The International Library of Environmental, Agricultural and Food Ethics

Volume 35

The ethics of food and agriculture is confronted with enormous challenges. Scientific developments in the food sciences promise to be dramatic; the concept of life sciences, that comprises the integral connection between the biological sciences, the medical sciences and the agricultural sciences, got a broad start with the genetic revolution. In the mean time, society, i.e., consumers, producers, farmers, policymakers, etc, raised lots of intriguing questions about the implications and presuppositions of this revolution, taking into account not only scientific developments, but societal as well. If so many things with respect to food and our food diet will change, will our food still be safe? Will it be produced under animal friendly conditions of husbandry and what will our definition of animal welfare be under these conditions? Will food production be sustainable and environmentally healthy? Will production consider the interest of the worst off and the small farmers? How will globalisation and liberalization of markets influence local and regional food production and consumption patterns? How will all these developments influence the rural areas and what values and policies are ethically sound?

All these questions raise fundamental and broad ethical issues and require enormous ethical theorizing to be approached fruitfully. Ethical reflection on criteria of animal welfare, sustainability, liveability of the rural areas, biotechnology, policies and all the interconnections is inevitable.

Library of Environmental, Agricultural and Food Ethics contributes to a sound, pluralistic and argumentative food and agricultural ethics. It brings together the most important and relevant voices in the field; by providing a platform for theoretical and practical contributors with respect to research and education on all levels.

Mbih Jerome Tosam · Erasmus Masitera
Editors

African Agrarian Philosophy

Springer

Editors
Mbih Jerome Tosam
Department of Philosophy
University of Bamenda
Bamenda, Cameroon

Erasmus Masitera [1979–2022]
SARChI in Higher Education and Human Development Research Group
University of the Free State
Masvingo, Zimbabwe

Department of Philosophy and Religious Studies
Simon Muzenda School of Arts, Culture and Heritage Studies, Great Zimbabwe University
Masvingo, Zimbabwe

ISSN 1570-3010 ISSN 2215-1737 (electronic)
The International Library of Environmental, Agricultural and Food Ethics
ISBN 978-3-031-43042-8 ISBN 978-3-031-43040-4 (eBook)
https://doi.org/10.1007/978-3-031-43040-4

This Springer imprint is published by the registered company Springer Nature Switzerland AG
The registered company address is: Gewerbestrasse 11, 6330 Cham, Switzerland

Paper in this product is recyclable.

Acknowledgements

I owe a special debt of gratitude to a number of people whose contributions and insights helped in the completion of this book. I would like to particularly thank Dr. Erasmus Masitera with whom I began the project before his sudden and untimely death. I am grateful to his colleague, Dr. Denis Masaka, for attending to reviewers' comments and revising Dr. Masitera's chapter, and for writing a biosketch on his behalf.

I am also highly indebted to Prof. Thaddues Metz of the University of Pretoria with whom I first discussed the book project and who gave useful advice and guidance on how to go about the project, and also supported me throughout the entire project.

I would also like to thank all the contributors to this volume for their insightful contributions, patience, and support. I am especially grateful to two anonymous reviewers for Springer for their constructive criticisms and suggestions. Their comments helped immensely in the improvement of the quality of the manuscript. I wish to thank my colleagues at the Department of Philosophy of the University of Bamenda for their Support. I also like to specially thank Dr. Pius Mosima and Dr. Richard A. Mbih for their encouragement and availability whenever their assistance was needed.

Finally, I would like to acknowledge, with thanks, my wife and children, Nkigoum Dolariste, Ankiambom Tosam Mbih, Peace V. L. Tosam, for their love, patience, and support.

Contents

Editors and Contributors

About the Editors

Mbih Jerome Tosam is Associate Professor of Philosophy at the University of Bamenda, Cameroon. He obtained his Ph.D. in Philosophy from the University of Yaoundé I, Cameroon, in 2011. He is former Chair of the Department of Philosophy at the Higher Teacher Training College (HTTC), Bambili (2011–2017) and the Faculty of Arts of the University of Bamenda, Cameroon (2017–2021). His research interests are in the areas of Bioethics, African Philosophy, and Intercultural philosophy. Some of his publications have appeared in the following journals: *South African Journal of Philosophy*, *Annali di studi religiosi*, *Medicine, Health Care and Philosophy: A European Journal*, *Developing World Bioethics*, *Journal of World Philosophies*, and *Polylog: Forum for Intercultural Philosophy.*

Erasmus Masitera (June 3 1979–March 1 2022) was a senior lecturer in Philosophy in the Department of Philosophy and Religious Studies at Great Zimbabwe University, Masvingo, Zimbabwe. At the time of his demise, he was a postdoctoral fellow at the University of the Free State, South Africa. His research areas revolved around the connections of Ethics, Ubuntu, land reform, and social justice.

Contributors

Daniel Banuoku Centre for Indigenous Knowledge and Organizational Development (CIKOD), Accra, Ghana

Birgit Boogaard Knowledge Technology and Innovation (KTI) Group, Wageningen University and Research, Wageningen, The Netherlands

Maduka Enyimba University of Calabar, Calabar, Nigeria

David Fletcher People Development Ltd, Antigonish, Canada

Bernard Yangmaadome Guri Centre for Indigenous Knowledge and Organizational Development (CIKOD), Accra, Ghana

Bernard P. Humbe Research Institute for Theology and Religion, College of Human Sciences, University of South Africa (UNISA), Mbombela, South Africa; University of Religions and Denominations (URD), Pardisan, Qom, Iran

Joseph Pardon Hungwe College of Education, University of South Africa, Pretoria, South Africa

Dorcas Hwati Nemangwe Secondary School, Gokwe, Zimbabwe

Alloy S. Ihuah Benue State University, Makurdi, Nigeria

Michael Kpughe Lang The University of Bamenda, Bamenda, Cameroon

David Ludwig Knowledge Technology and Innovation (KTI) Group, Wageningen University and Research, Wageningen, The Netherlands

Erasmus Masitera Great Zimbabwe University, Masvingo, Zimbabwe

Mbuh Tennu Mbuh Department of English, The University of Bamenda, Bamenda, Cameroon

Thaddeus Metz University of Pretoria, Pretoria, South Africa

Pius Mosima Department of Philosophy, University of Bamenda, Bamenda, Cameroon

Tasara Muguti Great Zimbabwe University, Masvingo, Zimbabwe

Victor C. A. Nweke University of Koblenz-Landau, Mainz, Germany

Babatunde A. Ogundiwin School of Geography, Archaeology and Environmental Studies, University of the Witwatersrand, Johannesburg, South Africa

Belko Ouologuem Department of Philosophy, University of Letters and Social Sciences, Bamako, Mali

David Anthony Pittaway University of the Free State, South Africa, Bloemfontein, South Africa

M. Rathbone Faculty of Economic and Management Sciences, North-West University, Potchefstroom, South Africa

Faith Sibanda Department of African Languages and Literature, Great Zimbabwe University, Masvingo, Zimbabwe

Elvis Tsvangirayi Siziva Great Zimbabwe University, Masvingo, Zimbabwe

Peter Takov Catholic University of Cameroon (CATUC), Bamenda, Cameroon

Mbih Jerome Tosam Department of Philosophy, Faculty of Arts, The University of Bamenda, Bamenda, Cameroon

Anthony Uzochukwu Ufearoh Department of Philosophy, Nnamdi Azikiwe University, Awka, Nigeria

Dominic Umoh Philosophy of Religion and Religious Ethics, Akwa Ibom State University, Mkpat-Enin, Nigeria

Chapter 1
Introduction: African Agrarian Philosophy

Mbih Jerome Tosam

Abstract This book explores indigenous sub-Saharan African agrarian beliefs, values, practices, institutions, as well as contemporary agrarian issues and challenges connected with a changing historical, economic, social, and political landscape in Africa. The book is hinged on the idea that wherever human beings have lived, they have been preoccupied with finding ways to ensure sustainable management of the natural resources at their disposal to take care of their basic needs: food, shelter, and security, and that agriculture is an essential, but generally neglected, determinant of the emergence and orientation of all philosophical traditions. In the quest for food security and subsistence of the group, different beliefs, knowledge systems, moral norms, cultural practices and institutions emerged to guide societies on how to manage the natural environment, to (re)produce and to nurture plants and animals. In this introductory chapter, I sketch out the general features of agrarian philosophy and African agrarian philosophy in particular.

Keywords African agrarian philosophy · Agrarianism · Agriculture · Nature · Farm ethics · Food ethics · Environmental management

This book explores indigenous sub-Saharan African agrarian beliefs, values, practices, institutions, as well as contemporary agrarian issues and challenges connected with a changing historical, economic, social, and political landscape in Africa. The book is anchored on the idea that wherever human beings have lived, they have been preoccupied with finding ways to ensure sustainable management of natural resources to take care of their basic needs: food, shelter, and security, and that agriculture is an essential, but generally neglected, determinant of the emergence and orientation of all philosophical traditions. Falvey posits that "once humans had settled in large communities, a philosophical class that could concentrate on esoteric thought arose. This relied on the sustained food supply that was eventually secured by agriculture

M. J. Tosam (✉)
Department of Philosophy, Faculty of Arts, The University of Bamenda, Bamenda, Cameroon
e-mail: mtosam2002@yahoo.com

M. J. Tosam and E. Masitera (eds.), *African Agrarian Philosophy*, The International Library of Environmental, Agricultural and Food Ethics 35,
https://doi.org/10.1007/978-3-031-43040-4_1

and its governance" (Falvey 2020: 4). In the quest for food security and subsistence of the group, different beliefs, knowledge systems, moral norms, cultural practices and institutions emerged to guide societies on how to manage the natural environment, to (re)produce and to nurture plants and animals.

As Paul B. Thompson has roughly sketched out, agrarian philosophy examines the "norms, values and social institutions that emerge from human beings' interactions with nature in the form of material subsistence practices such as obtaining food, clothing and shelter" (Thompson 2008: 528). Thus, agrarian philosophers critically reflect on the values and practices that inform and guide the way societies relate with and make use of nature to obtain food and shelter. As the different chapters of this book show, a philosophical investigation into African agrarian culture can cast new light on many aspects of the budding field of African philosophy.

There is an underexploration of agrarianism in the field of philosophy in general and African philosophy in particular. Even within mainstream environmental philosophy, agrarianism has been given scant scholarly attention. Agrarian philosophy is an appropriate domain for exploring the boundaries between traditional political theory on the one hand, and the emerging field of environmental philosophy, on the other (Thompson 2012: 214). In agrarian cultures, "cultivation of the soil provides direct contact with nature; through the contact with nature, the agrarian is blessed with a closer relationship to God" (Thompson 2008: 530). Also, farming instils in the farmer some canonical virtues such as "honor, self-reliance, courage, moral integrity, hospitality" (Thompson 2008: 530) and patience. Agrarian beliefs, norms, institutions, and practices that emerge therein, vary from one society to another as a result of differences in beliefs and customs about human relation to land and nature. These differences in cultural norms are also shaped by climatic, historical, economic setting, as well as by technological progress. In some societies, transformations in cultural, political, and economic relations to the earth and agriculture have sometimes resulted in social and economic developments, in others they have led to the loss of dignity which have provoked social, cultural, and ecological changes that degraded the natural environment which have affected human well-being (Sachs 2019). In other areas, the shortage or lack of rainfall has also influenced beliefs and customs related to farming and agriculture in general.

Although agrarian philosophy has not been distinctly and critically explored within mainstream philosophical discourse (Singer and Motter 2015: 2), it has been ubiquitous in all geographical, cultural traditions, and historical periods. Thompson underscores this idea when he argues that all human societies from primitive times have developed "moralities around group needs for sharing of provisions and protection" (Thompson 2008: 531). It is for this reason that "each culture has a unique way of thinking, unique value systems and beliefs, and different preferences determined and birthed by different factors" (Kugedera et al. 2021: 23), because each culture develops values and customs according to their perception of nature, beliefs connected with farming, food, and social organisation. How each society perceives nature determines the way they relate with it.

Techno-science, colonisation, globalisation, and climate change have also had serious ramifications on the structure of social, political, and economic life, and have contributed to agrarian changes in the world. In the Western world, scientific and industrial revolutions propelled modern man/woman's desire to conquer nature and other cultures and peoples for economic and political reasons. Colonialism and globalisation have also influenced agrarian reforms and changes across sub-Saharan Africa in that Euro-American agrarian values and practices were imposed at the detriment of local ones. Also, global climate change and changing weather patterns have influenced agrarian cultures across Africa. It is for the above reasons that agrarian perspectives vary from one region, country, community, to another depending on its historical and economic influences as well as its geographical location and ecological niche. For example, countries in the Sahelian region where there is less rainfall, and which are threatened by desertification and recurrent droughts, invariably adopt slightly different cultural approaches to agriculture from those in the tropical or equatorial regions where there is abundant rainfall. In areas threatened by the encroaching Sahara Desert, farming practices, beliefs, and customs are different. These natural and human-induced conditions which affect life in these regions also provoke socio-economic and political issues connected to land use and agriculture. Moreover, because the different regions or countries of sub-Saharan Africa were at some point in history under the colonial domination of one European country or another, and considering the de-Africanizing tendencies of European colonisation in Africa, this has, in subtle ways, influenced and shaped indigenous pre-colonial African agrarian cultures. More than half a century after independence, Western epistemological influences on the natural environment and on agricultural development in Africa have continued unabated (Boogaard 2019; Boogaard et al. 2023).

There is an urgent need to challenge and critically reappraise the onto-epistemic disrupting practices and policies introduced by the colonialists and amplified by globalisation. Roothaan argues that pre-modern ontologies concerning how indigenous people related to nature needs to be renegotiated and reconstructed in the postcolonial era (Roothaan 2019). This is because modernism with its attendant techno-scientific and capitalist outlook let to the banishment of spirit and non-material ontologies. She writes:

> We should look at the ways colonials introduced legislation of land ownership while denying already existing indigenous legal systems. Into the ways modern liberal democracies even today denigrate indigenous education of children while stressing all children should get modern education in schools. We should look into the food chains that sustain us, and how we disrupt the food chains of others, spoiling soils, river systems, seas, mountains and the climate that interacts with them. We should investigate why we reject traditional methods of hunting, of growing vegetables, of birth control, religious attitudes to nature, ways to communicate with spirits, and with trees, for that matter. The necessity to negotiate the environment, and to do it now, is put upon humanity by the inacceptable violence in the relations between modern and indigenous peoples in our times… (Roothaan 2019: 139).

It is necessary that while we strive to acquire modern education and the facilities of modernity, we must not lose sight of the alienating, dehumanizing, and above all, the environmentally degrading effects of modern life. A critical reflection in African agrarian beliefs and values is one of the ways through which indigenous African knowledge systems can be renegotiated and reconstructed.

While there have been growing literature on philosophy of nature, environmental ethics, land ethics and philosophy of agriculture during the last four decades or so, debates in the field of agrarian philosophy have, for the most part, focused on mainstream Kantian and utilitarian approaches (Boogaard 2019). Also, although African environmental philosophy is just beginning to be conceptualized, debated, and written, there is a dearth of scholarly work on non-Western, and especially indigenous African philosophical perspectives, intuitions, and moral orientations like the relational, care ethics, and spiritual approaches, which are relevant to African agrarianism.

1.1 What is Agrarian Philosophy?

The term agrarianism or agrarian philosophy has its origins from *lexagraria*, an ancient Roman law that authorised "the equal sharing and division of lands that were conquered by and thus belonged to the Roman Empire" (LeVasseur 2015: 3). It is for this reason that when the term was first used in conventional political literature in the 1700s and 1800s in the United States and Europe, it was over concerns about the interference of government with land ownership. Although this is still an important concern in contemporary agrarian philosophy, a worry which has always been fundamental to agrarianism from inception is "shared ownership, and … shared responsibility, in undertaking agriculture and in having farming act as the bedrock of civilization and society" (LeVasseur 2015: 3). There are different types of agrarian thought; there is ecological, religious, radical, literary (LeVasseur 2015), and agrarian pragmatism (Thompson 2008: 528). However, in spite of the variations in agrarian outlook, all agrarian thoughts share the idea that "soil, climate, and other geographical features will result in different material practices … and differences in norms, values and social institutions" (Thompson 2008: 529). Historically, agrarian philosophy can be traced back to the works of ancient Greek and Roman philosophers like Hesiod, Aristotle, Xenophon, Virgil, Cato, and Cicero.

In the literature on agrarian philosophy, the terms agrarian philosophy and agrarianism are often used interchangeably. In this book, we also use the terms interchangeably. Paul Thompson, one of the leading contemporary agrarian thinkers, defines agrarianism as:

> A class of philosophical views on human culture and practice as they relate to the broader environment …. Agrarian philosophies…offer formulations on the way that norms, values and social institutions emerge from human beings' interaction with nature in the form of material subsistence practices such as obtaining food, clothing and shelter (Thompson 2008: 528).

Central to any agrarian philosophy is human relation to nature and the moral norms, cultural and social institutions that emerge from this relation. Thompson underscores this idea when he posits that agrarian philosophy "stresses the role of nature, soil and climate in the formation of moral character as well as social and political institutions" (Thompson 2008: 527). Other commentators emphasize property ownership and the character formation towards becoming a responsible person that this relation to nature brings, as the core values of agrarianism. For instance, *The Encyclopaedia Britannica* defines agrarianism as "a social and political philosophy…that stresses the primacy of family farming, widespread property ownership, and political decentralization. Agrarian ideas are justified in terms of how they serve to cultivate moral character and to develop a full and responsible person" (Heath 2022).

On his part, Freyfogle contends that:

> Agrarianism's beginning point is with the land itself and people living on it. The practical necessity of gaining sustenance from local nature leads people to craft modes of production…tailored to the soils, waters, and other geographic features of their home lands. Out of these locally adaptive modes of production arise modes of social interaction, norms of behavior, and structures of living and governance. Nature… plays a key role in the ways people think, act, and interrelate….Nature… presses people to adapt, thereby, … "selecting" in evolutionary fashion for locally adapted land use norms and behaviors that enable a given people to flourish (Freyfogle 2008: 546).

Freyfogle thinks that individuals do not actually choose their moral values based on rational grounds. Somewhat, common behaviours by a particular group of people lead to common habits and moral attitudes, which members unconsciously enthral (Freyfogle 2008: 546). Whatever differences on how agrarian values and institutions emerge, one thing is common to all agrarian cultures: the use of land to satisfy the needs of the community has always been at the backbone of all agrarian philosophies.

Montmarquet has identified two approaches to agrarian philosophies based on two fundamental types of human relation to nature. The first approach is largely "focused on the relation to land as a source of value" (Montmarquet 1985: 12). In the history of philosophy, this approach is commonly associated with Hesiod. According to this view, our relation to nature results in "certain enduring human values" (Montmarquet 1985: 12). The first approach is authentic to the essential preoccupations of agrarian philosophy, because "insofar as the values traditionally associated with agriculture are… not economic, they find their expression in society whose goals are the classical ones not of gain but of living well" (Montmarquet 1985: 12). The second approach emphasises human relation to land as a basis of rights. According to this approach, "the most important moral aspects of human's relation to the land are defined by certain rights (Montmarquet 1985: 12). This agrarian approach is usually associated with Locke. However, one of the principal pitfalls of this form of agrarianism "is its tendency to degenerate into a mere concern with property and wealth, thus, a concern essentially indifferent to agrarianism" (Montmarquet 1985: 12) or agrarian ideals.

Wirzba captures, in a comprehensive sense, the general ideals of agrarian philosophy as follows:

> Agrarianism tests success and failure not by projected income statements or by economic growth, but by the health and vitality of a region's entire human and nonhuman neighborhood. Agrarianism … represents the most complex and far-reaching accounting system ever known, for according to it success must include a vibrant watershed and soil base; species diversity; human and animal contentment; communal creativity, responsibility, and joy; usable waste;

> social solidarity and sympathy; attention and delight; and the respectful maintenance of all the sources of life (Wirzba 2013: 3).

Although the idea that there is a nexus between cultural norms and the material environment only came to the fore in the social sciences in the second half of the twentieth century, philosophers who understand moral values, cultural practices, and social institutions as emanating from and "justified in virtue of the way that they facilitate material practices vital to a social groups' survival can be found throughout history" (Thompson 2008: 528). In summary, and as I have already stated, agrarian philosophy is an inclusive outlook that valorises the intimate and useful relations that exist between humans and the natural environment.

1.2 African Agrarian Philosophy

Indigenous, sub-Saharan African agrarian philosophy is an uncharted and largely overlooked area of study in the nascent field of African philosophy (Boogaard 2019: 3). Agriculture as a way of life is deeply connected to beliefs, values, and practices related to a wide range of issues connected to ecological ethics, food ethics, religion, medicine, political economy, social institutions, biological reproduction and species survival, indigenous knowledge, and property rights. In traditional African culture, it was difficult to separate agriculture from culture and agriculture from African onto-epistemology. Indigenous African agrarian philosophy is intimately connected to African communitarian and relational metaphysical and moral outlook. It was void of an anthropocentrism. Within African agrarian and environmental thought, the universe was seen as a "holistic community of mutually reinforcing natural life forces consisting of human communities… spirits, gods, deities, stones, sand, mountains, rivers, plants, and animals" (Ikuenobe 2014: 2). Everything in nature has a vital force such that the harmonious interaction amongst them strengthens the universe (Tempels 1959; Mbiti 1969). This interaction of forces forms the basis of indigenous African spirituality and religious worldview. Mbiti posits that "because traditional religions permeate all the departments of life, there is no formal distinction between the sacred and the secular, between the religious and the non-religious, between the spiritual and the material areas of life" (Mbiti 1969: 1). African agrarian philosophy can be a fertile area of study in the quest for epistemic justice in African philosophy.

As mentioned above, African agrarian philosophy refers to a set of beliefs, values, customs, and social institutions that inform and sanction the way humans relate to their natural environment, cultivate the land, ensure food security, and manage the resources of the Earth. In other words, African agrarianism is concerned with norms and practices related to agriculture and the proper management of nature to safeguard the survival of persons, plants, animals, and the ecological community. In the last three decades or so, there have been a number of publications that have explored issues which may be relevant to African agrarian philosophy. Some of these works include the following: Oruka and Juma (1994), Tangwa (1996, 2004),

Niekerk (2005), Bujo (2009), Maathai (2010), Mangena (2013), Kelbessa (2005, 2014), Chimakonam (2018), Chemhuru (2019), Tosam (2019) and Masitera (2021), Mosima (2021). Central to African agrarian thought are the following beliefs and values: There is no strict separation between humans and nature; the universe is both physical and supernatural; there is communal ownership of land and its resources, and the cultivation of good character towards other members of the community, including nature. Although some of these norms and customs have been changing as a result of the epistemologically disrupting and economically rapacious tendencies of colonisation and globalisation, some of the cultural beliefs and values related to the use of the Earth have survived and continue to influence the way people perceive and relate to nature (Roothaan 2019).

Communitarianism in African agrarian philosophy is anchored on African relational ontology. According to this outlook, the cosmos and everything in it is perceived as an interconnected and interdependent whole. This idea resonates in the writings of many African philosophers (Tempels 1959; Kagame 1969: 233; Mbiti 1969: 74; Wiredu 1994: 46; Tangwa 1996: 191; Bujo 1998: 20, 2009; Murove 2004: 185; Ikuenobe 2006: 63, 2014: 2; Udeani 2008: 67; Behrens 2014: 62; Chemhuru 2014: 75; Imafidon 2014; Tosam 2019: 177; Mosima 2021). In this universe, there is no strong subject-object, natural-supernatural split. According to Mbiti, "the spiritual universe is a unit with the physical, and … these two intermingle and dovetail into each other so much so that it is not easy, or even necessary, at times to draw the distinction or separate them" (1969: 74). Reality is a continuum and God who is perceived as the creator of the universe is considered to be found in all of His creatures. Hence, humans, plants, animals, the Earth, and the rest of nature [are] all infused with spirits and [vita] forces (Tempels 1959). For Tempels, African ontology is governed by communication of dynamic forces and "'force' is an essential component of 'being', and the "concept 'force' is inseparable from the definition of being. There is no idea among Bantu of 'being' divorced from the idea of 'force'" (Tempels 1959: 35). While sharing the idea of a dynamic universe with Tempels, Kagame insist that instead of force, being is the basic principle in the universe (Kagame 1956). Tempels contends that within the African universe, the aim of being:

> … is to acquire life, strength or vital force … Each being has been endowed by God with a certain force, capable of strengthening the vital energy of the strongest being of all creation: man. Supreme happiness, the only kind of blessing, is, to the Bantu, to possess the greatest vital force … Every illness, wound or disappointment, all suffering, depression, or fatigue, every injustice and every failure: all these are held to be, and are spoken of by the Bantu as, a diminution of vital force (Tempels 1959: 30–32).

For Tempels, the universe consists of forces which help to safeguard the link between forces. Hence, "force is the nature of being, force is being, being is force" (1959: 35). Reality, therefore, involves some intimate ontological relationships and interactions among beings. Tempels adopts a vitalist view of the universe in which everything is attached to and accounted for by vital force. It is force that connects humans to God, spirits, and the rest of nature. To emphasize the inherently interconnectedness and interdependence in the nature of things in African ontology, Tempels writes:

> Bantus hold that created beings preserve a bond one with another, an intimate ontological relationship, comparable with the causal tie which binds creature and Creator. For the Bantu there is interaction of being with being, that is to say, of force with force. Transcending the mechanical, chemical and psychological interactions, they see a relationship of forces which we should call ontological. In the created force (a contingent being) the Bantu sees a causal action emanating from the very nature of that created force and influencing other forces. One force will reinforce or weaken another. This causality is in no way supernatural in the sense of going beyond the proper attributes of created nature (Tempels 1959: 28–29).

In this integrated cosmos, health, fertility of the land, of humans, and of animals and plants are seen as harmonious balance between the various parts of the cosmos and disease (which may be reflected in infertility, natural disasters, poor harvest, and all types of misfortune that may affect humans, animals and plants) is the breakdown of cosmic harmony (Mbiti 1969). There were taboos against the abuse or irrational exploitation of the environment. In most African communities such taboos helped in the protection of the environment. There were sacred places, trees, and certain animal species, because it was believed that the violations of such taboos will have negative effects on both the wrongdoer and on the society such as disease, flood, drought, and death (Tangwa 1996; Masaka and Chemhuru 2010; Kelbessa 2015: 401). It is for this reason that there must always be expiatory rituals to repair and recover cosmic balance whenever and wherever it has been fissured.

Another important relational account of nature relevant to African agrarianism is that of Senghor. He argues that "as far as African ontology is concerned … there is no such thing as dead matter: every being, everything—be it only a grain of sand—radiates a life force, a sort of wave-particle; and sages, priests, kings, doctors, and artists all use it to help bring the universe to its fulfillment" (Senghor 1995: 49). Tangwa has described this indigenous African relational worldview as "eco-bio-communitarian" which involves "recognition and acceptance of inter-dependence and peaceful coexistence between earth, plants, animals and humans" (Tangwa 1996: 196). In this interconnected and interdependent conception of the cosmos, there is an extended interpretation of the community as a continuum of natural and supernatural forces. It is for this reason that in some parts of Africa, the Earth cannot be cut or farming carried out without obtaining the ritual permission of nature, the gods and spirits through ritual sacrifices (Tangwa 1996). For the Nso', for example, "the earth is a very potent force, to the extent that they do not cut the earth lightly (without ritual permission)" (Tangwa 1996: 190). From this perspective, it can be argued that African agrarian thought goes beyond the anthropological, ecological, and natural spheres. It is within this metaphysical and moral frame that African agrarianism operates.

Furthermore, there was communal ownership of land and its resources in pre-colonial Africa though traditional authorities, as custodians of culture and human representatives of the gods and the spiritual realm, were charged with the distribution and use of communal land.[1] Nyerere's *Ujama'a* (brotherhood), which can arguably

[1] However, in recent times, some traditional rulers have been abusing this moral, political, and spiritual authority vested in them by their communities and ancestors by usurping and selling communal land.

be considered as a pioneering work on indigenous African agrarianism, is based on traditional sub-Saharan African communal agrarian ethos or what is commonly referred in Southern Africa as ubuntu or in as *bom wul* (Kom) or *bir wir* (Nso') central Africa (Cameroon) which (all) denote the idea of 'because of a person.' Ramose considers "ubuntu as the root of African philosophy… it is the wellspring flowing with African ontology and epistemology. If these latter are the bases of philosophy then African [agrarian] philosophy has long been established in and through Ubuntu" (Ramose 2005: 35). In traditional African society, people owned property, especially land, in common. You could go to anybody's farm and harvest any crop provided you harvested only the quantity necessary to relieve your hunger (Tangwa 1996; Tosam 2019). Nyerere outlines three core values which define communal life in the traditional African community: "mutual involvement in one another"; communal ownership of "property", that is, "all basic goods were held in common and shared" (Nyerere 1968: 107) and "no one could go hungry while others hoarded food, and no one could be denied shelter if others had space to spare" (Nyerere 1968: 107). The third principle is that "everyone had an obligation to work. The work done by different people was different, but no one was exempt" (Nyerere 1968: 108). Within this communal context, people "are individuals within a community, we took care of the community and the community took care of us…Nobody starved, either of food or human dignity, because he lacked personal wealth; he could depend on the wealth possessed by the community of which he was a member" (Nyerere 1968: 162–171). The central moral structures of communalism, which include caring and human flourishing can be understood in terms of the fair production and distribution of goods, services and responsibilities in the community (Ikuenobe 2015: 1008).

Also of central import to African agrarianism is the cultivation of good character not only towards other members of the community, but also towards the ecologic and supernatural community. This may be considered as an indigenous African environmental virtue. How individuals use available natural resources tells whether or not they portray a good character towards others. In many African cultures, character is considered as the most important social good, even above all other goods (Ramose 2005; Gyekye 2010; Boogaard 2019; Metz 2022) such that "discourses or statements about morality turn to be discourses or statements essentially about character" (Gyekye 2010). In Kom culture, for instance, character is equated to one's humanity. It is good character (*nchinɨ ijuŋ*) that defines the humanity of a person (Tosam 2014). For the Nso', a person "is conduct" or character, *wir dze lii* (Mofor 2008: 208). A good person *(wir wo juŋ)* is one who loves, cherishes and promotes "truth, peace, uprightness, compassion, 'well-being' (*dze ye juŋi*) and other similar moral qualities and enduring cultural values" (Mofor 2008: 208). In African ethics, "one becomes more fully human through one's "way of life," by behaving more ethically" (Nkulu-N'Sengha 2009: 144). Sibanda shows that in traditional pre-colonial Africa, "agrarian metaphorical nuances in the Ndebele language had an effect of moulding individuals as well as the society into socially acceptable citizenry" (Sibanda, in this book). There were beliefs, values and customs which guided the regulation and exploitation of natural resources. The aim was to ensure that natural resources were not used irresponsibly or depleted. In some African societies, there were laws which regulated the proper relation between humans and nature. Kelbissa notes that in the Oromo community in Ethiopia, for example, "*Saffuu* regulates the relationship between various animals and human beings. The Oromo moral code does not allow irresponsible and unlimited exploitation of resources and human beings.... *saffuu* is based on justice. It reflects a deep respect and balance between various things (Kelbessa 2005: 24). These moral values ensured that there was no inconsiderate and wanton exploitation of nature.

In agriculture, there were/are practices that ensure(d) that the land remained fertile and species were not wantonly expended. It was morally wrong to completely expend a species. It is for this reason that the search for wholeness and interspecies care where humans care for nature and nature in turn cared for human well-being occupied an important place in indigenous African outlook towards the universe. Ramose notes that "the principle of wholeness applies also to the relation between human beings and ...nature. To care for one another, therefore, implies caring for physical nature as well. Without such care, the interdependence between human beings and physical nature would be undermined" (Ramose 2009: 309). Such care is also essential for cosmic equilibrium. Because nature is perceived as sacred and is considered dwelling place of spirits, and supernatural beings, some natural features like trees, bushes, rivers, mountains are believed to be inhabited by spirits. For this reason, such sacred places have to be cared for and protected (Tangwa 1996; Roothaan 2019; Tosam 2019; Mosima 2021; Ouologuem, in this book).

Another belief which is relevant to African agrarian and environmental thought is the belief in totems. Across the continent different ethnic groups believe they have affinity or spiritual link with different animals and plant species (Tempels 1959; Tangwa 1996: 191). A particular physical being serves as a symbol of a particular society recognized as its totem. "A human group and an animal species can occupy in their respective classes a rank relatively equal or relatively different. Their vital rank can be parallel or different. A chief in the class of humans shows his royal rank by wearing the skin of a royal animal" (Tempels 1959: 61). In such communities, there are taboos against killing or eating such animals or plants. Within African ontological worldview, respect for nature is respect for self (Mosima 2021:64).

A further important agrarian belief and practice which encouraged the protection and healing of nature was the practice of shifting cultivation. Tangwa argues:

> The practice of *shifting cultivation* which, unfortunately is fast disappearing, owing to both increasing population and increasing use of imported chemical fertilisers, was one way of ensuring that the soil, vegetation and the fauna recovered and reasserted themselves after a period of human exploitation. Strict taboos on eating or killing certain insects, reptiles, birds and mammals also ensured the survival of the fauna within this ecological niche (Tangwa 1996: 189).

These values and practices ensured that production methods did not cause pollution and the living did not exhaust available resources, and where there was an overuse of the Earth's resources, the area was allowed for some time (for one or two years) to heal before it was used for farming again. This method ensured that future generations were not deprived of the natural resources that God has blessed humankind with.

Indigenous African agrarianism is, no doubt, based on a profoundly spiritual and interconnected and symbiotic understanding of nature. This agrarian outlook is informed by a non-anthropocentric impulse, which holds that humans cannot survive by living in complete disregard for the natural environment with which they share the universe. Hence, African agrarianism promotes sustainable use of natural resources. Contrary to industrial agriculture, traditional agricultural methods ensured that there was less soil erosion, less pollution of soils and streams, and less agricultural poisons. To mitigate the negative effects of global climate change, modern industrial society has a lot to learn from indigenous attitude towards nature.

Another important contribution to African agrarian philosophy is that of Oruka (1993). Oruka discussed issues related to land ethics, consumerism, and global justice. In his "Parental Earth Ethics", Oruka contends that the earth or the world is a Family Unit, a global village in which members have kith and kin relationship with one another (1993). Oruka considers this as an ethic of duty towards humans and the natural environment. The earth is a commonwealth to all humanity in which we are all interconnected and interdependent. Oruka advocates for an eco-philosophical approach which perceives the universe as an interconnected whole in which everything is related to everything else. For this reason, there is need for solidarity and mutual care between members of this global community (Mosima, in this book). It is a kind of "organic unity" governed by the principles of interdependence and ethical responsibility (Graness and Kresse 1997: 257). For this reason, he argues, because "there is a need for a shift towards a new epistemological outlook in which humankind is viewed as part of a complex and systematic totality of nature" (Oruka and Juma 1994:115). It is from this eco-ethical outlook that Oruka challenges the anthropocentric approach in favour of the ecocentric ethical approach (1993). According to him, the anthropocentric approach is inadequate in proffering solutions to the current global environmental and health crisis that humankind is bedevilled with.

1.3 The Necessity of Studying African Agrarian Philosophy

There have been frantic efforts across Africa to promote agricultural development in order to ensure food security and in spite of huge investments in agriculture, the results have been dismally unsatisfactory as the problem of food insecurity seems to be worsening. African countries have failed to achieve food security as most African countries still depend on Western and Asian countries for most of their food supplies. The failure of the campaign against food insecurity may be explained by the fact that the various approaches employed have been predominantly anchored on Western industrial agricultural methods and ideologies which are not amenable to indigenous methods of agriculture and ways of being (Boogaard 2019: 273). Also, while this could be one of the rational ways of solving the problem, most of the theorists fail or ignore the negative impact of industrial farming on indigenous people, culture, and on the environment. Industrial agriculture is too economically oriented and farmers are less concerned with issues of environmental and ecological pollution. Industrial agriculturalists "are fully embedded in the market and as anxious as any economic group to curry favors with government" (Freyfogle 2008: 547). Industrial farmers "acquire their norms the same way most people do, not from surrounding nature but from the market, liberal values, and religious enthusiasms" (Freyfogle 2008: 547). According to Boogaard et al., the globalization of industrialized farming has been accompanied by the "globalization of social and environmental crises, ranging from loss of biodiversity to the erosion of rural livelihoods" (Boogaard et al. 2023). From the same perspective, Berry argues that the danger of modern industrial approaches to farming is that it "…prescribes an economy that is placeless and displacing. It does not distinguish one place from another. It applies its methods and technologies indiscriminately in the American East and the American West, in the United States and in India. It thus continues the economy of colonialism" (Berry 2013: 20). This destruction of rural livelihood has also ushered in an epistemological and metaphysical crisis, involving the destruction of indigenous knowledge systems, norms, customs, and ontologies (Boogaard et al. 2023). However, discussions about "sustainable development increasingly recognize the importance of indigenous knowledge in managing social and environmental change, mainstream agricultural development projects continue to marginalize [African perspectives] as irrelevant for sustainable agricultural practices" (Boogaard et al. 2023).

A further reason why it is imperative to study African agrarian philosophy is because at the heart of African agrarianism are fundamental ideas, values, and theories in African environmental philosophy. As Behrens has rightly argued, "no philosophy or ethic that does not account for how we should value the natural environment can be complete. Yet, relatively little attention has been given to these matters by African philosophers" (Behrens 2017: 192). African agrarian philosophy is therefore an important window through which African eco-philosophy can be contemplated. Moreover, there is a wealth of uncharted knowledge and values in indigenous African relation to nature that can profoundly deepen and expand the global debates on how to relate the natural environment. I agree with Behrens when he says that "if African philosophy is to be relevant to the needs and interests of the people of this continent, considerably more work needs to be done to develop African eco-philosophy and environmental ethics" (Behrens 2017: 192–193). In this direction, African agrarianism is an important field in want of critical exploration.

Closely connected to the idea of investigating indigenous pre-colonial African eco-philosophy is the need to find alternative approaches to resolving the environmental crisis that humankind is currently faced with. New developments in science and technology have led to remarkable progress and have resolved many problems for humankind including health, communication, etc. However, these advances in technology have engendered new and unfamiliar problems, which our current normative framework is unable to handle. Ecological crisis is one of these new problems provoked by technology. In the search for new ways of facing the environmental challenges that modern technologies have led us into, non-Western and indigenous ecological conceptions of nature and ways of being, which have been largely ignored in the quest for solutions to the global environmental crisis, have the potential to proffer new or alternative ideas that can "transform our popular conception[s] of self-interest, [our] human sense of home and our self-conception" (Riggio 2015: 1).

Beside the fact that much of the debates on agrarianism have focused on consequentialist and deontological models (Thompson 2008), in the history of philosophy, agrarianism is often discussed only in relation to the works of Xenophon, Plato, Aristotle, Rousseau, and liberal theorists like Locke and Jefferson, (Thompson 2008; Montmarquet 1985; LeVasseur 2015). According to Boogaard, the failure of Western-initiated agricultural policies in Africa can be attributed to the fact that the questions and solutions are framed from a Eurocentric perspective. African values, epistemologies, and outlooks on agriculture are usually being overlooked or simply derided and silenced. It is essential that agriculture be grounded on the ontology of the people. Boogaard thinks that:

> ...African philosophies can provide African perspectives on Africa's development and agriculture and pose critical questions about our global food system.... African philosophy seems highly relevant when thinking about and working on sustainable agricultural development and food production in Africa (Boogaard 2019: 274).

As I have shown above, there are many agrarian perspectives as there are forms of agricultural developments. For this reason, philosophies and outlooks "of sustainable agricultural development and their legitimacy can vary across contexts, cultures and stakeholders such as farmers, policy makers, business people, researchers, and practitioners" (Boogaard 2019: 274).

Moreover, and closely related to the views of Boogaard et al., is the problem of epistemic injustice in African philosophy. There has been little critical attention on indigenous African philosophical perspectives on agriculture and human relation to the nature, which are central concerns of agrarian philosophy. According to Thompson,

> ... the relational ontology existent between entities is so rich that humans are inseparable from nature or the world around them. Indigenous African Agrarians' cultivation of the soil provides a special direct contact with nature and the other beings around and above all is blessed with a closer nexus to the divine spiritual realm. (Thompson 2008: 538).

Indigenous African relational and spiritual attitude towards nature and the cosmos is a profoundly rich resource for African philosophy in particular and for higher education in general. It is therefore epistemologically and morally imperative to give voice to indigenous agrarian perspectives such as those of sub-Saharan Africa.

In recent years the question of epistemic injustice against African philosophy and African systems of knowledge has been expounded by a number of authors. The views, perspectives, and opinions of African and non-Western scholars are conspicuously ignored in many important global debates. This constitutes an epistemic injustice against these cultures, peoples and regions of the world (Graness 2012, 2015). Graness contends that "academic debate ... is still rooted in a system that has yet to adhere to such fundamental principles of justice as the recognition of and respect for the opinions, concepts, and systems of norms and values of different cultures and regions" (Graness 2015: 127).This Eurocentric approach which determines "the ethics of knowing", that is, imposing what should be taught and learned as philosophy (Chimakonam 2017: 125) or what should be considered as valid knowledge otherwise it is not philosophy or valid knowledge, has resulted in epistemic marginalisation of African and non-Western philosophical traditions in the mainstream philosophical discourse even though these cultures and peoples have a lot to contribute (Chimakonam 2017: 123). According to Chimakonam, "…any epistemological structure that does not balance all available perspectives is lopsided" (Chimakonam 2017: 125). The study of African agrarianism may help open new perspectives in indigenous African epistemologies.

Also, academic African philosophy in the universities has further heightened the epistemic marginalization of African agrarian philosophy because it "tends to focus mainly on academic theory, while being less engaged with agricultural practices and people's livelihoods in rural communities" (Boogaard et al., their contribution in this book). On his part, and in line with the call to give voice to indigenous agrarian knowledge systems, Ihuah proffers three reasons why it is necessary to consider indigenous African knowledge systems in agriculture and in the management of natural resources:

> (i) The long-term generation and transmission of knowledge of the local ecosystem offers a unique historical perspective into indigenous risk adjustment options; (ii) Modern scientists involved in the management and conservation of areas that may be ecologically fragile or marginal, or that contain genetically important plant or animal biodiversity, may benefit greatly from alternative knowledge; (iii) That African endogenous knowledge does not separate humans from natural environment; that the world is interconnected and interdependent" (Ihuah, in this volume).

There is a wealth of knowledge on indigenous African agrarian thought that is yet to receive critical philosophical attention within the field of African philosophy. The study of African agrarian philosophy will, as seen in some of the chapter of this book, help raise and clarify some fundamental philosophical questions like "What is the nature of the world around us (epistemological questions)?" "What is the nature of the non-human universe (metaphysical questions)?" And "what should be the appropriate relationship between human and non-human creatures (ethical questions)?" Any attempt to address these questions will, undoubtedly, be making novel contributions to African philosophy in particular and philosophy in general. It is by uncovering and critically scrutinizing different aspects of traditional African philosophy that the discipline can attain the universal relevance that it deserves. In this search for this universal relevance, no aspect of African cultural life should be ignored.

Finally, it may be argued that it is no longer necessary to rekindle indigenous pre-colonial African agrarian culture because African societies need to modernise or industrialise in order to achieve the much-needed development they yearn for. Such critics of agrarianism may argue that reawakening traditional African agrarian visions will be "anachronistic and dangerous" (Wirzba 2013: 1), because it may only help to thwart the much-desired development of the continent and further exacerbate the development gab between the industrialised affluent world and the African continent.

According to Wirzba,[2] such a critique will be out of place because industrial and technological enthusiasts "…rarely reveal the whole story. They do not tell us what the complete and long-term costs (to communities and ecosystems) of their visions are, when and where their visions fail, nor will they disclose the actual or potential profits they hope to realize. They hide the gap that often exists between promise and fulfillment" (Wirzba 2013: 1–2). While it is evident that modernity with its attendant scientific and technological advancements is not essentially bad, it should not be presented as if it is without blemishes. Techno-science has made life much easier; it has made production faster and in large scale; with new communication systems, it has reduced the world in terms of time and space, and it has also improved health care and education.

However, in spite of its putative and desired benefits, modernism and industrialization comes with great loss. Wirzba warns that we should not be uncritically carried away by these great developments. He opines that:

> There are good reasons to suggest that a culture loses its indispensable moorings, and thus potentially distorts its overall aims, when it foregoes the sympathy and knowledge that grows out of cultivating (*cultura*) the land (*ager*). Past cultures needed to be attentive to the requirements of regional geographies because for thousands of years human life and development were themselves firmly and practically oriented through multiple relationships with natural landscapes and the organisms they support. Most people, as a matter of practical necessity, understood themselves and their aspirations in terms of the limits and possibilities of the land. And so, whether we appreciate it or not, current widespread insularity from and ignorance about our many interdependencies with the earth represents an unparalleled development in human history (Wirzba 2013: 1–2).

In the context of Africa and most of the developing world, the quest for modernisation and industrialization is necessary for economic competitivity, however, "…in our haste to embrace technological improvements we must be careful not to overlook or degrade those elements of life—such as communal support, traditional wisdom, clean water, and nutritious food—that are fundamental" (Wirzba 2013: 3). What we need in today's modern world where indigenous ontologies and epistemologies have been silenced, is to engage in a genuine negotiation where we begin by "looking honestly at the ways modern society makes us live" (Roothaan 2019: 138). For Bujo, it is "only when the technological world listens to the symbolic language of nature will it become literate once more and able to promote life instead of death" (Bujo 2009: 288).

From the foregoing analysis, it is evident that indigenous African agrarian cultures provide a fertile ground on which latent and neglected aspects of traditional African philosophy can be explored.

---

[2] Although Wirzba's comments were directed against the critics of the call to return to some agrarian values in today's America, his comments may also be relevant to any critique against African agrarianism.

1.4 Structure and Organization of the Book

Part I focuses on African communitarian culture and how traditional African beliefs, values, and customs inform and shap indigenous African agrarian thought. There is consensus among agrarian philosophers that cultural norms, values and social institutions develop from human beings' interaction with the Earth in their quest for food, health, and shelter (Thompson 2008: 528; Freyfogle 2008: 546; Montmarquet 1985: 12; Nyerere 1968: 106–109). This part of the book is divided into five chapters. In Chap. 2, "Unpacking Ndebele Agrarian Metaphors for the Promotion and Preservation of Communal Social Development", Faith Sibanda argues that agrarian metaphorical nuances in Ndebele language had an effect on persons as these maxims morally and spiritually moulded them into socially responsible citizens. As I have shown above, character building is one of the cardinal canons of agrarian philosophy. From this standpoint, Sibanda uses the theory of *ubuntu/unhu* to analyse Ndebele proverbs and other related sayings to demonstrate that their agrarian exposure and practice provided a solid pedestal for the promotion and shaping of a socially just, humane and sustainable society. He interrogates the various metaphors, particularly proverbs, in explaining some of the cultural beliefs and standards transmitted through the people's experiences with agrarian farming tendencies.

In Chap. 3, "The Farm-Village Practice of Yorùbá in West Africa", Babatunde A. Ogundiwin uses the traditional Yorùbá farm-village practice (farming-workplace), to show its central role in this urban culture as well as the interconnection of the natural environment with agronomic practices and social institutions. This indigenous agrarian practice ensured urban sustenance for several centuries. Ogundiwin contends that the pace with which the continent is urbanizing, urban sustenance has become a growing concern in sub-Saharan Africa. For this reason, there is a critical need to rediscover agrarian values and practices which are inspired by indigenous African outlook on urban sustenance with the aim of protecting the environment. According to Ogundiwin, reviving traditional Yorùbá agrarian practice may provide an insight into the all-encompassing indigenous African agrarian custom which might provide guidance on sustainable urban development. To do this, we need to rethink emerging development notions in Africa which maintain that agriculture is a business which is unconnected to social values. He calls for need to pay close attention to holistic development by ensuring that in our attempt to modernise agriculture we do not render the environment unsafe for the local populations.

In Chap. 4, "On the Confluence of Permaculture and African Agrarianism", David Anthony Pittaway shows that there exist overlapping practices and values between permaculture and African agrarianism. Although he is quick to point out that there are differences between these two fields. In this chapter, Pittaway focuses essentially on the shared features in order to emphasise the "commonalities between people and practices" that portray our common humanity. He contends that the confluence of African agrarianism and permaculture demonstrates some features of "culturally-inclusive, ecologically-sensitive, and refreshingly simple SMART targets." Examples of commonalities between the two systems include the following: (1) both systems are profoundly land-based and not house-based; (2) in both cases, what happens on the land is part of what constitutes the home and indeed one's life, and the patterns of the seasons heavily influence one's behaviour; (3) in both systems of agriculture, there is no connection to a water-grid, electricity-grid, or sewage system; (4) animal manure plays a crucial role in both systems to fertilise the soil in which crops are grown; (5) and finally, there is companion planting.

In Chap. 5, "Dialogue between African agrarian philosophy and Adam Smith on underdevelopment and resource dependence in Africa", Mark Rathbone reflects on the principles of agrarian economics in Africa and uses these values to engage in a dialogue between African agrarian philosophy and some traces of agrarianism in Adam Smith's *The nature and causes of the wealth of nations* [1776] in order to highlight principles for agrarian economics in Africa. He argues that the importance of community and society in the work of Smith serves as an essential starting point for dialogue between African agrarian philosophy and classical economics. For Smith, agrarian economics was a broad form of economic activity that relied on raw materials, including agricultural and mineral resources. Agrarian economics that is contingent on limited raw materials must, therefore, also be rooted in normative and socially conscious value systems to ensure the sustainable use of economic resources. According to Rathbone, a mutually respectful engagement between African agrarian philosophy and classic economics like the free market economics of Adam Smith, can lead to fruitful debate for sustainable agrarian economics in Africa. Sustainable agrarian economics has to deal with two central problems of African economics, specifically underdevelopment and resource dependence. He thinks that it is precisely in an effort to find alternative solutions to these challenges facing African economics that Smith sets an important example for African agrarianism. There is need for a dialogue between the two economic systems because "both Smith and agrarian philosophy in Africa emphasise the role of ethics, community and agriculture although from divergent worldviews."

The focus of Part Two of this book is on the moral status of non-human nature in African agrarianism. Human relation to nature is a fundamental component of all agrarian traditions. Part II is divided into three chapters, which address, in different ways, the question of the moral worth of non-human environment. In Chap. 6, "Defending a Relational Account of Moral Status", Thaddeus Metz responds to critics of his approach to indigenous African relational account of the moral status. According to Metz's relational account of moral worth, "a being has a greater moral status, the more it is capable of communing (as a subject) or of us communing with it (as an object)." Basically, in this chapter, Metz responds to five objections to his modal account of moral status, namely: That it (1) "entails that we may rightly dominate mentally incapacitated human beings; (2) that it prioritizes mentally incapacitated human beings over animals with similar cognitive abilities without sufficient justification; (3) it entails that intelligent aliens lack moral status; (4) it cannot make sense of our duties towards the dead; and (5) it is unable to account for the standing of species as distinct from their members." According to Metz these criticisms do not hold water and should dissipate once terms are carefully defined and certain distinctions drawn.

In Chap. 7, "The Phenomenon of Male and Female Crops in Igbo Agrarian Culture: Implication for Gender Equality", Anthony Uzochukwu Ufearoh examines the metaphysical and ethical implications of the phenomenon of crop-gendering in traditional Igbo culture. He employs the hermeneutic approach to interpret the cultural symbols used in Igbo agrarian thought. Ufearoh reflects on the intersection between gendered agrarian practices which have onto-anthropological implications and gender equality in Igbo society. He focuses on the cosmogonic myths and religious feasts connected with some staple crops such as yam, cocoyam and pumpkin. According to him, the phenomenon of crop gendering in Igbo culture creates an equal and enabling space for complementarity and participation of both sexes not only in agriculture, but also in the socio-political and religious spheres. Ufearoh maintains that crop gendering encourages gender inclusiveness, complementarity, and autonomy. The quest for gender equality is propelled by principles that make for equal and fair treatment of both sexes.

In Chap. 8, "The Religious Significance of Mushrooms among the Shona People of Zimbabwe: An Ethnomycological approach", Bernard P. Humbe focuses on the religious importance of wild mushroom in indigenous Shona agrarian thought. The sacredness of mushroom is shown by paying attention to the place they sprout, their colour, shape, age, sex of the person who finds them, and how and when they are harvested. These beliefs and practices show a deep connection between African agrarianism and spirituality.

Part Three runs from Chaps. 9 to 14. The contributions in this part concentrate on African agrarian and environmental ethics. In Chap. 9, "The Consubstantiality of Living Things: Towards a Mandingo Cosmo-Anthropocentric Ethics", Belko Ouologuem discusses the pre-colonial Mandingo (also Mandinka), pre-colonial Sahelian environmental agrarianism. According to the Mandingos, nature is a precious partner in the cosmos because of its material and spiritual value. The Mandingos developed a relational ethic in which human welfare was intimately linked to environmental well-being. Grounding his reflection on pre-modern and pre-colonial Mandingo oral and written sources such as the Mandingo Hunters' Oath and the Charter of Kouroukan Fuga (1236),[3] Ouologuem shows that the Mandingos included nature in their onto-metaphysical, epistemic and moral conception of the universe.

In Chap. 10, "*Ɨwu-ɨ-Kom-ɨ-Twal:* Kom Agrarian-Environmental Ethics", Mbih Jerome Tosam uses the Kom triadic worldview as grounding for Kom agrarian-environmental philosophy. He argues that the Kom triad (*ɨwu-ɨ-kom-ɨ-twal*/the Kom three hands/arms) which includes "*wayn*" (a child), "*afo-aghina*" (food), and "*nyamngvin*" (communal flourishing) is the guiding principle of moral, social, spiritual, and ecological well-being in Kom culture. According to this triadic outlook, humans must procreate for cultural perpetuation; produce for subsistence, and strive for good health and for common well-being. From this outlook, the community includes non-human persons and Nature, which are related and interdependent on

[3] These documents show that not all pre-colonial sub-Saharan African societies were oral as it is often claimed.

each other. Hence, to count, morally, a being must not be necessarily real, rational or be able to defend their interest. Because humans share the cosmos with Nature, they cannot use its resources without consulting or obtaining permission from Nature. Tosam shows that according to the Kom agrarian worldview, there is deep a consideration for future generations.

In Chap. 11, "Land Ethics among the Traditional Annangs of Southern Nigeria: Traditional Environmental Ethics, Challenging Contemporary Hostilities towards our Planet", Dominic Umoh adopts a phenomenological and exploratory approach to demonstrate that traditional Annang religion has profound and far-reaching implications on the cultural life of the people. According to traditional Annang religion, the land, the mother Earth, is considered as a Deity who "sustains, nourishes, grooms, upholds, pampers, caresses and shelters her children in the warmth and tenderness of her bosom." In traditional Annang society, Nature was considered worthy of respect on the same level as humans. Nature had "rights and privileges and was allowed to flourish, mature, renew, heal and reproduce. Hence, as our (human) closest neighbours, any threat on the environment was tantamount to human self-destruction. Umoh advocates a return to indigenous African ancestral wisdom—to a new civilization of "ecological humanism", in order to save nature from extinction. For Umoh, although our forebears "were illiterate"… they were not fools."

Hwati in Chap. 12, "Shangwe Environmental Ethics: A Panacea for Agrarian Problems in Gokwe", discusses pre-colonial indigenous Shangwe agrarian beliefs, moral values and practices which were employed in the conservation of the natural environment, resolution of agrarian conflicts, and maintenance of harmonious interhuman relationship in the community. She argues that government policies should take traditional ethics seriously in the management of environmental issues.

Michael K. Lang in Chap. 13, "Agrarian Rituals, Food Security and Environmental Conservation in the Bamenda Grassfields of Cameroon", examines agrarian rituals in the Bamenda Grassfields of Cameroon and demonstrates their implications in food security and environmental preservation. Lang suggests that a new impetus be placed on agrarian rituals and practices because of their continuous bearing on the success of agricultural endeavours and their capacity to control ecological conditions in ways that contribute to environmental protection and food security. He uses ritual practices that are deeply entrenched in indigenous worldviews to show the intimate connection between religion, agriculture, and sustainability.

In Chap. 14, "Indigenous African Eco-communitarian, Agrarian Philosophy: Lessons on Environmental Conservation and Sustainability from the Nso culture of North West Cameroon", Peter Takov discusses the agrarian-environmental thought of the Nso' of North West Cameroon. The Nso' eco-bio-communitarian worldview considers humans as beings 'with others'; it emphasizes cooperative and communitarian values in living with others. Here, 'others' do not only refer to human others, nature and the supernatural are also included in the community. Takov argues that there are features of Nso' agrarian communitarianism which can help restrain the anthropocentric self-interest which has been responsible for the current global environmental catastrophe. Takov is convinced that indigenous Nso agrarian philosophy can be recuperated in spite of the threat these beliefs and values face as a result of epistemic colonization and globalisation on African cultures.

Part IV of this book focuses on indigenous African Knowledge Systems (IKSs) and how these can serve as a repository for Africa's development. In Chap. 15, "The emergence of a re-humanizing pedagogy for African agrarian philosophy", Birgit Boogaard, Bernard Yangmaadome Guri, Daniel Banuoku, David Ludwig, and David Fletcher argue that local people are knowledge authorities on African IKSs, values and practices, and have been regularly improving and adjusting practices in accordance with their changing environment. They argue that although African agrarianism existed before the advent of universities, contemporary African philosophy has tended to focus largely on speculative theories while neglecting agricultural practices and experiences of local people. Boogaard and her colleagues maintain that for African agrarian philosophy to be meaningful, it should not remain theoretical; it must be developed and put into practice by the local people. For them, African agrarian philosophy can contribute to the rediscovery and re-humanization of African values by revitalizing and re-establishing indigenous agricultural knowledge, beliefs and customs. Boogaard et al. propose seven themes for a pedagogy of African agrarian philosophy: "memory; a dialogical student–teacher relation; the value of lived experiences; intergenerational and spiritual methods of education; relationality of human beings and Mother Earth; unity between theory and practice; and critical consciousness about people's rights."

Alloy S. Ihuah in Chap. 16, "African Endogenous Knowledge and Sustainable Development: Evolving an African Agrarian Philosophy", illustrates how modern methods of living and farming are threatening traditional African lifestyles with virtual extinction through unresponsive development endeavours in which indigenous peoples' knowledge systems are not taken into consideration. According to Ihuah, by yielding to the dictates of modern technology, which is incompatible with indigenous ontologies, the human person is replaced thereby debasing Nature and the resources necessary for harmonious living. Ihuah argues that African endogenous knowledge in agriculture is a strategic resource to salvage the environmental problems that have been brought about by the irresponsible use of technology. He contends that African endogenous knowledge systems are a reservoir for environmental management and policies for resource development in Africa's complex and ageing ecological system. Ihuah advances three reasons why it is imperative to consider indigenous African knowledge systems in agriculture and in the management of natural resources: (i) "The long-term generation and transmission of knowledge of the local ecosystem offers a unique historical perspective into indigenous risk adjustment options; (ii) Modern scientists involved in the management and conservation of areas that may be ecologically fragile or marginal, or that contain genetically important plant or animal biodiversity, may benefit greatly from alternative knowledge; and (iii) African endogenous knowledge does not separate humans from natural environment;" because "the world is interconnected and interdependent."

In his Chap. 17, "The Shona People's 'Zunde raMambo' (King's granary) as a model for social responsibility: A task for higher education systems", Erasmus Masitera examines some humanising insights which were reflected in agricultural and communal practices of the Shona people of Zimbabwe. He demonstrates that through the 'Zunde raMambo' (King's granary), Shona communities ensured that food security, poverty and hunger were reduced, and good health and social well-being enhanced. The 'Zunde raMambo' activities led to a community welfare practice in which the underprivileged received assistance. Also, within this custom, ethical values such as communal responsibility, sharing and caring, cooperative work and common welfare were promoted. Masitera maintains that the informal safety nets that 'Zunde raMambo' provide demonstrates the indigenous African idea of community development which contemporary society ought to learn from. For him, such moral characters can be achieved through an educational system that reflects African indigenous moral norms that encourage communal well-being and cooperative living.

In Chap. 18, "The practice of African traditional medicine and agrarianism in Zimbabwe: The Quest for Karanga Agrarian Practice in Madamombe area of Chivi District", *Tasara Muguti* discusses the value of indigenous traditional medicine in agriculture among the Karanga people of Chivi in Zimbabwe, which has survived in spite of its denigration and outright rejection by the colonialists. The chapter explores the extent to which traditional medicine is continuously relied upon by the people of the Madamombe area in dealing with crop enhancement, processing, and preservation of seed and crop yield, storage of harvest, animal husbandry and other challenges in agriculture. Maguti argues for the need to preserve this fast fading, but relevant aspect of indigenous knowledge because the current challenges confronting Africa requires a multi-faceted approach, which will necessitate both indigenous and modern systems of knowledge.

Part V runs from Chaps. 19 to 23. In this part, we discuss some contemporary agrarian issues in Africa such as the problem of climate change and global justice, food security, the suppression of the rights of small scale farmers by governments and multinational corporations, and the decolonisation of agriculture. In Chap. 19, "Henry Odera Oruka's Parental Earth Ethics as Ethics of Duty: Towards Ecological Fairness and Global Justice", Pius Mosima discusses Henry O. Oruka's Parental Earth Ethics and notion of global justice as an important contribution to African agrarianism. He argues that there is an urgent need to consider marginalised and silenced voices in the Global South in the debate on global justice; a field which has been largely dominated by Euro-American perspectives. According to Mosima, the lack of a genuine intercultural and inclusive discourse on global justice raises doubts as to whether the current concepts of global justice can transcend regional and cultural boundaries. He maintains that Oruka's Parental Earth Ethics could be adopted as a new standard for global justice and as a foundation for global environmental ethics. Oruka's Parental Earth Ethics could be a genuine resource in our search for the promotion of global justice in the use and fair distribution of global wealth for the common good of all. Oruka's principle of global justice is essentially informed by the ethical principle of the human minimum, which ought to inspire us in our moral and political consideration of others including the natural environment.

In Chap. 20, "Rethinking Shangwe Traditional Philosophy in Resolving Agrarian Wrangles in Contemporary Gokwe Communities", Elvis Tsvangirayi Siziva investigates indigenous African (Gokwe) approaches to resolving agrarian disputes.

According to Siziva, although some of these traditional African methods, laws and customs in resolving agrarian conflicts were ignored and replaced by Western colonial legal and moral approaches, on the basis that traditional laws are ambiguous and unreliable in managing agrarian issues, indigenous ways of conflicts resolution related to land tenure, crop theft, livestock and other related agrarian disputes have not faded. Indigenous laws have not been adequately applied in dealing with agrarian issues.

Maduka Enyimba and Victor Nweke in Chap. 21, "Food Security as a Fundamental Human Right: A Philosophical Consideration from Africa", discuss the question of food security and human rights. They argue that among the three basic human needs, namely, food, shelter and clothing; food is the most significant. This is the case because food is a necessary requirement for the preservation of human life. According to Enyimba and Nweke, if food is a prerequisite for human life, then the right to life suggests the right to food security. And since the right to life is a fundamental human right, food security should therefore be a fundamental human right. According to them, there is a nexus between poverty, violent crime and food insecurity in Africa.

In Chap. 22, "*Murimi munhu*: A Quest for Decoloniality in Black African "Small Scale" Subsistence Farmers in Rural "Reserve" Zimbabwe", Joseph Pardon Hungwe uses the song, *Murimi munhu,* by the Zimbabwean musician, Oliver Mtukudzi to underscore the coloniality embedded in the treatment of "small scale" subsistence farmers in Zimbabwe. African small scale farmers are "systematically dehumanised through derision, negative stereotyping, squalid living conditions and government's general systematic marginalisation." Hungwe traces the origin of these dehumanising tendencies in the colonial arrangement in which black Africans were confined to rural areas (*ruzevha).* He presents two ways in which small scale farmers are marginalised: Firstly, the unfair market forces deny them the power to determine the prices of their produce. Secondly, the "urbanised" populace derides small scale subsistence farming as an engagement for the uneducated and "unemployable." Hungwe argues that the decolonial approach result in two things: Firstly, it permits him (the author) to confront some of the negative stereotypes and dehumanizing practices that are expressed against "small scale" subsistence farmers in rural "reserve" Zimbabwe, which do not only debase and deride African communal values, but also maintain coloniality. Secondly, by addressing coloniality as the basis of dehumanising tendencies towards subsistence farmers, the chapter seeks to conceptually restore the humanity of subsistent farmers in rural Zimbabwe.

In the final chapter (Chap. 23), "The Farm in Colonial and Postindependence Imagination: A Crisis of Continuity" Mbuh Tennu Mbuh argues that the establishment of colonial farms, which were essentially aimed at serving the subsistence interest of the colonialists and their home countries, undermined and distorted indigenous African agricultural values, beliefs, cultural practices, and economic interests. Mbuh argues that because neocolonial and postindependence African economies have failed to decolonise and displace colonial agrarian disarrangements, this has resulted in "a continuity of redundancy in replicating mega policy and even of stagnation." For this reason, he contends that the ongoing drive for "cultural glocalism" calls for the need to indigenize agriculture by giving voice to indigenous African agrarian values and practices.

References

Behrens, K. 2014. Toward an African Relational Environmentalism. In *Ontologized Ethics: New Essays in African Meta-Ethics*, eds. E. Imafidon, and J.A. Bewaji, 55–72. Lanham: Lexington Books.

Behrens, K. 2017. The Imperative of Developing African Eco-Philosophy. In *Themes, Issues and Problems in African Philosophy*, ed. I.E. Ukpokolo, 191–204. New York, Palgrave: Macmillan.

Berry, W. 2013. The Agrarian Standard. In *The Essential Agrarian Reader: The Future of Culture, Community, and the Land*, ed. Norman Wirzba, 22–32. Lexington, Kentucky: The University Press of Kentucky.

Boogaard, B. 2019. The Relevance of Connecting Sustainable Agricultural Development with African Philosophy. *South African Journal of Philosophy* 38 (3): 273–286.

Boogaard, B., D. Ludwig, B.Y. Guri, and D. Banuoku. 2023. A Reconsideration of African Spirituality in Agricultural Development Projects: Traditional Ecological Knowledge from Dagara Elders in Koro, Ghana. In *Beauty in African Thought—Critique of the Western Idea of Development*, eds. Angela Roothaan, Bolaji Bateye, Mahmoud Masaeli, Louise Müller. Lexington Books.

Bujo, B. 1998. *The Ethical Dimension of Community: The African Model and the Dialogue Between North and South*. Nairobi: Paulines Publications.

Bujo, B. 2009. Ecology and Ethical Responsibility from an African Perspective. In *African Ethics: An Anthology of Comparative and Applied Ethics*, ed. M.F. Murove, 281–197. Scottville, MI: University of KwaZulu-Natal Press.

Chimakonam, J. 2017. African Philosophy and Global Epistemic Injustice. *Journal of Global Ethics* 13 (2): 120–137.

Chemhuru, M. 2014. The Ethical Import in African Metaphysics: A Critical Discourse in Shona Environmental Ethics. In *Ontologized Ethics: New Essays in African Meta-Ethics*, eds. E. Imafidon, and J.A. Bewaji, 73–88. Lanham: Lexington Books.

Chemhuru, M., ed. 2019. *African Environmental Ethics: A Critical Reader*. Gewerbestrasse: Springer.

Chimakonam, J., ed. 2018. *African Philosophy and Environmental Conservation*. London: Routledge.

Falvey, L. 2020. *Agriculture and Philosophy: Agricultral Science in Philosophy*. Songkhla: Thaksin University Press.

Freyfogle, E.T. 2008. Fostering a Culture of Land Commentary on '"Agrarian Philosophy and Ecological Ethics."' *Science and Engineering Ethics* 14: 545–549.

Graness, A., and K. Kresse, eds. 1997. Sagacious Reasoning: H. Odera Oruka in Memoriam, Frankfurt, Peter Lang.

Graness, A. 2012. What is Global Justice? Henry Odera Oruka's Contribution to the Current Debate. *Journal on African Philosophy* 6: 31–46.

Graness, A. 2015. Is the Debate on 'Global Justice' a Global One? Some Considerations in View of Modern Philosophy in Africa. *Journal of Global Ethics* 11 (1): 126–140. https://doi.org/10.1080/17449626.2015.1010014.

Gyekye, K. 2010. African Ethics. In *The Stanford Encyclopedia of Philosophy*, ed. Edward Zalta. http://plato.stanford.edu/archives/fall2010/entries/african-ethics.

Heath, F.E. 2022. Agrarianism. *Encyclopaedia Britannica*, 2020, https://www.britannica.com/topic/agrarianism. Accessed 7 July 2022.

Ikuenobe, P. 2006. Philosophical Perspectives on Communalism and Morality in African Traditions, Lanham, Lexington Books.
Ikuenobe, P. 2014. Traditional African Environmental Ethics and Colonial Legacy. *International Journal of Philosophy and Theology* 2: 1–21.
Ikuenobe, P. 2015. Relational Autonomy, Personhood and African Traditions. *Philosophy: East and West* 65 (4): 1005–1029.
Imafidon, E. 2014. On the Ontological Foundation of a Social Ethics in African Traditions. In *Ontologized Ethics*: *New Essays in African Meta-Ethics*, eds. E. Imafidon, and J.A. Bewaji, 37–54. Lanham: Lexington Books.
Kagame, A. 1956. *La Philosophie bantu rwandaise de l'e^tre*. Bruxelles : Academie Royale des Sciences Coloniales.
Kagame, A. 1969. Le Fondement ultime de la morale bantu. *Au Coeur De L'afrique* 5: 231–236.
Kelbessa, W. 2005. The Rehabilitation of Indigenous Environmental Ethics in Africa. *Diogenes* 52 (3): 17–34. https://doi.org/10.1177/0392192105055167.
Kelbessa, W. 2014. Can African Environmental Ethics Contribute to Environmental Policy in African? *Environmental Ethics* 36: 31–61.
Kelbessa, W. 2015. African Environmental Ethics, Indigenous Knowledge, and Environmental Challenges. *Environmental Ethics* 37 (4): 387–410.
Kugedera, A.T., N. Sakadzo, T. Muiseva, E. Chivehenge, S.L. Kugara, F. Chimbwanda, and M.T. Decide. 2021. The Link Between Indigenous Knowledge Systems and Sustainable Agricultural Production in Zimbabwe. *African Journal of Religion, Philosophy and Culture* 2 (2): 20–39.
LeVasseur, T. 2015. Agrarianism. In *The SAGE Encyclopedia of Food Issues*, ed. Ken Albala. Thousand Oaks: SAGE Publications, Inc.
Maathai, W. 2010. We Are Called to Help the Earth to Heal. In *Moral Ground: Ethical Action for a Planet in Peril*, ed. K.D. Moore and M.P. Nelson, 275–278. San Antonio, TX: Trinity University Press.
Mangena, Fainos. 2013. Dis Cerning Moral Status in the African Environment. *Phronimon* 14 (2): 25–44.
Masaka, D., and M. Chemhuru. 2010. Taboos as Sources of Shona People's Environmental Ethics. *Journal of Sustainable Development in Africa*, 12 (7): 121–133.
Masitera, E. (ed.). 2021. *Philosophical Perspectives on Land Reform in Southern Africa*. Cham: Palgrave-Macmillan.
Mbiti, J.S. 1969. *African Religions and Philosophy*. London: Heinemann.
Metz, T. 2022. *A Relational Moral Theory: African Ethics in and Beyond the Continent*. Oxford: Oxford University Press.
Mofor, C. 2008. *Plotinus and African Concepts of Evil: Perspectives in Multicultural Philosophy*. Bern: Peter Lang.
Montmarquet, J.A. 1985. Philosophical Foundations for Agrarianism. In *Agriculture and Human Values*. Spring.
Mosima, P. 2021. Cosmic Interconnectedness: An African Exploration Towards a Dynamic Understanding of Nature. *Satya Niyalam Chennai Journal of Intercultural Philosophy* 39: 62–80.
Murove, F.M. 2004. An African Commitment to Ecological Conservation: The Shona Concepts of Ukama and Ubuntu. *Mankind Quarterly*, XLV: 195–215.
Niekerk, A. (ed.). 2005. *Ethics in Agriculture—An African Perspective*. Springer.
Nkulu-N'Sengha, M. 2009. Bumuntu. In *Encyclopedia of African Religion*, ed. M.K. Asante and A. Mazama, 142–147. Los Angeles: Sage.
Nyerere, J.K. 1968. *Ujama'a: Essays on Socialism*. London: Oxford University Press.
Oruka, H.O. 1993. Parental Earth Ethics. In *Quest*, vol. VII, no. 1 (June 1993), 20–27. Rejoinders. In: *Quest*, vol. VII, no. I, 106–109.
Oruka, H.O., and C. Juma. 1994. Ecophilosophy and Parental Earth Ethics. In *Philosophy, Humanity and Ecology*, ed. Henry Odera Oruka, 115–129. Nairobi: ACTS Press.
Ramose, M.B. 2005. *African Philosophy Through Ubuntu*. Harare: Mond Books Publishers.

Ramose, M.B. 2009. Ecology Through *Ubuntu*. In *African Ethics: An Anthology of Comparative and Applied Ethics*, ed. M.F. Murove, 308–314. Scottsville: University of KwaZulu-Natal Press.
Roothaan, A. 2019. *Indigenous, Modern and Postcolonial Relations to Nature: Negotiating the Environment*. London and New York: Routledge.
Riggio, A. 2015. *Ecology, Ethics and the Future of Humanity*. New York: Palgrave Macmillan.
Sachs, C.E. (ed.). 2019. *Gender, Agriculture and Agrarian Transformations: Changing Relations in Africa, Latin America and Asia*. Abingdon: Routledge.
Senghor, L.S. 1995. Negritude: A Humanism of the Twentieth Century. In *I Am Because We Are: Readings in Black Philosophy*, eds. Fred Lee Hord, and Jonathan Scott Lee, 45–54. Amherst: University of Massachusetts Press.
Singer, R., and J. Motter. 2015. Agrarian Environmental Rhetoric: A Theoretical Conceptualization. Presented at Bridging Divides: Spaces of Scholarship and Practice in Environmental Communication The Conference on Communication and Environment, Boulder, Colorado, June 11–14, 2015. https://theieca.org/coce2015.
Tangwa, G.B. 1996. Bioethics: An African Perspective. *Bioethics* 10 (3): 183–200.
Tangwa, G.B. 2004. Some African Reflections on Biomedical and Environmental Ethics. In *A Companion to African Philosophy*, ed. Kwasi Wiredu, 387–395. Malden: Blackwell Publishers.
Tempels, P. 1959. *Bantu Philosophy*. Paris: Présence Africaine Éditions.
Thompson, B.P. 2008. Agrarian Philosophy and Ecological Ethics. *Science and Engineering Ethics* 14: 527–544.
Thompson, B.P. 2012. *Nature Politics and the Philosophy of Agriculture*.
Tosam, M.J. 2014. The Relevance of Kom Ethics to African Development. *International Journal of Philosophy* 2 (3): 36–47. https://doi.org/10.11648/j.ijp.20140203.12.
Tosam, M.J. 2019. African Environmental Ethics and Sustainable Development. *Open Journal of Philosophy* 9: 172–192.
Udeani, C.C. 2008. Traditional African Spirituality and Ethics—A Panacea to Leadership Crisis and Corruption in Africa? *Phronimon* 9: 65–72.
Wiredu, K. 1994. Philosophy, Humankind and the Environment, in Henry Odera Oruka (ed.), *Philosophy, Humanity, and Ecology, Nairobi, Kenya, ACTS Press* 30–48.
Wirzba, N., ed. 2013. *The Essential Agrarian Reader: The Future of Culture, Community, and the Land*. Kentucky, The University Press of Kentucky: Lexington.

Mbih Jerome Tosam is Associate Professor of Philosophy at the University of Bamenda, Cameroon. He obtained his Ph.D in Philosophy from the University of Yaoundé I, Cameroon in 2011. He is former Chair of the Department of Philosophy at the Higher Teacher Training College (HTTC) Bambili (2011–2017) and the at the Faculty of Arts of the University of Bamenda, Cameroon (2017–2021). His research interests are in the areas of Bioethics, African Philosophy, and Intercultural Philosophy. Some of his publications have appeared in the folowing joiurnals: *South African Journal of Philosophy, Annali di studi religiosi, Medicine, Health Care and Philosophy: A European Journal, Deveolping World Bioethics, Journal of World Philosophies*, and *Polylog: Forum for Intercultural Philosophy*.

Part I
African Communitarian Agrarianism

Chapter 2
Unpacking Ndebele Agrarian Metaphors for the Promotion and Preservation of Communal Social Development

Faith Sibanda

Abstract Africans in general have had a long history of survival through agrarian means which are mainly agriculture and animal husbandry long before the advent of the so called "civilized" western/modern ways. Much as it is apparent that methods of farming which rely mostly on modern technological advancement have, in many instances, resulted in more production and harvest, it is also apparent that traditional agrarian methods of farming have neither been dislodged nor displaced by the same. This is evident in the current agro-ecosystem of Zimbabwe which has recently realized a bumper harvest by reverting back to the traditional *pfumvudza/intwasa* methods of farming which were promoted by the president of the country. While agriculture is commonly viewed as a physical or economic activity, this chapter seeks to espouse on the social, spiritual and even physiological benefits that Ndebele people have, and continue to enjoy from it. The chapter argues that agrarian metaphorical nuances in the Ndebele language had an effect of moulding individuals as well as the society into socially acceptable citizenry. The chapter makes use of the theory of *ubuntu/unhu* in order to examine Ndebele proverbs and other sayings so as to demonstrate that their agrarian exposure and practice provided a solid pedestal for them to promote and mould among themselves, a socially just, humane and sustainable society. Contemporary societies have a lot to benefit from the agrarian experiences not only economically but, socially since they promoted hard work, good and morally acceptable behaviour.

Keywords Agrarian · Ndebele language · Animal husbandry · *Ubuntu* · Good morals · Acceptable behaviour · Proverbs

F. Sibanda (✉)
Department of African Languages and Literature, Great Zimbabwe University, Box 1235, Masvingo, Zimbabwe
e-mail: sibandaf@gzu.ac.zw

M. J. Tosam and E. Masitera (eds.), *African Agrarian Philosophy*, The International Library of Environmental, Agricultural and Food Ethics 35,
https://doi.org/10.1007/978-3-031-43040-4_2

2.1 Introduction

Different societies around the world have got different survival methods and skills which have sustained them over the years. There are a variety of sociological, political, and spiritual aspects which have close similarities among many African nations so much so that their behaviours and conduct almost resemble this close connection. Apparently, what seems to be most common amongst them is the "culture" whose characteristics are almost similar all the way from Cape to Cairo. Most notably, their agriculture or agricultural practices have so much direct and indirect link with their culture spanning right from the way they plant, right through to harvests, even to consumption and storage of their various farm produce. It may not be very possible to talk about agriculture and exclude culture in the African context. In fact, according to De Rossi and Taylor (1985: 24), the word "culture" derives from a French term, which in turn derives from the Latin "colere," which means to tend to the earth and grow, or cultivation and nurture, thereby suggesting an intimate relationship between culture and agriculture whether agrarian or otherwise.

Ndebele people of Zimbabwe, like most African indigenous societies, have a very long history of undertaking agricultural activities particularly subsistence farming comprising land tillage as well as animal husbandry. Their long trail of close interaction with the land and what it offers is reflected in various cultural artifacts, songs, poems, religious activities as well as the language and literature of the people. This research focuses on how the various agrarian farming experiences have shaped the worldview of the Ndebele people in making them mould their members in order to produce acceptable citizens. The research is premised on the idea that, agrarian farming tendencies are intertwined with the culture of the people so much so that, even the perceptions and perspectives for the same people are shaped, determined and influenced by the same.

This chapter examines how Ndebele people used their agrarian farming experiences in cultivating noble characters among their members through formulation of linguistic metaphors whose tap roots go deep into the streams that have continued to water agrarian farm produce. The chapter interrogates the various metaphors, particularly proverbs in explaining some of the cultural expectations as well as cultural norms and values propagated through the people's experiences with the agrarian farming tendencies. Nyamnjoh (2021: 37) points out the cultural value of proverbs in understanding a people's life and behaviour when he says that:

> Proverbs, as condensed wisdom drawn from human experience, provide a rich resource for understanding, *inter alia,* how African communities have, through the ages, negotiated and navigated questions of being and belonging through a myriad of encounters with one another, as well as with people from elsewhere and those they have come to know and relate to through their own mobility. Proverbs are cherished repertoires of humans as dynamic and creative innovators in conversation with the geographies and environments that continually feed their individual and collective selves and their appetite for the nuanced complexities of being human.

The various perceptions about life in general, the interpretations about various aspects of life and phenomena, the explanations about the meanings of life and its complexities, can all be understood when one takes a deep contextual analysis of Ndebele proverbs, especially those which are informed by the traditional agrarian experience.

2.2 African Indigenous Knowledge: An Overview

African indigenous languages and knowledge systems (AIKS) are under threat due to various internal and external forces such as migration, cultural imperialism as well as globalization. Owusu-Ansah and Mji (2013: 7) argue that "… culture is the 'lens' through which a person perceives, interprets and makes sense of his or her reality, if we speak of the inclusion of African indigenous knowledge in any investigation, we would be speaking about the examination of African reality from the perspective of the African and not with the African on the periphery." The same observation has been noted with concern by both the United Nations and the World Bank who believe that African indigenous knowledge, and indeed the indigenous languages, which are the vehicles that carry and transpose indigenous cultures are under siege from the concept of globalization. The Report of the permanent Forum on Indigenous Issues read at the session of the United Nations observed that:

> Indigenous peoples are major stakeholders for UNESCO's mandate, as holders of rich and unique knowledge systems, traditions and languages that constitute an important segment of human cultural diversity. UNESCO recognizes that indigenous peoples can give a valuable contribution to its mandate in the fields of education, natural and social sciences, culture, communication and, in order to reflect their importance in its work, UNESCO has engaged with them in several manners.

This concern emanates from the fact that both African IKS and indigenous languages continue to be vilified as backward, barbaric and static yet, there is ample evidence pointing to the fact that Africans themselves in general, and Ndebele people in particular, have utilized these languages for survival through smooth and difficult circumstances. The negative portrayal of African languages has prompted this research whose thrust, among other things, is to demonstrate that Ndebele people have used their agrarian experiences to formulate linguistic metaphors which have assisted them in shaping the moral fibre which has kept their society together. In emphasizing the use and strength of agrarian metaphors in shaping the high moral fibre among African communities, Tutu (1999: 13) postulates that:

> These proverbs not only demonstrate that the Bantu people are generous people, they say something about the *Hunhu/Ubuntu* strand that runs through the traditional thought of almost all the Bantu cultures of Southern Africa whereby everything is done to promote the interests of the group or community. The proverbs show that the Bantu people are selfless people as summarised by the Nguni proverb which we referred to earlier, which says: *Umuntu ngumuntu ngabantu* (*"a person is a person through other persons."*) or, as they put it in Shona: *Munhu munhu muvanhu, translated to English as* (*"a person is a person through other persons."*) Without the attribute of *generosity*, it may be difficult to express one's selflessness.

Basically, the ethical codes which are reflected in various agrarian metaphors stand as testimony to the fact that African societies in general, and Ndebele people in particular, are guided by community views and standards of behaviour.

2.3 Conceptual Underpinnings

Carrying out a study about African cultures and languages requires an appropriate conceptual framework which adequately grounds the arguments in a manner that does not expose them to undue criticism and contempt. African culture, AIKS, African indigenous languages and literatures have been so much vilified and brutalized especially by a lot of scholars from the hostile North who have actually, in the past, condemned them as not worthy subjects of discussion. Some of the warped opinions about Africa and Africans with particular reference to the issues around their cultural activities, which, in all sense includes agrarian activites, for instance, such disturbing sentiments as the following from Trevor-Roper (1960: 8) which represents many others:

> "...at this point we leave Africa, not to mention it again. For it is no historical part of the world; it has no movement or development to exhibit. Historical movements in it – that is in its northern part – belong to the Asiatic or European World...".

These unfortunate ideas about the state of things among Africans has continued to guide most decisions and attitudes by those who have not taken their time to investigate for themselves what exacty is happening in and among Africans. Needless to mention that some detrimental decisions have been made in the past in the understanding that African cultures are so backward and have no need to be consulted even when making programs that affect them. In this case, the role of the local cultural activities on food security and nutrition have not been given enough cognition and thus aggraian activities have not been afforded the requisite status yet they have been a source of livelihood for local people for so many years. The effects of such an approach is highlighted by so many scholars including Mawere (2010: 213) who points out the following:

> The West considered Africa as a "dark continent", and hence despised its traditions, customs, belief systems, and indigenous knowledge systems as diabolic, and backward. This had a negative impact to Africa's own socio-economic and political development. Africa's valued traditions and knowledge systems had to change to fit in with the western scientism and modernity.

In light of such sad developments, it becomes imperative and necessary for a research of this nature to be grounded on the right and relevant conceptual grounding so that it can avoid to directly or indirectly contributing to the vile subjugation of African cultures and knowledge.

Nzewi (2007: 14) argues that it is the duty of all those who undertake the task of researching, writing, or making analyses pertaining African cultures, AIKS, or African languages and literature to make sure that they continue to fight the war of empowering and re-emancipating them from the horrific negative criticisms that they have been subjected to. He is of the view that, by emancipating the various African indigenous cultures and knowledge, in essense, it goes a long way in emancipating the local people themselves who have been the main unfair victims of the colonial system. He says that:

> Contemporary Africans must strive to rescue, resuscitate, and advance our intellectual legacy or the onslaught of externally manipulated forces of mental and cultural dissociation now rampaging Africa will obliterate our original intellect and lore of life (Nzewi 2007: 14).

In other words, writing, reading, analyzing African cultures can no longer be considered as 'business as usual' just because, as highlighted above, it has actually become a duty for scholars and researchers to do so in a manner that does not continue to injure the already injured. This approach finds meaning in the words of Ngugi (1982: 150) who advocates for an approach that provides us with "means of knowledge about ourselves … (and how we should relate to each other) and after we have examined ourselves, we radiate outwards and discover other peoples and worlds around us". Among the various Afrocentric and post-colonial theories that have emerged recently as a way of emancipating the sulbatan, this research is grounded on the theory of *ubuntu/unhu* (humanness) whose main thrust is about promoting and taking advantage of the humaneness which is inherently inert among various African societies.

Ubuntu is an ethic or humanist philosophy focusing on people's allegiances and relations with each other. The word has its origin in the Bantu languages of southern Africa. *Ubuntu* is seen as an exclusive African indigenous concept. The *Ubuntu* operating system was named for this principle. In other words, this conceptual framework, which is locally designed, is based on an idea which is meant to address peculiar African problems in an indigenous way instead of tackling them using a universal formula. The theory is meant to move away from looking at African people through an exotic lens because it is in doing so that they end up seeing an impaired vision of themselves. As a way of emphasising the need to adopt a theory that is relevant to our situation and problems, Hapanyengwi-Chemhuru and Shizha (2012: 23) go on to support the theory of *ubuntu* by suggesting that:

> In our context, the philosophy that comes to mind is the philosophy of *unhu/ubuntu* (humanity, good behaviour, respectfulness to others, pleasant and honest). *Unhu* forces us to come to terms with the fact that whether we are African, European, Shona, or Ndebele, *tose tirivanhu*. (In spite of our racial or ethnic diversity, the bottom line is that we are all human beings).

In line with the objectives of this chapter, which are, to examine the role played by agrarian farming experiences of the people in shaping their societies to be morally sound, the concept of *ubuntu* forms a good baseline on which to anchor the various arguments since it provides the yardstick by which to measure successes and failures of the society under study. At the centre of the Africans' life is the belief that we are all joined by a single umbilical cord whose length and breadth encircles all of us as the black indigenous community making it inevitable for us to look down upon each other, let alone fight one with another. The same feeling is echoed by Nussbaum (2003: 21) who postulates that:

> *Ubuntu* is the capacity in African culture to express compassion, reciprocity, dignity, harmony, and humanity in the interests of building and maintaining community. *Ubuntu* calls on us to believe and feel that: your pain is my pain, my wealth is your wealth, and your salvation is my salvation. In essence, *Ubuntu*, a Nguni word from South Africa, addresses our interconnectedness, our common humanity, and the responsibility to each other that flows from our connection.

When one takes a look at the various Ndebele metaphors which have links with their experiences in long years of agrarian farming, there is ample evidence pointing to the fact that Africans in general, and Ndebeles in particular, moulded their life and lifestyles around the concept of *ubuntu*. As a locally brewed concept, it forms the basis for most life decisions by members of the society and it gives meaning to a research of this nature whose thrust is to examine the link between agrarian farming experiences and the shaping of morally upright behaviour and standards.

2.4 Methodology

This chapter falls within the field of social sciences and, humanities and as such, intends to unpack the agrarian philosophy of the Ndebele people and how they use it to shape societal moral standards as well as achieve social development. The study utilized documentary analysis, observation and indepth interviews for data collection. Under the documentary analysis, the researcher made use of information from various literature on proverbs and other language aspects as well as electronic media such as television programs and the internet were consulted in order to access current trends in the perceptions of people as regards AIKS. This approach is in tandem with the dictates of the qualitative research design because it is the most appropriate design in analyzing human behaviour, ideas, ethics, among other aspects. Emphasising the importance of adopting the best research design, Bless, Higson-Smith and Sithole (2014: 130) say that:

> A research design relates directly to the answering of a research question. Because research is a project that takes place over an extended period of time, it is unthinkable to embark on such an exercise without a clear plan or design, a sort of blue print.

On the other hand, it was necessary to carry out strategic interviews with local people in a bid to gather information on the role and value of AIKS in the yesteryear in comparison with the current trends. Interview samples included the old and the young in order to vary the information depending on experiences in order to make analyses and conclusions that are fair and appropriate. This approach is in tandem with what is said by Chiromo (2006: 8) who views qualitative research as a systematic, interactive and subjective approach used to describe life experiences and giving them a meaning. In essence, qualitative research pre-occupies itself with investigating social phenomena which ordinarily is not quantifiable such as the ideas, ideologies and feelings of people whether as a group or as individuals. Putting it more explicitly, Denzin and Lincoln (2011: 4) postulate that:

> Qualitative research begins with assumptions and the use of interpretive theoretical frameworks that inform the study of research problems addressing the meaning individuals or groups ascribe to a social or human problem. To study this problem, qualitative researchers use an emerging qualitative approach to inquiry, the collection of data in a natural setting sensitive to the people and places under study, and data analysis that is both inductive and deductive and establish patterns or themes. The final written report or presentation includes the voices of participants, the reflexity of the researcher, a complex description and interpretation of the problem, and its contribution to the literature or a call for a change.

Effectively, the qualitative research design allows the researcher to interact closely with the subjects of research, thereby, affording the opportunity to even observe them practically as they undertake some of the issues under discussion through the research instrument of observtion.

2.5 Ndebele Agrarian Metaphors

As has been highlighted in the introduction, this chapter examines the various Ndebele language metaphors which have a direct or indirect link with agrarian farming practices in order to demonstrate that agrarian experiences have been, and continue to be manipulated in order to shape social and physical behaviour of citizens. The research argues that, while agrarian farming is basically a physical activity which involves physical labor, Ndebele people have used the various experiences gained from it so that they could shape the desired social characters among their children in order for them to become acceptable citizens when they are old. Not only have the agrarian experiences been used to deal with children, but adults have, and continue to have their share as well.

In other words, when one gleans at the close interactions which Ndebele people have with their agrarian farming, they can determine what type of a society they envisage because the various agrarian metaphors reflect the societal moral expectations. The sections below make an analysis of the agrarian metaphors as a way of determining their contribution to the social and moral fibre of the Ndebele societies. According to Achebe (2001) proverbs demonstrate that communication is a process in which meaning is multiple, layered and infinite, and where context is cardinal to understanding suggesting that, there is a lot about the history, the emotions, the joys, the desires of a people that one can learn from the metaphoric language that they use.

Isisu somhambi kasinganani, singangophonjwana lwembuzi ("a visitor's stomach is not very big, it is as tiny as a horn of a goat")

Greed or selfishness is a very unacceptable characteristic among most societies particularly among the Ndebele because, in most instances, people live as large entities rather than individuals. In most cases, traditional African societies lived as large or extended families rather than nucleus small families. This on its own is a sign that they never promoted individualism or greed. In fact, greed is discouraged as much as it is unacceptable so much so that there are various folktales which were narrated to the children as a way of discouraging them from such tendencies. Apart from staying as large families, Ndebele people were very generous to strangers or visitors whom they treated with care and love in the understanding that they themselves could also be strangers or visitors one of the days. It was believed that, if one fails to treat strangers properly, they themselves would be mistreated whenever they visited other areas especially where they would not be known. According to Samkange and Samkange, (1980: 45):

> In the case of the Shona/Ndebele communities in Africa where hospitality is given for free as when one provides accommodation and food to a stranger at his or her home, the magnitude is high. Coming to the idea of hospitality in Africa, it is important to note that in a traditional Shona/Ndebele society a person having traveled a long distance to visit a relative would sometimes get tired and hungry before they got to their relative's home. During their short stay (in transit), they would be provided with food, accommodation, and warm clothes (if they happened to travel during the winter season).

It should be bone in mind that before the advent of modern transport, local people used to travel on foot regardless of the distance. At times, there would be need for one to travel over a number of days thus, they would be required to sleep and rest along the way. In cases like that, the traveller would be compelled to spend some of the nights lodging at different homes until they got to their destinations. When a stranger came to a home and asked for a night lodging place, it was not acceptable to send them away or to make them sleep on an empty stomach, the host family was expected to do its best to make sure that the stranger's welfare is taken care of. Apparently, there was no room for people to count it as loss to share their food with strangers because, in the majority of cases, everyone had enough food and to spare. Strangers are not referred to as strangers, they are referred to as *umuntu wemzini*, ("a person from another household") which is a polite way of labeling a traveller who is ordinarily on their way from one place to another.

Feeding visitors was/is therefore, duty more than anything else because, in their eyes, a visitor would not stay for too long. Besides, even if they were given food, they woud not finish the whole granary where food is ordinarily stored. It is therefore, not surprising that, through their interaction and practice of animal husbandry, they came up with befitting metaphors to discourage people from withholding hospitality to travellers. *Isisu somhambi kasinganani, singangophonjwana lwembuzi* ("a traveler's stomach is not so large, it is just as tiny as the horn of a goat").

While there are so many animals, domestic and wild, that have horns, it is phenomenal that the imagery here is more emphatic when made in reference to the goat which, apart from its tiny physiological stature, it actually has small horns too. Taking it from a linguistic perspective, the very fact that the proverb uses an antithesis where, initially, it would appear as if the visitor has a very big stomach; */kasi-ngananil* ("it is not as big") as if it cannot be compared to anything, yet, the next statement which is in form of a rhetoric response, makes the absolute comparison; */singangophonjwana lwembuzil* ("Just as tiny as the horn of a goat"). Ironically, it should also be noted that the proverb converts the horn of the goat from */uphondol* ("horn") Which refers to a complete size of a horn, by making use of the diminutive suffix */-njwana/* ("tiny horn") which reduces the size of the horn to a minute size and deliberately intended to emphasise the fact that a visitor cannot finish your food, so no one is expected to be stingy, but, all are expected to share as this is a moral societal virtue. Smakange and Samkange (1980: 45) posit that:

> In the case of the Shona/Ndebele communities in Africa where hospitality is given for free as when one provides accommodation and food to a stranger at his or her home, the magnitude is high. Coming to the idea of hospitality in Africa, it is important to note that in a traditional Shona/Ndebele society a person having traveled a long distance to visit a relative would sometimes get tired and hungry before they got to their relative's home. During their short stay (in transit), they would be provided with food, accommodation, and warm clothes (if they happened to travel during the winter season).

Greed has a very negative effect on social, economic or even physical development of a society because it breeds selfishness and self-centered tendencies which have a bearing on the way one interacts and associates with others. Any society which wants to achieve social or economic development cannot do so unless it discourages its people from selfish motives or actions. In any case, greed violates the very ethos of *ubuntu* which by and large, promotes, propagates and prescribes members of a given society to practice generousity at all given opportunities. This principle is also reflected in another proverb which says *inkomo ehambayo kayiqedi tshani* ("a cow that is in transit cannot finish the grass").

The vast experience and daily interaction with cattle made the Ndebele people to closely observe the behaviour of cattle in relation to grazing as they pass through an area which is not their own. Sometimes cattle stray to neighboring villages for some reason or another. In such instances, they do not behave the same way they would have done had they been on familiar grounds. Apparently, as they try and retrace their steps back home, they go hungry and may need to feed whilst in transit. It is the behaviour of feeding which was spectacularly considered by the Ndebele people as it was observed to be different from the behaviour executed on familiar grounds as the cattle eat yet at the same time they move.

It was also noticed that no matter how much a moving cow may eat, it can never finish the entire pasture just because it is in transit. To the Ndebele, a visitor is just like a cow that is in transit, they can never finish the food. They cannot stay more than it is necessary just because they have to get to their intended destination on time. It is therefore, socially inappropiraite to deny visitors the requisite hospitality just because, apart from being morally right, it makes sure that society remains morally cohesive. In any case, being a visitor is not unique to certain individuals; anyone can find themselves in a situation where they are visiting, whether voluntarily, or otherwise. This possibility compels everyone to take good care of strangers because *unyawo alulampumulo* ("a foot has no rest") meaning that everyone could be a visitor the next day and they would always expect to be treated well regardless of wherever they are. In other words, the social expectations as far as hospitality to visitors is concerned actually empowers the visitors with some socio-legal right such as the following:

- Right to be treated with dignity
- Right to food and drink
- Right to good sleep
- Right to free passage on the journey.

The fact that everyone can be a visitor at one point or the other, makes it mandatory for the society to treat visitors with respect and to assist them so that people can commune as much as possible without any hinderances. Apart from making sure that visitors are well catered for, this principle helps to mould good character among individuals by encouraging them to place "human needs, interests, and dignity as of fundamental importance and concern" of others over thiers (Gyekye 1996: 158).

Ikhuba lomunye kalilandima ("someone else's hoe can not finish the job")

Agrarian farming itself is a physical activity which requires hard work, resilience, patience and perseverance if anything positive is to be realized. Traditional African families would undertake manual labour in the fields right from the time of planting, through harvesting and ultimately harvesting. All members of the family would engage in different activities depending on the size and structure of the family but, all would be done in order to maximize output so that the family would not have to beg for food during the dry season. Hard work, patience and perseverance have always been virtuous traits that are emulable not only among Ndebele societies but across the entire African continent. It is apparent that agrarian farming relies more on the natural rainfall such that, everyone has to carry out any farming activities as fast as possible while the rains are still around so that they are not left to starve due to droughts. It was therefore, essential that people be always alert and prepared for the farming season way before it arrived so that they are not caught unawares.

Having been into farming for quite a long time, Ndebele people noted that some people are never ready on time for the farming season or farming activities so much that they end up becoming victims of their lack of preparedness. The proverb which says *ikhuba lomunye kalilandima* (someone else's hoe can not allow you to finsh your job) emantes from the agrarian farming activities as a way of discouraging members of the society from over reliance upon other people's tools as that is detrimental to attainment of a fruitful harvest. A fruitful harvest does not only benefit an individual family entity, it benefits the entire community at large because the wealfare of the community comes before that of an individual. Eze (2011: 12) observes that the Africans have a "philosophy of *Hunhu/Ubuntu* which places the communal interests ahead of the individual interests" and as such, one is bound to operate in line with the dictates of the community. A productive community means that the disadvantaged, the vulnerable, especially orphans and widows will be taken good care of without any hustles.

It should be noted that *ikhuba* ("hoe") which is used in this case represents all farming tools, tangible and intangible as well as those aspects of farming life that are supposed to aid in the farming process. It was common practice that those who were well to do would loan ("*ukusisa*") some of their spans of cattle to those members of the community who were needy so that they also can find means with which to produce food for their families. In order to avoid conflicts in the event that the owner of the span of cattle has developed a need to have them back before the end of the farming season, he or she is empowered by this proverb in taking what belongs to him back. He can not be deemed to be insensitive or selfish because it has been socially accepted that "*ikhuba lomunye alilandima*" ("someone else's hoe can only be used on temporary basis") which means the owner has a right to take back his tools at any time.

Apparently, research shows that this proverb is not only restricted to farming activities alone, it is now applicable to anything that belongs to another person, whatever it is, as long it has been borrowed, the owner is permitted to ask for it back without hustles. In other words, drawn from the agrarian experience, this proverb serves as both a warnig for people to always prepare for farming on time as well as a conflict management strategy for swift exchange of materials from owner to borrower and the other way round so much that the interests of individuals and those of the entire community are protected. Yamamoto (1997: 52) puts it correctly in reference to the altruistic character of *Ubuntu* philosophy when he remarks that "*ubuntu* is the idea that no one can be healthy when the community is sick. *Ubuntu* says I am human only because you are human. If I undermine your humanity, I dehumanise myself".

While placing so much emphasis on the moulding of individual character of members, the proverb also makes an unapologetic demand for people to adequately prepare themselves on time for the farming activities by making sure that they are well equipped with all requisite physical and social tools. Failure to prepare on time leads to misery and puts the whole community in a compromising position as far as food security as concerned. If there is one family or more which are not well prepared for the farming activity at the same time with others, there are high chances that those particular families will starve and as such be an unnecessary burden to the entire village which might have to find means and ways to make sure that they are fed. Adequate and timely preparation for farming results in high yields which eventually ensures that the community welfare is well catered for. As has already been highlighted, the community's interests supersedes those of an individual, yet, on the other hand, an individual's welfare is of particular importance to the community so much that, it is the duty of the community to make sure that its members tore the line.

Apart form farming activities, the proverb *ikhuba lomunye kalilandima* ("someone else's hoe can not be relied upon") now applies to all facets of life where people are encouraged to be self reliant and not necessarily depend on loaned materials. Not withstanding the fact that there is no way people could have everything in life and not borrow from others, the bottom line is that each individual is encouraged to be self reliant and borrow only when it is necessary. Even when it comes to borrowing, it is fundamental that one who borrows is also able to help others to borrow something else from him.

What comes to mind is the other proverb whose etymology is closely associated with agrarian experiences of having observed the behaviour of cattle which says, *ikhotha eyikhothayo* ('a cow licks the one that licks it back'). Ndebele people had observed that when cattle are relaxed under a shade or in the pastures, they tend to use their togue to lick each other. They take turns to do so such that the one which licks first expects the other to reciprocate, and in the majority of cases, that is what happens. This behaviour complements the idea that members of the community should always be prepared for farming by having their own tools or equipment so that they do not get interrupted when owners come to claim their things, but, most importantly, if they should borrow, let them be in a position to reciproacate the favor.

Uthango ludla amakhomane ("the protective fence is eating up the gourd squash")

Diet and special food varieties have always been a concern for the Afrcan people in general as well as the Ndebele in particular. Food varieties have been used to complement the stapple meals depending on their availability, nutritional value as well as cultural needs. There are food stuffs that are seasonal such as water mellons, pumkins, wild fruits and insects whose availability depends on seasons since they appear and disappear. Apart form planting stapple foods such as maize, soghurm and millet, families would plant some of these vaieties which are not specifically for commercial purposes but are there to spice up things as far as diet is concerned. Beverages such as *amahewu* ("home brewed non-alcoholic beverage") are also popular since they work as energisers for people who would be working in the fields due to the nutritional value that they possess.

The various food varieties that are planted, usually in the traditional agrarian way, have got a variety of ways of planting and harvesting them. Some are just scattered in the field so that they grow on their own, some are buried in the ground while some grow on their own for instance mushrooms. The ways of cooking them also depend on the geography of the consumers as well as their cultural orientation which spells out the various cultural proscriptions concerning them. Some foodstuffs are not meant to be eaten during the day while some are reserved for men or women respectively. All in all, a collection of various foodstuffs explains how the people place value on their nutrition for the sake of good health and longevity of life.

Of special interest to this research is the food variety called *amakhomane* ("gourd squash") which is a leguminous type almost similar to pumpkins. This type of food crop is planted reservedly by the women alongside with stapple food crops such as maize. It is a type of food crop which is planted using the scattering method usually near the family homestead. When ripened, the fruit-like product is harvested while still green and soft and eaten either as a meal on its own or accompanying another main meal. Sometimes it is eaten with fresh or sour milk in order to add to its nutritional value and taste. *Amakhomane* ("gourd squash") are a delicacy among the Ndebele because they usually mark the beginning of the *kwindla* (autom) which is a time to start eating farm produce known as *ihlobo* ("spring season").

Apparently, since *amakhomane* ("gourd squash") are planted predominantly using the scattering method, their seeds may fall anywhere even where the sower does not intend. Most of the times, just because they are leguminous, the shrub that bears them usually flourishes well when stretching on the traditional fence ("*uthango*") which is made up of tree logs, thorns and bushes. The fence is the one that is used to keep both domestic and wild animals away from the fields where the crops are planted. The fence acts as both a barrier and a protection to the various crops that the family would have planted in their fields and as such, that fence is an important aspect of traditional farming practices. Actually, the fence is done with so much dexterity and skill so that, apart form the primary purpose of keeping away animals, it also dispays the ingenuity of the owner of the fields.

Be that as it may, there were times when people would observe that, while the fence is there, *amakhomane* ("gourd squash") seem to be disappearing from the field one by one yet the owner herself has not yet harvested them. This would be a cause for concern as it would be a clear reflection that there would be someone mischievious in the village, or even in the family, who would be stealing those *amakhomane* ("gourd squash"). It was then that proverbs like this one were deployed in order to express the observations and feelings of the people. Nyamnjoh (2021: 67) underscores the communicative nature of proverbs saying:

> Not only are proverbs universal in their use to express emotions, thoughts, experiences and challenges, they are also universal in the very fact of the mobility of humans, ideas and language across geographies. The power of the proverb lies in its eternal incompleteness of meaning, that constantly opens itself up to improvisation and creative innovation in usage with and across cultural communities.

If the culprit would not have been apprehended yet, the Ndebele would sarcastically say, '*uthango ludla amakhomane*" ("the fence is eating up the gourd squash"). In essence, they would be suggesting that, since *amakhomane* ("gourd squash") are disappearing from the fields, and since no one has been seen taking them officially or unofficially, then, the only explanation would be to say that the fence is the one which is eating them. In other words, apart from being a protective barrier for the crops, a fence is also viewed as the owner of those crops. The fence is given the mandate to physically protect whatever crops are in the field from harm and danger by any intruder so much that, if foul play is suspected, the fence is the first culprit. This metaphoric expression carries so much within itsef just because, everyone knows that the fence has no capacity to eat any crops, neither can it be held liable to their disappearcance by any means. Finnegan (1970: 389) linguistically describes proverbs as 'a rich source of imagery and succinct expression on which more elaborate forms can draw' while functionally, proverbs 'represent the creator's ability to create theories of existence from everyday observations. Each proverb is a part-truth about life, together the proverbs form artistic truths about the African worldview' (Bhebe and Viriri 2012: xiv).

However, far from lamenting the unexplained or unexplainable disappearance of *amakhomane* from the physical family field, this proverb is used as a social expression of the observation of the community where misdemeanor is being suspected. Ndlovu and Mangena (2014: 46) highlight the use of proverbs as conforminty to the United Nations Rights of Children. The UN Convention on the Rights of the Child (1989) is perhaps the most widely accepted Human Rights Convention and Zimbabwe is one of the 193 nations which have accepted to abide by the ethos of this convention. While the general belief is that that the human rights discourse is a modern concept, children's rights have always been moral imperatives for both the Shona and Ndebele of Zimbabwe so much that:

> An exploration of proverbs in both Shona and Ndebele cultures demonstrates that these people had ways of recognising the importance of children and ensuring that they are protected from various forms of abuse. A recourse into Shona and Ndebele oral literature is significant because for many pre-colonial African societies' civilizations were 'based on orality … religion, education, science, or ideology were all part and parcel of every type of activity' (Curtin et al. 1995: 469).

It was therefore, discovered during the process of carrying out this research that, when the Ndebele people suspect that someone who is in a vantage position is abusing their advantage and performing acts of misconduct on those who are in a vulnerable position due to their physical, social or economic condition, he/she, just like the symbolic fence, is deemed to be eating *amakhomane* ("gourd squash"). In essence, the proverb may be used as a diversion from direct confrontation with the perpetrator of the injustice, in case he is alerted and quickly conceals evidence in case of possible litigation. At the same time, it may be used as a way of telling him that his or her evil deeds are now in the open because the community would have unearthed them while he/she still thought the deeds were hidden.

More specifically, the proverb was used in a situation where a father would be sexually abusing minors, be they biological children or otherwise. The act is considered heinous and bizzare so much that, in other instances they use the proverb which says "*inkukhu idla amaqanda ayo*" ("a hen which eats its own eggs"). Again, the agrarian experiences with domestic birds which are a traditional part of family wealth, diet and culture, helps them to observe certain behaviour traits by these birds and as such use them to express those human attitudes that are considered to be both unerthical and uncouth. Procreation among both humans and animals is an invaluable expectation among Ndebele people since it assures them of a fruitful future which is full or production satisfaction. In fact, Ndlovu and Mangena (2014: 665) underscore the need to ensure that vulnerable members of the society are protected when they say that "in both Shona and Ndebele traditional cultures, as expressed in the proverbs, parents have an obligation to offer protection to their children".

In that case, no one, whether human or animal, is allowed to freely interfere with the prospects of production and continuity especially by intercepting life at its prime stage just like in the case of a hen and an egg. If the hen eats the eggs, where are we going to get more chickes? That is the principle. So, when a man violetes the children who are under his care, society does not take it lying down. A man who violates the rights, priviledges and dignity of his own children is just like a hen which intercepts life by eating its own eggs or worse still, a fence which eats *amakhomane* ("gourd squash") which are depending on it for protection and security. Speaking about the need to safeguard the rights and freedoms of children, Ndlovu and Mangena (2014: 663) make reference to the UN Convention on the Rights of Children, specifically Artice 6 which states that "no child shall be subjected to arbitrary or unlwoful interference with his or her privacy, family, or correspondence, not to unlawful attacks on his or her honour and reputation". By applying the proverbs whose etymology is grounded on agrarian experiences, the Ndebele are actually calling upon the society to be a custodian of the rights of the vulnerable both socially and physically.

It should also be noted that the imagery created by these proverbs is steeped in the cultural perceptions and expectations of the indigenous people in liaison with their agrarian farming experiences and as such, it exhibits their cultural inclinations and standards for measuring good and bad behaviour. Jenjekwa (2021: 15) speakes of the interconnectedness of the life and social environment of the people when he says that "culture is an obiquitious product of the intercourse between and amongst a people's social, historical, political, religious and economic values to produce a certain distinct way of life". Ultimately, the point to note is that, since agriculture is part of culture, it is therefore, understandable that indigenous people, due to their agrarian experiences, sought to shape the social, as well as cultural, behaviour of their societies for posterity.

2.6 Conclusion

This research has interrogated the relationship between Ndebele proverbs and agrarian farming experiences. It has established that, apart from engaging in agrarian farming for the purposes of food security and healthy wellbeing, indigenous Ndebele people formulated societal behaviour codes and expectations basing on the experiences they would have gained during their farming activities. Their dependence and reliance on agrarian farming for physical survival gave them an opportunity to study and explicate various farming experiences as fodder for social consumption by their members in the community for the sake of social cohesion, order and posterity. Agrarian farming itself, by its very nature, requires commitment to hard or manual labour, a social virtue among the Ndebele who view a hardworking person as a valuable resource for the community.

It should also, therefore, not come as a surprise that there are a plethora of proverbs which are inclined towards hard work as demanded by the agrarian type of farming such as some of those mentioned in the research discussion. It goes without saying that, Ndebele people, just like all other African cultural groups, have no room for deviant character traits, immoral behaviour or even misdemeanor, as these are viewed to be inconsistent with what the society expects among its members because they are always expected to live in line with the dictates of *ubuntu*. The very fact that life among Ndebele people depends on the agrarian activities, gives the Ndebele people a chance to locate their societal expectations on it just because it is the backbone of human life. Everybody in the society, one way or the other, finds themselves closely conforming to the expected standards just because they are embedded on the day-to-day social activity, agrarian farming, which is the source of life and welfare.

Therefore, by basing their principles of life on agrarian farming through associating and formulating metaphorical nuances such as proverbs, Ndebele people strongly communicate with both the inside and outside world about those aspects of life that matter to them. The world can glean at the way Ndebele people understand and interpret phenomena using the proverbs that they have formulated especially those that are a result of their years of experience with their interractions with agrarian farming such as those that have been discussed in this research.

References

Achebe, C. 2001. *There was a Country: A Personal History of Biafra*. London: Allen Lane.

Bhebe, N., and A. Viriri. 2012. *Shona Proverbs: Palm Oil With Which African Words Are Eaten*. Gweru: Booklove Publishers.

Bless, C., C. Higson-Smith, and S.L. Sithole. 2014. *Fundamentals of Social Research Methods: An African Perspective*, 5th ed. Cape Town: Juta.

Chiromo, A.S. 2006. *Research Methods and Statistics in Education: A Student's Guide*. Gweru: Beta Print.

Curtin, P., S. Feierman, L. Thompson, and J. Vansina. 1995. *African History from the Earliest Time to Independence*. London: Longman.

Denzin, N.K., and Y.S. Lincoln. 2011. *Introduction: The Discipline and Practice of Qualitative Research. The Sage Handbook of Qualitative Research*, 4th ed., 1–99. Thousand Oaks, CA: Sage.

De Rossi, C., and M.K. Taylor. 1985. Symbolic Dimensions in Cultural Anthropology. *Current Anthropology*, 26(2), 167–185.

Eze, M.O. 2011. I am Because You Are. A Philosophical analysis of the idea of *Ubuntu*. *The UNESCO Courier*, 10–13.

Finnegan, R. 1970. *Oral Literature in Africa*. Nairobi: Oxford University Press.

Gyekye, K. 1996. *African Cultural Values: An Introduction*. Accra: Sankofa PublishingCompany.

Hapanyengwi-Chemhuru, O., and E. Shizha. 2012. *Unhu/Ubuntu* and Education for Reconciliation in Zimbabwe. *Journal or Contemporary Issues in Education*. 17 (2): 120–132.

Jenjekwa, V. 2021. Sociolinguistics: An Overview. In *Sociolinguistics and Multilingual Education in Zimbabwe*, eds. W. Magwa, V. Jenjekwa, and Moyo, M. Gweru: MSU Press and Publications.

Mawere, M. 2010. Indigenous Knowledge Systems' (IKSs) Potential, For Establishing a Moral, Virtuous Society; Lessons from Selected IKSs in Zimbabwe and Mozambique. *Journal of Sustainable Development in Africa* 2 (7): 209–221.

Ndlovu, S., and T. Mangena. 2014. Reflections on How Selected Shona and Ndebele Proverbs Highlight a Worldview that Promotes a Respect and/or a Violation of Children's Rights. *The International Journal of Children's Rights* 22 (3): 660–671.

Ngugi wa Thiong'o. 1982. *Moving the Centre: The Struggle for Cultural Freedoms*. James Currey Publishers.

Nussbaum, B. (2003). Ubuntu: Reflections of a South African on Our Common Humanity. In *Reflections, The Society for Organisational Learning and the Massachusetts Institute of Technology*, vol. 4, no. 4, 21–26.

Nyamnjoh, F.B. 2021. *Being and becoming African as a permanent work in progress: inspiration from Chinua Achebe's proverbs*. Deliverd as a Public Lecture in the Department of Social Anthropology, University of Cape Town.

Nzewi, M. 2007. *A Contemporary Study of Musical Arts: Informed by African Indigenous Knowledge Systems*. Volume Four Illuminations, Reflections and Explorations. Ciima Series. Report submitted by UNSECO to the 19th Session of the United Nations Permanent Forum on Indigenous Issues (UNPF11) 13–24, April 2020.

Owusu-Ansah, F.E., & G. Mji. 2013. *African Journal of Disability* 2(1).

Samkange, T., and S. Samkange. 1980. *Unhuism or Ubuntuism: A Zimbabwe Indigenous Political Philosophy*, 1980. Harare: Graham Publishing.

Shutte, A. 2008. African Ethics in a Globalizing World. In *Persons in Community: African Ethics in a Global Culture*, ed. R. Nicolson, 15–34. Scottsville: University of KwaZulu Natal Press.

Trevor-Roper, H.R. 1960. Discussion of H. R. Trevor-Roper: "The General Crisis of the Seventeenth Century." *From past and Present* 18: 8–42.

Tutu, D. 1999. *No Future without Forgiveness*. New York: Doubleday. *United Nations Convention on The Rights Of Children*. 1989.

Yamamoto, E.K. 1997. Race Apologies. *Journal of Gender, Race and Justice* 1: 47–88.

Faith Sibanda is a Senior Lecturer at the Great Zimbabwe University in the Department of African Languages and Literature specialising in African indigenous knowledge systems and cultural studies. His research interests are in the area of Languages, Linguistics and Culture where he has published a variety of research articles in international accredited journals. He has presented research papers at national and international conferences on Culture and African Indigenous Knowledge Systems. He is also a newly appointed member of the National Intangible Cultural Heritage Advisory Committee.

Chapter 3
The Farm-Village Practice of Yorùbá in West Africa

Babatunde A. Ogundiwin

Abstract The Yorùbá ethnic group of West Africa, an urban people, resided in walled towns in pre-colonial times but farmed the surrounding areas in villages devoted to agriculture. In this chapter, I argue that this traditional, pre-colonial agrarian culture informs contemporary agrarian visions of urban sustenance. This chapter explores the Yorùbá agrarian practice of urban-hinterland relations as exemplified in the farm-village practice using the notion of indigenous geographies. The outlying farming practice reveals a spatial thought that drew attention to the importance of the physical environment, agronomic practices and social institutions in human–environment interactions. Its impact on the indigenous political system is evident in the agricultural food system, political identities and territorial formation. Conversely, it informs emerging agrarian ideas on the post-colonial landscape. The chapter concludes that the satellite farm-village as a cultural lens offers a spatial interpretative framework to rethink eco-social balance for the sustenance of urban populations and the protection of the physical environment in contemporary modernizing attempts on the agricultural landscape.

Keywords Farm-village practice · Yorùbá · Agrarianism · Indigenous geographies

3.1 Introduction

The Yorùbá farm-village practice is a fundamental part of the Yorùbá urban culture. Although there have been discussions on the farm-village, there is still the need to draw attention to the spatial character of this indigenous agrarian practice. As a rapidly urbanising continent, the challenge of urban sustenance is a growing concern in sub-Saharan Africa (Mabogunje 2008). Hence, the need to engage agrarian thought that stress on indigenous African outlook on urban sustenance while protecting the

B. A. Ogundiwin (✉)
School of Geography, Archaeology and Environmental Studies, University of the Witwatersrand, Johannesburg, South Africa
e-mail: 1807179@students.wits.ac.za

M. J. Tosam and E. Masitera (eds.), *African Agrarian Philosophy*, The International Library of Environmental, Agricultural and Food Ethics 35,
https://doi.org/10.1007/978-3-031-43040-4_3

physical environment. The notion of agrarianism used here refers to a set of philosophies that place agricultural practices within the broader concerns of the biophysical environment and social values (Montmarquet 1985; Thompson 2008; Tosam 2020; Wirzba 2003). Drawing upon the notion of indigenous geographies, this chapter explores the Yorùbá agrarian philosophy of farm-village practice and its associated conservational practices. Indigenous geographies are critical frameworks that destabilise and denaturalise the taken-for-granted knowledge of modernist agriculture and the neoliberal logic of profit motive and its environmental degradation, which is considered a necessary trade-off (Wirzba 2003).

The key feature of Yorùbá agrarianism is the farm village practice. This village is a farming workspace closely related to its immediate urban settings but remains an outlying production space. It represents the principal agricultural spatiality apart from the nearby farms of the towns. In the urban hinterlands of Yorùbáland, farm-village practice evolved sometime in the fifteenth century, becoming a well-established social practice some centuries after. For instance, farm-villages were the principal feature of the agricultural landscape of Ibadan during the nineteenth century. The idea of farming with two modes of inner and outer farmlands was a well-known practice in Western Africa (Hopkins 2020). It was practiced amongst farmers in Eastern and Northern Nigeria (Chisolm 1962; Floyd 1969). Also, in the Asante Region of Ghana and Burkina Faso, in West Africa, the use of nearby and far away farms are employed in agricultural activities (Gray 2003; Hopkins 2020; Oyeleye 2001; Steel 1948). However, in contrast to these similar agrarian systems, the indigenous Yorùbá farm-village practice was intertwined with the sustenance of their urban culture.

The purpose of this chapter is to explore this farm-village practice from an indigenous geographical context demonstrating the agrarian conceptualisation of space within socio-environmental practices on the landscape. I argue that the emergence of the notion of *Agbe Oko d'Oko*, a reworking of the modernist settlement scheme, arose in the cultural encounter with the indigenous agrarian conceptualisation of space. As Ogar and Bassey (2019: 76, original emphasis) observe, "Studying *societal structures and values* may also be instructive". An insight into the indigenous agrarian practice might highlight the possibilities of agrarian ideas for urban sustenance, and rethink emerging notions, in sub-Saharan Africa, that insist that agriculture is a business unconnected to social values (Green 2013).

The discussion in this chapter is as follows: the next section conceptualises this study through the notion of indigenous geographies drawing attention to the importance of place-based knowledge; in the third section, I briefly discuss the Yorùbá ethnic group and their urban culture. The fourth section examines the indigenous agrarian practice, making a distinction between the nearby farm and the outlying farming practice. Thereafter, the fifth section explores the farm village practice to show the interconnection of the natural environment, agronomic practices and social institutions. The sixth section examines the impact of this agrarian practice in nineteenth-century Ibadan showing its influence on indigenous politics. The seventh section explores a postcolonial reworking of the agrarian practice. The eighth section engages insights from the farm-village as an agrarian spatial thought. Finally,

I conclude by drawing attention to contemporary challenges of urbanisation and environmental ethics in Sub-Saharan Africa.

3.2 Indigenous Geographies: A Conceptual Approach

Indigenous geographies, a concept with varied meanings and situated within different contexts, emerged from the application of critical theory in geographical thought. This critical approach seeks to decolonise taken-for-granted Western modernist geographical narratives (Sidaway 2000). For instance, in southern Nigeria, the colonial administration described indigenous agrarian practices as "their system of agriculture, as practised in the more wooded portions of the country, is one of the most wasteful known" (Colonial Office 1908: 37). Such Western assumptions became normative theorisations of indigenous agricultural knowledge. These meta-narratives and grand-theories of Western modernity projected a universal fixed truth, based on secured ontological and epistemological foundations. Conversely, the concept of indigenous geographies deconstructs these modernist geographical narratives by engaging and rediscovering indigenous ways of seeing, knowing and thinking (Blomley 2006; de Leeuw and Hunt 2018; Nayak and Jeffery 2011). As Mbembe (2015: 17), following Ngugi, observes, "Africa has to be placed at the centre". Therefore, Indigenous geographies contest this taken-for-granted geographical knowledge of dominant worldviews by exploring silenced voices and experiences of indigenous forms of knowledge. The notion of indigeneity offers some insight into indigenous worldviews, for instance, their relational conceptualisation of agrarian space and its associated conservation practices.

"Indigeneity", Radcliffe (2015: 1) explains, "can be defined as the socio-spatial processes and practices whereby Indigenous people and places are determined as distinct (ontologically, epistemologically, culturally, in sovereignty, etc.) to dominant universals." Critical geographies of indigeneity seek to examine and understand the production of place-based knowledge. Therefore, Radcliffe (2015: 2) observes "as an analytical concept, indigeneity attends to the social, cultural, economic, political, institutional, and epistemic processes through which the meaning of being indigenous in a particular time and place is constructed." This involves the historicising and spatialising of indigenous subjects to identify and theorise their contingent and relational positions. As a contingent and relational approach, the concept draws attention to indigenous epistemologies of knowledge production and the diverse positioning in its encounter with the colonial past and the (post)colonial present (Yeh and Bryan 2015). For instance, the pre-colonial *Ebi* social system (Akinjogbin 1966) was reflected in Yorùbá spatial organisation of agricultural production. This spatial organisation includes the village and the agricultural landscape. While the village is a farming workspace (Oyeleye 2001), the landscape refers to the cognitive framing of farm villages at a larger spatial scale. This urban-hinterland relation continues to have (with varied) effects on the ways the Yorùbás organise agrarian production in remote hinterlands of the post-colonial space.

Indigenous geographical knowledge requires an engagement with these place-based conceptualisations of space and the ways of thinking, not only in the past but also in contemporary times. One way of engaging these indigenous geographies, Radcliffe (2015) suggests, is through the map metaphor. Following Bryan (2009), she employs the map metaphor as a framework of indigeneity, for "Indigeneity is to Indigenous peoples as cartography is to the earth's surface". Hence, geographies of indigeneity like maps are "selective, interested, highly codified, co-produced through routines and technologies, and entails traceable consequences" (Radcliffe 2015: 2). Mapping place-based, indigenous geographical knowledge can be approached in diverse ways (Larsen and Johnson 2012). Wise (2018: 139) contends, "building geographical and historical studies out of Indigenous epistemological foundations provides one way of accomplishing this objective". In this study, the objective is to describe and explain the farm village practice and its effect on indigenous environmental thought. Therefore, employing and engaging indigenous epistemologies involved oral histories, proverbs, songs, and language that are firmly rooted in geographically specific places (Gbadegesin 1991; Moseley et al. 2014; Wise 2018). Conversely, written records on local environmental knowledge in the pre-colonial period, mostly authored by Western Europeans, also exist. Although there are limits to the use of such records (Clayton 2003; Hopkins 2020), sources without modernist bias are valuable. Hence, indigenous geographies, as attempted here, employ reconstruction and deconstruction strategies through a critical lens to engage these ethnographic records.

3.3 Yorùbás

The Yorùbá ethnic group is situated in West Africa roughly between the middle and lower course of River Niger and Lake Volta (Atanda 1980; Usman and Falola 2019). This geographical location is in the present-day Republics of Nigeria, Benin and Togo. The Yorùbá country lies between Longitude 2° 30″ and 6° 30″ East and Latitude 6° and 9° North of the equator, covering about 181,300 km^2. The River Niger bound Yorùbáland in the north and northeast and to the south by the Atlantic Ocean. It is dominated by a plateau running from northwest to the southeast. However, there are hilly areas dotted with inselbergs in the south-western, north-eastern and central parts of Yorùbáland. There are several rivers around the country while there is a 274 kms coastline (Obateru 2005: 44). There is two river-system in Yorùbáland, one set of rivers draining northwards into the Niger River, and the other set of rivers flowing southwards into the Lagoon and the Atlantic Ocean (Ogundiwin 2009). There are three lagoons, Lagos, Lekki and Mahin, inter-spaced with swamp, marshes and tidal mudflats, behind this coastline. Beyond the coastal mangrove vegetation, the southern parts are forestlands whilst the northern parts are savanna grasslands.

Since the pre-colonial period, the Yorùbá ethnic group of West Africa has been town-dwellers (Akintoye 2010: 175). As Fadipe (1939: 167) observed every peasant was a townsman. The *Ìlú*, an urban settlement, in pre-colonial Yorùbáland, had an internal structure, where the elites resided in the urban centre, and the urban quarters were organised along occupational lines (Mabogunje 1976). At the centre of the town were the *Àfin*, the king's palace, important shrines and the *Ọjàba*, the king's or central market (See Fig. 3.1). Flanking the core zone in a concentric pattern were *Àdúgbò* or *Itún*, the wards or quarters. These wards consisted of *Agbo-Ilé*, compounds. Each compound was a rectangular building of varied sizes. In addition, residents of the compound were usually members of this same lineage (*Ẹbi*). These lineage compounds could house between thirty to more than one hundred persons (Clarke 1972: 236). Often the residents of these compounds were engaged in primary and traditional occupations like farming, dying, blacksmithing and cloth weaving in addition to petty trading (Bascom 1955; Atanda 1980). Fadipe (1939: 154) observed, "a noticeable feature of Yorùbá economic organisation is that of specialisation by compounds. While there is hardly any compound that is not without its quota of farmers, there are whole compounds to be found in large towns whose male population is made up of people following the same trade. There are compounds of blacksmiths, of weavers, of calabash dressers, and so on". However, since farming was the basic occupation the lineage also served as a production unit.

Although the primary spatial organisation was the urban settlement, there were also villages in Yorùbáland. Mabogunje (1974: 27) observed three criteria of distinguishing a town from a village (*Abúlé*) in the Yorùbá cultural context. These criteria are an elaborate system of administration; a greater complexity of economic activities; and intensive trading activities. He considered a population of 5000 usually fulfilling these criteria. However, several towns had a population larger than 5000. For instance, in the nineteenth century, several European explorers and visitors to Yorùbáland in the 1820–1850s reported, based on estimates, that several Yorùbá towns had populations between twenty-five thousand and seventy thousand (Bowen 1857; Burton 1863; Clarke 1972; Lander 1830; Tucker 1855). In addition, most pre-colonial Yorùbá urban settlements had a system of town walls (Ajayi and Smith 1971: 23–28). Outside the towns were the spatial patterns of Yorùbá agriculture. As aforementioned, agriculture was the main occupation. Even the craftsmen who are townsmen had farms tended by their children, dependents and servile workers (Fadipe 1939: 167). These farms were located within and outside the urban space (Ogundiwin 2009). Bascom (1958: 191) observed, "Yorùbá cities were non-industrial, being based on farming, craft specialisation, and trade involving large markets, true money and true middlemen. Farming is not only a rural occupation; city dwellers work outlying farms which surround the city". Administrative and economic activities were integrated in Yorùbá settlement geographies.

This urban spatial organisation also integrated the indigenous agrarian practice and a multipurpose agroforestry practice of the home forest. The surrounding land immediately behind the first town wall was employed usually for subsistence farming (See Fig. 3.1). For instance, in the 1850s, Clarke observed, "as people live altogether in towns, the surrounding country is first cultivated until badly worn and then is left to reinstate itself when another field is cleared. Hence, the fields around a town are oftentimes in a wild uncultivated state" (1972: 259). These fields were the farmlands immediately before the *Igbó-Ilé*, the home forest. The home forest was primarily for

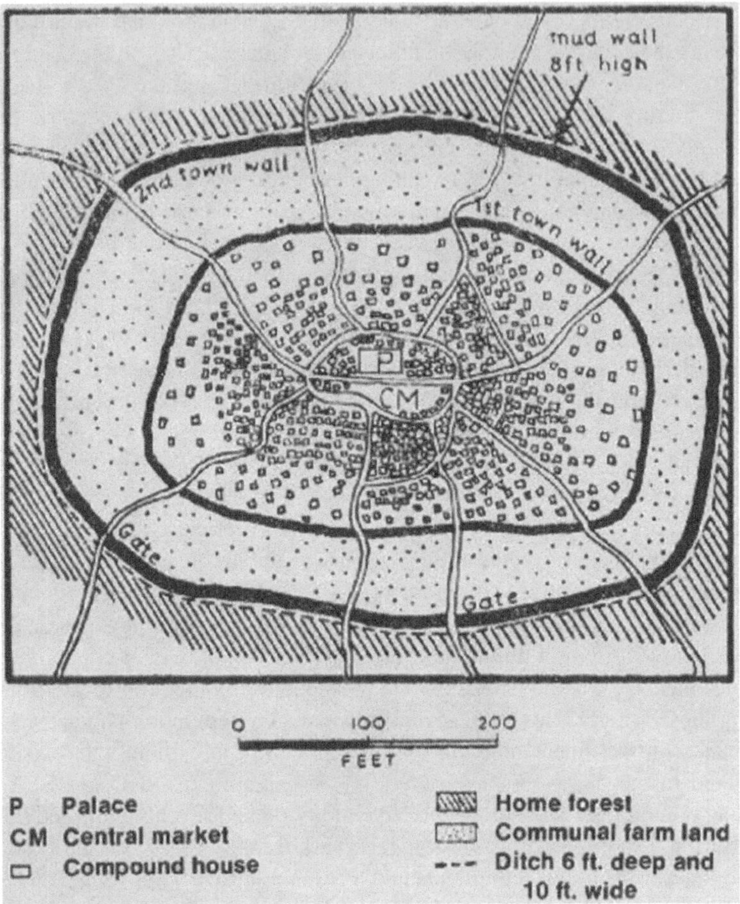

Fig. 3.1 Typical Yorùbá Town Plan in the pre-colonial period (*Source* Masood 1978: 311)

the defence of the settlement, an additional defence feature to the town walls (Ajayi and Smith 1971: 23). As Johnson (1921: 91) indicated "for the greater security of smaller towns a bush or thicket called Igbo Ile (home forest) is kept, about half to one mile from the walls right round the town". Nonetheless, the *Igbo-Ile* besides a defence feature also served as an environmental shield and was occasionally a source of firewood. Beyond the home forest was the second form of agrarian practice—the farm village.

3.4 Agrarian Practice: Nearby Farms and Outlying Farms

The indigenous spatial organisation of agriculture involved two forms of farming practice. The Rev. Samuel Johnson (1921), a Yorùbá cleric, writing in the late 1880–1890s, observed that there are two types of farms operated by all farmers of importance; first, the home farm, known as '*Oko Etilé*' and second, the far-away farm, known as '*Oko-Egàn*'. He wrote that "when engaged in the nearer one, they work from 6 or 7 am to 5 pm, with intervals for meals, and then return home; but at the distant farm, they invariably remain there for weeks and months before returning home" (Johnson 1921: 118). In contrast to Johnson, other scholars identified three modes of this agrarian spatiality but the dual spatial practices of farming remained the usual focus (Adeoye 1979: 106–116; Ojo 1970: 466). The nearby farms were often located either within the town walls or just outside the town walls. Cultivation on these nearby farms involved mixed cropping patterns using the rotational bush fallow (Ojo 1970: 466). The main crops grown were yam, maize, and cassava. The workforce was usually the farmer and his family members. This farming family worked from sunrise to sunset with intervals for meals and rest. Thereafter, the farmers returned home. This daily routine was so common in the typical Yorùbá town of the pre-colonial period. Such was the thrill of this daily routine that Rev. Clarke in the 1850s observed,

> A description of a farmer's life would serve to give a clear idea of the extent of cultivation and the mode of culture. It has already been stated that the natives dwell altogether in towns and cities around which they cultivate the soil, thereby necessitating the daily return of every man who visits the farm. A farmer therefore rises early, breakfasts, equips himself for the field with such implements as may be needed and starts for the country. In the course of an hour or more he reaches his farm where he labours until late in the afternoons, when taking a load of provisions or wood, he beats a retreat and reaches town in time for the evening market. As the country over which he travels is level or beautifully undulating he can readily walk twelve or fifteen miles and work from six to eight hours per day. This mode of life is kept up during the whole year by thousands of persons who seem to be delighted with the idea of going to the farm if for nothing else other than to make the visit. (Clarke 1972: 260–261)

The daily practice of working on the nearby farms was the basis of the culture of urban–rural migration. This agrarian migratory culture had a particular effect on the urban landscape, as it temporarily became a gendered space. For example, the English traveller, Daniel J. May at Iwo, "was much struck with the apparent paucity of men at this place; the preponderance of females is always evident, but it seemed here to be greater than usual" (May 1860: 213–214). This gender difference in the town can be explained as the menfolk were away at the nearby farms during the daylight. In contrast, farther from the town, about ten to twenty kilometres were the far-away farms (Oyeleye 2001). As Burton (1863: 67) noted at Abeokuta, "if you ask for the farms, you are told that they are distant five to twenty miles". These far-away farms had villages attached to them. These farm villages were different from the usually large residential villages with a farming community: rather, they are committed farming spaces. Likewise, they were the location of large-scale cultivation of crops and animal rearing for commercial purposes. Again, these farm villages were seen as "places of work and temporary residence", hence, they were satellites of the urban settlement (Fadipe 1939: 147; Goddard 1965). As a Yorùbá proverb states that *Ilé la bọ simi oko* (The town household is the place of rest after the wandering and work at farm) (Atteh 1980). Hence, the farm-village was an integral part of the urban culture.

3.5 The Outlying Farm: The Farm-Village Practice

The origins of the outlying farm-village practice were entwined with the Yorùbá urbanisation. The practice grew as towns were settled and farming hamlets were sited farther away from the agglomerated population. It probably became well-established around several towns by the late fifteenth century (Morgan 1959: 52). However, the value of this agrarian practice seems more decisive during the nineteenth-century wars and geographic shift in population around Yorùbáland (Ajayi and Smith 1971; Crowder 1966: 116; Usman and Falola 2019). Several farm villages became the nucleus of new towns, (Johnson 1921: 225) while new farm villages were established to carry on the work of urban sustenance as more settled conditions emerged from the wars and migrations. The character of this agrarian practice was evident in and associated with certain representations of geographic space, agronomic practices and social institutions.

The geographic space of the farm village, *Oko*, consists of two spatial configurations: *Abúlé*, the village layout and *Oko*, the actual farm layout (See Fig. 3.2). The village layout consisted of the buildings and the farmlands. The layout had an identical geometric form with the urban plan at a smaller scale or size. The two main types of buildings in the farm village were the *Ahéré*, farmhouse and the *Àká*, the barn. The farmhouse, a key material structure of the farm-village, loomed large in the mental image of the farming population. As a Yorùbá proverb attests that *Ahéré ni yio kehin oko, atta ni yio kehin ilé* (The farmhouse remains to the last, [upon the ground], and the ridge of the roof completes the building). The farmlands were located around the village interspaced with forestland, scrub or parkland. Sometimes these scattered fields seem like a rough radial pattern around the village. The notion of *Abúlé* was a general topographic descriptor of the farm village, but there was another term to distinguish its size. The satellite farm village often began as *Abà*—a small size farming hamlet. Similarly, the entire farm village was referred to, as aforementioned, as *Oko*, farm, in contradistinction to the actual farmlands.

The *Oko* specifically referred to the total farmlands as well as individual cultivated fields near the farm village. Since the farm village represented a production unit of a lineage in the urban settlement, the cultivated fields belonged to lineage members. Indeed, the scattered farms of the farm village were like a spatial diagram of the *mọlẹbi* system. Therefore, there were boundaries in and around the farm villages. Besides, there were other farm villages around. Boundaries were essential as some farmlands were adjoining one another, as "many persons resided on their own separate

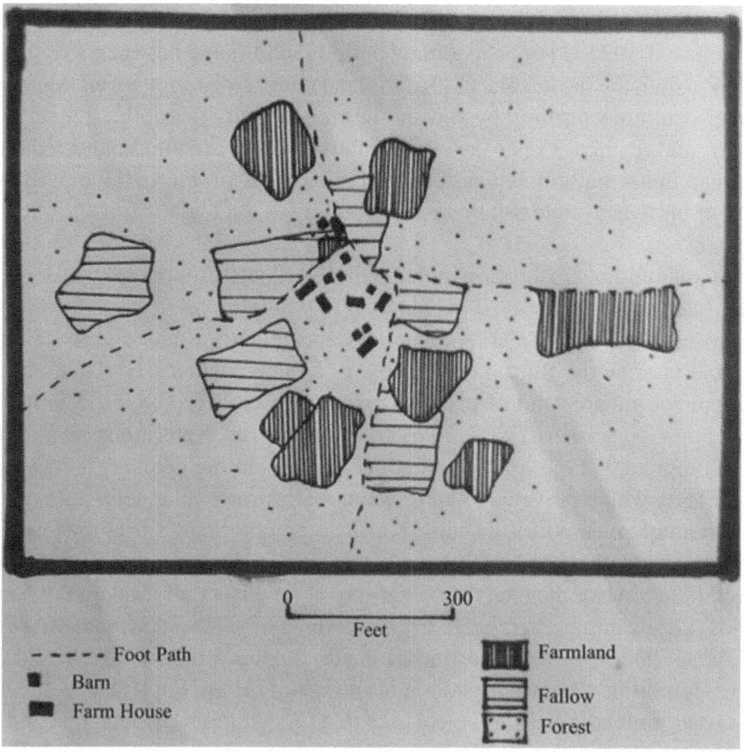

Fig. 3.2 Sketch plan of a farm village (*Source* Private Collection, Redrawn from an Original Sketch)

farms" (Tucker 1855: 13). Although *Ọ̀na*, paths or roads, sometimes served as boundaries, other methods of spatial identification were employed for boundary purposes when farmlands of different individuals are close to one another. These farmlands might seem indistinguishable, but in fact, 'small paths' and natural features are used as *Àla*, boundary. Such natural features include streams, rock outcrops, trees and even deliberately planted trees, such as the *Pẹrẹgun*. The type of crop production usually determines the organisation of the farm layout. The farm layout could have zero-tillage or soil tillage patterns. A common tillage pattern made use of *Ebè*, heaps. As Burton (1863: 62) observed in outlying farms, "the land is carefully prepared with the hoe, and dwarf heaps, in whose summit a hole is drilled receive the seed". Whether it was a zero-tillage or heaped tillage, the farm pattern often reflected a subjective geometry of planting certain crops. The heaps were used as a measure of size and distance within the micro-geography of individual farmland (Atteh 1980).

This indigenous geographic space emerged from the modification of the physical environment through agronomic practices. These material practices centred on the field system, environmental care and the role of weather. The field system used on the individual farmlands was the rotational bush fallow. Bush fallow refers to the agrarian practice in which farmlands, after some years of cultivation, are left to lie fallow, so the land and soil can regenerate naturally to regain their fertility during the fallow period (Johnson 1921: 118). This fallow period could range between five to fifteen years. Thereafter, the farmer places the farmland under cultivation again. Meanwhile, the farmer continues farming by shifting farming activity to another parcel of land in the surrounding area. Hence, around the farm villages are interconnected patches of cultivated fields and fallow lands. Indeed, the spatial structure of the farm village was that of cultivated fields, fallow areas, buildings, secondary forests, and virgin forests.

Environmental care on the farm includes the use of intercropping, agroforestry and soil care. Intercropping involved the planting of two or more crops on the farmland. For instance, there was the common practice of planting maize or beans across cassava and yam fields. Burton (1863: 62), as he approached Abeokuta, observed, "here… the people are fond of planting together several (crops), especially beans, maize and manioc (cassava)". Leguminous crops such as beans are known to fix the nitrogen component of the soil. In addition, cassava planting on many field was done by zero tillage, which prevented soil erosion. Agroforestry, the deliberate retention of trees on a field cleared for farming, was a common feature of indigenous farms. These trees provided several ecological functions like the leguminous plants on the cultivated fields. Also, they served as anti-erosion barriers on the large expanse of farmlands. For example, there were trees such as *Amùja*, *Bujẹ*, *Sapo*, *Ìyeyè* (Yellow Plum), *Akòko* and *Epin* retained on farmlands for these reasons. Alongside their ecological importance, they had economic and social values. For instance, the *Amùja* tree was used for medicinal care; the *Akòko* tree for ceremonial, medicinal and rope-making purposes; and the *Bujẹ* tree were used in producing a kind of colour for tattooing (RBG 1891: 218). Furthermore, several lone-standing trees are used as cartographic labels for describing locations on the farmlands. Soil care on farmland involves several placed-based methods. One of such methods depends on the natural regeneration of the field by worms. Alvan Millson (1891: 584–585) gave an insight into organic fertiliser used by Yorùbá farmers in the 1890s, when he wrote,

> In the dry season …..The whole surface of the ground among the grass is seen to be covered by serried ranks of cylindrical worm casts. These worms casts vary in height from a quarter of an inch to three inches, and exist in astonishing numbers. …. For scores of square miles they crowd the land, closely packed, upright, and burnt by the sun into rigid rolls of hardened clay. There they stand until the rains break them down into a fine powder, rich in plant food, and lending itself easily to the hoe of the farmer….This work goes on unceasingly; year after year…. Where the worms do not work, the Yoruba knows that it is useless to make his farm.

The weather played a fundamental role in agrarian practices. The farming operations of planting, weeding and harvesting were mainly determined by the rainfall pattern (Hopkins 2020: 72). There are two seasons in Yorùbáland, as in other parts of West Africa, the dry and wet seasons. The dry season occurs between October and March, and the wet season occurs between April and September. Indeed, rainfall pattern was a paramount factor not only in the farm village but also interconnected to the entire ancient Yorùbá culture. The ancient Yorùbá began counting the year (*Odún*), made up of fourteen months, at the end of the rainy season sometimes within October and November (Dennett 1910: 130). This mensuration was directly related to farmwork. As the ancient year begins, during the dry season, the farmer clears and burns the field in preparation for the next farming season. At the commencement of the rains between April and May, known as *Aṣerọ Òjo*—the first rains, planting of crops begins. At height of the rains between June and July, usually referred to as *Ộwãrà Òjo*—heavy downpour, weeding of the farm intensifies; these were the months of the most demanding farm work, particularly in the farm village, as the vegetation was at its peak. By the end of the heavy rains, the first harvest was ready.

This was the effect of the rhythm of rainfall pattern on farming that the Yorùbá has a saying that, *Oṣù mẹ̀ta ni ebi ipa àgbẹ̀* (A farmer's period of hunger is three months), the basis of which was the length of time that maize and some vegetables will grow and ripen (Fadipe 1939: 177). The first maize planting begins with the early rains. After a brief dry season, occurring in August–September, the crop of the second harvest was ready as the *Arộkúrò-Òjo*—latter rains approached. The dry season comes again, and the preparation for another year begins. The re-roofing of farmhouses and some other activities are carried out only in the dry season. This ancient Yorùbá year system was superimposed upon, sometimes later in the mid-sixteenth century, with the thirteen lunar months (Dennett 1910). However, the routine of farm activity remains the same and apply to both the home farm and the farm village.

These agronomic practices led to some distinctive normative agrarian institutions associated with the farm village and indigenous agriculture in general. For example, regulations of fieldwork, urban–rural migration and rest house services. The labour force on individual farms usually consists of the farmer and his immediate family. On the farms of chiefs, it involved extended family members. However, there are two forms of collective labour available to the farmer. These are the *Ộwệ* (obligatory labour) and the *Ãró* (reciprocal labour). The obligatory labour and the reciprocal labour were employed on the home and satellite farms, but "in either case, the immediate remuneration was a good meal provided by the convener of the Owe or Aro" (Atanda 1980: 60). These collective labour systems were used in agricultural activities for clearing land, bush or forest growth, as well as planting. As Fadipe (1939: 150) noted, "a member of an aro association is entitled to call upon the entire group to come and help him on his farm, either to clear the land of weeds preparatory to planting, or to reciprocate in kind". Besides, there was also the use of servile workers and the *Ìwộfà*—peonage system of labour before the colonial period. Again, farm labour was gendered but with variations across Yorùbáland. For instance, women did not usually till the soil in western Yorùbáland. However, towards the eastern part of the country, women were involved in regular farming. Either involved in regular farming or taking care of their kitchen gardens, women were also employed in the more demanding task of housekeeping. Besides, the raising of livestock—chicken, ducks, goats—were done by women on the farm. The farm village practice was also associated with urban–rural migration, but of a longer duration unlike the daily routine of farmers to the nearby farms. For example, the Rev. Bowen (1857: 135–136, my emphasis) observed,

> To acquaint myself with the language, intellect, feelings, and every-day life of the natives, I used to visit them on their farms, ten or fifteen miles from town; and remain two or three days. In the spring of 1851, I went with Sam and Shumoi, the cook, to the most distant farms

> up the Ogun river, whence it is not far to the [Oyo] Yoruba line. .....Here we found one or two hundred men, women and children, in long open shanties, surrounded by goats and chickens. They lived in Abbeokuta, but were in the habit of *coming and going* as occasion required.

These urban–rural migrations involved distances that were relatively similar in most towns. For instance, in Eastern Yorùbáland in 1858, D. J. May observed, "I left my quarters in Ilesha, accompanied by a chief's messenger, and travelled north-westerly 4½ miles over a fair but rugged road to a farm village—numerous roads branching to farms indicating much cultivation in the Vicinity" (May 1860: 219). On the other hand, in Western Yorùbáland, D. J. May, the same English traveller wrote, "the next day's journey was to a farmstead, Obagba…a farm to Iwo, which… I found to be but about 5 miles distant" (*Ibid*: 213–214). Hence, the farm-villages serving Iwo and Ilesa were 6.4–8 km (4–5 miles) from the towns respectively. However, some farm villages were farther, between 16 and 12.8 km (10 and 8 miles) from the town, as Thomas B. Freeman encountered in 1843 when travelling to Abeokuta (Tucker 1855: 50–51). Some farm villages were situated on the frontier of the town or the kingdom (Akintoye 2010). Therefore, they served as guest-house to long-distance travellers. Here water and food provisions are made available for the passing traveller without charge or "at exceedingly low prices". For example, Millson (1891: 582) observed,

> There is indeed a custom among the Yorubas of the interior that the passing traveller may stop at any farm or field and cook sufficient food from the standing crops for one meal, but it is considered a very heinous offence for him to carry any away with him.

3.6 Farm Village Practice in Nineteenth Century Urban-Hinterland Relations: The Ibadan Experience

The farm village, which was a visible physical object at a lower geographical scale … [had] 'become an attribute at another' (Downs and Stea 1973: 315). At a larger geographic scale, the notion of the farm village was a way of framing or conceptualising the cultural landscape. However, the basic features of the farm village described above are subject to variations. The ecological differences in different geographical areas of Yorùbáland produced these variations. These variations were evident in the material practices of several Yorùbá sub-ethnic groups (Usman and Falola 2019: 10). Hence, farm village practices in the dry savanna in northern Yorùbáland had some dissimilarities from the same outlying farming practices in the moist forestland of the southern Yorùbás. The practice of satellite farm villages that best reflects the urban–rural hinterland economic relations of the Yorùbá in the nineteenth century, were found near the biggest towns. For example, the towns of Abeokuta, Ijebu-Ode, Ilesa, Ondo, Oyo, Ogbomoso, Ilorin and Ibadan. Ibadan, situated between the two ecological zones, became the largest city in the pre-colonial period (Hopkins 2020: 63) and had a highly developed practise of satellite farm-village culture.

The cultural geography of Ibadan reveals the influence of this agrarian idea on its spatial organisation of the political economy. Ibadan was founded about the first decade of the nineteenth century (Awe 1965: 221). An army of Yorùbá warriors who re-established the settlement as a war camp overran a small town of the same name. The initial town, later known as *Ọjàba*, became the core zone of the new settlement (Akinyele 1911: 44–46). Several wards were created, as the new town expanded. Unlike in the ancient Yorùbá urban setting, several lineages reside in each compound in the wards. These lineages, having fled the ravages of war, and slavery attacks, often settled together in the same compound under the protection of a warlord (*Ibid*: 210). Although the primary occupation of the residents was military service, agricultural work also played an important role in the development of the town and the formation of the Ibadan state. In 1853, Anna Hinderer (1872: 59– 60) observed,

> the immense town of Ibadan…is enclosed by mud walls, eighteen miles in circumference, beyond which there is a broad belt of cultivated land, five or six miles in breath, reclaimed from the bush…. a lively traffic is carried on in the produce of the farms and native manufactures. ….But the principal occupation of the people is farming, in which everyone is engaged, whatever other calling he follows, each having a right to such land as he chooses to occupy outside the walls, provided only that it be not already appropriated.

The basic agricultural production unit for this expanding urban population was the farm village. The pioneering settlers of Ibadan in the 1810s and 1820s resided on the hilly areas while establishing their nearby and satellite farms in the immediate and farther areas respectively (Afolayan 1994: 136). As the town boundary expanded so also the re-location of nearby farms and the farm-villages. Several farm villages were shifted several times outside the town walls, as it was rebuilt thrice before the colonial period began in 1893 (Akinyele 1911: 44; Johnson 1921: 327). Millson (1891: 583) noted,

> Some three miles to the north of Odo Ona Kekere, from the crest of a rising in the undulating land, the great city of Ibadan – the London of Negroland – comes full in view, extending for over six miles from east to west, and for more than three from north to south. Surrounded by its farming villages, 163 in number, Ibadan counts over 200,000 souls, while within the walls of the city itself at least 120,000 people are gathered. Its sea of brown roofs covers an area of nearly 16 square miles, and the ditches and walls of hardened clay, which surround it, are more than 18 miles in circumference.

The Ibadan agricultural landscape constituted several farm villages. Every major compound in the town had one or more farm-village(s) at the outskirt of Ibadan, like in other Yorùbá towns (Goddard 1965). The farm villages were identified by place names. Oftentimes the place name used was the name of the founder of the compound. Therefore, several farm villages on the Ibadan agrarian landscape in the nineteenth century had the same names as that of the residential compound in the town. However, there were other types of place names. For instance, there was the use of civil rank, military titles, events, names of animals or the occupation of the founder of the village. Owing to the impact of high population density, expanding external trade and political-economic influence, this agrarian spatiality was modified from the simple ancient lineage-based farm village practice. Hence, the satellite farm village in Ibadan territory had a larger village layout of several buildings, with surrounding larger farmlands employing a larger workforce. As Awe (1973: 69–70) observed,

> This … type of farm was colonised by a labour force made up of women, children and slaves. The average size of the farm is not known, but the fact that some *oko egan* could boast of as many as six hundred or more slaves indicates that each of these distant farms must have covered a fairly large area.

This agrarian practice had a direct influence on urban political economic sustenance, geopolitical identities and indigenous state territorial formation. First, the farm villages were crucial in urban food sustenance, a foremost element of the political economy of the state. Although there was a rapid influx of migrant population in the mid-nineteenth century into the town, the agrarian production system sustained the indigenous social order. From the founding of Ibadan, there has been a rapidly expanding population over the decades of the nineteenth century, which increased the concern for domestic sustenance. For example, in 1851, Rev. Hinderer estimated the population of the town as between 60,000 and 100,000 (Hinderer 1872: 20). By 1890, Mathews noted it must have had almost a quarter of a million inhabitants (Findlay and Holdsworth 1922: 228). As a military state, Ibadan was often in a state of emergency, which made large-scale agricultural production for domestic survival in the farm villages crucial to state policy. "These extensive plantations not only support their huge establishments but also supply the markets, so that a military state though Ibadan was, food was actually cheaper there than in many other towns" (Johnson 1921: 325).

Although the state was involved in internal commerce in Yorùbáland there was also a concern for external trade (Falola 1984). For example, the palm oil trade by European merchants on the seacoast led the Ibadan state to organise palm plantations to meet this external demand (Webster and Boahen 1980: 60). In 1858, Daniel J. May observed, "Between Iwo and Ibadan palm-trees are particularly numerous" (May 1860: 213). Similarly, in 1890, Millson pointed out, "Northward from Ibadan, [were] ….many villages, and a landscape dotted far and near with oil-palms (*Elais guineensis*)" (Millson 1891: 583–584). The satellite farms in the Ibadan hinterland were integral to achieving this agricultural commercial success. Indeed, this indigenous farm system in Ibadan and other Yorùbá towns created and sustained the palm oil trade, which had value amounted to £1.1 million in the first half of 1888 in the Port of Lagos (Crowder 1966: 203). Therefore, this agrarian practice was embedded in the political-economic sustenance of the urban polity.

Second, farm villages contributed to the sense of political identity in the politics of the indigenous Ibadan state. The 163 farm villages served as an additional political representational device in the heavily populated settlement. An adage in the late-nineteenth century bears this out *Ọmo Ibadan ti o ni oko, ẹrú ni* (An Ibadan person who does not have a farm village is of servile origin). The citizenry included the farm village(s) of their household in the framing of their sense of identity. Indeed, the place-name of a farm village, as a crucial symbolic device in the internal geopolitics of the state, manifested political identities amongst the Ibadan citizenry.

Third, farm villages were landmarks in defining the territorial space of the Ibadan state. Farm villages served as geographical boundaries on the frontier of indigenous Yorùbá states (Akintoye 2010: 220). For instance, an Ibadan saying states that *Lalupon ni ilú ti bẹrẹ* (it is from Lalupon that the Ibadan settlement commences).

This saying reveals the farm-village of Lalupon, on the northern outskirts, as a landmark in the indigenous cognitive maps and mappings. It was an important landmark in the layers of territorial boundaries of the physical space. In the 1850s, the Ibadan army invaded and subdued the greater part of central Yorùbáland to establish Ibadan Empire leading to layers of territorial boundaries of the town, indigenous state and Empire—on the cultural landscape. As specific farmlands used roads, trees, rocks as boundaries, so the farm villages served as boundary landmarks on the cultural landscape at a larger spatial scale. For example, in 1851, Hinderer coming from the Egbaland sought to "reach some of the Ibadan farms before night-fall" (Tucker 1855: 234). These were frontier farm villages of Ibadan near the Egba boundary. Therefore, the 163 farm villages served as the visible mark of the frontier of Ibadan state.

3.7 The Farm Village—An Integrated Model

The cultural encounter of Yorùbá agrarianism with Western scientific agriculture, particularly state-owned land colonisation, has encouraged innovation in indigenous agro-spatial thought. The modernist farm settlement scheme was conceived as '*Agbé Oko dá Oko*' [living and working on the farm]—a notion very similar but different from the traditional farm-village and the regular state farm. The same notion of *Agbé Oko d'Oko*, an indigenous agro-spatial thought, which emerged from the farm village practice of Yorùbá urbanism, have been re-employed as an interpretative framework to reconceptualise this post-1945 modernist settlement scheme (A. Gbadamosi, personal communication, March 23, 2019). This reworking of the state-owned farm scheme through indigenous farm village practice offers a hybridised integrated model in exploring possibilities for sustainable agriculture in contemporary sub-Saharan Africa.

Incorporating agricultural production with environmental care, as agrarian thinking at the grassroots, provides an alternative vision. These indigenous-based agrarian imaginaries could take different spatial forms. For example, it could encourage the creation of community-driven farm settlement schemes (CFS). As indigenous farmers often prefer and define, a modern farming village as a 'loose-type' group or collective farming, "rather than the joint tenancy co-operative which is based on the Israeli moshav" (Oshuntogun 1980: 145). These locally defined schemes of *Agbé Oko d'Oko* will provide recognised spatial production units as a key measure in responding to the challenge of diminishing farmland in Sub-Saharan Africa (Jayne et al. 2014). Conversely, environmental care principally revolves around the questions of limiting agrochemical use, reforesting and revitalising dead soil (Fabricant 2012). For instance, to reduce or avoid the use of agrochemicals, operators and participants of the CFS must embrace natural pesticides and organic fertilisers. This self-defined agrarian space, if legally backed, can offer indigenous knowledge-based sustainable agriculture in rapidly emerging urban fringes.

Thus, the elements of indigenous agro-ecological practice should manifest in the envisioned sustainable agrarian landscape: soil maintenance, crop rotation, natural fertilisers, forest conservation, water sources, wildlife management, weather observation and human social values. For example, soil care should be paramount in knowledge production and agronomic practice. In 1904, George Hazzledine (1904: 183) commenting on Yorùbá farmers at Ilorin, observed, "they understand their soil well". Likewise, Alan Millson's insight into organic fertilisers emphasise the regeneration of soil fertility in agronomic practice. Thus, crop production using the rotation bush fallow method "maintain an excellent average, and the same plot of ground serves for generations to support its owners" (Millson 1891: 584). As Lal (2009: 8) observed, "Soil—the essence of all terrestrial life—embodies Mother Nature in that it has the capacity to meet all human needs without the greed".

3.8 The Farm Village—An Agrarian Thought Amongst the Yorùbá

The cognitive representations of this Indigenous agricultural space highlight the material practices dependent on human–environment interaction for sustainability and question the geo-narratives of the wasteful system of subaltern agriculture in colonial knowledges (Blomley 2006). Agricultural production in modernist spatial thought aims at high productivity with little or no serious concern for the natural environment. In contrast to modernist Cartesian space, indigenous spatial thought conceives the landscape as an evolving cultural phenomenon, which acknowledges history and tradition in environmental ethics (Olwig 2009: 245). From the above study, the conceptualisation of Indigenous agricultural space as near and far workplaces reflects the understanding and reliance of this urban culture on the natural environment as an integral part of society. This Indigenous environmentalism is instructive on the cautious use of space to ensure ecological security. As Ogar and Bassey (2019: 78) observe, following Ogungbemi (1997), African societies "recognize the importance of water, land and air management, particularly the traditional ethic of not taking more than you need from nature." Indigenous agricultural space framed by sustainability manifests the continued importance of local geographies.

These place-based knowledges, at all spatial scales, reflected a continuous state of actively constructed indigenous geographies (Radcliffe 2015). The geographies of indigenous agriculture are not static or fixed, but as relational and contingent, they always respond to environmental, social and political changes. In the above study, there were changes from the ancient lineage-based farm village practice to the household-based practice of a larger urban settlement consequent to the social upheavals of the nineteenth century. These modifications of local spatial knowledge were also evident in the colonial period. They were responses to the colonial restructuring of the agricultural economy. These indigenous spaces of agricultural work, which adjusted to the realities of capitalist modernity continued to serve the needs of the subaltern populations (Hopkins 2020: 103). Recently, this outlying agrarian practice has been a cultural lens to reinterpret agricultural modernity in response to economic stagnation arising from the crisis of capitalist development. This reflects the adaptive capability of traditional and indigenous spatial thought. As Mackenzie et al (2017: 43) assert, "Traditional knowledge is adaptive and adaptable. Due to the grounding of traditional knowledge in experience, those bearing and utilising traditional knowledge are constantly adjusting and readjusting their understanding of the landscape and seascape. Traditional knowledge does not exist in a vacuum, and continues to evolve and adapt". Adaptive qualities of local spatial knowledge are the cultural lens of alternative modernity to rethink agrarian development (Fabricant 2012: 126).

Although several African governments have consistently sought agricultural modernisation, these modernisation programmes have had limited impact on the general farming population. Rather, the adaptions of traditional agriculture to Western cultural encounters have mainly come from subaltern populations through their indigenous agrarian skills. In this study, I have shown the emergence of the notion of *Agbé Oko d'Oko* in the twentieth century that reconceives a modern agricultural spatial organisation: the settlement scheme. Indeed, there are previous adaptions in the agricultural economy evident in the Yorùbá cocoa industry (Berry 1975, 1986; Faluyi 1995). However, the physical control and impact of capitalist modernity have meant challenges to these indigenous conservational skills. Their co-option into the capitalist market economy has had some positive but mostly negative effects on indigenous agrarian practices.

This is evident in the stereotypical representation of an inferior African agricultural system, which reflects capitalist modernity's conception of tradition and nature (Mbembe 2015: 9; Hopkins 2020). Within this current denigration of indigenous agrarian practice, a legacy of colonial knowledge systems in modern state administration (Loomba 2005; Wirzba 2003), alterations of this conception of indigenous agrarian space still exists in remote hinterlands (See Fig. 3.3), though characterised by several limitations in the modern world. Therefore, amidst the elusive rhetoric of sustainability from political actors of African administrative geographies, these indigenous agricultural geographies actively, provide, within their limitations, the necessary moral fabric and eco-social ethic to ensure the environmental sustenance of the agricultural space.

3.9 Conclusion

The Yorùbá people of West Africa are mainly urban dwellers, but like most peoples of Sub-Sahara Africa, they depended on agriculture for their sustenance. Also, like in other parts of West Africa, they operated an agricultural system with a geographical pattern of inner and outer farming units. The outer farming practice of the farm-village was the bulwark of urban sustenance through the centuries. This practice reveals that in Yorùbá societies, there is a connection in eco-social interaction with the sustenance of livelihood, urban political identity and political geography. This was exemplified in nineteenth-century Ibadan and remerges in alternative agricultural

Fig. 3.3 A farm village of Isundunrin town, along the Iwo-Oke—Isundunrin road, Ejigbo Local Government Area, Osun State, Southwest Nigeria, 2021 (*Source* Author, 2021)

visions in the post-colonial landscape. Self-defined agrarian practices ought to be given much consideration in the sustenance of African cities, as these cities are experiencing rapid urbanisation. In addition, the sustenance of the African population meant employing developmental policies in line with environmental concerns that embed indigenous sustainability discourses in the grassroots. This requires rethinking modern economic development from the African perspective and the need to draw attention to agrarian politics, eco-social relations and environmental ethics in post-colonial governance. Consequently, the farm village offers an indigenous spatial framework to reinterpret the sustenance of African cities through outlying agrarian production in the urban hinterlands.

References

Adeoye, C.L. 1979. *Àṣà and Ìṣe Yorùbá.* [Traditions and Customs of the Yorubas] 2014 Reprint. Ibadan: University Press.

Afolayan, A. 1994. Migration, Social links and Residential Mobility in the Ibadan Region. In *Ibadan Region*, eds. Michael Filani, Festus O. Akintola, and Chris O. Ikporukpo, 136–143. Ibadan: Rex Publications.

Akinjogbin, I. 1966. It History, Its People and Its Culture. In *An Introduction to Western Nigeria: Its People and Culture and System of Government*, ed. Adedeji, Adebayo, 8–19. Ife: Institute of Public Administration, University of Ife.

Akintoye, S. 2010. *A History of the Yoruba People.* Dakar: Amalion Publishing.

Akinyele, I. 1911. *Iwe Itan Ibadan [History of Ibadan]*, 1981st ed. Ibadan: Board Publications.

Ajayi, J., and R. Smith. 1971. *Yoruba Warfare in Nineteenth Century.* Ibadan: Ibadan University Press.

Atanda, J. 1980. *An Introduction to Yoruba History.* Ibadan: Ibadan University Press.

Atteh, D. 1980. *Resources and Decisions; Peasant Farmer Agricultural Management and its Relevance for Rural Development Planning in Kwara State, Nigeria.* PhD Thesis. Geography Department, University of London.

Awe, B. 1965. The End of an Experiment: The Collapse of the Ibadan Empire 1877–1893. *Journal of the Historical Society of Nigeria* 3 (2): 221–230.

Awe, B. 1973. Militarism and Economic Development in Nineteenth Century Yoruba Country: The Ibadan Example. *Journal of African History* 14: 65–77.

Bascom, W. 1955. Urbanisation Among the Yoruba. *American Journal of Sociology* 60 (5): 446–454.

Bascom, W. 1958. Yoruba Urbanism: A Summary. *Man* 58: 190–191.

Berry, S. 1975. *Cocoa, Custom and Socio-economic Change in Rural Western Nigeria* Oxford: Clarendon Press.

Berry, S. 1986. *Fathers Work for their Sons*. Los Angeles: University of California Press.

Blomley, N. 2006. Uncritical Critical Geography? *Progress in Human Geography* 30(1): 87–94. https://doi.org/10.1191/0309132506ph593pr.

Bowen, T. 1857. *Central Africa: Adventures and Missionary Labours in Several Countries in the Interior of Africa from 1849 to 1856*. Charleston: Southern Baptist Publication Society.

Bryan, J. 2009. Where Would we be Without Them? Knowledge, Space and Power in Indigenous Politics. *Futures* 41: 24–32. https://doi.org/10.1016/j.futures.2008.07.005.

Burton, R. 1863. *Abeokuta and the Cameroon Mountains: An Exploration*. London: Tinsley Brothers.

Chisolm, M. 1962. *Rural Settlement and Land Use*. London: Hutchinson University Library.

Clarke, W. 1972. *Travels and Explorations in Yorubaland (1854–1858)*, ed. Jacob A. Atanda. Ibadan: Ibadan University Press.

Clayton, D. 2003. Critical Imperial and Colonial Geographies. In *Handbook of Cultural Geography*, eds. Kay Anderson, Mona Domosh, Steve Pile, and Nigel Thrift, 354–368. London: Sage Publications.

Office, Colonial. 1908. *Colonial Reports-Annual, Southern Nigeria, Report for 1906*. London: HMSO.

Crowder, M. 1966. *The Story of Nigeria*, 2nd ed. London: Faber.

De Leeuw, S., and S. Hunt. 2018. Unsettling Decolonizing Geographies. *Geography Compass* 12 (7): 1–14. https://doi.org/10.1111/gec3.12376.

Dennett, R. 1910. *Nigerian Studies or The Religious and Political System of The Yoruba*. London: Macmillan.

Downs, R., and D. Stea. 1973. Cognitive Maps and Spatial Behaviour: Process and Products. In *The Map Reader Theories of Mapping Practice and Cartographic Representation*, eds. Martin Dodge, Rob Kitchin, and Chris Perkins, 2011, 312–317. Chichester: Wiley-Blackwell.

Fabricant, N. 2012. *Mobilizing Bolivia's Displaced: Indigenous Politics and the Struggle over Land*. Chapel Hill: University of North Carolina Press.

Fadipe, N. 1939. *The Sociology of the Yoruba*, ed. Francis Okediji, and Oladejo Okediji,1970. Ibadan: Ibadan University Press.

Falola, T. 1984. *The Political Economy of A Pre-Colonial African State: Ibadan 1830–1900*. Ile-Ife: University of Ife Press.

Faluyi, E. 1995. *A History of Agriculture in Western Nigeria 1900–1960*. PhD. Thesis. History Department, University of Lagos, Nigeria.

Findlay, G., and W. Holdsworth. 1922. *The History of the Wesleyan Methodist Missionary Society* (IV). London: The Epworth Press.

Floyd, B. 1969. *Eastern Nigeria: A Geographical Review*. London: Macmillan.

Gbadegesin, S. 1991. *African Philosophy: Traditional Yoruba Philosophy and Contemporary African Realities*. New York: Peter Lang.

Goddard, S. 1965. Town-Farm Relationships in Yorubaland: A Case Study from Oyo. *Africa* 35 (1): 21–29.

Gray, L. 2003. Investing in Soil Quality: Farmer Responses to Land Scarcity in Southwestern Burkina Faso. In *African Savanna: Global Narratives and Local Knowledge of Environmental Change*, ed. Thomas J. Bassett and Donald Crummey, 72–90. Oxford: James Currey.

Green, A.R. 2013. *Agriculture is the Future of Nigeria*. Skoll World Forum www.forbes.com/sites/skollworldforum. Accessed 10 Mar 2020.

Hazzledine, G. 1904. *The White Man in Nigeria*, 2018th ed. London: FB Ltd.

Hinderer, A. 1872. *Seventeen Years in the Yoruba Country: Memorials of Anna Hinderer, Wife of the Rev. D. Hinderer, C. M. S. Missionary in Western Africa.* With An Introduction by Richard B. Hone. London: The Religious Tract Society.

Hopkins, A. 2020. *An Economic History of West Africa*, 2nd ed. London: Routledge.

Jayne, T., A. Chapoto, N. Sitko, C. Nkonde, M. Muyanga, and J. Chamberlin. 2014. Is The Scramble For Land In Africa Foreclosing a Smallholder Agricultural Expansion Strategy? *Journal of International Affairs* 67 (2): 35–53.

Johnson, S. 1921. *The History of the Yorubas: From the Earliest Times to the Beginning of the British Protectorate.* 1960 Reprint, ed Dr. Obadiah Johnson. Lagos: CMS Bookshop.

Lal, R. 2009. Mother of Necessity: The Soil. In *Organic Farming, Pest Control and Remediation of Soil Pollutants*, ed. Eric Lichtfouse, 5–9. Heidelberg: Springer.

Lander, R. 1830. *Records of Captain Clapperton's Last Expedition to Africa (1).* London: Colburn and Bentley.

Larsen, S., and J.T. Johnson. 2012. In Between Worlds: Place, Experience, and Research in Indigenous Geography. *Journal of Cultural Geography* 29 (1): 1–13. https://doi.org/10.1080/08873631.2012.646887.

Loomba, A. 2005. *Colonialism / Postcolonialism.* London: Routledge.

Mabogunje, A. 1974. The Precolonial Development of Yoruba Towns. In *The City in the Third World*, ed. Denis Dwyer, 26–33. London: Macmillan Publishers.

Mabogunje, A. 1976. Cities and African Development. In *Studies in the Development of African Resources No. 3*, eds. Geoffrey C. Last, and Akinlawon L. Mabogunje. Ibadan: Oxford University Press.

Mabogunje, A. 2008. The Inclusive City: Popular Empowerment of Local Governments in Rapidly Urbanizing Africa. *Journal of Nigerian Institute of Town Planning* 21 (1): 1–16.

Mackenzie, K., W. Siabatob, F. Reitsmaa, and C. Claramun. 2017. Spatio-temporal Visualisation and Data Exploration of Traditional Ecological Knowledge/Indigenous Knowledge. *Conservation and Society* 15 (1): 41–58. https://doi.org/10.4103/0972-4923.201391.

Masood, M. 1978. The Traditional Organisation of a Yoruba Town: A Study of ljebu-Ode. *Ekistics* 45 (271): 307–312.

May, D. 1860. Journey in the Yoruba and Nupe Countries in 1858. *Journal of the Royal Geographical Society of London* 30: 212–233.

Mbembe, A. 2015. *Decolonizing Knowledge and the Question of the Archive: Africa is a Country.* https://wiser.wits.ac.za/system/files. Accessed 6 Nov 2019.

Millson, A. 1891. The Yoruba Country, West Africa. *Proceedings of the Royal Geographical Society and Monthly Record of Geography* 13 (10): 577–587.

Montmarquet, J. 1985. Philosophical Foundations for Agrarianism. *Agriculture and Human Values.* 2: 5–14.

Morgan, W.B. 1959. The Influence of European Contacts on the Landscape of Southern Nigeria. *The Geographical Journal* 125 (1): 48–64.

Moseley, W., E. Perramond, H.M. Hapke, and P. Laris. 2014. *An Introduction to Human–Environment Geography: Local Dynamics and Global Processes.* Chicester: Wiley-Blackwell.

Nayak, A., and A. Jeffery. 2011. *Geographical Thought: An Introduction to Ideas in Human Geography.* London: Routledge.

Obateru, O. 2005. *Planning Regional and Rural Development.* Ibadan: Penthouse.

Olwig, K. 2009. Landscape, Culture and Regional Studies: Connecting the Dots. In *A Companion to Environmental Geography*, eds. Noel Castree, David Demeritt, Diana Liverman, and Bruce Rhoads, 238–252. Chicester: Wiley-Blackwell.

Ogar, J., and S. Bassey. 2019. African Environmental Ethics. *RAIS Journal for Social Sciences* 3 (1): 71–81.

Ogundiwin, B. 2009. *Geographical Thought and Methodology among the Yorubas*, Unpublished Manuscript, Geography and Planning Department, University of Lagos, Akoka.

Ogungbemi, S. 1997. An African Perspective on the Environmental Crisis. In *Environmental Ethics: Readings in Theory and Application*, ed. Louis J. Pojman, 330–337. Belmont, CA: Wadsworth Publishing Company.

Ojo, A. 1970. Some Observations on Journey to Agricultural Work in Yorubaland, Southwestern Nigeria. *Economic Geography* 46 (3): 459–471.

Oshuntogun, A. (1980). Cooperative and Small Farmers in Nigeria. In *Nigeria Small Farmers: Problems and Prospects in Integrated Rural Development*, eds. Olajuwon Olayide, Julius A. Eweka, and Victoria E. Bello-Osagie, 133–148. Ibadan: CARD.

Oyeleye, D.A. 2001. *Settlement Geography*. Akoka, Lagos: University of Lagos Press.

Radcliffe, S. 2015. Geography and Indigeneity I: Indigeneity, Coloniality and Knowledge. In *Progress in Human Geography Progress Reports*. 1–11. https://core.ac.uk. Accessed 11 Nov 2019

Royal Botanical Gardens (RBG). 1891. Indigenous Plants of Yoruba-Land. In *Bulletin of Miscellaneous Information*, vol. 56, 206–219. Kew: Royal Botanic Gardens.

Sidaway, J. (2000). Postcolonial Geographies: an Exploratory Essay. *Progress in Human Geography*. 24 (4): 591–612. https://doi.org/10.1191/030913200100189120.

Steel, R. 1948. Some Geographical Problems of Land Use in British West Africa. *Transactions and Papers (Institute of British Geographers)* 14: 27–42.

Thompson, P. 2008. Agrarian Philosophy and Ecological Ethics. *Science and Engineering Ethics* 14: 527–544.

Tosam, M.J. 2020. Negotiating and Overturning the Othering of Indigenous Epistemologies. *Journal of World Philosophies* 5 (1): 282–286. https://doi.org/10.2979/jourworlphil.5.1.18.

Tucker, S. 1855. *Abeokuta; or Sunrise within Tropics: An Outline of the Origin and Progress of the Yoruba Mission*. New York: Robert Carter and Brothers.

Usman, A., and T. Falola. 2019. *The Yoruba from Prehistory to the Present*. Cambridge Cambridge: Cambridge University Press.

Webster, J., and A. Boahen. 1980. *The Revolutionary Years: West Africa Since 1800*. London: Longman Group.

Wirzba, N. 2003. Why Agrarianism Matters—Even to the Urbanites. In *The Essential Agrarian Reader: The Future of Culture, Community and the Land*, 2010th ed., ed. Norman Wirzba, 21–44. Lexington: University Press of Kentucky.

Wise, M. 2018. The Ties of Historical Geography and Critical Indigenous Studies. *Historical Geography* 46: 239–284. https://doi.org/10.1353/hgo.2018.0033.

Yeh, E., and J. Bryan. 2015. Indigeneity. In *The Routledge Handbook of Political Ecology*, ed. Tom Perreault, Gavin Bridge, and James McCarthy, 531–543. London: Routledge.

Babatunde A. Ogundiwin is currently a Ph.D candidate in the School of Geography, Archaeology and Environmental Studies, University of the Witwatersrand, Johannesburg, South Africa. He received his B.Sc and M.Sc from the University of Lagos, Akoka, Nigeria. He is currently completing his thesis on the interlinkage of cartography, the modern political state, subaltern groups and agricultural development, emphasizing the role of socio-economic maps in providing visual insights into human spatial thought. His research interests include the cartographic history of Africa, economic cartographies, topographic mapping, landscape studies, participatory planning and development geographies.

Chapter 4
On the Confluence of Permaculture and African Agrarianism

David Anthony Pittaway

> "Indeed, many permaculture techniques have been adopted from indigenous farmers around the world."—Charles Eisenstein (2012)

Abstract This chapter offers a sketch of some practical components of a small but arguably typical permaculture homestead, with attention given to possible motivations that permaculturalists may have for some of their endeavours. A brief generalised sketch of an African agrarian homestead is offered, with some conceptual factors coming into view, and overlap between permaculture and African agrarianism is highlighted. The issue of centralisation in South Africa is raised, with the argument being that it is inimical not only to permaculture and African agrarianism, but also to the well-being of ecological systems in general. The example of extant waste-water and sewerage management systems in South Africa is briefly given the spotlight for the purpose of illustrating what is at stake ecologically if decentralised systems are not embraced more widely. The chapter ends with a call for action: despite transformation rhetoric in South African politics, centralisation continues to give impetus to homogeneous forms of development at the cost of cultural and ecological diversity. The confluence of African agrarianism and permaculture illustrates some characteristics of culturally-inclusive, ecologically-sensitive, and refreshingly simple SMART targets.

Keywords African agrarianism · Permaculture · Centralisation · Decentralisation · Externalities · SMART targets.

D. A. Pittaway (✉)
University of the Free State, South Africa, Bloemfontein, South Africa
e-mail: pittaway@gmail.com

M. J. Tosam and E. Masitera (eds.), *African Agrarian Philosophy*, The International Library of Environmental, Agricultural and Food Ethics 35,
https://doi.org/10.1007/978-3-031-43040-4_4

4.1 Introduction

My decade-long permaculture journey impels me to write this introduction in the first-person. The decade of permaculture-esque living began after my own academic study of philosophy culminated in an MA, and also after lecturing philosophy in various contexts. Thereafter, during the years of trying to practice permaculture as a way of life (inspired by philosophy as a way of life as explored by Pierre Hadot 1995), I lectured philosophy at an African university (which involved tailoring the syllabus to be more suitable for an African context) while completing a philosophy Ph.D that ventured partly into ecological territory thematically. It is my hope that my time and experience as an inclusivity-focused and ecology-focused philosopher and permaculture practitioner in South Africa count as grounds from which intimate first-hand experience on and with the land may yield some observations relevant to the subject of African agrarianism on one hand, and ecological problems and solutions on the other.

I am not an expert in African agrarianism (AA), but even if I were, as a white South African man it may be inappropriate for me to comment on the subject in any but a few oblique ways. One of these ways is, I suggest, to draw attention to the areas in which there seems to be overlaps between what I have seen of permaculture and AA. Without a doubt, there are differences between them, but in the spirit of emphasising commonalities between people and practices to foreground "our common humanity" (Roberts 2015), my focus will be on what look to me like shared characteristics and features. This approach lends itself well to criticism. First, because it may legitimately be asked if what appear to me to be similarities, resonances, and overlaps are indeed those. Second, because my glimpses of African agrarian practices have always been from the periphery, which happens when a cultural outsider tries to look in. Caveats therefore apply, and charitable interpretation is requested.

My ongoing study and implementation of permaculture have led me to believe that there is such a thing as a generalised small-scale example of permaculture implementation. I describe this example in the next part of the chapter, and I address what I consider to be some reasons for why permaculture practitioners may embark on permaculture projects. Thereafter, I describe some practical and conceptual aspects of AA that appear to resonate with some aspects of permaculture, leading to the analysis that centralisation (e.g. a municipal sewerage system) is inimical to both AA and permaculture. I follow with an example of centralised waste-water treatment methods to illustrate ecological short-comings typical in centralised systems, with focus on insights that AA and permaculture may offer towards remedy. The overall picture painted in the exploration of the confluence of AA and permaculture leads to an example set of SMART targets that offer some down-to-earth solutions to some of the problems of centralised sewerage systems. Finally, It is my hope that the process that allows for arrival at these targets may be used in a broader process towards addressing some of the problems of centralisation in its other forms.

4.2 Permaculture Example and Potential Motivations

4.2.1 Example

In this generalised land-based[1] example of the application of various permaculture principles, the following system components are among those present: a small house, rain-water storage tanks, a small solar energy system, a vegetable garden, some fruiting trees, indigenous trees, chickens and a chicken coop, a compost toilet, a compost-making zone, and a wood-burning cooking stove. The small house is constructed out of as many low-impact building materials and reclaimed/upcycled materials as possible, for example locally-felled timber and adobe (clay/sand/straw), straw-bales, earth-bags, and/or wattle-and-daub, with various store-bought materials also present, for example a corrugated-iron roof. While the house provides shelter to the people who live on the land, the house is not the main feature of the space as it would be in a typical urban and suburban area. Instead, the outdoor environment is as much a living space as the indoor environment, with the house providing a roofed and enclosed area for people to take shelter in at night and during inhospitable weather.

In this example, there is no connection to grid electricity or water. Potable water and irrigation water both come from the water stored in storage tanks. The tanks are connected via a gutter system to the roof of the house. When it rains, water flows into the tanks, and the water is accessed by way of a passive gravity water system—there is no electric pump present to create water-pressure in this example. Electricity is present for other uses, but its usage is conditional. During sunny daylight hours, one can power numerous devices because the power is not being drained from the deep-cycle batteries but is being sourced more-or-less directly from the solar panels. At night in this sketch, the amount of electricity drained from the batteries must be 'capped' or they may become damaged, which in practice may mean occasionally having to pull the plug on some electric devices. Correct design of the photovoltaic system should prevent such outcomes, but adapting to environmental factors such as electricity constraints is par for the permaculture course.

The vegetable garden contains a variety of vegetables as appropriate for the season and the amount of rain-water that can be taken from the water tanks without emptying them. Vegetables are 'companion planted': vegetables that grow well together are planted alongside each other to create a mutually beneficial relationship. A good example of companion planting is the 'Three Sisters': corn, squash, and beans. The beans fix nitrogen into the ground and thus enhance fertility; the corn provides the support poles that support the beans; and the big leaves of the squash cover the soil and keep it cool and moist. Seeds from each crop are saved for future crops. There are no grains grown in this scenario due to the large space that is required to grow them. Instead, potatoes and sweet potatoes serve as the primary starches.

[1] Permaculture can be applied in various contexts, not just land-based ones. For example, my permaculture teacher, whose name is Aranya, applied permaculture principles when re-designing his music studio. However, the vast majority of his permaculture portfolio is constituted by land-based examples.

The soil in the vegetable garden is kept fertile by regularly topping it up with home-made compost. Where possible, the ingredients for the compost come from the patch of land in focus in the forms of tree leaves, grass cuttings, kitchen scraps, harvest by-products, chicken manure, as well as human manure. The 'humanure' is incorporated into the compost pile sparingly and carefully, i.e. by placing it at the centre of a hot pile of compost in order to kill potential pathogens. Other animal manure is brought onto the land from nearby horse-stables where the incoming organic matter is a mix of horse manure and horse-stable wood-chip floor-cover. The contents of a few buckets of cow manure slurry are also incorporated into the compost pile in order to encourage the growth of useful fungi and micro-organisms. A variety of carbon-rich organic materials is used to mulch the garden beds to retain moisture in the system—this is important because it minimises the use of valuable and often limited water resources.

Chickens provide eggs for the human diet, and a chicken is occasionally killed for its meat. The birds also act as a natural pest control on the land and in the vegetable garden. In this example, chicken-feed is occasionally purchased and fed to the chickens because what is grown on the land is insufficient to fully feed the chickens. The people in this example also eat store-bought food because insufficient produce is grown on site. This is due to the need for the people to spend much of their time earning a wage that pays the inevitable bills and the various other expenses that are part and parcel of life in a capitalist system. In this example, however, the people do not aim to be self-sufficient, but rather to produce a good quantity of healthy eggs and vegetables as an ethical imperative while also developing their relative-autonomy that comes with the proverbial 'life on the land'. Some of the produce is bartered in the local community.

The house, vegetable garden, and chicken coop occupy approximately 30% of the land-space in this generalised example. Most of the land is dedicated to trees: largely endemic trees, but also with some fruiting trees in the mix. The trees provide forage for birds, insects and bees, while also serving as a natural privacy barrier for the humans. Exotic trees in the area are felled and their wood used in the small cook-stove for the purpose of cooking meals and heating water for dish-washing and body-washing. Keeping the human 'footprint' on the land to a minimal size is an ethical imperative in this example. There is a commitment to the custodianship of the land rather than mere ownership of it, as well as to the nurturing of a system that incorporates and respects the non-human life and energies that largely constitute this basic permaculture homestead, an ethic that arguably applies not only to the homestead but also to broader contexts in which the permaculturalist interacts.

4.2.2 Potential Motivations

While it may seem presumptuous to speak for the motivations of others, permaculture theory and practice are largely grounded on the work by Bill Mollison, most notably in *Permaculture: A Designer's Manual* (1988). From the first paragraph of the book,

Mollison (1988: i) positions permaculture as being in opposition to "modern agriculture, monoculture forestry, and thoughtless settlement design", all of which he refers to as damaging, and he advocates for a movement "towards regional or village self-reliance". The *Designer's Manual* is, says Mollison (1988: 1), "about design", but it is "also about values and ethics, and above all about a sense of personal responsibility for earth care."

Mollison takes aim at the "madness of uncontrolled industrial growth" (Ibid), noting that "[o]ur consumptive lifestyle has led us to the very brink of annihilation" (Ibid), and he comments that the "experience of the natural world and its laws has almost been abandoned for closed, artificial, and meaningless lives" (1988: 2). Keeping in mind that Mollison hails from Australia, he states that he "believes that unless we adopt sophisticated aboriginal belief systems and learn respect for all life, then we lose our own". He is of the view that the more "we depart from communal permanence, the greater the risk of tyranny, feudalism, and revolution and the more work for less yield" (1988: 6).

Mollison (1988: 7) identifies the following as one of two "responsibilities to pursue": "to get our house and garden, our place of living, in order, so that it supports us", noting that "truly responsible conservationists have gardens which support their food needs, and are working to reduce their own energy needs to a modest consumption" (Ibid). He calls for "a society in which we are all designers, based on an ethical and applied education, with a clear concept of life ethics" (1988: 10). He speaks highly about the Gaia hypothesis (in which life on Earth is seen to function as a single complex system), and after mentioning it for at least the second time, he states that the "reaction of the earth is to restore equilibrium and balance. If we maltreat, overload, deform, or deflect natural systems and processes, then we will get a reaction, and this reaction may have long-term consequences" (1988: 11).

From these general observations, an ethic of 'earth care', 'people care', and 'fair share' arises, with twelve permaculture principles listed in order to guide permaculture design. The first of these principles is 'observe and interact'; the third is 'obtain a yield'; the sixth is 'produce no waste'. These are three examples from the twelve principles that together offer guidance that differs from what seem to be the imperatives of industrial growth systems that have given rise to a global economy that, as Joel Kovel (2002) and others have gone to great lengths to emphasise, has catastrophic consequences due to the amplified scale of its activity. The tiny permaculture homestead is, in part, something of an alternative to large-scale systems. Furthermore, by embedding themselves in spaces that are designed according to natural and ecological principles, and by being outdoors for much of their time and getting their hands into the soil, permaculturalists tend to find value and purpose in their deeds and immediate environment, which they may battle to find in typical large city-home style scenarios that can often feel alienating and 'disconnected' from nature.

In general, permaculture resonates with the conservationist's plea to 'tread softly' (as per Thomas Princen's observations 2010). Permaculturalists often frame themselves as custodians over a small patch of land, guided by the commitment to decrease reliance on centralised systems while nurturing spaces that are more in harmony with the rhythms of nature and thus foster biodiversity and organic soil fertility. While permaculturalists do participate in capitalist industrial growth systems, they often do so to the extent that they can work on and with the land, make fertile soil, grow vegetables, maintain off-grid water and electricity systems where possible, create and maintain spaces for birds and insects and micro-organisms to share, and interact in nature in a mindful and respectful way.

4.3 African Agrarianism: Resonances

4.3.1 Example

As in the generalised sketch of a permaculture homestead, a small house, or several small houses, are present on the given patch of land that hosts this brief sketch of AA. The house or houses are constructed from locally-available materials such as adobe bricks (made from clay/sand/straw), wattle-and-daub, and thatch. As in the permaculture homestead example, the house is not the primary living area, with most activities occurring outside during daylight hours. Already here there is a strong overlap with permaculture because both systems are heavily land-based and not house-based; in both cases, what happens on the land is part of what constitutes the home and indeed one's life, and the patterns of the seasons heavily influence one's behaviour.

In this example of AA, no electricity is available, nor is there a connection to a water grid. As in the case of the permaculture homestead example, the cooking of food and heating takes place on wood fires. Wood for the fires is collected from nearby areas. Rainwater collects in a pond on the land, with buckets of water occasionally carried in from a river or lake. The toilet is an out-house, i.e. a basic structure built over a hole in the ground, and as in the permaculture homestead example, humanure is kept in the system. So in neither the AA example nor in the permaculture homestead example is there a connection to a water-grid, an electricity-grid, or a sewerage system.

As in the permaculture homestead example, chickens are kept at the AA site, providing eggs and meat for the human diet. In the AA scenario, there is generally much more of a focus on herd animals than at the permaculture homestead example, with cattle, sheep, and goats playing a major role in the diets and bartering systems of traditional agrarians. The animal manure, however, plays a crucial role in both scenarios: to fertilise the soil in which crops are grown. In AA, herd animals often graze in a field that later becomes the area where crops are grown. The animals defecate in the field, and they mix the fertile manure into the ground as they walk over it. In the permaculture homestead example, a similar outcome is achieved by making compost at a dedicated site and adding the fertile soil to the vegetable garden. Bigger permaculture systems do incorporate grazing animals.

Companion planting is common in AA in the same way that it is in the permaculture homestead example, with the Three Sisters being popular in both contexts. A relatively small permaculture homestead generally does not have enough space for the production of grain crops such as wheat, while an AA scenario may unfold in an area large enough to accommodate such crops. With no artificial fertilisers and pesticides being used in the AA context or permaculture context, birds and other wildlife perform the function of pollinating crops and trees, as well as keeping pests at bay.

Seasonality, as already touched upon, is a central feature in both AA and the permaculture homestead example. Different crops are grown at different times of the year. Harvesting of a crop is an important event, with people coming together in seasonal celebration more so than in the permaculture homestead example due to the strong community setting of the former. Responding to the changes of the seasons, for example the changing number of daylight hours available in which to do manual work outside, is unavoidable in both AA and the permaculture homestead example. Bartering occurs in both contexts, although far more so in AA, with this trend diminishing in all cultural contexts as industrial capitalism spread via globalisation.

Both traditional AA settings and the permaculture homestead example appear aesthetically 'messier' in comparison to a city or village that has colonial roots. Grass is not uniformly mowed, trees are generally not neatly manicured, and organic materials are often piled in several areas for later incorporation into the fertility of the systems. Garden beds are covered in mulch rather than cleared of all but a few flowers, and weeds are generally not removed unless they compete with some other more desirable plant or crop. A strong intention in both AA and the permaculture homestead example is to produce a crop (one of the twelve permaculture principles is indeed 'obtain a yield'), and to do so with resources that are locally available, versus typical city-style homes and gardens where no food or energy is produced.

4.3.2 Conceptual Aspects of African Agrarianism

In his 2008 essay, "Agrarian Philosophy and Ecological Ethics", Paul B. Thompson (2008: 538) points out the following:

> Like his friend and one-time mentor Ralph Waldo Emerson, Thoreau believed that practical experiences such as planting beans, building a home, gathering huckleberries, or fathoming the depth of a pond are formative for human personality and character. Emerson and Thoreau both believed that "nature," by which they meant not wilderness but simply a world not totally given over to urban rhythms and the built environment, was the most reliable guide to character and a balanced, self-reliant personality. Thus it is agrarian philosophy [...] that will provide the most useful connections to ecological ethics.

The sketches of both AA and permaculture provide scenarios in which people engage in exactly the types of agrarian activities described by Thompson. Whether the aim of the activities is explicitly to form the human character, a part of the outcome is almost certainly that the human character is developed by the activities. Anyone to have aimed for self-reliance will attest to the fact that it is a near-impossible goal (after all, no person is an island), and self-sufficiency can only be achieved in actively cooperative communities (which AA certainly had), which is to suggest something other than self-reliance! There is therefore a difference between self-reliance and the "self-reliant personality", but in building basic structures, planting crops and tending to them, managing the fertility of the soil, engaging in bartering, etc., practitioners of AA and permaculture alike do cultivate a self-reliant personality.

The personality nurtured by agrarian practices is further elaborated on by Thompson (2008: 530) when he quotes the following: the farmer "has a sense of identity, a sense of historical and religious tradition, a feeling of belonging to a concrete family, place, and region, which are psychologically and culturally beneficial." Perhaps this mix of factors conveys to some extent why land is so sacred to so many African cultures. The land is the space on which vegetables are grown and animals grazed; the space where family works and plays together, and where people celebrate, worship, barter and trade, and conduct most of their lives' activities (notably, this is not to suggest that AA is a backward-looking or unchanging system). As Thompson (2008: 538–539) comments, "agrarian views saw norms and character traits as products of the interaction between an organism (a person) and its environment"—this observation is true of AA as well as permaculture.

Continuing with the theme of the interaction between the farmer (agrarian) and the land, Thompson (2008: 541) again quotes from elsewhere to highlight a

> three-way co-production. The farmer produces crops and livestock, to be sure, but in doing so adapts to soil and climate in producing the farm. The farm (and the life of farming) in turn produces the farmer, meaning that the goodness of the farmer's work is more evident in the character of the farm family and the farm community than in the quantity of farm commodities.

Adapting to the soil and the climate is relevant to both AA and permaculture, while industrial agriculture adapts the soil via the addition of artificial fertilisers and pesticides. Industrial agriculture has become driven by the economic imperative to produce as much of a yield as possible in the shortest period of time, and to do so relentlessly. While AA and permaculture both aim to serve various human needs, they do not do so exclusively, and their farming practices are widely regarded as regenerative rather than purely destructive. This resonates with the following (Thompson 2008: 538): "agrarian [...] views have tended to focus attention on elements of human conduct that are [...] arguably more congenial to people who think in ecological terms."

Thompson (2008: 530) quotes from elsewhere again to incorporate a spiritual dimension to agrarianism, which is particularly relevant for AA considering that strong links are often made between ancestors and the land:

> Cultivation of the soil provides direct contact with nature; through the contact with nature the agrarian is blessed with a closer relationship to God. Farming has within it a positive spiritual good; the farmer acquires the virtues of 'honor, manliness, self-reliance, courage, moral integrity, and hospitality" and follows the example of God when creating order out of chaos.

Furthermore, cultivation of the soil fulfils these spiritual functions 'while obeying the rhythms of nature" (Ibid). Practitioners of AA on the one hand, and permaculture on the other hand, may not all agree on the point about a closer relationship with God, but the former often streamline the spiritual aspect of their practices in ancestral directions to varying degrees, while the latter do often venerate nature in a way that positions it as a giver of life. Whether the intention of venerating nature as upper-case Nature is explicit, the effect of practising AA or permaculture does provide direct contact with nature, and to various extents the practices do create order out of what might appear to be chaos.

4.4 The Tragedy of Centralisation: The Waste-Water Treatment Example

4.4.1 Illustration/Problem

The practical sketches of an African agrarian (AA) scenario and a basic permaculture homestead, when viewed against the backdrop of some agrarian ethics, do suggest that both arenas resonate to some extent at the levels of practice and theory. As will become clear in this section of the chapter, AA and permaculture resonate further in their classification as decentralised systems. Decentralised systems are very different to centralised ones, with both sketches painting pictures of basic decentralised systems, and with a city being the typical example of a conglomeration of centralised systems.

Take centralised food production as an example. In a city, food is produced almost entirely on far-away farms that almost unanimously produce one kind of crop or 'meat product', for example, corn, wheat, beef, or chicken. This is most often achieved via the use of chemical inputs into the farms—chemical fertilisers, pesticides, herbicides, and large quantities of fossil fuels. The crops and products eventually appear as items on supermarket shelves, usually after having gone through a multi-phase process involving chemical applications, transportation, processing, and packaging. Each one of the steps entails more fossil fuels and various other resource inputs. This is by default how industrial agriculture works.

At the conceptual level, centralised agricultural systems are arguably inimical to the ethos of AA and permaculture. As already demonstrated, Mollison is unambiguously critical of what he refers to as the madness of uncontrolled industrial growth, which simultaneously enables and is enabled by centralised systems that concentrate power and resources into the hands of bureaucrats and corporations. These centralised entities tend to preside paternalistically and often one-dimensionally over the autonomy of individuals and small groups of people wishing to put into practice negative freedoms rather than prescribed and state-endorsed positive freedoms (Pittaway 2018). The political pandemic context of 2020 and 2021 surely shows these remarks to be fitting.

Conceptually again, it is difficult to see how there is room in the ethos of AA (as defined for the purposes of this chapter, i.e. in its pre-industrial phase) for what could loosely be called the ideology of centralisation. A farmer cannot strive towards the ideal of self-reliance, or various other qualities noted in the previous section of this chapter, in their practical or conceptual formats if the reality is that centralisation makes reliance on the State and on Capital unavoidable either by design or by accidental outcome. In theory, the practitioner of AA may be 'free' to decline the services and products of centralised systems, but in practice, these systems radically altered all arenas of society and nature, with the outflow of millions of once-were agrarians from rural areas to cities being an example of a consequence of the structural development or design of centralisation. Even in historically-advantaged contexts, permission needs to be granted by the State for land-use endeavours, but ever-present rules and regulations encourage resource-heavy developments.

While industrial agriculture provides many people with never-before seen conveniences, various physical "externalities" (Princen 2010: 41) occur in the production and distribution of resources and products. For example, as explained by William H. Schlesinger (2020), nitrogen runoff occurs in the process of fertilising monocrops, with the nitrogen (as well as various other chemicals sprayed onto crops and fields) draining into rivers and water-systems, ultimately ending up in the ocean, causing massive problems for wildlife, ecosystems, and human activity too seeing as we depend for our survival on the integrity of healthy ecosystems. Mollison's observation, already quoted above, is indeed accurate: "If we maltreat, overload, deform, or deflect natural systems and processes, then we will get a reaction, and this reaction may have long-term consequences" (1988: 11).

Nitrogen runoff is just one example of an externality. There are many examples of externalities caused by centralised systems, and nature is plagued by them. The production of electricity at power-plants requires huge quantities of fossil-fuels to be mined and burned, and externalities such as the release of carbon dioxide (and other greenhouse gases) into the atmosphere occur, as well as the loss of biodiversity that accompanies large mining projects and city-style developments that are enabled by most large industries. Centralised water systems require the construction of large dams that completely alter the dynamic of natural systems downstream, and the infrastructure required to put water in the taps of every city-home has obvious impacts on nature, for example the laying of pipes and concrete across countless kilometres of landscape. These kinds of externalised consequences of centralised systems are, as Charles Eisenstein (2011) points out, disastrous in the context of ecology, and he adds that they reflect a mindset of separation between people and planet, which (as explored in previous sections) is completely out of place in AA and permaculture.

To bring the notion of externalities closer to most people's personal habits, consider the example of a flushing toilet. Having just seen the example of far-away dams that pipe water into cities and into the taps of homes, it should be noted that the first externality of flushing toilets starts there, i.e. at the dam. Large dams upstream are not sensitive to what happens downstream, and it does not take any stretch of the imagination to see that an animal or plant species that once thrived on or near a healthy river or wetland is radically impacted when the animal's or plant's water source either drastically reduces or disappears altogether due to upstream damming. While African agrarians may have created some small dams, the small scales at which they did so are incomparable to the massive scales of industrial endeavours.

Much of the water that flows from dams to the taps of city homes serves the function of flushing 'away' human urine and faeces, both of which are rich in nitrogen and can be used as organic fertilisers. These substances flow into city sewerage systems and get mixed into the cocktail of waste-water that is constituted not only by faeces and urine, but by everything that human beings throw into drains. The waste-water eventually ends up at waste-water treatment sites. In 2018, I played a role in the facilitation of three outings of university staff and students to the waste-water treatment site at Driftsands, Port Elizabeth/Gqeberha. What was seen and learned there alarmed all participants.

The waste-water that enters the site flows at a strong rate of flow. The outings occurred during a province-wide water crisis, so it was rather surprising and worrying that a 'river' of toxic waste-water flowed constantly into the site—I found myself saying, "no wonder that we have a water-shortage". With most attendees of the outings holding their noses closed due to the stench in the air, we saw how the first stage of filtration involved removing larger debris such as plastic packets, tampons and condoms, and even items such as discarded children's toys, that found their way into the sewerage system. Our guide commented that we would be surprised and alarmed to know the full range of materials that folks flushed down their toilets and into their drains.

Several stages of filtration and treatment occur at the site, with the end result being at least two-fold. First, large piles of organic matter are accumulated on site after being sifted out of huge concrete treatment troughs; and second, the filtered water is allowed to pass through the site and into the ocean. Our guide informed us that the engineers at the site test for approximately 100 chemicals and heavy metals in the water, and that their treatment protocols aim to neutralise or remove those chemicals and heavy metals. Uncertainty exists around the questions, first, of how many chemicals and heavy metals that are not tested for are present in the water that enters and leaves the site. Second, how effective the treatment methods are at fully neutralising or removing the known chemicals and heavy metals. Third, how viable is it to add more chemicals to water in the attempt to rid chemicals from it.

The huge piles of organic matter on site are apparently made available to anyone who wants to collect some of the contents for free, for example farmers who may wish to use it for agricultural purposes. Judging by the size of the piles during all three site visits (which occurred three months apart from each other), few people were jumping at the offer. This is understandable because it does not take a large leap of logic to comprehend that the organic matter contains known and unknown toxic chemicals. This is true of the waste-water that leaves the site and flows into the ocean. Our guide gave us the impression that while the engineers and other site employees do the best they can with limited budgets, they are up against a juggernaut of waste-water inflow. The filtration process (whose efficacy is questionable to begin with) is often bypassed completely during power outages, of which there are many in South Africa. Heavy rains that cause large surges of water inflow also render the filtration process completely obsolete during and after the surges. This is when large swathes of plastics, rubbers, chemicals, heavy metals, and various other substances are flushed out to sea.

The United Nations Environment Programme (UNEP 2021) website states the following, which supports my observations from the visits to the waste-water treatment site: "Pollution from inadequately managed sanitation and untreated/partially treated wastewater seeps into the ocean every day all around the world, smothering underwater ecosystems like coral reefs and degrading coastal fisheries and water sources that communities rely on." However, with new chemicals being invented and flushed into sewer systems annually, and with inevitably limited resources and staff to manage the process that could presumably purify waste-water before it enters the ocean, it is not unreasonable to suggest that waste-water will by default remain "inadequately managed" (Ibid). History is replete with examples of people assuming that their large-scale systems are safe, non-toxic, and manageable, but where these assumptions proved to be tragically off-the-mark, e.g. the Fukushima Daiichi nuclear disaster.

The example of nitrogen-rich and chemical-rich waste-water that enters the ocean constantly, i.e. every second of every hour of every day, is not an example of a problem that can be solved by doing more of the same thing, i.e. flushing fertility 'away' down the toilet and continuing to use chemical-heavy products whose usage ultimately sends chemicals, plastics, heavy metals, etc. into the sewers and into the oceans. A completely different, decentralised approach to solutions is needed, and the confluence of AA and permaculture holds some insight in this regard.

4.4.2 Remedy

A breathtakingly simple solution to the externality of polluted waste-water is apparent in the sketches of AA and permaculture. In a nutshell, the solution is not to use flushing toilets in the first place. This is the case in both sketches above: urine and humanure are kept on site. In both sketches, the 'wastes' are incorporated into the land for the purpose of creating fertility, which is to say that there is no externality in this regard—'wastes' become internalised resources.

The obvious criticism in response to the suggestion of using a compost toilet or AA-style out-house can be raised via the question, how can the widespread use of compost toilets and out-houses be implemented at a large scale in extant city contexts? While there is no one-size-fits-all answer to the question (which is entirely fitting for decentralised approaches), it is perhaps the case that AA and permaculture do provide glimpses of answers. They both display how keeping humanure within a system is a virtue rather than a vice at the level of theory and practice. Stated in colloquial terms, in theory it is admirable and character-forming to deal with one's own shit, while practically it can serve the dual function of adding fertility to the soils and preventing externalities.

This suggestion towards a simple remedy raises the issue of behaviour change. Throughout the history of environmentalism, ecological problems and solutions have been identified, but clear routes towards changing the problematic behaviour of citizens have not arisen towards an ecologically-respectful global dispensation. What has been seen in this chapter of permaculture and AA praxis is that some simple agrarian practices either intentionally or unintentionally close the loop on the inputs and outputs for small agricultural systems. By establishing agrarian/permaculture nodes in city and village spaces, 'small and slow solutions' (a permaculture principle) can be implemented that embody basic agrarian practices and change people's behaviour towards internalising resources rather than externalising wastes.

Once city households begin using compost toilets, something will have to be done with the fertile organic matter that will no longer be flushed down toilets. The wastes that were once externalised will need to become internalised resources. Considering the high levels of unemployment that plague African nations, it is not unreasonable to call for State-funded teams of well-compensated compost-material collectors, teams that would look similar in organisation to the teams of garbage collectors that operate throughout various nations. Sites would need to be identified for the processing of the humanure wastes, and in a short period of time there would be mounds of compost that can be used for food-growing purposes or for the replacement of topsoils that have been lost due to short-sighted human activity (usually industrial agriculture and development).

Questions might arise about the purity of the compost considering that many people use medications that might find their way through the human system and into faeces and urine. Several responses can be given to this concern. First, the composting systems will need to be arranged so that incoming receptacles of humanure are labelled with a 'medication' or 'no-medication' label, and different batches of compost will need to be tested for their chemical constituents. Second, it should not be forgotten that traces of medications in faeces and urine end up going somewhere in the present centralised system—specifically, to rivers and/or the ocean, in the form of an externality that damages non-human life and planetary health, which ultimately comes back to haunt human beings via the destabilisation of ecological systems on which we depend for our well-being. The solution towards remedy here changes the situation so that this issue is internalised rather than externalised.

Third, it should also not be forgotten that industrial agriculture makes use of a variety of chemicals that do find their way into the food that we all eat, with pressing questions remaining unanswered about the safety of the phenomenon—Joël Spiroux de Vendômois et al. (2010) contextualise this concern in a manner that shows clearly how industry rigs the research process that supposedly guarantees the safety of genetically modified foods that are grown with chemical fertilisers, pesticides, and herbicides. Fourth and finally, large heaps of almost-unusable chemical-laden organic matter are already piled up at waste-water treatment sites; the suggested solution aims to position piles of carefully-managed piles of usable fertility at 'source' (i.e. the start of the process) rather than near the sea (the end of the process).

Once the problem of waste-water treatment externalities is solved via the method proposed here, various other problems can be solved in manners that further resonate with AA and permaculture. For example, if members of a citizenry can comprehend why their fertility (in the form of humanure) is valuable as an 'internalised' organic fertiliser but detrimental as an externalised 'waste', their ecological consciousness may be tweaked towards other pro-ecological behaviours that are desperately needed in the context of the ecological crisis. Another multi-pronged example is that organic fertiliser in various city spaces can be used to grow healthy organic food without industrial chemicals (or at least with a remarkable reduction in them), with drastically-reduced or eliminated transport requirements, and with consumers ultimately receiving organic produce that in turn will keep a citizenry healthier. These are just a few examples sourced from a broad review of the confluence of AA and permaculture that demonstrate a knock-on effect towards making a bad situation much better. In the spirit of achieving this outcome, i.e. making a bad situation much better, the following SMART targets are offered, ones that resonate to varying degrees with what has so far been explored about the confluence of AA and permaculture.

4.5 SMART Targets

SMART stands for: Specific, Measurable, Achievable, Realistic, and Time-bound. From the confluence of AA and permaculture discussed so far, the following SMART targets are offered as examples to assist policy-makers in the crucial process towards sustainable decentralised agricultural systems, and perhaps even multi-cultural ones too considering that agrarianism lies at the roots of many cultures. Time-frames are being excluded here because it will have to be up to the discretion of policy makers and policy implementers to decide on them. Considering that South Africa has an Act called the Protection, Promotion, Development and Management of Indigenous Knowledge Act 6 of 2019, but that little appears to be in process to, for example, "provide for the facilitation and coordination of indigenous knowledge-based innovation", these SMART targets should be of use to municipalities throughout South Africa:

- Create education programmes for citizens of all cultural backgrounds that inform them about AA practice and theory.
- Include in the educational material information on the overlap between AA, permaculture, and other compatible ways of organising a decentralised homestead.
- Include in the educational material information about how AA and permaculture (and other compatible methods) mitigate ecological problems and create local jobs that serve the purpose of adding to the local economy and local food security.
- Establish sites throughout cities that demonstrate organic methods of composting and food production. These sites can be ones that were left to fall into disrepair via neglect or a multitude of other reasons.
- Establish teams of 'fertility collectors' in a model not unlike that of city garbage collection.

- Establish sites where humanure can be processed with other organic materials in the compost-making process.
- Establish testing procedures for the organic matter so that it can be adequately managed.
- Promote decentralised AA, permaculture, and other compatible approaches, as responsible and ethically commendable 'hands-on' methods that facilitate resilient, sustainable, and multi-cultural community spaces.

4.6 Conclusion

I have not addressed many limitations or shortcomings of either permaculture or African Agrarianism. Legitimate questions can be asked about the ability for either to feed the astronomically large number of people that inhabit Africa, and the many more people that in the next several decades will be added to the present number. No claims are being made here about the potential for AA or permaculture to achieve that goal. However, it should be clear that humanity now faces problems such as the 'externality' caused by a centralised waste-water and sewerage system. This system is one example of many centralised systems that by default cause some of many externalised problems.

In practice, current economic and political trends generally push for more technology, more growth, and more centralisation, all of which have arguably accelerated the human species in the direction of ecocide. What if citizens could put into practice some basic ecological principles that AA and permaculture adhere to in a manner that actualises some of the goals established in legislation promoting indigenous knowledge systems? What if tried-and-tested methods of organic food production can become national pass-times that increase resilience in the domains of food-security and employment in a multi-cultural and ecologically-sensitive manner? There may be many answers to these questions, but I will offer a brief answer: more citizens will interact with each other and with their environment in environmentally sustainable ways that empower the participants as per the agrarian ethic, with the outcome partly being healthier, happier, and more engaged community-members and citizens. This appears to be a win–win scenario. Ignoring it would be to look the gift-horse of simple solutions in the mouth.

References

Eisenstein, C. 2011. *Sacred Economics: Money, Gift and Society in the Age of Transition*. USA: North Atlantic Books.

Eisenstein, Charles. 2012. 'Permaculture and the Myth of Scarcity.' https://charleseisenstein.org/essays/permaculture-and-the-myth-of-scarcity/. Accessed 13 Sept 2021.

Hadot, Pierre. 1995. *Philosophy as a Way of Life*. USA: Blackwell.

Kovel, Joel. 2002. *The Enemy of Nature: The End of Capitalism or the End of the World?* New York: Zed Books.

Mollison, Bill. 1988. *Permaculture: A Designer's Manual.* Australia: Tagari Publications.

Pittaway, David. 2018. Placing the Ecological Crisis in a Broader Context: The Orphic and the Promethean. *Phronimon* 19 (1): 1–16. https://doi.org/10.25159/2413-3086/4508.

Princen, Thomas. 2010. *Treading Softly: Paths to Ecological Order*. USA: MIT Press.

Roberts, Gregory David. 2015. *Gregory David Roberts: 'Final' Interview on The Mountain Shadow by Shantaram author*. https://www.smh.com.au/entertainment/books/gregory-david-roberts-final-interview-on-the-shadow-mountain-by-shantaram-author-20151005-gk1o20.html. Accessed 13 Sept 2021.

Schlesinger, William H. 2020. *Nitrogen in Runoff*. https://blogs.nicholas.duke.edu/citizenscientist/nitrogen-in-runoff/. Accessed 13 Sept 2021.

Thompson, Paul B. 2008. Agrarian Philosophy and Ecological Ethics. *Science and Engineering Ethics* 14: 527–544. https://doi.org/10.1007/s11948-008-9094-1.

UNEP (United Nations Environment Programme). 2021. *New Global Collaboration to Tackle Impacts of Wastewater Pollution on the World's Coral Reefs*. https://www.unep.org/cep/news/editorial/new-global-collaboration-tackle-impacts-wastewater-pollution-worlds-coral-reefs. Accessed 13 Sept 2021.

Vendômois (de), Joël, Dominique Cellier, Christian Vélot, Emilie Clair, Robin Mesnage, Gilles-Eric Séralini. 2010. Debate on GMOs Health Risks after Statistical Findings in Regulatory Tests. *International Journal of Biological Sciences* 6(6):590–598. https://doi.org/10.7150/ijbs.6.590.

David Anthony Pittaway Ph.D, is a Research Associate with the Philosophy Department at the University of the Free State, South Africa.

Chapter 5
Dialogue Between African Agrarian Philosophy and Adam Smith on Underdevelopment and Resource Dependence in Africa

M. Rathbone

Abstract The purpose of this chapter is to highlight principles for agrarian economics in Africa based on the dialogue between agrarian philosophy in Africa and agrarian traces in Adam Smith's work *The nature and causes of the wealth of nations* [1776]. I will argue that the importance of community and society in the work of Smith is a vital point of departure for dialogue between agrarian philosophy in Africa and classic economics. The dialogue attempts to provide perspective on the issues of underdevelopment and resource dependence in Africa that are barriers to sustainable agrarian economics. Further, agrarian economics may provide alternative views to the sometimes exploitative and narcissistic tendencies associated with capitalism that are wrongly associated with the work of Smith. The analysis of the interlocutors is appreciative and value-driven, which highlights that the role of ethics, multi-sectoral development and justice are core aspects of sustainable agrarian economics in Africa.

Keywords Agrarian philosophy · Africa · Adam Smith · Community · Underdevelopment · Resource dependence

5.1 Introduction

In this chapter, the focus is on dialogue between agrarian philosophy in Africa and agrarian traces in the book *An inquiry into the nature and causes of the wealth of nations* [1776] (WN) of Adam Smith in order to highlight principles for agrarian economics in Africa. This is a novel dialogue that could expand scholarly knowledge due to the cultural divergence between Smith's work and African philosophy

---

M. Rathbone (✉)
Faculty of Economic and Management Sciences, North-West University, Potchefstroom, South Africa
e-mail: mark.rathbone@nwu.ac.za

M. J. Tosam and E. Masitera (eds.), *African Agrarian Philosophy*, The International Library of Environmental, Agricultural and Food Ethics 35,
https://doi.org/10.1007/978-3-031-43040-4_5

and provide insightful perspectives for Afro-agrarian economics. Smith is an important interlocutor as the father of economic science that can contribute fundamental principles in the discussion about agriculture due to his extensive writing about the topic but also due to his socio-ethical framework for economics that resonates with the more communal sentiments in Africa, although different from the worldview of the Scottish Enlightenment. In other words, the aim of the dialogue is not to smooth out differences in the worldviews of Smith and African philosophy. It is rather a robust dialogue that deals with the tensions as points of critical reflection.

The agrarian traces in Smith's work underscores the role agriculture and social engagement played in his economics that will secondarily be explored in the dialogue with Afro-agrarian philosophy. For Smith, agrarian economics was inclusive of all forms of economic activity that are dependent on raw materials. This includes raw materials from produce (for example barley, maize, fruit) and mineral resources (for example gold, coal, iron ore). Agrarian economics that is dependent on limited raw materials must therefore also be embedded in normative and socially conscious value systems for sustainable use of these materials. It will be argued that Smith's economics, although influenced by agrarianism to an extent and appreciative of the work of contemporaries like Francis Quesnay (WN IV.ix.5 1981:516), hold that a single sectorial approach to economics at the expense of manufacture and trade will limit economic development. This is due to Smith's non-interventionist approach to economics and the value he placed on principles such as justice, liberty and equality, for instance, free market economics.

In African culture society, norms and values are rooted in communal ethics, which is a crucial interlocution point for dialogue between agrarian economics in Africa and Smith. The free market economics that Smith is known for today highly underscores the individual, society and ethics although divergent from African concepts of community and land. These facets of Smith's economics place priority on principles that include liberty, fair treatment of others, harmonious interrelations, and also respect for individual differences. The implications of these principles are that economic activities should not be constrained by interventionist policies that suppress free and fair economic activity. In this way, Smith envisions a balance between self-interest and sympathy (Rathbone 2015). Self-interested and unrestricted economic activity in the various sectors will be to the benefit of the economy. However, unfair restraints placed on certain sectors to promote other preferred sectors will be met with disapprobation because the unjust suffering imposed on people involved in the curtailed sector, will ignite the sympathy of people. In this way, there is an important socio-ethical aspect present in Smith's understanding of commerce, rooted in instinctive moral development and economics, such as business ethics.

Agrarian economics in Africa, which is highly dependent on agriculture and resources, must grapple with two critical issues, namely underdevelopment and resource dependence (prominent in agriculture and mining). These are important aspects that Smith's work addresses and therefore he is a valuable interlocutor (although the philosophies are inherently divergent) that may provide alternatives for agrarian economics in the African context. In this chapter, we will first analyse underdevelopment in Africa and its possible relationship with single-sector dependence on resources from the perspective of the interaction between Smith and African philosophy on these issues. Secondly, the agrarian legacy in Africa and its communal value system will be placed in dialogue with agrarian traces in Smith's work to explore possibilities for sustainable economics in Africa. In this section, the focus is on underdevelopment and resource dependence in Africa based on the preceding section on the dialogue between African philosophy and Smith. In this discussion, the African worldview of land and community is a focal aspect that grounds the dialogue. Finally, ethical economics, sectoral development and justice will be highlighted as salient aspects of sustainable African agrarian economics.

5.2 Economics in Africa

5.2.1 Underdevelopment

The etymology of the word *development* can be traced back to the French word *developer* in the 1650s, which means "to unfold". The later 19th-century meaning that was informed by a more utilitarian understanding of development as progress and economic growth, added a more figurative dimension that incorporated the idea of releasing a person's potential. The foil of this perspective—stagnation—denotes a situation where the potential of a person, nation, or economy is not unfolding into a state of flourishing and thus not growing as a function of the various dimensions of personhood. Underdevelopment can therefore be understood as unfolding that is not at its optimal level—in other words, development is taking place but at a rate that is not resulting in economic growth to reduce socio-economic issues such as poverty and inequality. It is true that this view of the common good can be interpreted in different ways and from an African perspective the notion of economic growth because of unfolding may be criticised as a capitalist and Western ideal. In other words, the unfolding of society can be understood from the African worldview as the promotion of self-interest and subjection to the market, or it can also be viewed from a communal perspective in which the growth of a few can be to the benefit of society except for cooperatives. The problem is the function of the cooperative is direct market-related to gain a pricing advantage. However, in general, these views are not mutually exclusive because from a Smithian perspective, the economy must be ethically and socially embedded with justice as a measure of fairness (see discussion in Sect. 5.3.2 on the impartial spectator and sympathy of Smith). Therefore, the extreme wealth of a few at the expense of others points to the failure of the economy and its unsustainability (Rathbone 2015, 2018). The sharing of wealth from a communal perspective can be equally problematic if it led to the continued injustice of underdevelopment of many and the increase of patrimony of others.

Under-development in the African context can therefore be assumed to refer to the untapped potential of African economies to reach their full growth potential. The problem is that the word *potential* can mean different things for different people, which in turn has implications for policy changes and government programs. In Africa, framed by colonialism and imperialism, this question regarding potential is loaded and it can mean one of two things. Firstly, it may refer to the advancement of the developed, advanced and powerful to determine what is needed for Africa to reach its full potential. Rodney (2012) notes that European nations contributed to Africa's underdevelopment due to the establishment of dependency between Europe and Africa, with the development of the former and the deprivation of the latter. Alternatively, potential has become more internalised and complex. Post-colonial studies note that potential in Africa can negatively imply the continuation of colonial mentalities and mechanisms imposed by the colonial rulers of the past, which limits the opportunity and development of the people in general. The collusion of the elites of the previously colonised people and colonisers (that today include multinational companies) of earlier times depletes the resources of nations and undermines the distribution of wealth (for example through corruption) and development, so that the potential of the nation does not unfold (Mbembe 2015).

Another dimension to this is the global village, or—more ominously—globalisation or colonisation on a grand scale. In more ironical terms, it may also be described as the WN deconstructed (for example Smith's criticism of mercantilism), Smith's masterpiece debunked and overrun by multinational corporations, or corporations changing the way we function, our culture and probably souls (if such a thing exists). Underdevelopment then starts to mean something totally different: it becomes a game of commodification and failure to respect what is sacred and valuable. However, modern perspectives infused by globalisation, postcolonial mentalities and global markets have a different perspective on Africa. Chabal and Daloz (1999) note that Africa "works" and that despite numerous challenges, there is economic transformation underway that can be seen in the rise of the middle class, opportunities in African markets for foreign investors, and so forth (Beresford 2016: 1). The dilemma is that this optimism is based on intensified resource exports, which extracts the wealth of a nation without economics development and only increases the dependence on the sometimes-volatile foreign markets of the import nations. This resources dependence of African economies only increases underdevelopment, as will be discussed in the next section. These problematic areas focus on structural problems of African economies. However, these structural perspectives, although important, sometimes ignore the personal or human face of underdevelopment.

A more nuanced and human perspective is to start with the individual person and her potential in a particular place in Africa; and not with the utility or calculation of the needs of elite's or CEOs of global conglomerates (Beresford 2016: 3). The starting point should be a person in a particular place and relevant socio-economic issues like sanitation, health, education, transport and so forth (Mkandawire 2014). Amartya Sen's (1999) capability theory approaches the question of underdevelopment in a more human manner, by giving priority to the person and not the utilitarian statistics. To be poor involve much more than not having enough money for basic services in comparison to people in other places. It is rather a matter of not having the same potential to be a complete human being as others, because there are very few accountable structures to ensure the development of people, even though the country may have enough resources and wealth to accomplish this goal. Often, accountable structures have been replaced with corrupt systems that benefit the elite.

Underdevelopment is therefore also about the past, namely colonialism and its continuation through post-colonialism. Colonialism was about the land, minds and bodies of people who were sacrificed for the development (or rather the greed) of others. Smith was correct in that greed is a human instinct that transcends culture. It involves self-interest and poisons civilisation. Colonialism is an inhumane display of the saving grace instinct gone wrong—caring for oneself and one's own at the expense of others (often resulting in xenophobia)—and it creates a world of resentment and anger that feeds on itself and the insecurities that inhabit our psyche. Therefore, even the notion of change and liberation cannot shake itself free from this narcissistic impulse.

After all the possible philosophical perspectives of underdevelopment have been exhausted, the realism and experience of poverty remain a fundamental fact—and not only poverty but all the secondary aspects that come with poverty, like a failure of health care, education, governance and justice (Atkinson 2015), as well as. Also, the demise of government infrastructure in all sectors of society increases poverty and resentment. This becomes a catalyst for social, political and societal frustration, tension and violence. The net results and effects of underdevelopment highlight that change is essential. The problem is complex and the answers will also have to focus on the different dynamics of the situation, for example, economic policy, education, social development, the legal system and so forth.

Education is one of the most pressing issues in Africa (Bloom et al. 2006: i) and represents an important opportunity for people to change the trajectory of their lives and provide the means for new possibilities. Therefore, education is vital for social and economic change through new technologies that can enhance the lives of people and provide more productive and environmentally sensible methods for agriculture, production and growth. The failure of the education system (as with individuals) also has a devastating effect on the trajectory of an economy because the economy hinges on singular resource sectoral dependence and manual labour. This is in the form of agriculture and mining, which are two of the most prevalent and exploited domains of the resource sector (Beresford 2016).

5.2.2 Resource Dependence

Resource dependence must not be viewed as an impediment to African economics. It is an unquestionable fact that Africa has numerous and abundant resources that must be cherished. Africa also has plentiful geographical space and fertile soil for agriculture. These natural resources must be celebrated and managed responsibly for the benefit of the population of the continent and generations to come. These resources provide ample opportunities for labour and income for the people of Africa. It becomes problematic, however, when this sector remains the only sector. This, coupled with limited education and sometimes rampant corruption, means that the resources will remain the only source for GDP growth and wealth of a country (Beresford 2016: 2). Dependence on resource exports leaves African markets and economics vulnerable to the growth trajectory of resource consuming and manufacturing economics such as China, which has stimulated GDP growth in Africa (Beresford 2016: 2). This dependence has a more daunting effect in that the imports of manufactured goods are far costlier than the exported resources, and this erodes the capital available for development. The problem is that the elite continues with extraction economics similar to previous colonial rulers, which restricts investment for the sustainable development of agriculture, mining and other sectors. The result is that producers will limit investment because the expectation is that a "large share of future output will be appropriated by the elite" (Mizuno and Okazawa 2009: 409).

The implication of this type of misappropriation of capital (for instance through corruption) is that the developmental agenda of a country is stripped of investment, and the resources that are available are mismanaged and/or exhausted (Rathbone 2020). Because of the growing poverty, poorer nations are dependent on others and do not have strong negotiating power. Moreover, the exploitation of natural resources and land leads to devastating environmental problems and exhaustion of resources, thus limiting the future income of a nation.

Education and technology are important to cultivate other sectors of the economy, like manufacturing. Bloom et al. (2006: i) make a convincing case for the link between education and development, specifically in terms of tertiary education and technological advancement when they argue that "African higher education can assist countries with technological catch-up and thus improve the potential for faster growth". Technology is also a tool to enhance agriculture and other resource-dependent activities. Manufacturing is important because the resources can be used to produce goods for own consumption and export. Agriculture can also be a source of sustainable food security and exports. Although technology and manufacture are often associated with progress, environmental exploitation and ecological damage, it is clear that contemporary economics is increasingly underscoring the need for sustainable economic development and environmental awareness, such as green products, manufacture, marketing, and so forth.

In the next section, an appreciative and dialogical analysis will be presented of the African context in terms of the land, people, culture and philosophy. Similarities with the work of Adam Smith will then be identified. In addition, the role of agriculture and multi-sector economics as a prominent resource for development will be accentuated.

5.3 Agrarian Philosophy in Africa and Adam Smith

5.3.1 Agrarianism, the Land and Community

Agrarian philosophy focusses on agriculture as the basis for economic prosperity. It, therefore, involves the land, labour, community, the environment, and ethical values (Thompson 2008: 528). The interaction between the environment and humans is viewed as the basis for the development of culture and norms. Thompson (2008: 527) notes that agrarian philosophy "stresses the role of nature, soil and climate in the formation of moral character as well as social and political institutions". In other words, it has a profound moral and ethical aspect that underpins agriculture, community, and soil. As mentioned earlier, from the perspective of Smith agrarianism includes any form of raw material because these substances—whether grown and harvested or mined—are related to the land and can have an impact on communities, as the environment influences values. Therefore, the interaction between people and the geographical space has an influence on behaviour, traditions, values and norms.

In the African context, agrarian philosophy generally highlights the unique challenges faced by Africans such as food security, poverty, exploitation of the land and legacy issues relating to colonialism and imperialism, and so forth (Ndaguba and Hanyane 2018; Gassner et al. 2019). Additionally, agrarian philosophy in Africa is complemented by African culture and philosophy that have a deeply community-orientated anthropology and values that are closely connected to the land (Masitera 2020; Rathbone 2014). However, Boogaard (2019: 274) laments that "agriculture has not gained much attention in academic African philosophy so far". Consequently, African philosophical values associated with agriculture need to be explored in more detail because insight into the culture, identity and worldview of people is essential for sustainable practices as agriculture (Munyaka and Motlhabi 2009: 70–73; Rathbone 2014). In this regard, the importance of collective harvesting in Southern Africa inspired by the notion of *letsema* (voluntary work together), that is far removed from Smith's idea of private ownership of property, require further consideration. As pointed out in the previous section cooperatives that follow a collective approach to obtain market advantage can probably benefit from dialogue with African experiences of collective harvesting. However, Twala (2004) cautions that *letsema* may also threaten full employment and must therefore be administered properly not to take advantage of workers. From this perspective, Smith's contribution to the development of labour can provide important insights for the administering of collective harvesting.

Agrarian philosophy and agriculture have implications for economics because economics is a social science in which management and trade of resources do not exclude the close connection between the values, norms, land and community. These are important values that must not be underestimated and that provide an important normative dimension for African economics, specifically in relation to the challenges of underdevelopment and resource dependence (Rathbone 2018). Therefore, Boogaard (2019) points out that agricultural and economic development in Africa is hampered because of a general absence of knowledge of African philosophy among agricultural scientists and practitioners in general. Boogaard (2019: 273) notes that mutually "respectful and transformative dialogues can lead to mutual understanding and a more equal relation between Africa and the West, in the sense that this relation becomes more open to African philosophies and less Eurocentric". Therefore, engagement between agrarian philosophy in Africa and classic economics (for instance, Adam Smith's free market economics) can lead to fruitful debate for sustainable agrarian economics in Africa. This does not mean that differences are ignored but that mutual themes can be discussed with respect for the other. This can lead to the creative unlocking of alternatives on sensitive issues in Africa such as land, community, development and so forth.

Consequently, the work of Smith can be an important interlocutor with agrarian philosophy in Africa and economics because it provides insights from classic economic science that looks at basic human instincts (for instance, self-interest and sympathy) and cognitive processes involved in ethics, society and economics. In other words, it is not a rationalist approach but rather focuses more on an empirical analysis of human passions and sentiments and their social embedded character represented by the impartial spectator. The impartial spectator assists the subject in cognitively considering the most appropriate action as judged by others (Rathbone 2018). This aspect of Smith's work underscores the importance of the community where a person engages with others. It also implies that any activity or decision has ramifications for others. This aspect is not in complete alignment with the African communal philosophy, but it does provide a point of contact with the marked difference that Smith also argues in WN that self-interest with the mediation of the invisible hand directs action to the greater good of society (WN.IV.II.9 1981: 349). This aspect is clearly influenced by the values of the Enlightenment that cherries individuality and freedom.

Further, there is a more poignant reason for dialogue with Smith and it is that Smith's work has been vilified as a source of narcissistic capitalism and colonial exploitation (Forje 2020: xli). What is not mentioned by Smith's critics, is that he fiercely disapproved of corruption and injustice by the British government that colluded with the mercantile classes (Rathbone 2020). Therefore, his specific empirical approach to economics and society adds an alternative perspective as an interlocutor with the communal approach of agrarian philosophy in Africa, specifically because there are salient agrarian and social traces (Dwyer 2005: 662). Thus, for Smith the land and communities are important as a source of economic well-being and wealth (Caton 1985: 835). Smith (WN V.II.12 1981: 644) uses the reference to the community directly for the tenants that cultivate the land and sometimes experience unfair increases in rent. The community of farmers seems to embody the communal values of agriculture but also the injustice of greedy landowners. Although Smith's ideas are influence by the Enlightenment the reference to the community of farmers is closer to the African conception of community because of the mutual interdependence of this group of people. Furthermore, the labour and goods provided by communities are the basis for the well-being and wealth of nations. Smith's agrarian traces and the importance of social engagement are similar with agrarian philosophy in Africa and a basis for dialogue.

Smith notes that agriculture is decisive for economic prosperity because the produce of a farmer provides a profit and rent from the land (WN IV.ix.10 1981: 517). From this, the continued surplus produce from the land creates capital to improve agriculture and the development of manufacture (WN IV.ix.23 1981: 521). Smith (WN IV.ix.24 1981: 521) concludes that the increase in resources such as agriculture and goods will create more capital for both agriculture or manufacture: "According to this liberal and generous system, therefore, the most advantageous method in which a landed nation can raise up artificers, manufacturers, and merchants of its own is to grant the most perfect freedom of trade to the artificers, manufacturers, and merchants of all other nations." The problem is that if agriculture is exclusively promoted, it will be to the detriment of agriculture because the cost of imported goods would deplete the capital required to invest in agriculture and enhance the technology for more sustainable agrarian economics, since "[t]hose systems, therefore, which, preferring agriculture to all other employments, in order to promote it, impose restraints upon manufactures and foreign trade, act contrary to the very end which they propose, and indirectly discourage that very species of industry which they mean to promote" (WN IV.ix.49 1981: 533).

The point that Smith makes, is that land and agriculture are salient sectors for economic growth and wealth. The main reason for this is that nature supports agriculture: the sun and soil produce a great harvest, and livestock produce meat and milk (Caton 1985: 835). The natural resources, therefore, benefit agricultural endeavours to sustain communities. In addition, for Smith agriculture is closely linked to the community, or rather the community of farmers who are daily working on the land and improving the potential of the land to produce. Their way of life is prudent and patient, as they work with nature for the prosperity of society. In this regard, farming is a way of life that encapsulates the values required for the flourishing of society. The value system of the farming community and the excess that is produced is a source of wealth that may enhance development, as will be discussed in the next section.

5.3.2 Agrarianism and Development

In agrarian economics in Africa—as in Smith's economics—development is embedded in the culture and values of the context. This is noted by Newenham-Kahindi (2009), who investigated the role of communal values and the workplace. It was evident that the sustainability of the business is directly linked to the socio-cultural values of a particular context. In this case, team-building, negotiation and consensual leadership were among the important business and managerial practices for business in Africa (Newenham-Kahindi 2009: 101). Therefore, agrarian economics that is not mindful of the communal culture of Africa will find it difficult to survive. Furthermore, this communal mindset, as is the case with Smith, includes a wider sectoral developmental perspective that provides people the opportunity to develop themselves through training, collaboration, and new expertise. This focus on development is for the benefit of society and to increase the possibility to provide the goods and services that are desperately needed in Africa. As mentioned in the previous section Smith's reference to the role of self-interest and market interaction is different from African communal modes of engagement. However, Smith's view of the market was not limited to cold calculated rationalism. He envisioned a more socially embedded market dynamic in which the impartial spectator and sympathy were in tension with self-interest (Rathbone 2015). This was evident in his free market economy that encouraged trade across various sectors of the economy as a manifestation of the values of liberty and equality. The reason for this is that people engage in economic activities and must be allowed to express themselves in a dignified way without government intervention and control. Obviously, in the African context, communal values will form the basis of a socially embedded market dynamic for economic development.

Many critics argue that Smith adheres to a stadial development approach that privileges industrialisation, manufacture and commerce as the epitome of development, as opposed to agriculture, pastoral and hunter-gathering approaches that only represent stages followed to the full commercial approach (Meek 1976; Campbell 1976; Evensky 2005). This stadial approach has been refuted by Paganelli (2020) and others by noting that Smith provides examples of development that contradict and discredits the logic of the stadial approach. Preferably, development for Smith is a historical fact that does not follow a prescribed pattern. History consists of random events, inventions and the discovery of new technologies that change the future of nations, for example, the development of gunpowder. Paganelli (2020) argues that the stadial approach is rather used by Smith to classify different types of societies. Paganelli (2020: 3–4) explains that the "four stages are a taxonomy of different relations between means of production and social, moral, political, and legal institutions, not a model of development from one stage to another".

For Smith, the developmental potential of agrarianism is closely related to the growth and wealth it generates and the values system that underpins it (WN IV.iv.24 1981:326). One of the main aspects of development is the wealth it creates, but wealth is not an outcome in itself. The wealth and access created by agriculture improve the land and technology, and it also ignites the potential of people and the freedom to create new possibilities and opportunities. Although Smith does not promote a model for development, he is of the opinion that multi-sector development is an expression of liberty and that agriculture is an important foundation for sustainable economics. The logic of sustainable agrarian economics does therefore include the creation and development of other sectors of the economy (WN IV.ix 1981: 521)—manufacture, trade and so forth are therefore a logical consequence of agrarianism as Smith highlights (WN IV.ix.24. 1981: 326).

Smith (WN I.i.4 1981: 1) is of the opinion that multi-sectorial development is restricted in agrarian economics due to the nature of the labour involved in agriculture:

The nature of agriculture, indeed, does not admit of so many subdivisions of labour, nor of so complete a separation of one business from another, as manufactures.

It is impossible to separate so entirely the business of the grazier from that of the corn farmer as the trade of the carpenter is commonly separated from that of the smith. The spinner is almost always a distinct person from the weaver; but the ploughman, the harrower, the sower of the seed, and the reaper of the corn, are often the same. The occasions for those different sorts of labour returning with the different seasons of the year, it is impossible that one man should be constantly employed in any one of them. This impossibility of making so complete and entire a separation of all the different branches of labour employed in agriculture is perhaps the reason why the improvement of the productive powers of labour in this art does not always keep pace with their improvement in manufactures.

Therefore, the developmental limitation of agriculture must be complemented by education, technology and multi-sectoral collaborations. Smith (WN I.i.4 1981: 1) uses the example of wealthy nations as an example of the benefits of multi-sectoral development.

The most opulent nations, indeed, generally excel all their neighbours in agriculture as well as in manufactures; but they are commonly more distinguished by their superiority in the latter than in the former. Their lands are in general better cultivated, and having more labour and expense bestowed upon them, produce more in proportion to the extent and natural fertility of the ground. But this superiority of produce is seldom much more than in proportion to the superiority of labour and expense. In agriculture, the labour of the rich country is not always much more productive than that of the poor; or, at least, it is never so much more productive as it commonly is in manufactures.

For Smith, the instincts of self-interest and sympathy are the framework for human moral development. These instincts are in many respects in a state of tension (Rathbone 2015) or disjunction (Rathbone 2021). This implies that the instincts have distinct loci of control and directions, namely self-preservation in the case of self-interest, and harmonious community relations and fellow-feeling in the case of sympathy. The former focusses focuses on the person and the advancement of the individual and the latter on the place of the subject in a community that contributes to the wealth fare of society. Furthermore, the contribution of the individual is met with approbation and fellow-feeling, like social cohesion from appreciation (Rathbone 2018). A dominant expression of self-interest clearly has the danger of excessive self-preservation at the expense of others, while overly demand for harmonious relations can move to the opposite pole of determinism and social control. The tension between these instincts is therefore important for healthy individuals and communities. These instincts are paramount in Smith's economics as the basis of ethical business practices that are sustainable. Smith's ethics is thus the foundation of his economics: development cannot take place without rigorous ethical deliberation that is situated in community and context. The context includes the community, the land and economics.

However, all the economic benefits will lead to short-term prosperity if it is not rooted in a value system that is respectful of the land and the community, as is argued by Smith and agrarian philosophy. Respect and accountability are vital for the responsible management of the land and resources for a sustainable future. The social, anthropological, and environmental focus of development addresses key contemporary economic problems. Firstly, people and communities must not be sacrificed for profit. Secondly, ecological systems cannot be ruined because of mismanagement or the greedy agendas of governments and multinational companies. The importance of community and land are core to African philosophy and can also be viewed in the work of Smith.

In the next section, the aspects highlighted by the dialogue between Smith and agrarian philosophy in Africa will be discussed as the basis of sustainable agrarian economics in Africa.

5.4 Sustainable Agrarian Economics in Africa

5.4.1 Ethical Economics

Agrarian philosophy in Africa and Smith both agree that economics requires a normative ethical core. Although the cultural background of both perspectives differs, there is common ground in that Smith's individualism is placed in the context of sociability, whilst African philosophy's communalism accentuates healthy respect for human dignity. Newenham-Kahindi (2009) emphasises that the communal values of Africa should be contextualised for business and implemented with practical programs and activities. Bush (2013: 51) concurs that sustainable economics in Africa should include more "participation in local decision-making and control of international capital". Ethical economics can be greatly enhanced by the communal ethic of Africa. However, participants can only reap benefits if patriarchal and traditional structures do not limit the dialogue.

The agrarian traces in Smith's work also underscore the importance of land and responsible management of land in his economics. Land is not only a productive means to an end but is interwoven by community and values of prudence (Johnson 1997). Smith is certainly not an environmentalist or activist in the modern sense of the word in terms of the issues facing humanity in the twenty-first century. Land is also not viewed as sacred by Smith, but he encourages respect for others and an approach where land cannot be reduced to commercial aims only. The interrelation between land, nature and people is more embedded in the mysteries of existence than in pure production and profit motives.

African philosophy's holistic ontology that informs agrarian philosophy in Africa does indeed go much further than Smith's atomism by linking land, culture, and community (Rathbone 2014). However, there are enough similar themes (although from divergent cultural perspectives) that can critically complement one another and support sustainable economics. In other words, the tensions between the worldviews can highlight weaknesses that can be critically evaluated. For example, the danger is that Smith's approach can become functionalist and reduce the land to a means of production—African philosophy provides a far more sacred view of land that will not tolerate any reductions that can be restrictive or harmful to the land and/or community. At the same time, Smith highlights the productive capacity of land and agriculture as key for economic and sectoral development that is important for agrarian economics. A critical ethic of land that scrutinises cultural and economic aspects associated with the mining of resources and cultivation of the soil, is required (Rathbone and Verhoef 2021). Dialogue between divergent positions is not a technique to suspend the differences but to move forward without offending people or increasing underdevelopment. For example, land in Africa is linked to ancestors and other spirits which makes dialogue extremely sensitive and the utmost respect should be exhibited. Dialogue opens divergent worldviews up for greater understanding and also creative possibilities to be explored. Therefore, the development of land for agriculture or industry may encounter problems related to the belief of the community inhabiting the land. Dialogue is not an attempt to provide a quick answer. It rather encourages robust engagement to explore alternatives in a respectful manner.

5.4.2 Sectoral Development

In the dialogue between African philosophy and Smith, the focus of the latter on multi-sectoral development is a necessary contribution to sustainable Agrarian economics in Africa. On the other hand, the value-driven nature that is supported by the interaction between African philosophy and Smith is vital for multi-sectoral development not to fall prey to human greed and exploitation by multi-national conglomerates.

Another important aspect of the relationship between land and community is the fact that agriculture can be enhanced and developed by collaborating with other sectors to preserve the land and ecology by refraining from over-exploiting the land. Mismanagement of Africa's resources and land will be detrimental to the identity of the communities and sustainability. The care and welfare of the community are more important than individual self-interest. It also means that no form of greed or corruption will be tolerated. Sectoral development means that underdevelopment is challenged because the value is placed on the land and community. More efficient and sustainable farming through technology and education is beneficial to the community and will result in sustainable management of the land.

At this point, it is important to acknowledge dissenting voices that reject industrialisation as a Western form of economic and developmental imperialism. In the article *The Classical Agrarian Question: Myth, Reality and Relevance* (2013) the authors argue that industrialisation as the basis for development is a flawed myth because "such a view fails to acknowledge the historic importance of the national question and its land and peasant components, which are irreducible to industrialization" (Moyo et al. 2013: 93). The premise is that Marxism views industrialisation as a remedy for the "backwardness" (*sic*) of some economies. It is imperative to state that Smith in no way suggests that agrarian economics is *not* important or should be replaced by industrialisation. Smith is appreciative of the important role of agriculture in the development of free market economies and consequently underscores a more robust economy that will not be at the mercy of manufacturing economies. Smith also appreciates the value system of agricultural communities and does not see them as backward at all. As mentioned earlier, recent research also emphasises Smith's criticism of stadial development that praises commerce and manufacturing as the pivot of development (Paganelli 2020).

Finally, Gassner et al. (2019:309) argued that development within the agriculture sector must distinguish between subsistence and commercial farming: "It is important to understand where and for whom agriculture will have the main purpose of ensuring food and nutritional security and where and for whom there is the potential for significant increases in incomes and a contribution to wider economic growth." This is a valuable distinction between developmental effort and government policy. Subsistence farming does not necessarily have to be linked to poverty but in many cases (due to droughts and/or other natural disasters, among others) many of these farmers struggle to survive and are economically challenged to obtain other goods. Commercial farming is more intensive and requires more capital to remain competitive in the market.

5.4.3 Justice as a Relational Concept

Justice is a pressing issue for both agrarian philosophies in Africa and Smith. Smith views justice as a limiting mechanism to avoid unfair and harmful practices (Rathbone 2020). However, justice is crucial for society to function in harmony and sustainable economics. Smith was outspoken against corruption and broken interpersonal and institutional trust, such as Poor laws (Rathbone 2015). Government intervention and collusion with certain classes, like merchants, created an elitist sector and disenfranchised the general population, thereby reducing the prosperity and development of the nation.

In the African context, justice—against the backdrop of colonialism, imperialism and globalisation—is viewed as a means to bring about restitution and healing. In the case of colonialism and globalisation, this aspect is clearly of great importance. However, justice can also be viewed from an African communal perspective, specifically in the case of corrupt practices and activities such as nepotism and exploitation by the elite. Broken trust and failure to provide services because of incompetence and greed, lead to the breakdown of community structure and morale. More tangibly, it also contributes to underdevelopment and increased poverty. This injustice is a serious debilitating factor in the development of sustainable agrarian economics and well-being (Metz 2009). Masitera (2020: 35) adds that in Africa, "humaneness and communitarian life are centered on values such as relationality, cooperation, common good and equal distribution of communal goods". The African values and norms, therefore, represent a perspective on redistributive justice that is based on relational rather than rights principles, in which case the land is not owned but equally shared. The difference between individual and collective ownership of property is an important distinction between Smith and African philosophy. However, collective ownership in Africa has also led to similar abuses of Smith's time where tenants are unfairly treated and exploited by landowners, where African leaders abused their powers in the unfair appropriating land, or its produce. In this regard, dialogue is important for a just dispensation without overruling private ownership or privileging any other.

5.5 Conclusion

The purpose of the study was to highlight salient aspects of sustainable agrarian economics in Africa through robust and critical dialogue. This was accomplished by placing the agrarian traces in the work of Smith in dialogue with agrarian philosophy and its implications for economics in Africa. Sustainable agrarian economics has to grapple with two major problems of African economics, namely underdevelopment and resource dependence. It is precisely in an attempt to find alternatives to deal with these issues that the classic economics of Smith is an important interlocutor with agrarian philosophy in Africa because Smith's empirical approach offers an analysis of the cognitive processes involved in decision-making and moral development. The dialogue is based on the fact that both Smith and agrarian philosophy in Africa emphasise the role of ethics, community and agriculture although from divergent worldviews. The aspects that surfaced from the discussion are the ethical foundation of economics, multi-sectoral development and justice as a relational concept. These aspects draw on the strengths and innovations that the dialogue between Smith and agrarian philosophy highlighted for sustainable agrarian economics.

References

Atkinson, A.B. 2015. *Inequality What Can Be Done?* Published by Harvard University Press. https://www.tony-atkinson.com/new-book-inequality-what-can-be-done. Accessed 11 Nov 2019.

Beresford, A. 2016. Africa Rising? *Review of African Political Economy* 43 (147): 1–7. https://doi.org/10.1080/03056244.2016.1149369.

Bloom, D., D. Canning, and K. Chan. 2006. *Higher Education and Economic Development in Africa.* Human development sector: African region.

Boogaard, B.K. 2019. The Relevance of Connecting Sustainable Agricultural Development with African Philosophy. *South African Journal of Philosophy*, 38(3): 273–286

Bush, R. 2013. Making the Twenty First Century its Own: Janus Faced African (Under) Development. *Afrika Focus* 26 (1): 51–65.

Campbell, R.H. 1976. Introduction. In *An Inquiry into the Nature and Causes of the Wealth of Nations* Indianapolis: Liberty Press.

Caton, H. 1985. The Preindustrial Economics of Adam Smith. *The Journal of Economic History* 45 (4): 833–853.

Chabal, P., and J.-P. Daloz. 1999. *Africa Works: Disorder as Political Instrument*. Oxford: James Currey.

Dwyer, J. 2005. Ethics and Economics: Bridging Adam Smith's Theory of Moral Sentiments and Wealth of Nations. *Journal of British Studies* 44 (4): 662–687.

Evensky, Jerry. 2005. *Adam Smith's Moral Philosophy*. Cambridge: Cambridge University Press.

Forje, W. 2020. *Unravelling the Mysteries of Africa's Underdevelopment : Changing Africa, One Idea at a Time*, Langaa RPCIG, 2020. ProQuest Ebook Central, https://ebookcentral.proquest.com/lib/northwu-ebooks/detail.action?docID=6215662.

Gassner, A., D. Harris, K. Mausch, A. Terheggen, C. Lopes, R.F. Finlayson, and P. Dobie. 2019. Poverty Eradication and Food Security Through Agriculture in Africa: Rethinking Objectives and Entry Points. *Outlook on Agriculture* 48 (4): 309–315.

Johnson, D.G. 1997. Agriculture and the Wealth of Nations. *The American Economic Review* 84 (2): 1–12.

Masitera, E. 2020. Indigenous African Ethics and Land Distribution. *South African Journal of Philosophy* 39 (1): 35–46. https://doi.org/10.1080/02580136.2019.1706383.

Mbembe, A. 2015. *On the Postcolony*. South Africa: Wits University Press.

Meek, R.L. 1976. *Social Science and the Ignoble Savage*. Cambridge University Press: Cambridge; New York.

Metz, T. 2009. Africa Moral Theory and Public Governance: Nepotism, Preferential Hiring and Other Partiality. In *African Ethics: An Anthology of Comparative and Applied Ethics*, ed. F.M. Murove. Pietermaritzburg: University of KwaZulu-Natal Press.

Mizuno, N., and R. Okazawa. 2009. Colonial Experience and Postcolonial Underdevelopment in Africa. *Public Choice* 141: 405–419. https://doi.org/10.1007/s11127-009-9461-8.

Mkandawire, T. 2014. Can Africa Turn from Recovery to Development? *Current History* 113 (793): 171–177.

Moyo, S., P. Jha, and P. Yeros. 2013. The Classical Agrarian Question: Myth, Reality and Relevance Today. *Agrarian South: Journal of Political Economy* 2 (1): 93–119.

Munyaka, M., and M. Motlhabi. 2009. Ubuntu and its Socio-moral Significance. In *African Ethics: An Anthology of Comparative and Applied Ethics*, ed. F.M. Murove. Pietermaritzburg: University of KwaZulu-Natal Press.

Ndaguba, E.A., and B. Hanyane. 2018. Exploring the Philosophical Engagements for Community Economic Development Analytical Framework for Poverty Alleviation in South African Rural Areas. *Cogent Economics & Finance* 6: 1. https://doi.org/10.1080/23322039.2018.1539942.

Newenham-Kahindi, A. 2009. The Transfer of Ubuntu and Indaba Business Models Abroad: A Case of South African Multinational Banks and Telecommunication Services in Tanzania. *International Journal of Cross Cultural Management* 9 (1): 87–108.

Paganelli, M.P. 2020. Adam Smith and Economic Development in Theory and Practice: A Rejection of Stadial Model? *Journal of the History of Economic Thought (forthcoming)*. Preprint at SocArXiv, osf.io/preprints/socarxiv.

Rathbone, M. 2014. Interpretations of the Tower of Babel Narrative in the African Context. *Acta Theologica* 34 (1): 173–196.

Rathbone, M. 2015. Love, Money and Madness: Money in the Economic Philosophies of Adam Smith and Jean-Jacques Rousseau. *Southern African Journal of Philosophy* 34 (3): 379–389.

Rathbone, M. 2018. Adam Smith, the Impartial Spectator and Embodiment: Towards an Economics of Accountability and Dialogue. *Religions* 2018 (9): 118.

Rathbone, M. 2020. Capitalism, the Book of Amos and Adam Smith: An Analysis of Corruption. *HTS Theological Studies* 76 (4): 1–9.

Rathbone, M. 2021. Laughter and the Economic Philosophy of Adam Smith. *Southern African Journal of Philosophy* 40 (3): 242–253.

Rathbone, M., and A.H. Verhoef. 2021. Towards a Critical Ethic of Land in the Southern African Context. In *Philosophical Perspectives on Land Reform in Southern Africa*, ed. E. Masitera, 267–284. Cham: Palgrave MacMillan.

Rodney, W. 2012. *How Europe Underdeveloped Africa*. Oxford: Pambazuka Press.

Smith, A. [1776]. 1981. *An Inquiry into the Nature and Causes of the Wealth of Nations*. Indianapolis: Liberty Fund.

Sen, A. 1999. *On Ethics & and Economics*. London: Blackwell

Thompson, P.B. 2008. Agrarian Philosophy and Ecological Ethics. *Science Engineering Ethics* 14: 527–544.

Twala, C. (2004). The 'Letsema/Ilima' Campaign: A Smokescreen or an Essential Strategy to Deal with the Unemployment Crisis in South Africa. *Southern Journal for Contemporary History*, 29(1): 184–198. Retrieved from https://journals.ufs.ac.za/index.php/jch/article/view/409.

Mark Rathbone is a professor and chair of the Research Ethics Committee of the Faculty of Economic and Management Sciences at the North-West University. Prof Rathbone has published numerous articles and book chapters on topics that range from business ethics and classic economic philosophy to phenomenology. He is also the editor of various books that focus on similar topics and is currently busy with a research project on the work of Adam Smith that focuses on contemporary themes such as economic inequality, trust, commodification, corruption, among others.

Part II
Moral Status of Non-human Nature in African Agrarian Thought

Chapter 6
Defending a Relational Account of Moral Status

Thaddeus Metz

Abstract For the more than a decade, I have advanced an account of what makes persons, animals, and other beings entitled to moral treatment for their own sake that is informed by characteristically African ideas about dignity, a great chain of being, and community. Roughly according to this account, a being has a greater moral status, the more it is capable of communing (as a subject) or of us communing with it (as an object). I have mainly argued that this characteristically African and relational approach to moral status is a better account than salient Western approaches, especially individualist views associated with utilitarianism and Kantianism. Over the years, several commentators have raised criticisms of my approach, including that it objectionably: entails that we may rightly dominate mentally incapacitated human beings; prioritizes mentally incapacitated human beings over animals with similar cognitive abilities without sufficient justification; entails that intelligent aliens lack moral status; cannot make sense of our duties towards the dead; and is unable to account for the standing of species as distinct from their members. In this chapter I provide a comprehensive response to these and related objections, defending the initial account as an attractive way to understand what makes a being matter morally for non-instrumental reasons. For many animals to have a moral status would have important implications for the practice of agriculture, for instance farming animals for food and expanding crops.

Keywords Animal rights · Dignity · Direct duties · Indirect duties · Marginal cases · Moral status · Relationality · Species

---

T. Metz (✉)
University of Pretoria, Pretoria, South Africa
e-mail: th.metz@up.ac.za

M. J. Tosam and E. Masitera (eds.), *African Agrarian Philosophy*, The International Library of Environmental, Agricultural and Food Ethics 35,
https://doi.org/10.1007/978-3-031-43040-4_6

6.1 Introduction

For about 15 years I have been interpreting the African moral-philosophical tradition in an analytical manner, drawing on it to construct a basic moral principle that could clearly entail and powerfully explain a wide array of intuitions about which actions are right and which are wrong (culminating in Metz 2022a). My aim has been to advance a normative ethical theory with a characteristically African content that rivals standard Western views such as utilitarianism and Kantianism. As part of this project, I have also proposed and defended companion accounts of distributive justice (e.g., Metz 2015), criminal justice (e.g., Metz 2019), moral virtue (Metz 2022b), and moral status (initially Metz 2010, 2012a). In this chapter I focus strictly on moral status, which primarily concerns which beings on the Earth we owe duties for their own sake. We do not owe a pen a duty for the sake of the pen; if we have moral reason to treat it certain way, it is not ultimately because of facts about the pen, but more because of the human person who owns it or otherwise might be affected by our treatment of the pen. It is common to say that while the person is an object of a 'direct duty', meaning that it has a moral status, the pen is at best an object of an 'indirect duty' and lacks a moral status.

Which other things located on the land besides human persons have a moral status, and in virtue of what? Are rocks and rivers similar to pens in lacking moral status, even if we have many indirect duties towards them? Do we owe direct duties to human embryos and early foetuses? Which animals have a moral status, and how does it compare to that of human persons or human non-persons? Such questions are central to philosophical debates about moral status, and I have sought to develop a comprehensive principle that would provide plausible answers to them and similar kinds of queries. Roughly according to my account, a being has a greater moral status, the more it is capable of communing (as a subject) or of us communing with it (as an object). Such a view has been described as a 'modal relationalism' (first by me in Metz 2010 and subsequently by Horsthemke 2015: 85–92, 2017; Maj 2020; and Samuel and Fayemi 2020). On the one hand, my principle appeals to a being's *relational* properties, as opposed to intrinsic properties such as sentience, autonomy, or life, and, on the other, it does not appeal to a being's actual relational properties and instead focuses on its *possible* ones, i.e., ways in which it *could* interact in the relevant way.

Over the years, several commentators have raised criticisms of this modal relational approach to moral status, including in several essays that have been devoted to it (e.g., Horsthemke 2015: 85–92, 2017; Ikuenobe 2016; Molefe 2017; Samuel and Fayemi 2019, 2020; Maj 2020).[1] Amongst their criticisms are that my principle objectionably: entails that we may rightly dominate mentally incapacitated human beings; prioritizes mentally incapacitated human beings over animals with similar cognitive abilities without sufficient justification; entails that intelligent aliens lack

[1] That said, a number of commentators have also found this approach to moral status to be appealing and have applied it in variety ways, including Behrens (2017), Samuel and Fayemi (2019), and Cordeiro-Rodrigues and Ewuoso (2022).

moral status; cannot make sense of our duties towards the dead; and is unable to account for the standing of species as distinct from their members.[2] In this chapter I provide a comprehensive response to these and related objections, defending the initial account as an attractive way to understand what makes a being matter morally for non-instrumental reasons. I maintain that modal relationalism would be plausible to employ when addressing agricultural ethical controversies, including whether and, if so, how to farm animals for food and when it is permissible to expand crops at the expense of wildlife and biodiversity.

In the following I begin by recounting the essentials of my approach to moral status (Sect. 6.2), so as to facilitate the debates between myself and my interlocutors. I then dispatch what I take to be some of the weaker objections to my view, for instance that suggest that my approach involves human beings choosing which other beings have moral status or entails that all value is contingent on humanity (Sect. 6.3). After pointing out that these and other criticisms rest on mischaracterizations, conflations, and the like, I use more space to address criticisms that are prima facie stronger. Specifically, I address objections according to which my account of moral status is *incomplete* (or exclusionary), failing to include certain beings that purportedly have a moral status, such as alien persons or dead human bodies (Sect. 6.4). After that I take up objections that my view is *elitist* (or anthropocentric) in that, while it includes the right sorts of beings, it does not accord them the proper degree of status compared to other beings, where the key topic concerns 'marginal cases' of human beings with severe mental incapacitation (Sect. 6.5). I conclude by indicating some other issues that need to be addressed elsewhere in order to provide a full defence of modal relationalism (Sect. 6.6).

6.2 An Overview of Modal Relationalism

The account of moral status that I have developed counts as African insofar as it is informed by three moral ideas that have been salient in the literate sub-Saharan philosophical tradition, viz., those of human dignity, community, and a great chain of being. I have reinterpreted these ideas and integrated them in a way that is meant to be attractive, not merely to contemporary African philosophers, but also to many philosophers and related enquirers around the world. In this section, I briefly recount the traditional understandings of these three notions, after which I indicate how I have reconstructed them into a principled understanding of moral status and what some of the key philosophical motivation for the resultant principle is.

---

[2] In this chapter I set aside criticisms that my view grounds an implausible account of the dignity of human persons, on which see Ikuenobe (2016). Instead, I address issues pertaining to the moral status of mainly non-human persons, human non-persons, and non-human animals. In addition, I have already responded to Horsthemke (2015: 85–92) in Metz (2017a), and so do not repeat facets of our debate here.

First off, it has been common for African philosophers to hold that human beings have a dignity, that is, a superlative non-instrumental value that is inherent to us. Common has been the view that, while groups such as peoples or nations might well matter morally, individuals also matter in their own right, a perspective enshrined in the African ('Banjul') Charter on Human and Peoples' Rights (Organization of African Unity 1981). Standard views amongst African philosophers about what confers a dignity on a human individual include the ideas that we have a particularly strong or complex vital force that has come from God (a clear statement of which is in Wiredu 1990: 244), that we are members of a clan or other relationship (Bujo 2001: 88), and that we have treated others in certain morally correct ways (Ikuenobe 2016).

A second salient theme in African philosophy concerns the importance of community. Probably the most frequent invocation of this value is the idea that a good life involves relating to others in certain communal ways (with salient examples including being part of an extended family that includes the not-yet-born, human beings, and the living-dead), participating in collective harvesting and other kinds of labour, upholding long-standing traditions and rituals, and seeking unanimous agreement to resolve disputes about how to resolve conflict.[3] In all four of these examples, people come together in certain cohesive ways or can be said to 'enter into' community with one another.

A third view frequently espoused by African philosophers is that the world forms a great chain of being or hierarchical ordering of entities (e.g., Magesa 1997: 39–51; Imafidon 2014: 40–42). Traditionally the thought is that all concrete objects are infused with a divine energy, but that some have a stronger or more sophisticated form of it than others. 'Life-force varies quantitatively (in terms of growth and strength) and qualitatively (in terms of intelligence and will)' (Anyanwu 1984: 90). The mineral kingdom is at the bottom of the ranking for having the least quantity or poorest quality of life-force. Plants are thought to have a greater life-force than rocks, say, for being capable of reproduction and movement. Animals have a greater one than plants, e.g., for being capable of self-motion and some pattern-making (consider a bird's nest or beaver's dam). Finally, '(o)ver all visible beings, in terms of intensity of vital force, stands humanity' (Magesa 1997: 51), particularly for having self-awareness and genuine creativity, while ancestors, who are imperceptible persons living on Earth, are deemed to have an even greater vital force than human beings. Beyond the metaphysical claim that beings can be ranked according to their life-force is the axiological claim that moral status tracks this ranking, such that plants matter morally more than rocks, animals matter more than plants, and humans matter most (of perceptible beings) (see, e.g., Etieyibo 2017).

Now, I have sought to develop a theory of moral status that is recognizably African but that could appeal to those from a wide array of philosophical backgrounds. I have drawn on sub-Saharan perspectives to construct moral principles that could be

[3] Much more controversial (and in my view implausible) claims are that the community, understood as a group, is ontologically prior to the individual in some way that gives it a moral priority or that one's foremost obligation is to abide by a community's extant norms, whatever they happen to be.

justified to a global audience. That has meant setting aside metaphysical claims that are difficult to back up with evidence or at least that lack receptors in other major traditions around the world. Specifically, this approach has prescribed developing an account of moral status that does not invoke claims about imperceptible agents and forces such as God, the not-yet-born, the living-dead, ancestors, and vital force. My view is instead strictly secular or focused on the perceptible.[4]

Despite being shorn of contested metaphysical views, certain strains of African axiology retain an attractive logic and can ground a plausible account of moral status. One prima facie appealing strategy would be to develop a naturalist account of vitality,[5] but in my work I have instead mainly articulated and defended a naturalist account of relationality. Specifically, I have drawn together the three strands of dignity, community, and hierarchy this way: a being has a greater moral status, the more it is markedly capable of being party to communal relationships with us. In the rest of this section, I briefly analyze the central concepts of this principle and indicate some of why I believe it is philosophically promising.

By 'communal relationship' I mean a way of interacting characterized by identifying with others and exhibiting solidarity with them. Identifying with others means enjoying a sense of togetherness with them and participating with them on a voluntary, trustworthy basis to support their goals. Exhibiting solidarity with others means going out of one's way to meet their needs and otherwise improve their quality of life and doing so out of sympathy and for their own sake. The combination of identity and solidarity does a good job of capturing what is morally appealing about extended family, collective harvesting, sustaining customs, and seeking consensus. What is desirable about these kinds of interaction is plausibly the sharing of a sense of self, the cooperation, the aid, and the altruistic motivation. All together, those constitute communal relating.

An individual might be 'capable' of being party to a communal relationship in two different ways. On the one hand, it might be able to be a subject of it, that is, could by its nature identify with others and exhibit solidarity with them. On the other hand, it might be able by its nature to be an object of communal relationship, i.e., we could identify and exhibit solidarity with it.

According to my account, a being that in principle can be a subject and object of communal relationship has the highest moral status, which entails that it has a dignity. Its urgent interests, e.g., in remaining alive and living well, must be satisfied before those of any other being in cases of conflict. A being that can be merely an object of a communal relationship with us has a partial moral status; it does matter morally for its own sake, but lacks a dignity. Its urgent interests must be satisfied

---

[4] I am therefore seeking neither to represent African philosophical beliefs in anthropological or ethnophilosophical fashion nor to provide an 'authentic' African ethic. I am instead interested in determining which beliefs merit acceptance, at least given a certain audience of philosophers. I also maintain that authenticity is much more of a European value (on which see Taylor 1992) than an African one. Those who want a 'pure' African philosophy are often (ironically) appealing to a value is that not characteristically African, where the latter prescribes interdependence and relationality, not independence and isolation.

[5] Which I have pursued a bit elsewhere, e.g., in Metz (2012b).

before attending to anything without a moral status, but not if doing so would prevent one from satisfying the urgent interests of a being with dignity. Finally, a being that can be neither a subject nor an object of communal relationship lacks a moral status, such that any moral reason to treat it a certain way is indirect, ultimately grounded on facts about some other being.

Although there are more nuances to the principle, particularly about there being degrees of the ability to be an object of communal relationship and how to balance the interests of entities with differential moral status (which I discuss below), what I have spelled out so far is enough to grasp the core of the view and to see some of its appeal. Nearly all human persons are capable of being a subject and object of communal relationship and hence have a dignity. That is, just about all of us human persons can enjoy a sense of togetherness with others, participate with them on a cooperative basis, help meet others' needs, and do so altruistically, and, furthermore, other persons can do these things with just about all of us.

A very large majority of animals are capable of being an object of communal relationship with us and hence have a partial moral status. Animals such as apes, dolphins, elephants, cows, dogs, cats, and mice all matter morally for their own sake such that we have direct duties towards them. Although they cannot exhibit identity and solidarity with us (even if apes might have the rudimentary ability to do so), we can clearly do so with them, meaning they are able to be objects of communal relationship with us. We can enjoy a sense of togetherness with these creatures, interact with them in trustworthy ways that foster their ends, do what meets their needs, and do so out of sympathy and not merely for our own long-term self-interest.

Finally, beings such as rocks, shrubs, and mosquitos can be neither the subject nor the object of communal relationship with us and hence lack a moral status altogether. There of course could well be moral reasons to treat them in a certain ways, but, by my account, not because of facts about these beings and instead because of facts about other beings such as humans or animals. So, for example, if there is reason not to destroy an ecosystem that includes rocks, shrubs, and mosquitos, it is not directly because these beings would be negatively affected, but rather indirectly because negatively affecting them would wrong other kinds of beings, roughly those with intentions and a quality of life.

In all these respects, I submit that the modal relational account of moral status accounts for widely held intuitions about which beings matter morally for their own sake and to what degree.[6] In particular, it is quite common for professional philosophers in the African, Western, and many other traditions to think that human persons matter more than animals, which, in turn, matter more than plants. If you had to choose between running over a person or an animal with one's car (there being no third alternative), surely you should avoid the person and instead strike the animal. And if you had to choose between running over an animal or a shrub (again, there being no third alternative), surely you should mow down the shrub. The modal

[6] I also think the theory does a good job of providing accounts of how to rank the moral importance of various kinds of animals, on which see Metz (2012a: 400), and of how to judge the status of embryos, foetuses, and newborns (Metz 2022a: 182–186).

relational account easily accommodates these intuitive judgements, since the various entities have differential abilities to relate communally with us and since the urgent (but not trivial) interests of those with a higher standing must be satisfied before the urgent interests of those with a lower standing (in cases when not all interests can be satisfied).

Note that rival approaches in the Western tradition cannot easily capture the above intuitions. Kantian views have difficulty ascribing moral status to animals at all since they lack rationality or autonomy. Utilitarian or welfarist views naturally ascribe moral status to animals, insofar as they feel pain and have preferences, but these views have difficulty explaining why human beings have a greater standing and they also eschew the concept of dignity. Biocentric and subject of a life views have the same problem, entailing that the lives of persons and animals are of equal standing, and hence unable to entail that the urgent interests of the former must be satisfied before those of the latter in cases of conflict.

Although the urgent interests of human persons should be satisifed if at the necessary expense of the urgent interests of animals, modal relationalism does ground some strong duties towards animals. Consider the principle that the urgent interests of beings with partial moral status should not be sacrificed for the *trivial* interests of those with full moral status. It follows that we should not greatly subordinate or harm animals for the sake of, say, the mere pleasure of taste. While factory farming chickens for meat, for example, would be justified in the absence of other sources of protein for human persons, it would not be if adequate plant-based sources were available. In addition, while expanding crops and thereby destroying the habitats of many animals would be justified if necessary to feed human persons, it would not be if it were done merely to give consumers a version of maize differing only slightly from current ones.

My primary aim in articulating modal relationalism has been to draw on resources from the African tradition pertaining to dignity, community, and a great chain of being to construct an account of moral status that could be found attractive without relying on contested metaphysical claims and that improves on salient Western approaches. However, many of those in the literature have tended to criticize it, not so much in favour of one of the Western rivals, but instead more absolutely, as providing an inaccurate account from a more comprehensive philosophical perspective. These kinds of criticisms, addressed in the Sects. 6.4 and 6.5, are fair and welcome. However, I first address some criticisms that are based on what I deem to be misinterpretations or conflations.

6.3 Weaker Objections

In this section I take up concerns about modal relationalism expressed in the literature that I think clearly miss their mark. They can be readily seen not to apply, once terms are carefully defined and certain distinctions are drawn. I address them because responding to them might help to avoid misunderstanding on the part of readers.

To start, Filip Maj (2020) suggests at times in his essay that I am to be read as suggesting that actual relationships are necessary for moral status, a view that he (aptly) finds counterintuitive.

> (Metz) mentions that animals that are objects of our care 'matter in their own right'….This seems like an inconsistency as how are animals considered to 'matter in their own right' if their moral status comes down to us choosing them as being so….It is confusing when Metz says 'we have direct duties toward many animals with which we can by nature commune'….Does this mean that we do not have a direct duty to all the animals that we choose to identify with and care for, just those that we can in principle? Can the 'many' be changed to 'all'?....Metz's theory could also cause a problem for the diversity of animals existing in the world, e.g., wild animals, as there is no apparent direct obligation to secure species, if they do not have a considerable input as being objects of human interest, that is, communion (Maj 2020: 344, 345).

This analysis neglects the essentially modal dimension of my approach, according to which it is a being's natural *capacity* to be communed with by characteristic human beings that confers a (partial) moral status on it, not its actually having been communed with by a given one of us. The question is whether it is a kind of being with which we *could in principle* identify (i.e., enjoy a sense of togetherness and foster its ends) and exhibit solidarity (i.e., enhance its quality of life and do so out of sympathy and for its own sake).

Once the modal approach is kept in view, several of the above queries and claims are seen to fall away. So, it is not beings that *are* objects of our care that matter for their own sake, but rather beings that by their nature and ours *could be* objects of our care, for instance beings such as pigs, rabbits, and whales. (Of course, actually being an object of care is sufficient for it to be able to be such an object, but actually being an object of care is hardly a necessary condition for that.) It is not a matter of us 'choosing' which beings have a moral status, as moral status for me does not depend on any humans electing to interact with anything. It is not my view that we have a direct duty to all and only the animals we *choose* to identify with and care for, but rather to all and only (not merely many of) those animals we *could* identify with and care for. Finally, my view indeed accords many wild animals a moral status insofar as we could in principle commune with them (note that we could not with, say, bacteria or fruit flies); it is not the case that by my account we must take an interest in animals and commune with them first, in order for direct duties to them to arise. It is really the other way round, such that, roughly, one ought to commune with a being because one can.

The remaining objections addressed in this section more firmly grasp the modal dimension of my account, but are claims that it has counterintuitive implications that I think it clearly does not have. For some, it appears that if for an animal to have moral status it must be the kind of thing with which we could commune, then all its value is dependent on us. However, so their objection goes, the value of an animal does not entirely depend on us; it has some value in its own right. Motsamai Molefe remarks,

> Metz imagines human beings as an essential furniture of the world, so much as to make them a basis upon which the value of other things like animals depends. In this light, this theory implies that human beings are an essential furniture of the world to the effect that without them there would be no concerns of value at all (Molefe 2017: 199–200).

Maj approves of Molefe's reasoning: 'Metz would have to commit to the fact that there is no value in the world without a human valuer' (Maj 2020: 344).

Modal relationalism neither expresses such a viewpoint nor is otherwise committed to it, which can be readily seen upon distinguishing the two concepts of moral status and final value. For something to have final (or non-instrumental) value is for it to be good for its own sake in some respect; it is valuable not merely as a means. In contrast, for something to have moral status (at least in respect of human agents) is for us to owe it moral treatment for its own sake; it is owed moral treatment not merely for indirect reasons. Having moral status is not necessary in order for something to have final value; something might lack a moral status and yet have a non-instrumental value. If so, then, even if moral status is logically dependent on us, final value need not be.

To appreciate how it is possible for something to have final value without having moral status, consider a species of intelligent and sentient beings who live in another galaxy. Supposing that, living so far away, we could not affect them in any way whatsoever, they lack a moral status in relation to us. We have no direct duties to them since we cannot influence them at all. We do not owe them anything since we simply cannot do anything for them, whether negative or positive. Despite lacking a moral status in relation to human agents, such beings would nonetheless plausibly have a final value.

For another way to recognize how it is possible for something to have final value without having moral status, think about certain scientific laws or musical compositions. Scientific laws that are true and do much to unify a wide array of phenomena are plausibly good for their own sake, or at the very least the state of understanding them is. Musical compositions that are original and beautiful are also plausibly good for their own sake, or, again, at least the states of creating and apprehending them are. Even so, it would be quite unusual to ascribe moral status to scientific laws or musical compositions, or even to the states of us interacting with them in various ways. We do not owe principles or songs (or our interactions with them) direct duties for their own sake, and yet they have some non-instrumental value.

So, one must distinguish sharply between a thing being owed a duty for its own sake by a human agent and a thing being good for its own sake. Upon doing so, one can see that, even if a thing's being a owed a duty on our part were to depend on its capacity to relate to us, it need not follow that a thing's being good would have to depend on that.

It is true that I have in my work suggested that if a being has a full moral status then it has a dignity, i.e., a superlative non-instrumental value that does not depend on contingencies such as what a society thinks of the being or which choices it makes. However, I do not recall ever having suggested that a being has a dignity only if it has a full moral status, and I do not see that I have to hold such a view. Those creatures in a far away galaxy mentioned above might well have a dignity, even if, because we cannot interact with them at all, they do not have a moral status in respect of us.

That is part of the explanation of why I do not believe that modal relationalism is vulnerable to a third criticism that I rebut in this section, viz., the 'last man objection' that Molefe has also advanced against my approach (Molefe 2017: 200). According to this objection, a theory of moral status is counterintuitive if it cannot entail and plausibly explain why it would be wrong for the very last human person to do what would destroy the natural world after his demise. 'If one would find it horrendous for the last man to destroy all of nature, this points to the direction that probably nature matters for its own sake' (Molefe 2017: 200).

One thing to say in reply is that it is perfectly coherent for me to hold that parts of the natural world have a non-instrumental value and that, even if human beings died out, they would retain it. They would not have a moral status from the perspective of humanity if no human agents existed, but they would still be good for their own sake, as per the reply to the previous objection.

Another thing to say is it is precisely my view that a large part of nature 'matters for its own sake' when it comes to morality; that is what is involved in claiming that it has a moral status or we owe it direct duties. I do not think that a large part of nature matters morally in itself in the sense of merely in virtue of intrinsic properties that it has. However, it is a mistake to suppose that something can matter *for its own sake*, morally or otherwise, only in virtue of what it is *in itself* or its intrinsic properties, as I believe many philosophers in the analytic tradition have demonstrated (for just two examples, see Korsgaard 1983; and Kagan 1998). The means/end and intrinsic/relational distinctions are orthogonal, such that a being can be valuable as an end in itself (or have moral standing for non-instrumental reasons), and not be a mere means, in virtue of non-intrinsic properties. A work of art is plausibly good for its own sake, and not for merely instrumental purposes, but only when it is in relation to a perceiver. Something can have more non-instrumental value when it is rare, but rarity is a relational property. These and a myriad of other examples that philosophers have advanced over the years are strong evidence that relational properties can make something good for its own sake or enhance its final value; founding value on a thing's relational properties need not mean that the thing is good solely as a means or indirectly.

A final reply to the last man objection is that the last remaining human agent, by my theory, should view a large part of nature as having a moral status and so would be wrong to do what would lead to its destruction. My theory implies that something has a moral status for a human agent if characteristic human beings are capable of relating communally to it. Roughly speaking, if a being has intentional states and a quality of life, and if we are in a position to affect it, then it has a moral status, even if only a partial one (entailing that its urgent interests must be protected at the cost of trivial interests of beings with a full moral status). The last man should apprehend the fact that many animals meet these criteria, such that it would wrong them if he were to do what would greatly subordinate or harm them. There might be *other* parts of nature that also would be wronged, but that kind of objection is different, and is addressed in the next section.

6.4 Charges of Incompleteness

In this section I address claims that the modal relational account of moral status I have developed entails that certain beings lack a moral status that in fact intuitively have it. I first address the claim that it cannot account for the moral status of aliens, then that of dead humans, and finally that of species as distinct from their individual members. On all three counts, I maintain that there are reasonable responses to be made on behalf of modal relationalism.

In respect of aliens, Molefe imagines that a race of Martians exists who are very much like us for being capable of exhibiting identity and solidarity with others. The Martians are similar to humans in that they are persons capable of relating communally. Now, Molefe urges us to suppose

> that these two entities, for some strange biological reason, I *stipulate*, cannot enter into any kind of interaction with each other. If Metz is truly committed to the view that moral status is accounted for by an essential reference to some *human* feature (ability to commune with human beings), then it should follow that Martians have no moral status (Molefe 2017: 200).

However, the objection continues, such aliens would indeed have a moral status. Perhaps that would be because they are capable of relating communally with one another, even if not with human beings (Molefe 2017: 200), or it might be because of certain intrinsic properties they exhibit such as intelligence and sentience (Molefe 2017: 198–199). In any event, for Molefe, the mere fact that it is impossible both for us to commune with the Martians and for them to commune with us does not disqualify these aliens, whom we imagine to have relational capacities, from having moral status.

In reply, it is, in fact, enormously plausible to deny that these aliens would have a moral status, if indeed we 'cannot enter into any kind of interaction with each other'. If we simply could not affect the pursuit of their ends or their quality of life for better or worse, no matter what we did, then why should we think that we would have any direct duties towards these beings? What would we owe them, if we could neither interfere with them nor support their goals and if we could neither harm them nor help them?

That is not to say that these Martians would be valueless. Recall from the previous section that it is perfectly coherent to suppose that a being can have a final value, even a dignity, but not have a moral status. The Martians case is similar to the aliens case from the previous section, in which I imagined a race of intelligent and sentient beings living in a different galaxy. In both cases, since we cannot affect these alien persons in any manner whatsoever, they lack a moral status as it concerns the direct duties of human agents. We would owe them nothing, if there really would be nothing that we could do for them. I believe this is a compelling reply.

Molefe offers another version of the case, however, that does require me to clarify and acknowledge a limit to the scope of my claims. In the second version of the case, Molefe imagines that human beings have died out but that Martians have remained on Earth, and he maintains that not only Martians, but also many animals, could have a moral status in that world (Molefe 2017: 201). Martians could have direct duties towards other Martians and also towards beings such as goats, turtles, and pigeons.

In reply, I accept the point. In my work, I have been strictly interested in the moral duties of human agents and hence moral status from within a human linguistic and conceptual scheme. I have not been aiming to provide an account of moral obligations for any agent whatsoever, whether it is a Martian or God. I believe that this restriction has been fair, for two reasons, one pragmatic and the other more theoretical.

Pragmatically, it is sensible for philosophers to develop accounts of what merely *we* owe to others since, for all we know, we are the only agents with whom we are in contact. Concerns of the day for philosophers are what human agents owe to each other and what else they can affect, particularly in regards to the land on which we live. Concerns of the day for philosophers do not include what a multitude of different hypothetical agents (or agents believed to be actual on mere faith) might owe to others for their own sake.

Theoretically, and more deeply, I accept that the causal theory of reference applies to many moral terms (cf. Metz 2013a: 91–93). According to that view, part of the meaning of moral language is determined by the objective nature of behaviour that a certain group of language users have dubbed with words such as 'right', 'unjust', 'base', or the like. Similar to 'Moral Twin Earth' thought experiments (e.g., Horgan and Timmons 1991), a group of language users with different biological, social, and environmental backgrounds from ours could dub with moral terms features that differ substantially from those that we have dubbed with such terms. In that case, what counts as 'wrong', for instance, would differ between human agents and these others agents. If the causal theory of reference is true, then we have no alternative but to leave open the nature of moral status from the standpoint of other kinds of agents, in the absence of any information about how they deploy moral terms. Since I do not believe that we can address moral status for non-human agents, I have not done so, and have instead addressed only the question of what has moral status given our linguistic and conceptual frameworks.

The second charge of incompleteness on the part of modal relationism that I address here is that it cannot account for the moral status of a certain class of human beings. Specifically, Molefe (2017: 203–204) points out that my account of moral status entails that dead human beings lack a moral status, where 'dead' means the permanent termination of a person's existence (and hence we do not suppose that what are often called the 'living-dead' continue to exist). After all, the dead do not have the ability to be subjects of communion, that is, they cannot identify with others or exhibit solidarity with them. These require mental states involving an awareness of others, amongst other features lacking in a dead body. In addition, the dead do not have the ability to be objects of communion with us, for they lack goals or a good that we could affect. It does seem to follow that the dead lack a moral status on my account, having the same standing as a rock, viz., none, which Molefe finds counterintuitive. Furthermore, Molefe makes the dialectical point that I myself have at times claimed that there are regarding duties prescribing certain ways of treating the dead (e.g., Metz 2012a: 389).

One thing to note up front is that, as the literature stands, there is no comprehensive theory of moral status that accounts for the idea that we have direct duties towards dead bodies. Welfarism does not since the dead are incapable of a high or low quality of life. Rationalism does not since the dead cannot reason, exhibit intelligence, have autonomy, or the like. Subject of a life theories do not since there is no life. The same remark applies to biocentric theories, according to which all living beings have a moral status; the dead are not alive. If, therefore, modal relationalism cannot account for the moral status of the dead, it is not a strike against the theory relative to competitors. There is no reason to reject it in favour of any extant rival.

However, do dead human bodies have a moral status? At present, I am not sure.[7] I do believe that we have duties *in respect of* dead human bodies, but, *contra* what Molefe suggests (Molefe 2017: 203), that need not mean that we have duties *to* them, for the duties might be indirect instead of direct. For instance, it is plausible to hold that the moral reason not to eat or have sex with the dead is not that it would wrong the dead body, but rather that it would wrong the *living relatives and other associates* of the person who used to inhabit the body. It might also be that if we were to mistreat a dead body we would wrong, not the dead body, but rather *the person who used to inhabit it*, i.e., agent who once had the ability to exhibit identity and solidarity. If we would wrong a person who once lived by disregarding her will posthumously, it is analogously plausible to hold that we would wrong a person who once lived by failing to treat her dead body in certain ways. I find these indirect accounts of our duties in respect of the dead pretty convincing. Molefe does not consider them when mounting criticisms of my approach, but they are prima facie strong.

However, I want to note two ways to reinterpret modal relationalism to account for direct duties to the dead, if one is firm in holding that they exist. One would be to hold that moral status inheres in any being capable of being party to a communal relationship with us *or* a being that recently had that capacity but no longer does. Something like 'recently' must be part of the formulation in order to avoid strong counterexamples; for there is a moral difference of some kind between the body of a human being that died a week ago and the disintegrated carbon, i.e., more or less oil or soil, of one that died many thousands of years ago. In addition, even when it comes to a human body that has recently died, it would be implausible to ascribe it a full moral status, one comparable to that of a living person able to commune with others. Surely, if you had to choose between saving the life of a such a person and maintaining the integrity of a dead body, one should opt for the former, the best explanation of which is that the former has a higher standing (supposing the latter has a standing at all).

Another strategy, and one more inherent to modal relationalism as I have normally expounded it, would be to maintain that perhaps there is in fact a sense in which we can commune with a dead human body. It can be the object of communal relationship, not by virtue of us being able to advance its ends, improve its quality of life, or sympathize with it, as Molefe rightly points out. However, perhaps we can think of it as part of a 'we'. One part of a communal relationship, as I have most thoroughly analyzed it

[7] For a firm 'No', and on grounds of modal relationalism, see Matisonn and Muade (2023).

(e.g., Metz 2013b), includes a sense of togetherness or thinking of oneself as bound up with another. Such a thin sense of the ability to identify with a dead human body might be invoked to ground a partial moral status on its part (cf. discussion of the moral status of human embryos in Metz 2022a). It would, however, follow from this approach that a living animal has a greater (albeit still partial) moral status than a dead human body, since the former would be much more able to be the object of a communal relationship with us than the latter.

Let me turn now to a third way in which modal relationalism has been accused of failing to include certain things with a moral status. Samuel and Fayemi (2020) together have argued that, despite its relationality, my theory is ultimately still individualistic in a certain way. It ascribes direct duties only to individuals and not also to groups or wholes. Although modal relationalism is not individualistic in the sense of basing moral status on the properties intrinsic to an individual, it does base moral status on an individual's relational properties. That renders it unable to ascribe a moral status to 'species populations of non-humans' (Samuel and Fayemi 2020: 36), where a species is a collection or group as something distinct from its individual members (Samuel and Fayemi 2020: 42). Furthermore, since modal relationalism does not accord moral status to species, Samuel and Fayemi maintain that it cannot account adequately for the moral importance of biodiversity.

On this score I believe that indirect duties are a quite plausible way to account for the moral significance of both species and biodiversity. If one acted in a way that foreseeably caused the death of an entire species of beetle and did so for a trivial benefit, one would likely feel guilty most clearly about these two things: what one did to individual members of that species, and what the loss of the species means for other individual animals and human beings (perhaps including future generations), say, because of damage to an ecosystem. It is not so clear, I submit, that there would be a third instance of guilt in respect of what one did to the group as something distinct from the members that had composed it. If not, then it is implausible to think that the species is itself owed duties for its own sake, as opposed to what that group can do for individual members of other groups, animals and humans included.[8]

Unfortunately, Samuel and Fayemi fail to acknowledge the distinction between direct and indirect duties, making their key argument for ascribing moral status to species a *non sequitur*. They say this:

> The holistic view posits that a species population (say, of humans, animals, trees, plants) has standing. The holistic framing is a promising approach because it is morally implausible to confer moral standing on every individual animal as we currently do across the globe regarding individual humans, at least for prudential reasons – *we eat and feed on animals, and humans also use animal skins for clothing and leather,* and animals are a reason why we flourish (Samuel and Fayemi 2020: 38).

[8] One might well feel shame for having obliterated a species; such an action plausibly suggests that one is lacking good character in certain respects. However, shame is not nearly as reliable a marker of moral wrongness and of moral status than guilt, for shame is often sensibly triggered by non-moral considerations, such as one's poor appearance or financial poverty (Metz 2012a: 401).

This reasoning does nothing to suggest that a species or other group has a moral status, understood as being the sort of thing that is owed duties for its own sake. Pointing out that we have prudential reasons to protect species hardly means that species are owed duties for their own sake! Instead, Samuel and Fayemi themselves are in effect suggesting that merely indirect duties apply to species, meaning that there is no objection to my view whatsoever.

And what goes for species applies with the same force to biodiversity: it, too, is morally important to protect, not because it has a moral status, but rather because individual animals and humans have a moral status and we need to protect biodiversity for their sake. Samuel and Fayemi remark, 'There are situations when intervening in the ecosystems to sustain biodiversity is morally defensible. For example, we may protect certain species populations such as the beluga whales if human actions threaten their diversity' (2020: 37). I believe that too, and it follows straightaway from modal relationalism! There would be two-fold reason to protect the whales on my view: one, the individual whales matter for their own sake, and it would be wrong to frustrate their urgent interests for our trivial ones, and, two, reducing species diversity is likely to cause harm to individual humans and other animals, who also matter of their own sake. Pointing out the need to protect biodiversity does not, for all Samuel and Fayemi have said, provide reason to doubt modal relationalism.

Now, there are occasions in their essay when Samuel and Fayemi suggest that animal interests provide greater reason to act than my theory can recognize (an objection that I address in the following section). However, that is quite different from the sort of claim, considered in this section, that modal relationalism excludes a certain kind of thing, viz., an animal species from having moral status at all.

6.5 Charges of Elitism

In this last major section I address criticisms that modal relationalism, while not excluding certain beings from the domain of direct moral consideration altogether (as per those in the previous section), fails to accord them the degree of standing that they in fact have. In particular, critics have suggested that my theory objectionably prizes the interests of human beings over those of non-human animals and so counts as 'anthropocentric'.

One version of this criticism is based on an inaccurate representation of my theory, and so I begin with it. Samuel and Fayemi recognize that I ascribe moral status to animals, but they believe that I hold, or am committed to holding, 'that human interests must always come before the interests of nonhumans' (Samuel and Fayemi 2019: 91).

However, that is simply not true. Yes, I believe that beings capable of exhibiting identity and solidarity have a full moral status, while those that are merely capable of us identifying and exhibiting solidarity with them have a partial moral status. Beings who can commune with others, such as typical human persons, have the highest standing, while beings with which we can commune (but that cannot themselves commune), e.g., giraffes, sharks, and otters have a standing but a lower one. What follows from this ranking is not that 'human interests must always come first when doing so might lead to the destruction of the environment' (Samuel and Fayemi 2020: 39). Instead, in my work I have maintained that two other corollaries follow (already appealed to in previous sections). First, it would be wrong to sacrifice the urgent interests of a being with a partial moral status for the trivial interests of a being with a full moral status (Metz 2017a: 167, 169, 2017b: 275, 285–286, 2022a: 160, 162–163). Second, it would be wrong to sacrifice the urgent interests of a being with a full moral status for the urgent interests of a being with a partial moral status (Metz 2017a: 169, 172, 2017b: 274–275, 286, 2022a: 161).

Samuel and Fayemi do not acknowledge these corollaries regarding wrongness, but they make an enormous difference to understanding the implications of modal relationalism. I have suggested in various places that the first corollary entails that hunting merely for sport, killing just for the taste of meat, and inflicting pain for cosmetic testing are seriously immoral. Since the urgent interests of animals, beings with a partial moral status, may not be sacrificed for the trivial interests of humans, despite having a full moral status, we are obligated to refrain from subordinating and harming animals in these ways.

It is true that when the urgent interests of animals and humans come into conflict, the second corollary entails that those of the latter must be prioritized. However, that is quite intuitive. To invoke a case I have often appealed to, surely the reader would not think it right to shoot a person if necessary to feed a pig, but would think it right to shoot a pig if necessary to feed a person. The best explanation of the case is that persons have a higher moral status than pigs, such that, in cases where there is unavoidable conflict between the urgent interests of human persons and those of non-human animals, normally the former should be satisfied if at the necessary expense of the latter.

Having clarified some of the implications of modal relationalism, I now turn to objections that do not rest on incomplete descriptions of it. They concern the standing of human beings who lack the capacity to be subjects of communal relationship, i.e., who have mental deficiencies that prevent them from exhibiting other-regarding mental states of the sort requisite for identifying and exhibiting solidarity with others. On the one hand, some object that I do not accord them the same status as human persons, while, on the other hand, some object that I do not accord animals the same status as these human beings.

I have argued that human beings who can be merely objects of communal relationship with us have a partial moral status, albeit one greater than the moral status of animals (Metz 2012a: 397–398, 2022a: 163–165). According to my account, their standing is in between that of persons and animals. Since they lack the ability to be a subject of communal relationship, they do not have a full moral status. However, I have contended that their ability to be an object of communal relationship with characteristic human beings is markedly greater than that of animals, which means that their moral status is greater. One piece of evidence that we have a greater ability to commune with human non-persons than with animals is that we in fact commune more frequently and more intensely with them.

Given the two corollaries above, here is what follows when it comes to decision-making: the urgent interests of a human non-person should come before the trivial interests of a human person; the urgent interests of a human person should come before those of a human non-person in cases where there is an unavoidable conflict between them; and the urgent interests of a human non-person should come before those of an animal in cases of conflict. I submit that these implications are plausible, although not, I accept, downright uncontroversial.

Now, Molefe questions some of the evidence I present in favour of the claim that we are more able to commune with human non-persons than with animals. Molefe points out that '*doing more* for some being does not necessarily and always imply that she has greater moral status' (2017: 202), sensibly pointing out that doing more for one's child does not mean that she has a greater standing than other people's children.

In reply, first note my claim is not that *a particular agent* doing more for *a specific being*, such as me doing more for my child or my cat, is direct evidence that *its moral status is greater*. It is rather that *characteristic human beings* doing a lot more for *a certain kind of being* compared to another is evidence that we are *markedly more able to do something for that sort of being* relative to the other (which, in turn, entails that the former's moral status is higher, by modal relationalism). Molefe's point does not directly target my view.

Secondly, there is additional evidence that I can appeal to in order to show that we can commune to a greater degree with human non-persons than with animals, beyond the fact that generally we in fact do commune with them more often and more intensely. It is that the biological, psychological, and social nature of human non-persons is more like ours than is any animal's nature. Maj and Molefe both find this point 'speciesist', but I doubt that, since, if there were a non-human being with a nature like ours, such that we could commune with it as well as we characteristically can with human non-persons, it would follow from my view that they would have the same moral status.

Let me turn away from those suggesting that animals should be on a par with human non-persons and towards those who are naturally read as holding that human non-persons should be on a par with human persons. In particular, Samuel and Fayemi object that my approach would 'support the domination of humans who are incapacitated' (2020: 35, see also 40).

Samuel and Fayemi unfortunately do not spell out what 'domination' means. However, notice that in the case of animals my theory entails that hunting for sport, eating meat for the taste, and harming animals for cosmetic testing are all wrong. If we may not 'dominate' animals in these ways, it would be even more wrong to do similar kinds of things to human non-persons. Their urgent interests must be satisfied at the expense of at least our trivial ones, and human persons can be particularly responsible to satisfy the interests of human non-persons since the former have created the latter and the latter are unable to meet their own needs. A being does not have to have a full moral status in order to deserve real moral consideration.

It is true that if a public hospital had to make the unfortunate choice between saving the life of a human person who is able to commune as a subject and a human non-person who utterly lacks the awareness of other people altogether, I believe it should choose the former. Now, would the reader truly flip a coin in this case? If not, the best explanation is probably that there is differential moral status between these kinds of beings.

6.6 Concluding Statements on Remaining Concerns

In this chapter I have not here addressed literally all the criticisms of modal relationalism that have been made in the literature, not even by those whose essays I have discussed. I bring it to a close by mentioning some of the ones that remain and that I believe deserve a reply at some point.

One suggestion, from Maj (2020: 342, 344), is that some animals might display the ability to commune as a subject more than I acknowledge in my work. Chimpanzees and gorillas loom especially large. If they indeed can robustly act for the sake of others' ends and good, then they have a full moral status, and not human persons alone. Note that this point undercuts Maj's repeated description of my view as 'speciesist'; my view is capacity-based, and if some non-human being displays the relevant capacity, then it has the corresponding moral status. There is nothing in my view according to which human beings *as such* matter the most. That said, although from what I know of the literature there are rudimentary displays of other-regard amongst primates and some other animals, typical human persons have a markedly richer capacity for it. So, while I am currently inclined to retain the title of 'dignity' for human persons alone of Earthly beings, I accept, as I have said in my work (e.g., Metz 2012a: 400), that these animals should count as the 'highest' members of the animal kingdom (as they are often called), precisely because they approximate the ability to relate communally as subjects. Perhaps more scientific research (or awareness of it) will reveal all the greater ability in this respect and hence require a change in how to understand the implications of modal relationalism.

Other objections, or at least queries, concern how to make kinds of trade-offs that at present I find difficult to appraise—at least because of the variety of moral considerations at stake and sometimes because they seem to balance out. For instance, 'if I was driving in a brakeless car and I had to run over a psychopathic murderer or my good companion Max, my dog,' Maj wonders which should be spared (2020: 340). The case is complex, since some of the factors include moral status, but others include a partial obligation to one's pet as well as a positive obligation to prevent harm to strangers. Maj says he would save his dog, as I guess would I, if the psychopathy were extreme and more killing were on the cards; however, one need not suppose that the moral status of the killer and the dog are the same to support that judgement. Another case that I find hard to resolve concerns situations in which people's cultural practices cause harm to animals, something Maj (2020) also raises. Although human persons have a full moral status and animals a partial one, I have said above that the urgent interests of the latter may not be traded off for the trivial interests of the former. So, much depends on the details, of precisely how important the practice is to a people's self-conception and how harmful it is to the animals (for more reflection, see Metz 2017b). A conception of moral status cannot be expected to provide all the information relevant to deciding whether certain actions are wrong, which can be a complicated matter. Such a conception is, however, essential for making decisions about wrongness, and in this chapter I have argued that modal relationalism is much more promising than several critics have maintained.

References

Anyanwu, K.C. 1984. The Meaning of Ultimate Reality in Igbo Cultural Experience. *Ultimate Reality and Meaning* 7 (2): 84–101.

Behrens, K. 2017. A Critique of the Principle of 'Respect for Autonomy', Grounded in African Thought. *Developing World Bioethics* 18 (2): 126–134.

Bujo, B. 2001. *Foundations of an African ethic*. Translated by B. McNeil. New York: Crossroad Publishers.

Cordeiro-Rodrigues, L., and C. Ewuoso. 2022. An Afro-Communitarian Relational Approach to Brain Surrogates Research. *Neuroethics* 14 (4): 561–574.

Etieyibo, E. 2017. Anthropocentricism, African Metaphysical Worldview, and Animal Practices. *Journal of Animal Ethics* 7 (2): 145–162.

Horgan, T., and M. Timmons. 1991. New Wave Moral Realism Meets Moral Twin Earth. *Journal of Philosophical Research* 16: 447–465.

Horsthemke, K. 2015. *Animals and African Ethics*. New York: Palgrave Macmillan.

Horsthemke, K. 2017. Biocentrism, Ecocentrism, and African Modal Relationalism. *Journal of Animal Ethics* 7 (2): 183–189.

Ikuenobe, P. 2016. The Communal Basis for Moral Dignity. *Philosophical Papers* 45 (3): 437–469.

Imafidon, E. 2014. On the Ontological Foundation of a Social Ethics in African Tradition. In *Ontologized Ethics*, ed. E. Imafidon and J.A.I. Bewaji, 37–54. Lanham, MD: Lexington Books.

Kagan, S. 1998. Rethinking Intrinsic Value. *The Journal of Ethics* 2 (4): 277–297.

Korsgaard, C. 1983. Two Distinctions in Goodness. *The Philosophical Review* 92 (2): 169–195.

Magesa, L. 1997. *African Religion: The Moral Traditions of Abundant Life*. Maryknoll, NY: Orbis Books.

Maj, F. 2020. The Animal Other in Thaddeus Metz's Modal Ubuntu Ethics. In *Handbook of African Philosophy of Difference*, ed. E. Imafidon, 333–346. Cham: Springer.

Matisonn, H., and N.E. Muade. 2023. Research on Dead Human Bodies: African Perspectives on Moral Status. *Developing World Bioethics* 23 (1): 67–75.

Metz, T. 2010. For the Sake of the Friendship: Relationality and Relationship as Grounds of Beneficence. *Theoria* 57 (4): 54–76.

Metz, T. 2012a. An African Theory of Moral Status. *Ethical Theory and Moral Practice* 15 (3): 387–402.

Metz, T. 2012b. African Conceptions of Human Dignity. *Human Rights Review* 13 (1): 19–37.

Metz, T. 2013a. *Meaning in Life*. Oxford: Oxford University Press.

Metz, T. 2013b. The Western Ethic of Care or an Afro-Communitarian Ethic? *Journal of Global Ethics* 9 (1): 77–92.

Metz, T. 2015. An African Egalitarianism. In *The Equal Society*, ed. G. Hull, 185–208. Lanham, MD: Rowman & Littlefield.

Metz, T. 2017a. How to Ground Animal Rights on African Values. *Journal of Animal Ethics* 7 (2): 163–174.

Metz, T. 2017b. Duties Towards Animals Versus Rights to Culture. In *Animals, Race, and Multiculturalism*, ed. L. Cordeiro-Rodrigues and L. Mitchell, 269–294. New York: Palgrave Macmillan.

Metz, T. 2019. Reconciliation as the Aim of a Criminal Trial. *Constitutional Court Review* 9: 113–134.

Metz, T. 2022a. *A Relational Moral Theory: African Ethics in and Beyond the Continent*. Oxford: Oxford University Press.

Metz, T. 2022b. Virtue in African Ethics as Living Harmoniously. In *Harmony and Virtue*, ed. C. Li, and D. Düring, 207–229. New York: Oxford University Press.

Molefe, M. 2017. A Critique of Thad Metz's African Theory of Moral Status. *South African Journal of Philosophy* 36 (2): 195–205.

Organization of African Unity. 1981. African ('Banjul') Charter on Human and Peoples' Rights. https://au.int/en/treaties/african-charter-human-and-peoples-rights.

Samuel, O., and A. Fayemi. 2019. Afro-Communal Virtue Ethic as a Foundation for Environmental Sustainability in Africa and Beyond. *South African Journal of Philosophy* 38 (1): 79–95.

Samuel, O., and A. Fayemi. 2020. A Critique of Thaddeus Metz's Modal Relational Account of Moral Status. *Theoria* 67 (1): 28–44.

Taylor, C. 1992. *The Ethics of Authenticity*. Cambridge, MA: Harvard University Press.

Wiredu, K. 1990. An Akan Perspective on Human Rights. In *Human Rights in Africa*, ed. A.A. An-naim and F.M. Deng, 243–260. Washington D.C.: Brookings Institution Press.

Thaddeus Metz is Professor of Philosophy at the University of Pretoria. He is known for drawing on the African philosophical tradition analytically to address a variety of contemporary moral/political controversies. Metz has had more than 300 books, chapters, and articles published, including: 'Recent Work in African Philosophy' *Mind* (2021); 'Traditional African Religion as a Neglected Form of Monotheism' *The Monist* (2021); and *A Relational Moral Theory: African Ethics in and Beyond the Continent* (Oxford University Press 2022). Metz was once designated one of 'The World's Top 50 Thinkers' by *Prospect Magazine* for having helped bring African philosophical ideas to global audiences.

Chapter 7
The Phenomenon of Male and Female Crops in Igbo Agrarian Culture: Implication for Gender Equality

Anthony Uzochukwu Ufearoh

Abstract This chapter is an investigation into the phenomenon of gendering of crops in traditional Igbo-African agrarian culture in order to determine the extent to which it promotes gender equality. The approach is both phenomenological and analytical. The philosophical tool of hermeneutics is equally employed in interpreting the cultural symbols in the study. The focus is on the stable crops, their cosmogonic myth and the religious feasts associated with the crops: yam and New Yam Festival, for male crops and, cocoyam/Pumpkin with their attendant feasts for female crops. The paper X-rays the intersection between this gendered agrarian praxis imbued with onto-anthropological elements and gender equality in Igbo society. It is discovered that the phenomenon of male and female crops creates equal and enabling space for complementarity and participation of both sexes not only in the agro-economic but also in the socio-political and religious arenas. The paper therefore submits that the phenomenon of gendering crops enhances gender equality.

Keywords Crops · Igbo · Gender · Equality · Complementarity · Festival

7.1 Introduction

The present chapter examines the phenomenon of gendering of crops in traditional Igbo agrarian culture in order to determine the extent to which it could fecundate better relationship between the male and female gender in today's society. Amidst the eloquent feminist cry of purported gender inequality and polarization, it would seem, prima facie, that the female gender is marginalized in Igbo patriarchal culture. It has also been observed that the "effect of patriarchy and gender, and indeed the contributions of African women, have up until now been largely ignored by critics" (Stratton 1994: i). While gender can be a means of differentiation, it is most often

A. U. Ufearoh (✉)
Department of Philosophy, Nnamdi Azikiwe University, Awka, Nigeria
e-mail: au.ufearoh@unizik.edu.ng

M. J. Tosam and E. Masitera (eds.), *African Agrarian Philosophy*, The International Library of Environmental, Agricultural and Food Ethics 35,
https://doi.org/10.1007/978-3-031-43040-4_7

believed that the binary division it creates privileges one half. The attendant inequality tells negatively on the status, self-determination and overall wellbeing of the under-privileged gender. This, understandably, is a malady that plagues many societies. In search of possible panacea to the above predicament, the present chapter examines the phenomenon of gendering in Igbo agrarian culture. It aims at distilling therefrom, principles that could enhance gender equality.

The practice of binary classification or gendering of stable crops in Igbo agrarian culture, the cosmogonic myth and concomitant religious feasts associated with these crops say much about the status of both the male and female gender. This derives from the fact that the crops, as buttressed by their myth of origin, in a way serve as metaphor for the human. What significance or implication does this classification have for gender equality in the society? This is the major engagement of this chapter.

The chapter adopts the philosophical methods of phenomenology and analysis. The work builds on personal experiences of the researcher himself, having been born and reared in a rural agrarian Igbo village in Eastern Nigeria. This qualifies the work as an "insider account". The chapter equally relies on written account of life in the traditional Igbo community. African literatures such as novels that reflect the Igbo agrarian culture also serve as fodder for the present research. As part of the phenomenological approach, the authors are sometimes allowed to speak for themselves. The Igbo myth of origin provides the (meta) historical background for the practice of gendering of crops. The paper goes beyond phenomenological presentation to interpretative (hermeneutic) excursus in order to draw the undergirding principles and its implications on the lived experiences of the people in the Igbo society.

Accordingly, in the first section, I consider the concept of gender equality/equity. The Sect. 7.3 presents the practice of crop gendering in Igbo culture. The praxis, cosmogonic origin and the agricultural festivals and ritualization of the crops are here highlighted. In the Sect. 7.4, I sieve out the principles that undergird the gendering of crops in Igbo agrarian culture. The implications for gendering of crops and gender equality are discussed in the Sect. 7.5. The Sect. 7.6 is the conclusion.

7.2 The Concept of Gender Equality

Gender equality, in a way, can be said to be an off-shoot of unequal binary created by cultural and historical factors between male and female sexes. Understandably, when a new born is delivered, it is either a girl or a boy. Period! Each of these two comes with complexities of biological, hormonal and psychological trappings. Gradually, come gender roles and expectations which are mostly dictated by socio-cultural values and environment and are subject to change with time. Hence, the assertion that, "gender is socially constructed" (Butler 1999: x). It is believed that gendering as a means of classification easily slips into or serves as springboard and tool of oppression by way of privileging one of the binary components over the other. The imbalance comes in various incarnations such as social, political, economic and available opportunities. All manner of marginalization, exclusion and denigration based on sex are manifestations of the above inequality. It has been variously argued that women, more than men, have historically borne the brunt of the systematic gender inequality. This therefore calls for a process or means that guarantees fairness for all. This is gender equity. Can all genders be equal? The answer is yes and no! Human beings, considered as a species of animal gifted with rationality, are ontologically equal. Conversely, ontological equality does not translate to biological equality. All fingers are not equal. It is the opinion of the present researcher that ontological equality serves as the ground norm, indeed, the justification for equality of the sexes.

The thesis that traditional patriarchal societies or culture privileges the male gender and marginalizes the feminine gender suffers from hasty generalization. The submission simply overlooks what Nkiru Nzegwu calls, "cultural specificity and the historicity of societies" (2004: 559). In Igbo traditional setting one finds complexities of gender ideologies and roles. The Igbo agrarian culture as we shall see, presents a melting pot of plurality, unique but complementary existence of female and male crops which metaphorically stand for the mutual sharing of spaces, labour, representation and co-existence between man and woman in the Igbo world. Indeed, the search for equity or equality here might seem superfluous given the rootedness of the representation in the cosmogonic myths and the metaphysics of balance that undergird every relation in the Igbo worldview.

7.3 Male and Female Crops in Igbo Agrarian Culture: The Praxis, the Myth and Ritualization

This section X-rays the myth, the ritualization and the praxis of gendering of crops which is part and parcel of the traditional and existential Igbo agrarian culture. Gendering of crops, in a literal language, means that, while some crops are designated as male, others are regarded as female. Yam, for instance, is regarded as male crop and King of crops whereas some other crops like cocoyam and pumpkin, are seen as female. The gendering is observed more in the taxonomy of Igbo mystical trees whose symbols and their implicit meanings may not be apparently clear to uninformed observers. For example, *Chrysophyllum Albidum* (*udala* tree) stands as symbol of fertility and motherhood in most parts of Igbo land. This is a quintessential feminine tree, treated with great respect. Yam tuber (*Dioscrea*) stands as symbol of providence. The gendering extends to inanimate entities. Rain stone which is an indispensable instrument in rain making has both female and male counter-parts and each has got its respective functions. A few words about the Igbo seem ad rem here.

The Igbo is one of the three major ethnic nationalities in Nigeria. Though highly adventurous and enterprising, inclined to traveling far and wide in search of greener pasture, they predominantly occupy the five South Eastern states of Nigeria where they are the aborigines. Theirs is largely and fundamentally an agricultural community comprising mainly of subsistent farmers with a few exceptions that engage in fishing and wine tapping in riverine areas, smiting, handicraft etc. A sizeable portion of the Igbo population today engage in business merchandise and white-collar jobs (Nwoye 2011: 306). As in many traditional societies, agriculture here serves as the basic means of sustenance. Remarkably, success in agriculture is largely believed to be dependent on the Transcendence or supernatural influence. Good harvest is inextricably premised on fidelity to these unseen forces. Gregory Adibe writes that "the Igbo primordial times were agrarian and pastoral before the arrival of Western civilization. The economic life and existence of peasants were dependent on nature, which are dependent on supernatural forces" (2008: 5). Agriculture thus becomes the pivot around which economic and political and somehow, religious power is structured and measured.

Religion remains highly significant in the Igbo society today. The influence of Christianity notwithstanding, the resilient indigenous occupational or agricultural festivals still hold overwhelming sway in contemporary Igbo society and beyond the borders of Nigeria. The Igbo celebrate New Yam Festival in countries of Europe, America, and other areas where they reside. As we shall see, the Igbo believe in the complementary roles of opposites. On this note, Innocent Asouzu avers that anything that exists is a missing link of reality (2011: 41). The farm crops are not exempted.

Accordingly, one of the phenomena that confront one in traditional Igbo agrarian culture is the gendering of crops in such a way that they co-exist in a perfect combination in the farm, thanks to share-cropping and are taken care of by their respective human counterpart. While men cultivate male crops, women plant female crops. As the crops complement each other in the farm, so too on the dinner table! A combination of pounded yam (male crop) with cocoyam soup (female crop, with bitter leaf or similar vegetable) makes a perfect Igbo dish widely sought after. A tantalizing account of this pre-eminent food in Igbo culture is given in chapter five of Achebe's *Things Fall Apart*. Similarly, among the African literatures that captured the praxis of gendering of crops, Chinua Achebe's *Things Fall Apart* and *Arrow of God*, seem to be the most outstanding. The texts reflect the raw and unadulterated life in Igbo society. The title of the present chapter was informed by a line from Achebe's *magnus opus* (*Things Fall Apart*) where he wrote that, Okonkwo's mother and sisters "grew women's crops, like cocoyam, beans and cassava. Yam, the king of crops, was a man's crop" (1958: 16).

On another note, from an anthropological perspective, one can tacitly draw the inference that, using the gendered crops, Achebe metaphorically presents the communal, mutual and complementary human functioning and division of labour in traditional Igbo agriculture. The complexity of human existence is such that human beings need the services of each other at one point or the other. Each has got areas of strengths and weakness, hence the need for complementarity. Achebe wrote that while the men undertake the exacting task of cultivating yam, the women plant maize, melons, beans and weed the farm at intervals (1958: 24). The above presentation offers a bird's eye view of the practice of share-cropping and division of labour in a typical Igbo farm. Moreover the division of labour as presented above is not done in cast iron. It rather presents what is fashionable in Igbo society. Any crop can be cultivated by either men or women. Karen Warren is therefore wrong to have described extensive involvement of African women farmers in cultivation and processing of cassava as "plight" (2000: 9). The arrangement is simply guided by the spirit of gendering of crops and division of labour in African agrarian cultural setting.

The practice of gendering of crops is deeply rooted in the Igbo cosmogonic myth. Adiele Afigbo (1981: 41–42) describes the origin of the stable foods: yam, cocoyam, oil palm and bread fruit, in these lines:

> While Eri lived, *Chukwu* [God] fed him and his people with *azu-igwe!* But this special food ceased after the death of Eri. Nri his first son complained to *Chukwu* for food. *Chukwu* ordered Nri to sacrifice his first son and daughter and bury them in separate graves. Nri complied with it. Later after three-Igbo-weeks (*Izu atọ* = 12 days) yam grew from the grave of the son and cocoyam from that of the daughter. When Nri and his people ate these, they slept for the first time; later still Nri killed a male and female slave burying them separately. Again, after *Izu Ato*, an oil palm grew from the grave of the male slave, and a bread fruit tree (*ukwa*) from that of the female-slave.

The above account shows that the stable food did not emerge *ex nihilo*. Gendering plays out in the myth in such a way that the immolation of male child gave rise to male crops, while the immolation of the female child yielded the female crops. The male and female are equally involved and represented. Igbo agricultural festivals commemorate the above primordial event and serve as avenue for thanksgiving for the gift of the stable crops. And more, for a group that relies heavily on the supernatural for good harvest, appreciation for good harvest.

Agrarian festivals include New Yam Festival, Cocoyam Festival, Festival of Pumpkin Leaves, etc. Ritually speaking, by virtue of these crops (yam, cocoyam, oil palm, breadfruit) serving as the indispensable matter for the festival, their ontological status ambivalently changes from mere edible crops to sacramental objects. For instance, Achebe's account of the Feast of Pumpkin Leaves describes how "the women waved their leaves from side to side across their face, muttering prayers to Ulu" (2010). The leaves of pumpkin thus become ritual objects serving for spiritual purification. The New Yam festival seems to be the most widely celebrated agricultural festival. According to Punch Newspaper, "The age-long festival is celebrated across Nigeria and the world wherever the Igbo people live" (2019). The annual New Yam festival is celebrated in thanksgiving to the yam deity for the gift of yam and good harvest. Its corollary is Cocoyam Festival. This is popularly known as Festival of Ede Opoto or Ede Aro in Dunukofia and Njikoka councils of Anambra state, Nigeria, from where the present researcher comes. In Abba community of Nijikoka Council, the Cocoyam Festival is dedicated to the most portent goddess called *Oyi*. Though cassava is a female crop and widely eaten in Igbo land, it does not have any festival. The reason may not be unconnected with the fact that it is not indigenous to the Igbo, having originated from South America and not dedicated to any traditional Igbo goddess. According to Josef Pieper (1963: 33), a festival without gods is simply inconceivable.

7.4 The Metaphysics of Gendering Crops in Igbo Agrarian Culture

For a better appreciation of the implications for gender equality, one must X-ray the principle(s) that undergird the phenomenon of male and female crops or gendering of crops. As already stated above, the Igbo believe in the metaphysical principle of complementary duality and balance. This is the principle of mutual co-existence and complementary roles of opposites. This metaphysics is informed by the Igbo understanding of reality as force, dynamic and interdependent. This is unlike Western ontology that perceives being as static and isolated. No reality exists in isolation. Entities do, in fact, most times exist in copula and complementary configurations such as: *muo na madu* (spirit and human); *elu na ala* (the sky and the land); *nwoke na nwanyi* (male and female) et cetera. This calls for balance. On the need for balance, Nwoye (2011: 309) states that, "Igbo cosmology places emphasis on the importance of striking a balance between masculine and feminine principles." The dynamic ontology of complementary duality and balance therefore informs the gendering of crops.

The above principle equally plays out in the cosmocentric configurations of Igbo deities where some parts of the earths' crust are deified in Igbo traditional religion. The sky is a male god whereas the land is female (goddess). The earth goddess plays a significant role as the custodian of morality and fertility. She is inextricably linked with productivity and harvest. According to Nwoye, "in Igbo religious worldview, key areas, such as land, river, hills, forests, caves, are believed to be controlled by female deities. Such sites are also connected with agriculture, fertility, morality, mores, beauty, and blessings" (2011: 308). The intersection between the gods, morality and productivity in African agrarian culture is an area that calls for further research which is beyond the scope of the present work. As it relates to agriculture, it is believed that the male and female cosmological forces (the sky and the land) must necessarily harmonize to facilitate better yield in the farms. Meanwhile, it is quite informative to know that unlike some other religions such as Christianity and Islam that abysmally lack female representation in their divine configurations, Igbo theogony has pride of place for the female and there also exist priestesses in Igbo traditional religion. A total eclipse of female presence makes for an imbalanced theogony which invariably, tells negatively on the status of the female in the society. Little wonder why Okonkwo, the protagonist of *Things Fall Apart*, queried the missionaries: "You told us with your own mouth that there was only one god. Now you talk about his son. He must have a wife, then" (Achebe 1958: 103). Male and female ought to be represented so as to create the balance needed for gender equity.

Given the above metaphysics of complementary duality and balance which undergirds relationships in Igbo worldview; and the involvement of the male and female in the cosmogonic origin of the various stable crops in Igbo land, it is logical enough to infer that the phenomenon of gendering of crops has deep foundations in Igbo traditional agrarian culture. This praxis not only gives room for proper representation, but also enhances division of labour. This leads to economic empowerment in a society that thrives predominantly on an agrarian economy. Since economic power is in a way political power, the gendering of crops gives political voice or power not only to the male but also to the female. Religion portrays the culture of a people in its nakedness. The use of the crops as ritual objects at the agricultural festivals elevates the ontological status of the crops. Inclusion of the male and female crops in the Igbo festal calendar not only makes for balanced representation but also affords the rest and leisure which humans direly need to balance the long periods of work.

7.5 Male and Female Crops: Implications for Gender Equality

The quest for gender equality is propelled by principles that make for balanced and fair treatment of both male and female sexes. From the above exposé on the phenomenology of male and female crops, one can glean the following implications:

Balanced Representation (Inclusiveness): Reality is concatenated. There is equal representation of both male and female sexes in the cosmogonic myth, the actual praxis in the farms, the festal calendar. This makes for a sense of belonging. Inclusiveness is a *conditio sine qua non* for gender equity. Discriminatory representation leads to gender inequality.

The Principle of Complementarity: This plays out in the phenomenon of gendering of crops in Igbo agrarian culture. The male crops serve to complement the female crops. Division of labour in the field, as we saw above equally toes the line of complementarity of efforts. This implies that there ought to be combination and complementarity of efforts from the male and female gender in order to achieve set goals and objectives in the society.

Autonomy: creation of sense of autonomy is imperative in actualization of gender equality. Equal space must be given for the genders, firstly, to be whom they are. In the language of Martin Heidegger, this entails, "letting being be" (2002). Secondly, there should be adequate space and motivation for self-actualization without undue interference and constraints.

Empowerment: None of the binary components (male and female) should be unduly disadvantaged by virtue of sex. The phenomenon of gendering of crops in Igbo agrarian culture makes for empowerment of both genders via agriculture which is the mainstay of the traditional economy. It simultaneously offers both sexes economic, political and religious voices. The lesson for the modern society is that, adequate platform should therefore be created for empowerment of men and women without discrimination.

7.6 Conclusion

This chapter has examined the phenomenon of gendering of crops in Igbo agrarian culture in search of principles that could fecundate gender equality. The work X-rays the cosmogonic myth, the praxis and principles that undergird the gendering of crops. From the above discourse, the chapter brings together the values of inclusiveness, complementarity, autonomy, and empowerment as core principles derivable from the phenomenon of male and female crops in Igbo agrarian culture. These are vital ingredients for gender equality. A good application of these principles and values in the social, economic, political, and religious arena can serve to enhance gender equality.

References

Achebe, C. 1958. *Things Fall Apart*. Ibadan: Heinemann.

Achebe, C. 2010. *Arrow of God*. London: Penguin Books.

Adibe, G.E. 2008. *Igbo Mysticism: The Power of Igbo Traditional Religion and Society*. Onitsha: Trinity Press.

Afigbo, A.E. 1981. *Ropes of Sand: Studies in Igbo History and Culture*. Nsukka: University of Nigeria Press.

Asouzu, I. 2011. *"Ibuanyidanda" and the Philosophy of Essence (Philosophy The Science of Missing Links of Reality) (Inaugural Lecture)*. Calabar: University of Calabar Press.

Butler, J. 1999. *Gender Trouble: Feminism and the Subversion of Identity*. New York: Routledge.

Heidegger, M. 2002. *The Essence of Truth: On Plato's Cave Allegory and Theaetetus*. London: Bloomsbury.

Nwoye, C.M.A. 2011. Igbo Cultural and Religious Worldview: An Insider's Perspective. *International Journal of Sociology and Anthropology* 3 (9): 304–317.

Nzegwu, N. 2004. "Feminism and Africa: Impact and Limits of the Metaphysics of Gender." *A Companion to African Philosophy*, edited by Kwasi Wiredu, Blackwell, Malden. 560–569.

Pieper, J. 1963. *In Tune With the World: A Theory of Festivity*. Harcourt, Brace & World, New York.

Punch-Online. 2019. How Ekweremadu was Attacked with Stones, Yam in Germany. https://punchng.com/how-ekweremadu-was-attacked-with-stones-yam-in-germany/. Accessed 14 Jan 2022.

Stratton, F. 1994. *Contemporary African Literature and the Politics of Gender*. London: Routledge.

Warren, K.J. 2000. *Ecofeminist Philosophy: A Western Perspective on What It Is and Why It Matters*. Toronto: Rowman & Littlefield.

Anthony Uzochukwu Ufearoh Ph.D is a senior lecturer recently transferred from the Department of Philosophy, University of Calabar, Cross-River State to Nnamdi Azikiwe University, Awka, and Anambra State, Nigeria. He is also an adjunct senior lecturer in philosophy at Chukwuemeka Odumegwu Ojukwu University, Igbariam, and Anambra State, Nigeria. Dr. Ufearoh has published a number of articles in peer-reviewed journals. His areas of specialization/interest include metaphysics, African philosophy, environmental philosophy and the philosophy of Hannah Arendt. He is a member of Association of Philosophy Professionals of Nigeria (APPON) and Association for the Promotion of African Studies (APA).

Chapter 8
The Religious Significance of Mushrooms Among the Shona People of Zimbabwe: An Ethnomycological Approach

Bernard P. Humbe

Abstract One of the world's biggest challenges in contemporary times is to secure sufficient, healthy, safe, and high quality food for all in an environmentally sustainable manner (Zeleke et al. in Forests 11(875):1, 2020). This challenge is especially noticed in Africa, a continent characterised by hunger, food insecurity and under nutrition (Fernandes et al. in Appl. Sci. 11, 4221:1–27, 2021). As a response to this problem, this chapter focuses on wild mushroom (*howa*) in indigenous Shona culture. However, the importance of mushrooms in Zimbabwean indigenous religious and agrarian culture has remained underexplored in religion studies. The chapter employed African environmental ethics to explore the emergence and endurance of alternative food systems like wild mushroom. In this chapter, an ethnomycological work was done using a phenomenological approach, interviews and observations in rural areas of Bikita and Buhera. It showed that mushrooms have a religious significance. The sacredness of mushrooms was unpacked based on the beliefs and values connected to the place where they sprout, their colour, shape, age, sex of the person who discovers them, how they are harvested and what drives human beings to preserve the environment with wild mushrooms. I contend that in the discourse on African agrarian thought, it is necessary to rethink indigenous African environmental values in the consumption of wild mushrooms and how this contributes to the sustainability of the country's agricultural economy.

Keywords Ethnomycology · Ethnobotany · Fungi · *Howa* · Sacred world · Taboos · Wild mushroom

---

B. P. Humbe (✉)
Research Institute for Theology and Religion, College of Human Sciences, University of South Africa (UNISA), Mbombela, South Africa
e-mail: bhumbe@gzu.ac.zw

University of Religions and Denominations (URD), Pardisan, Qom, Iran

M. J. Tosam and E. Masitera (eds.), *African Agrarian Philosophy*, The International Library of Environmental, Agricultural and Food Ethics 35,
https://doi.org/10.1007/978-3-031-43040-4_8

8.1 Introduction

One of the world's biggest challenges in contemporary times is to secure sufficient food that is healthy, safe and of high quality for all in an environmentally sustainable manner (Zeleke et al. 2020: 1). This is especially noticed in Africa, a continent characterised by hunger, food insecurity and under- nutrition (Fernandes et al. 2021). Agriculture and food production is the main sector of occupation for the majority of indigenous people in Zimbabwe. Agriculture is shaped and sustained by African environmental values which are based on holistic and all-encompassing (Tosam 2019) outlook about the cosmos. In religious and philosophical studies, nutritional attributes of indigenous cuisine remain extensively underexplored. It is against this background that this chapter concentrates on the indigenous cultural importance of food in Zimbabwe. This is because in contemporary times, there has been a steady loss of content of food-culture relationship. Using an ethnomycological perspective, the chapter identified wild mushrooms as vital components of the livelihoods of indigenous people in Zimbabwe. A wild mushroom, also known as *howa* or *hohwa* in Shona, is both an edible and inedible fungus. It grows naturally during the rainy season in areas of high humidity and rainfall (https://www.pindula.co.zw ›Mushroom_in_Zimbabwe). Ethnomycology is the study of the use of fungi by humans and the relationship between traditional societies and fungi (Thangaraj et al. 2017). From an ethnomycological standpoint, there are religio-cultural beliefs and practices associated with wild mushrooms in the Shona indigenous communities (Ndemanu 2018: 71). Such beliefs and practices are also deeply-rooted in African environmental holism which can be perceived in African metaphysico-religious and moral outlook regarding nature (Tosam 2019). There is an inescapable connection between traditional African religions and the people's ways of being and relating to the environment to a point that it is nearly impossible to extricate oneself from it without strong feelings of stripping off a major part of one's cultural identity. Accordingly, any attempt at studying indigenous African peoples and their environment without considering their religions, the bedrock of their cultural view of nature, would be shallow and futile (Ndemanu 2018: 71).

Since earliest times, the spiritual value of wild mushrooms has been recognised; they are considered as an extraordinary kind of foodstuff. For example, the Greeks believed that mushrooms provided strength for soldiers in battle, while the Romans considered mushrooms as the "Food of the Gods" and served mushrooms only on festive occasions (Singha et al. 2020: 3161). In Ancient Egypt, mushrooms were regarded as plants of immortality which were given to people by the god Osiris. Because of their unique taste, mushrooms were proclaimed as a food reserved only for Egyptian royalty. Common people were not only prohibited from eating them, but even forbidden to touch them (Kotowski 2019). It is in this line of thinking that wild mushrooms' edibility and spiritual properties make them socially important for investigation in Zimbabwean agrarianism. However, it should be pointed out from the outset that as the country becomes more economically developed, it extemporaneously swings to Western- oriented foods understood to be rich in saturated fats, salt, sugar and sweeteners. Despite this growing acceptance of "Western" eating habits (Fernandes et al. 2021: 2), the traditional food choice of wild mushroom in Zimbabwe has prevailed and still considered beneficial in terms of nutrition, health outcomes and ecological sustainability. People in the two different districts under study (Bilita and Buhera) are in contact with the same natural resources and have a similar traditional knowledge of wild mushrooms. This similarity is the result of their religio-cultural exchanges, coexistence, and shared historical experiences which are hinged on some indigenous African environmental norms. This chapter argues for the need to rekindle indigenous African environmental values as a means of achieving a fertile human-food relationship across generations for sustainable development. This is because each society's perception of nature profoundly influences the way it treats it, and how a society relates with nature can either enhance or degrade the health of the environment (Tosam 2019). The case of wild mushrooms is an example.

This research was carried out using a poly-methodical approach. An explorative ethnomycological survey was qualitatively conducted between January 2020 and December 2021. Because of the nature of this study, a site-based approach was adopted designed to generate a representative or stratified sample for qualitative research on wild mushrooms in large community-based studies (Arcury and Quandt 1999: 129). Purposive sampling method was utilized in the identification and selection of 32 households from 4 villages in 2 Wards from Buhera and Bikita districts. The selected participants who spoke Shona were aged between 20 and 85. They served a specific purpose of finding the religious significance of wild mushrooms among the Shona people of Zimbabwe. Entrenched in this site-based approach and purposive sampling was the notion that the study placed great importance on who a person is and where that person is located within a group (Arcury and Quandt 1999: 129). The ideal target audience for the research had what Bourdieu (1986: 105) calls 'cultural competence' to understand the traditional religious perceptions of wild mushrooms in Zimbabwe's rural communities. Cultural competence included the 'forms of skill and knowledge which enabled participants to make sense' of the religious significance of mushrooms among the Shona people.

Data from the two chosen districts of Buhera and Bikita was gathered using semi-structured interviews and observations. These two districts are located in the same geographic region. The reason for choosing them was that the climatic condition of the region is favourable for wild mushroom development and diversity. The interview guide had loosely planned open-ended questions to get a preliminary list of wild edible and non-edible wild mushrooms in Buhera and Bikita districts. In a bid to employ a grounded approach during the study, forest excursions were also undertaken to investigate the habitat of wild mushrooms and their uses. The interviewees provided information about the identification methods used when harvesting wild mushrooms. They also required participants' knowledge about the different uses of wild fungi and their religious significance which are linked to the Shona traditional worldview. In addition to interviews, observations were made during the field work, paying attention to identification of the wild mushrooms, edibility, preparation and medicinal applicability in the context of participants' cultural practices. In this study, the framework of African environmental ethics was built on the interpretation of anthropocentrism and ecocentrism. While an anthropocentric mindset predicts a moral obligation only towards other human beings (Rülke et al. 2020), ecocentrism includes all living beings where people are inseparable from inorganic/organic nature that encapsulates them (Fang et al. 2023). The two approaches influence the perception of nature and its protection and, therefore, have an effect on the nature-related attitude (Rülke et al. 2020). During the study, observations did not solely involve watching subjects and objects of study, they also involved asking questions to ensure that the interpretation of what was observed about wild mushrooms was what really transpired in the traditional religious worldview of the Shona people. Though there are various types of edible wild mushrooms in Zimbabwe, detailed focus was paid on *dindindi, firifiti, howachuru, dare* and *nhedzi*. The interpretive paradigm allowed the study to view the traditional religious significance of the wild mushrooms through the perceptions and experiences of the participants.

Although wild mushrooms are highly valued in Zimbabwe, not all are suitable for human consumption. Within local culture, there is a well-established criterion for determining the edibility and non-edibility of wild mushrooms which is informed by the people's cultural beliefs and customs. This is embedded in their language and naming systems. In some instances, edible mushrooms are just called *howa* while poisonous mushroom is identified as *howamupengo*, translated to mean dangerous or harmful mushroom. To be more precise, all non-edible mushrooms are simply known as *howamupengo*. Some of the known poisonous wild mushrooms include the following: *mutekeramavu* (*Russula*), *davupfu* and *chivandikira* (https://www.pindula.co.zw ›Mushroom_Poisoning_in_Zimbabwe). The indigenous ethnomycological knowledge or skills to identify wild mushroom take cognisance of the following traits: All participants confirmed that edible mushrooms are usually found during the rainy season, particularly between November and March. The elderly respondents revealed the habitat of the wild mushrooms as a determining factor to consider when distinguishing edible from non-edible mushrooms. There are certain places which are suitable for the growth of edible mushroom; for example natural forests, mountains, anthills, farmlands, grazing lands and home gardens. Some participants were very specific in stating that when they are in the forest, they only collect wild mushrooms which grow near Muzhanje (Uapaca kirkiana) and Msasa (Brachystegia spiciformis) trees. However, other participants clarified that mushrooms which sprout in gum tree plantations are poisonous even if the type of mushroom looks familiar and non-poisonous.

Varied critical opinions have been expressed about the importance of wild mushrooms. Discourses on wild mushrooms reveal that scholars have shown profound interest on their commercial and medicinal benefits. A history of mushroom consumption and its impact on traditional views on mycobiota focusing on Poland is explored by Kotowski (2019). Kotowksi's study acknowledged that the 2015 study conducted by the research group connected with *Max Planck Institute for Evolutionary Anthropology* in Leipzig proved the presence of mushrooms in human diet as early as the Upper Palaeolithic Period. The discovery emerged from the examination of dental calculus from teeth found in the Lower Magdalenian burial of a woman, also known as the "Red Lady", where ochre-covered remains were found in 2010 at El Mirón cave in Cantabria (Nothern Spain). The inclusion of wild mushrooms is also a common feature in African diet. In the African context, Ndifon (2022) has done a systematic review to determine the state of mushroom-related indigenous knowledge in Southern Africa. This indigenous knowledge has been seen to be vital in determining the importance of wild mushrooms as a source of food and medicine. This is attested in Dube et al. (2021) which evaluated the importance of wild edible mushrooms as food in three villages of Binga, Zimbabwe. The understanding of wild mushrooms is enhanced by Piearce and Sharp's (2000) stipulation of indigenous names of Zimbabwean fungi. In the context of this study, the practice of giving names to wild mushrooms constitutes an existential script that underscores the sanctity of this natural food resource. The indigenous names have appeared in a number of publications dealing with edible and poisonous fungi in Zimbabwe. Despite this wild mushrooms naming system, there is always a problem in differentiating edible from non-edible mushrooms in Zimbabwe. Elsewhere in the United States of America, a very penetrating contribution on this issue is made by Ostry et al. (2010) whose advice is that mushrooms should always be dug, not picked, in order to detect the cap feature of potentially poisonous mushrooms.

Zeleke et al. (2020) documented the traditional knowledge and uses of wild mushroom species from three ethnic groups in Ethiopia which are the Amhara, Agew and Sidama. However, Zakele et al. recommended that more studies are needed to ensure that much of the potential value of wild mushroom species and the ethnomycological knowledge of local communities is not lost. Such knowledge is part of the identity of these communities; knowledge of wild mushroom uses, linguistic, and harvesting knowledge can prevent their loss as modernisation deepens due to the dominance of hegemonic cultures. Another study in the same country was carried out by Sitotaw et al. (2020) in the Menge District. The authors found out that this district is rich in wild mushroom diversity and associated indigenous knowledge. However, they cited anthropogenic factors together with loss of indigenous knowledge and very poor conservation efforts as threatening the survival of economically and ecologically important mushrooms in the area. Thus, the researchers highly recommended the adoption of complementary in situ and ex situ mushroom conservation strategies.

Wendiro et al. (2019) examined artisanal practices for the cultivation of six wild saprophytic mushroom species including Volvariella speciosa (akasukusuku), two Termitomyces sp. (obunegyere and another locally unnamed species), Agaricus sp. (ensyabire) and Agrocybe sp. (emponzira), and one exotic Pleurotus sp. (oyster) that are used as food or medicine. The researchers' descriptions of artisanal mycoculture methods that respond to conservation and utilisation pressures demonstrate the value of addressing traditional knowledge to improve ethno-biology and mycoculture industry practice. Traditional communities engage in multiple technological and organisational innovations and practices for sustainability and, in the case of mushroom production to conserve the environment and culture, ensure variety, food and nutrition security and income.

Although the majority of literature produced on wild mushrooms has dealt with indigenous knowledge on how they are used as food and medicine, the ethical and religious significance within indigenous African and Shona societies of modern

Zimbabwe have not been explored. This is despite the claim made by Kotowski (2019) that currently many researchers suggest the importance of fungi in the creation of early religious practices. There are many examples that suggest ancient use of fungal fruiting bodies containing psychoactive compounds in order to achieve metaphysical experiences. But this work uses African environmental values as a starting point to interpret the religious significance of wild mushrooms among the Shona people in Zimbabwe. There are agrarian beliefs, values, and practices connected with wild mushrooms in contemporary Zimbabwe. These beliefs and values which show the spiritual and moral dimension of wild mushrooms are expressed through myths and rituals practices.

8.2 A Conceptual Understanding of Wild Mushroom in African Religious Traditions (ARTs)

It has been argued that Africans are notoriously religious (Mbiti 1969). Ndemanu (2018: 71) avers that Africans are first and foremost members of traditional religions before any other religion. In African worldview, it is believed that human behaviour towards the natural environment also defines their relation with God. This is true about the religiosity of the Shona people. This group constitutes the largest ethnic group in Zimbabwe. Their belief system includes myths and rituals that explain the social and religious order in relation to the supernatural. From the same perspective, Magesa argues that "… the sacrality of life; respect for the spiritual and mystical nature of creation, and, especially of the human person, the sense of the family, community, solidarity and participation; and an emphasis on fecundity and sharing in life, friendship, healing and hospitality" (Magesa 1997: 52–53). The people believe in a Supreme Being called *Mwari,* who is also known by various anthropomorphic attributes. Between *Mwari* and humanity are ancestral spirits who serve as conveyor belts. What is understood as religion in the life of a contemporary indigenous African is a product of long-standing traditions about *Mwari* and ancestral spirits (Matenda 2018). *Mwari* and the ancestral spirits belong to the spiritual domain which superintends the physical world, a sphere of human beings and nature. This physical world is also known as Mother Earth/Nature. Among the indigenous Shona people, the Earth/Land is sacred. Tempels contends that in the African universe, "after the category of human forces come the other forces, animals, vegetables and minerals" (Tempels 1959: 63). Another realm is the supernatural world which is home to the interred dead people (varipasi). The natural environment sustains water sources, minerals, traditional medicine and fertility of the soil. So nature is sacred in the sense that it is a nourishing force and a place of burial. Through nature, the spiritual world and the supernatural world provide life, fertility, rain, health, food and other necessities needed for sustaining the physical world (Roothaan 2017). It is also from this perspective that Ikuenobe avers that in African ontology the universe was seen as a "holistic community of mutually reinforcing natural life forces consisting of human communities… spirits, gods, deities, stones, sand, mountains, rivers, plants, and animals" (Ikuenobe 2014: 2).

The Shona cosmology is replete with observed religious ceremonies, customs and rituals, myths, taboos and prescriptions, stories and proverbs which carry religious symbolism with the aim of safeguarding the welfare of the community and nature (Mbiti 1975: 12). This worldview shows that religion and food are symbolically intertwined. Taking this into consideration, it can be noted that the consumption of food in African religious traditions is embedded in values that have been transmitted across generations orally (Matenda 2018). For example, in African outlook towards nature, to ensure the sustenance of flora and fauna before the start of rains, a rain inducing ceremony is conducted. It is a religious act of renewing, sanctifying, and reviving life, for both human beings and other creatures (Mbiti 1975: 131). The centrality of such rituals is based on the understanding that the life of the Shona people revolve around food production (Matenda 2018). In this line of thinking, their economic activities show that indigenous Shona are a community of agriculturalists, fruit and mushroom gatherers. It is the availability of rains, a life-giving force which necessitates the sprouting of wild mushroom on sacred mother earth. So, the presence of mushrooms on sacred places like natural forests, mountains and anthills is a manifestation of the spiritual world on earth. It explains why there is a myth which specifies that people emerged from fungus on a tree (Kriel 1989: 24).

The use of wild mushrooms in ARTs is attached to cultural, religious and traditional knowledge symbolism that needs to be unpacked. For example, in the indigenous districts of Buhera and Bikita, the processing of mushrooms in times of abundance for times of scarcity is done using indigenous techniques such as drying which is exclusively the task of women. In ARTs women are regarded as sacred figures that perform religious tasks to enhance the spirituality of their families and the communities. In this study, food patterns reflect indigenous Shona people's religious beliefs, including their social and economic organisation about the nutritional and health properties of mushrooms.

8.3 Human-Mushroom Relationship in Indigenous Communities: An Enduring Ecocentric Heritage

This section explores the relationship between human beings and wild mushrooms in Bikita and Buhera indigenous communities. The strong bond between the two is an enduring heritage among the indigenous Shona people. Some of the responses had a sense of quest for identity as was espoused in the following reaction from an elderly participant. "Though modernity has dominated the culinary items in Zimbabwe, I still eat wild mushrooms because I grew up eating them. Therefore, I highly regard good environmental practices to ensure the sustainability of wild mushrooms." The participant thought that eating mushrooms has become his family's heritage where the mushrooms are perceived as a family cultural food for their unique flavour and taste. So, for them, eating wild mushrooms fosters their African identity which values the inseparability of people and inorganic/organic nature. Among the types of wild mushrooms mentioned was *dindindi* (Boletus Edulis). It is the largest tubed fungus in Zimbabwe with a thick brownish cap which can weigh up to 2.5 kg and measures more than 35 cm in diameter. Most of these large sizes were observed in Buhera district during the research. One female participant emphasised that "dry *dindindi* resembles dry meat in appearance with the taste being as good as meat." This traditional cuisine is passed down from one generation to the next as an expression of indigenous Shona religio-cultural identity. This includes how they act towards the preservation of the environment where *dindindi* is found. Consuming this traditional food is a way of preserving their religio-cultural values even when they move to new places.

It was discovered that the spiritual world communicates with human beings through nature and such a link is strengthened is also through wild mushrooms. The idea of the interconnectedness and interdependence between humans and nature has been underscored by many African philosophers (Tangwa 1996; Bujo 1998; Murove 2004; Behrens 2014; Tosam 2019; Mosima 2021). In some instances, Mwari and ancestral spirits use nature to communicate with human beings. This is the reason why Taringa argues that "The Shona, like many other African people, recognise that spirits operate in the human world through animals, birds and fish" (2006: 205). Although Taringa was referring to animals, he showed that the spiritual world communicates to the Shona people through nature. Today climate change is increasingly becoming a critical challenge to ecological, health, human well-being and people's livelihoods and sustainable development in Africa (Chapungu and Sibanda 2015: 27). Among the indigenous Shona people, weather prediction is recognised as a conventional way of enabling the preparedness of a community to adapt to problems associated with climate change. If the community suspects an impending drought, it puts in place means and ways of surviving during the drought period. On this note, informants pointed out that an edible wild mushroom called *firifiti* (Cantharellus longisporus) which has a scarlet and yellow colour is used by indigenous people to make predictions about the approaching farming seasons. This means that human beings also rely on the environment to get knowledge about weather forecasts. So, it is imperative for them to keep it safe. One elderly woman posited that if there is an abundance of *firifiti* in the forest when rains start falling, villagers are assured of getting plenty of food in that particular season. The abundance of foods would be realised in terms of crop yields, fish in the dams and pools, fruits and flourishing of livestock. The idea of large quantities of *firifiti* as an indicator of food abundance contradicts findings of a study by Chapungu and Sibanda (2015: 27) which shows that affluence of wild fruits predicts low rainfall and subsequently drought.

There are designated places where wild mushrooms sprout and one such place is the anthill. When wild mushroom are found on anthills, it is called *howachuru*. The mushrooms grow from termite nests deep underground. They have a characteristic root-like extension of the stem and a hard, sometimes pointed, centre of the cap. They are predominantly white and mostly appear in the early rains. They are said to have exceptional nutritional value. They deteriorate quite rapidly compared to other types of mushrooms and must, therefore, be eaten only when very fresh (https://www.pindula.co.zw ›Mushroom_in_Zimbabwe). An anthill is revered as a provider of wild mushroom. Therefore, its preservation is sometimes based on indigenous African religious beliefs and values. It is believed that *howachuru* appears on sacred anthills. A female mushroom forager illustrates that harvesting *howachuru* is associated with some injunctions. For example, some should be left as a reserve for the anthill when collecting them. A male traditional leader clarified that when this is done, the spiritual world becomes satisfied and ensures that the anthill continuously provides people with mushrooms. Indigenous families recognise that this knowledge is acquired progressively since there has to be recognition of the locations where there are mushrooms during the rainy season given the fact that not all mushrooms grow and develop uniformly (Lara-Vázquez et al. 2013). If the anthill is in someone's field, it is the prerogative of the owner to harvest and share them.

According to some indigenous belief, *howachuru* is sacred in ARTs because of the high spiritual status accorded to anthills in indigenous Bikita and Buhera communities. If someone wants to spill the beans against his adversary (*bembera*), he stands on an anthill and shouts on top of his/her voice, mentioning the name of a person he accuses of being a witch and who they suspects has caused his/her family members to be sick. He tells the person to stop bewitching his family. The moment this *bembera* is proclaimed publicly, the witchcraft activities done by the offender come to be exposed. More often than not, the sick member becomes well. Because of this, the participants believed that an anthill has some spiritual powers associated with it and mushrooms which sprout from it have mystic powers and serve as medicine. This view is supported by Kriel (1989: 38) who thinks that an anthill is seen as the door to the supernatural which provides healing for some health problems. The anthill does not only serves in removing evil influences, but also in serving as a source of spiritual powers (Kriel 1989: 38). Responses from traditional healers showed that medicines prepared from *howachuru* are used to cure diseases which are of spiritual rather infectious nature such as bad dreams, spirit possession, impurity and insanity. However, another participant added that *howachuru* is used to deal with problems of palpitation.

8.4 The Indigenous Shona People Understand Wild Mushroom as a Vehicle of Indigenous Soteriology

There is a wild mushroom species which was found to be of high utility value in Shona soteriology belief system. It is called is *nhedzi* (Russula heterophylla). This is the most common wild mushroom in Zimbabwe found across the national territory and throughout the rainy season. The young caps of *nhedzi* are hemispherical, sticky and golden-brown in the centre and fading to shiny white at the finely striated edge (https://www.pindula.co.zw ›Mushroom_in_Zimbabwe). There is a symbiotic relationship between the environment and human beings as attested by the spiritual interaction between *nhedzi* and human beings. This was confirmed by participants in Masvingo district who witnessed this spiritual interaction, according to an experience of a certain woman who was doing mushroom foraging in the forest of the Masvingo district. As she was picking the *nhedzi* and putting them in a red basket, the harvested mushrooms started speaking in a human male voice. The voice notified the woman that he was a dead person whose remains were hidden at the site where the woman had picked the mushrooms. She was then to go and notify his relatives in the same district. Once the relatives gathered, the mushroom voice narrated how he was abducted, murdered and buried in a shallow grave in the forest. Guided by the voice, the dead person's remains were exhumed and given a decent burial on a family cemetery.

The religious significance of this incident might confirm the belief that spiritual interaction may exist between wild fungi and humans. In the indigenous Shona worldview, failure to get a proper burial especially by relatives makes the deceased fail to attain ancestorhood. So, relatives of the deceased later performed a bringing back ceremony (*magadziro*), which was the final ritual to make the departed relative eligible for ancestorhood. Osarenkhoe et al (2013: 49) share the same sentiments when they say that mushrooms were reported to contain stimulatory, toxigenic, hallucinogenic and lethal properties that could create mycophobia in humans. What is interesting to note is the tripartite connection of the dead person, the woman forager and *nhedzi* mushroom. It was through *nhedzi* that the deceased manifested himself to the human beings. The wild mushroom made an African spirituality complete in terms of burial and post burial rites. In the Shona cosmology, the spiritual world usually communicates with the human world through women as mediums. In this case, a female medium with mushrooms knowledge is identified and given a task by the spiritual world. Missing a family member and spending five years without a clue of his whereabouts had resulted in serious psychological distress in the community. The anxiety disturbed their socio-economic life because more time was spent in searching for the missing person. So, the mediumship of this female forager and her mushroom brought peace and tranquillity to the family and community of the deceased. The above incident also shows the influence of gender and environmental interaction, especially on the heritage of local knowledge of mushrooms.

Wild mushrooms also have medicinal value. One participant, a traditional herbalist, pointed out that one of the worst problems in ARTs is impotence. For example, *Nhedzi* is a medicinal mushroom species used for the treatment of infertility. Procreation is a fundamental in African culture (as in all cultures); it is the main reason why people marry. In most communities, those who are unable to sire children are socially ostracised. Problems caused by impotence are detrimental to the wellbeing of families. It is drunk by the impotent person, and the liquid was believed to have some mystic healing powers. One participant underscored that the efficacy of roasting the mushrooms facing upwards was symbolic. "It is a good sign in the healing process in the sense that all providences in the patient's pursuit of potency have been opened." The position of the caps ensured the restoration of potency to the patient. Both men and women can take the medicine. This shows that wild mushrooms are essential for the health and well-being of indigenous people.

Apart from being soteriological vehicles as pointed out above, mushrooms are believed to confer fortunes and blessings in Shona culture. During the research, participants were in agreement that a wild mushroom specie called *dare* (*Termitomyces*) is very emblematic in indigenous communities. It is Zimbabwe's largest gilled fungus. *Dare* mushroom has some spiritual powers and it is believed to be identified mostly by lucky people. For example, it is a sign of good omen for pregnant women as well as for those who venture in enterprises like mining, farming, hunting and well digging. One participant believed that seeing *dare* signifies *humambo* (kingship). She illuminated that this might not be kinship in the literal sense, but the opportunity to amass wealth. Then harvesting *dare* reinforces a person's bond with nature which happens to be the provider. So, picking *dare* mushroom is accompanied by blessings. This is the reason why a person who has seen it does not assign someone to harvest it on his or her behalf because that will result in the transfer of the blessing to the person who does the actual picking.

8.5 Discussion and Ethical Implications of the Study Findings

Based on what has been discussed in the preceding sections, the research has the following as its findings: It was established during the study that there are some gendered ethnomycological practices embedded in traditional religion among the indigenous Shona people. In ARTs, most of the collecting practices and culinary uses of wild mushrooms among the Shona people are mostly handled by women. This is supported by a study done by Osarenkhoe et al. (2013: 42) who argues that women have the best capacity to distinguish edible from poisonous mushrooms, have knowledge of their spatial distribution in terms of habitat, phenology and associated substrate(s), possess the skills for processing (handling, drying and cooking) and appropriate local cuisine uses. This indicates a gendered role of household food security using locally available food resources such as wild mushrooms. Among the Shona people, women are understood as the best environmental keepers. This supports Chemhuru's (2018: 11) notion of the African ecofeminist argument for environmental ethical thinking which is centred on trying to understand environmental ethical issues from a feminist standpoint. Women occupy a central religious role as they are perceived to be *Mwari*'s (the Supreme Being of the Shona people) wives (Humbe, 2017: 216). Thus, they are also understood as sacred figures. These sacred figures are those who participate in wild mushroom vending. Although it is correct to argue as Warren does when she argues that "among white people, people of colour, poor people, children, the elderly, colonised peoples, so-called Third World people, and other human groups harmed by environmental destruction, it is often women who suffer disproportionately higher risks and harms than men" (Warren 2000: 2). In some African cultures, women still play a fundamental role in environmental conservation.

Another finding of the study was that the indigenous Shona people's knowledge of ecological dynamics is often reflected in prescribed ways of harvesting mushrooms. Determining the relationships between ecological dynamics and the Shona cultural practices is especially important for managing the environment where resources like wild mushrooms sprout for generations. Ecologically, all sacred places where wild mushrooms sprout are revered and well conserved. The conservation practices are sustained by beliefs and myths about mushrooms which constitute part of the rich culture of the indigenous Shona people in Zimbabwe. This finding correlates with Kelbessa (2005) who confirms that like other multi-cultural traditions, African indigenous traditions contain symbolic and ethical messages that are passed from generation to generation in order to ensure respect and compassion for other living creatures. People in various indigenous communities adopt what Kelbessa has called 'ethical messages' for the benefit of the environment. For example, the most common mythical uses are those meant to bring luck/fortunes (*Dare*), or to bring unity between family members as well as a remedy for infertility (*Nhedzi*). In health matters, the research agrees with Singha et al. (2020: 3168) when they say that there are strong ritualistic beliefs among the local tribal communities regarding the medicinal uses of wild mushrooms.

From the above, one can note that there is a symbiotic relationship between religion and agrarian entrepreneurship in indigenous communities of Zimbabwe. In some instances, the motive to preserve the environment is influenced by the commercial value of wild mushrooms. In other instances, it is purely for spiritual reasons the environment is preserved. From this view, it can be said that since food is a portal into a people's culture, they have a duty to reciprocate in keeping the environment safe as it provides them with wild mushrooms. However, if the indigenous people need the environment to be protected in order to ensure the continuous supply of wild mushroom, this is typically anthropocentric environmentalism. Conversely, an ecocentric view should be also noted in that there are some religio-cultural implications of wild mushrooms. Mbiti argues that "because traditional religions permeate all the departments of life, there is no formal distinction between the sacred and the secular, between the religious and the non-religious, between the spiritual and the material areas of life" (Mbiti 1969: 1). Indigenous Shona people include the sacred world in foraging practices. It is believed that wild mushroom is an ancestrally given resource meant to sustain the lives of people and, therefore, should be well preserved. This is a religio-cultural belief which links them with their departed elders of the family. The ecocentric view is well corroborated in that what makes the indigenous Shona people remain bonded with mushrooms is their connection to their roots. In view of these two centrisms, this study's findings are in accordance with Rülke et al. (2020) who found that the majority of the people have an anthropocentric perspective than an ecocentric one in terms of protecting biodiversity. Ramose notes that "the principle of wholeness applies also to the relation between human beings and …nature. To care for one another, therefore, implies caring for physical nature as well. Without such care, the interdependence between human beings and physical nature would be undermined" (Ramose, 2009: 309).

8.6 Conclusion

This chapter has argued for the need to revisit indigenous African environmental ethics in resolving agrarian and environmental issues. So, there is an immediate need to preserve indigenous knowledge systems regarding the utilisation of wild mushrooms in Zimbabwe (Rülke et al. 2020). In line with anthropo-ecocentrism, the chapter establishes the symbiotic relationship between human beings and nature, which is the basis for understanding the ecological significance of wild mushrooms in indigenous African culture (Osarenkhoe et al. 2013: 39). African environmental ethics is hinged on the nexus between the health of the natural environment and the well-being of humans. Wild mushrooms have material, cultural, and spiritual importance. There is the need for more research on wild mushrooms to explore its multidimensional value. Wild mushrooms can be elevated to cash crops status that are well priced as food, medicine, and a wide range of other potential uses as myco-fungicides, biofertilizers, novel drugs, animal feed supplement and bioremediants and tools in the healthy management of agroforests (Osarenkhoe et al. 2013: 39).

References

Arcury, T.A., and Quandt, S.A. 1999. Participant Recruitment for Qualitative Research: A Site-Based Approach to Community Research in Complex Societies. *Human Organization* 58 (2): 128–133. Society for Applied Anthropology.

Behrens, K. 2014. Toward an African Relational Environmentalism. In *Ontologized Ethics: New Essays in African Meta-Ethics*, ed. E. Imafidon, and J.A. Bewaji, 55–72. Lexington Books, Lanham.

Bourdieu, P. 1986. 'Forms of Capital' in Richardson J.,G. (ed). In *Handbook of Theory and Research for the Sociology of Education.* London: Greenwood Press.

Bujo, B. 1998. *The Ethical Dimension of Community: The African Model and the Dialogue between North and South.* Nairobi: Paulines Publications.

Chapungu, L., and F. Sibanda. 2015. Effectiveness of Conventional Indigenous Practices in Climate Change Adaptation and Mitigation in Masvingo District, Zimbabwe. In *Indigenous Knowledge in Zimbabwe, Laying Foundations for Sustainable Development*, ed. J. Mapara and M. Mazuru, 22–39. Gloucestershire, United Kingdom: Diaspora Publishers.

Chemhuru, M. 2018. Interpreting Ecofeminist Environmentalism in African Communitarian Philosophy and Ubuntu: An Alternative to Anthropocentrism. *Philosophical Papers.* https://doi.org/10.1080/05568641.2018.1450643.

Dube, P., Madamombe, G., Tapfumaneyi, L., Ngezimana, W., and Simango, K. 2021. *Collection And Consumption of Wild Edible Mushrooms In Three Villages of Binga, Zimbabwe*, (https://www.researchgate.net ›publication ›355723959_...) pp 1–19.

Fang, W.T., A. Hassan, and B.A. LePage. 2023. The Living Environmental Education Sound Science Toward a Cleaner, Safer, and Healthier Future. *Springer.* https://doi.org/10.1007/978-981-19-4234-1.

Fernandes, T., Garrine, C., Ferrão, J., Bell, V., and Varzakas, T. 2021. Mushroom Nutrition as Preventative Healthcare in Sub-Saharan Africa. Applied Science 11 (4221): 1–27.https://doi.org/10.3390/app11094221.

Humbe, B.P. 2017. African traditional religion in post-colonial Zimbabwe: A sustainable heritage for water resources management. In Green, M.C., Hackett, R.I.J., Hansen, L., and Venter, F. (eds).

Religious Pluralism, Heritage and Social Development in Africa. Stellenbosch: AFRICAN SUN MeDIA:205–219
Ikuenobe, P. 2014. Traditional African Environmental Ethics and Colonial Legacy. *International Journal of Philosophy and Theology* 2: 1–21.
Kelbessa, W. 2005. The Rehabilitation of Indigenous Environmental Ethics in Africa. *Diogenes* 52: 17–34.
Kotowski, M.A. 2019. History of Mushroom Consumption and its Impact on Traditional View on Mycobiota—An Example from Poland. *Microbial Biosystems* 4 (3): 1–13.
Kriel, A. 1989. *Roots of African Thought 2 Sources of Power*. Pretoria: University of South Africa.
Lara-Vázquez, F., Romero-Contreras, A.T, and Burrola-Aguilar, C. 2013. Traditional Knowledge Regarding Wild Mushrooms, in The Otomí Communities of San Pedro Arriba, Temoaya, Estado De México, *Agricultura, Sociedad Y Desarrollo,* Julio–Septiembre, pp 305–333.
Magesa, L. 1997. *African Religion: The Moral Tradition of Abundant Life*. New York: Orbis Books.
Matenda, J. 2018. *The Cultural and Religious Significance of Indigenous Vegetables: A Case Study of the Chionekano-ward of the Zvishavane-district in Zimbabwe*, A thesis submitted in fulfilment of the requirements for the degree of Master of Arts in the Department of Religion and Theology, University of the Western Cape.
Mbiti, J.S. 1969. *African Religions and Philosophy*. Oxford: Heinemann.
Mbiti, J.S. 1975. *Introduction to African Religion*. London: Heinemann.
Mosima, P. 2021. Cosmic Interconnectedness: An African Exploration Towards a Dynamic Understanding of Nature. *Satya Niyalam Chennai Journal of Intercultural Philosophy* 39: 62–80.
Murove, F.M. 2004. An African Commitment to Ecological Conservation: The Shona Concepts of Ukama and Ubuntu. *Mankind Quarterly* XLV: 195–215.
Ndemanu, M.T. 2018. Traditional African Religions and Their Influences on the Worldviews of Bangwa People of Cameroon: Expanding the Cultural Horizons of Study Abroad Students and Professionals. *Frontiers: The Interdisciplinary Journal of Study Abroad* XXX (1): 70–84.
Ndifon, E.M. 2022. Systematic Appraisal of Macrofungi (Basidiomycotina: Ascomycotina). *Biodiversity of Southern Africa Journal of Asia-Pacific Biodiversity* 15 (1): 80–85.
Osarenkhoe, O.O., O.A. John, and D.A. Theophilus. 2013. Ethnomycological Conspectus of West African Mushrooms: An Awareness Document. *Advances in Microbiology* 4: 39–54.
Ostry, M.E., Anderson, N.A., and O'Brien, J.G. 2010. *Field Guide to Common Macrofungi in Eastern Forests and Their Ecosystem Functions*. Delaware: U.S. FOREST SERVICE.
Piearce, G., and Sharp, C. 2000. Vernacular Names Of Zimbabwean Fungi: A Preliminary Checklist, in *Kirkia*, vol. 17, no. 2, 219–228. National Herbarium & Botanic Garden.
Ramose, M. B. 2009. Ecology through ubuntu. In M, F. Murove (ed.). *African ethics: An anthology of comparative and applied ethics*, 308–314. Scottsville: University of KwaZulu-Natal Press.
Roothaan, A. 2017. Hermeneutics of Trees in an African Context. Enriching the Understanding of the Environment 'for the Common Heritage of Humankind'. in *African Philosophy and Environmental Conservation,* ed. Jonathan Chimakonam. London: Routledge.
Rülke, J., Rieckmann, M., Nzau, J.M., and Teucher, M. 2020. How Ecocentrism and Anthropocentrism Influence Human–Environment Relationships in a Kenyan Biodiversity Hotspot. *Sustainability* 12: 8213. doi:https://doi.org/10.3390/su12198213.
Singha, K., S. Sahoo, M. Roy, A. Banerjee, K.C. Mondal, B.R. Pati, and P.K.D. Mohapatra. 2020. Contributions of Wild Mushrooms in Livelihood Management of Ethnic Tribes In Gurguripal, West Bengal, India. *International Journal of Pharmaceutical Sciences and Research* 11 (7): 3160–3171.
Sitotaw, R., Lulekal, E., and Abate, D. 2020. Ethnomycological study of edible and medicinal mushrooms in Menge District, Asossa Zone, Benshangul Gumuz Region, Ethiopia. In *Journal of Ethnobiology and Ethnomedicine*, 16:11. https://doi.org/10.1186/s13002-020-00361-9:1-14
Tangwa, G.B. 1996. Bioethics: An African Perspective. *Bioethics* 10 (3).
Taringa, N. 2006. How Environmental is African Traditional Religion? *Exchange* 35 (2): 3. https://doi.org/10.1163/157254306776525672.
Tempels, P. 1959. *Bantu Philosophy*. Paris: Présence Africaine Éditions.

Thangaraj, R., Raj, S., and Renganathan, K. 2017. Wound Healing Effect of King Alferd's Mushroom (Daldinia Concentrica) Used By Tribes Of Sirumalai Hills, Tamilnadu, India, in *International Journal of Pharmacy and Pharmaceutical Sciences* ISSN: 0975-1491, vol. 9, Issue 7, 161–164.

Tosam, M.J. 2019. African Environmental Ethics and Sustainable Development. *Open Journal of Philosophy* 9: 172–192. Scientific Research Publishing.

Warren, K.J. 2000. *Ecofeminist Philosophy: A Western Perspective on What It Is and Why It Matters*. Lanham: Rowman and Littlefield.

Wendiro, D., Wacoo, A.P., and Wise, G. 2019. Identifying indigenous practices for cultivation of wild saprophytic mushrooms: responding to the need for sustainable utilization of natural resources. In *Journal of Ethnobiology and Ethnomedicine*, 15:64. https://doi.org/10.1186/s13002-019-0342-z:1-15

Zeleke, G., Dejene, T., Tadesse, W., Agúndez, D., and Martín-Pinto, P. 2020. Ethnomycological Knowledge of Three Ethnic Groups in Ethiopia. *Forests* 11 (875): 1–18. doi:https://doi.org/10.3390/f11080875.

Bernard Pindukai Humbe holds a PhD in Religion Studies from the University of Free State, South Africa. Currently, he is a Research Fellow at the following institutions: University of South Africa (UNISA), South Africa and University of Religions and Denominations (URD), Iran. His areas of research interest include: Religion and Covid-19, African Indigenous Religious Knowledge Systems (AIRKS), Traditional Law and Social Development, Religion and Entrepreneurship, Religion and Social Transformation, and Religion and Power.

Part III
African Agrarianism and Environmental Ethics

Chapter 9
The Consubstantiality of Living Things: Towards a Mandingo Cosmo-Anthropocentric Ethics

Belko Ouologuem

Abstract The Socratic shift in philosophical thinking from nature to the human person marked the birth of ethical thinking in the history of Western philosophy. It was also the starting point of anthropocentric ethics. African philosophy, on the other hand, has developed a relational conception of human beings in intimate and permanent interaction with the natural environment. It is from this interconnected and interdependent outlook that a cosmo-anthropocentric ethic, which consists considering human well-beings only in relation with other constituent elements of the cosmos, has emerged. In this chapter, I critically explore the (African) Mandingo agrarian environmental ethics. Anchoring my reflection on medieval and pre-colonial Mandingo oral and written texts such as the Mandingo hunters' Oath and the Charter of Kouroukan Fuga, I show that the Mandingos included the natural environment in their moral conception of the universe.

Keywords Human welfare · Ethics · Anthropocentric ethics · Cosmo-anthropocentric ethics · Mandingo · The Mandinka hunters' oath · The Charter of Kurukan Fuga

9.1 Introduction

The modern world was built on mechanical and experimental rationality by granting human beings a privileged position in their relation with nature and its constituent elements. This self-assigned privilege has given humans the power to dominate and exploit nature for their own interest, excluding animals and plants and the ecosystem

This paper was originally written in French under the title: «De la Consubstantialité des êtres vivants: Pour une éthique Cosmo-anthropocentrée Mandingue.» It was translated into English by Mbih Jerome Tosam with the assistance of Denis Ghislain Mbessa.

B. Ouologuem (✉)
Department of Philosophy, University of Letters and Social Sciences, Bamako, Mali
e-mail: belko_wologueme@yahoo.fr

M. J. Tosam and E. Masitera (eds.), *African Agrarian Philosophy*, The International Library of Environmental, Agricultural and Food Ethics 35,
https://doi.org/10.1007/978-3-031-43040-4_9

from the field of morality. However, recent developments in environmental ethics have brought back nature to the centre of contemporary ethical thinking.

The anthropocentralisation of modern ethics is rooted in the Socratic epistemological shift of the object of philosophical reflection from nature to human concerns. Cicero gives an account of this shift in the following words:

> Moreover, from Ancient philosophy to Socrates [...] it was numbers and movements that were studied; the stars, their distances, their orbits and, in general, all celestial phenomena were also studied. Socrates was the first to invite philosophy down from the sky, to install it in the cities, to introduce it into the homes, and to impose on it the study of life, of morals, of good and evil (Tusculanes V 4, 10, trans. Humbert).

This invitation of philosophy to descend from the heavens to deal exclusively with morals, good and evil, in human life is based on the assumption that nothing ethical can be constructed from the "*phusis*" (nature). This is what Louis-André Dorion remarks: "The lack of interest in the study of nature stems from the conviction that one cannot deduce ethics from physics, that is to say ethical reflection has its own exigencies and has nothing, or very little, to learn from the study of nature" (Socrates, 2004: 5). If this anthropocentricalisation of ethics was not followed by Socrates' most famous followers, it nevertheless exerted an unprecedented influence on the whole history of Western ethical thought. It is only with the serious ecological and climatic threats of the 20th century that the need for an environmental ethics has emerged, following the observation that a science that would make mankind master and possessor of nature, as precribed by Descartes, is suicidal.

In contrast to anthropocentric ethics, African philosophy developed a cosmo-anthropocentric ethics variously described by African ethicists as eco- bio-communitarian ethics (Tangwa 1996) or simply relational ethics (Mbiti 1969; Bujo 2009). What is a cosmo-anthropocentric ethics? What are its foundations? How does it manifest itself? What is its relevance for our contemporary society? The ansers to thses questions will form the lineaments of the body of the text.

9.2 On the Consubstantiality of the Elements of the Cosmos

The teachings of the Komo[1] convey the idea that the cosmos is made up of visible and invisible worlds, the handiwork of the unique creator, *Mangala*. These cosmogonic teachings relate a story of the creation of the universe, told by Fodé Moussa Sidibé, a Komo initiate:

> *Mangala* created the Cosmos through the "fresh Nothingness", *sumaya madalen*, Glan, the original void, imprinted the universal movement, rolling up on itself in two spirals of opposite directions, releasing a force, an energy called *Zo* from which the *Yo* spirit proceeds. It is the *Yo* spirit, spinning at the four cardinal points of the Universe, which conceives by vibration,

[1] Komo is a society for the initiation and education of the Bamanan/Mandinka or Mandingo, the komo is a training centre for knowledge and character formation, the guarantor of the transmission of ancestral traditions, order and social cohesion. The komo consists of thirty three classes.

> yèrè yèrèli, four worlds: the first is the original world; the second, the mythical and past world; the third, the present world; the fourth, the future world.
>
> The four worlds, created at the beginning, starting from Glan, at the four cardinal points, include the following symbols:
>
> The first world is the Western world, the original world. It is the symbol of the preservation in the secret of uncreated things. It is the world of initial calm. Its constituent is air, *fignè*, which in turn is the symbol of intelligence, invention, skill, subtlety;
>
> The second world is that of the North, the mythical and past world. It is the symbol of fixedness, precision and the limit of all things. Its constituent is the earth, *dugu* or *dugukolo*, which is the symbol of completion, of the final result, of the end of cycle;
>
> The third is the world of the East, the actual world, symbolising intelligence, light, energy, heat, and all that is necessary for the birth of life. Its constituent is breath or fire, *goni* or *ta*, which is the symbol of ardour, motivation, enthusiasm;
>
> The fourth is the world of the South, the future world, that of and perpetuation. Its component is water, *jii* or *ji*, symbol of organic matter, suffering, pain and hindrance (De donsoya à donsologie 2020: 54–55).

It is evident from this account that *Mangala*, the supreme creator, created the cosmos from the original void by endowing it with a universal movement whose permanent activity generated the energy from which emerged the *Yo* spirit. The spirit in its spinning conceived the four worlds whose elements are air, earth, fire and water, whose intense stirring called *Glanzo*, led to the creation of beings and things. The following account relates the origin of things and being on earth:

> A heavy mass called Penba or Fenba, detaches itself from Glanzo by swirling and gives birth to the earth. The 22 elements which are the general characters of beings are ordered. At the same time, Fãro, a portion of the Yo spirit of Glan, builds the sky. And then it descends to earth, in the form of water, and brings life in *gnoron gnoron*, the muddy mud. From this matter, fonio, grass, plants, turtles, fish, scorpions, crocodiles, and other aquatic animals and humans successively emerged (Idem, 57–58).

The Fãro, the portion of the *Yo* spirit, by way of permanent and intense stirring, generated a living mud that we call original substance from which emerged plants and animals including human beings, hence the idea of consubstantiality of living things. The universal movement of intense mixing ensures the birth, growth and individual and collective degeneration of all beings in the plant and animal kingdom and maintains inter-activity between them which guarantees the equilibrium of the ecosystem. Thus, any human attitude, including moral and ethical, which seeks to make human beings independent and special entities in the cosmos, ignoring their interrelation and interdependence with the rest of nature, would lead to an imbalance of the ecosystem, the consequences of which will be unforseeable for humanity.

9.3 The Reaction of Nature in Relation to Human Attempts at Omnipotence

The consubstantiality of living things, with the corollary of an interaction whose knowledge and preservation guarantees ecological balance, is a concept that is common among the Dogon[2] people of Mali. As a farming people, the Dogon have developed a specific relationship with nature, and particularly with the earth, which they consider as the nourishing and living mother, hence the cult of the earth, which is at the basis of almost all spirituality in the Dogon society. The founding agrarian rite, which constitutes the cultural and territorial identity of the Dogon people, is *Ougourou.*[3] *Ougourou* is a cyclical sacrificial rite of purification and recognition of the agreement between the different clans that make up the Dogon people and the spirits of nature in relation to the occupation of the space which has become the Dogon Country.

In Dogon cosmogony, *Amba*, the Supreme Being, created the world by populating it with visible and invisible beings, *yèben,* who live and share the earth in harmony for the balance and stability of their respective environments. The visible beings are the human beings and the entire animal and plant world, and the invisible world which includes the spirits of the waters, the spirits of the ancestors and the *antoumbouloum.*[4] The latter can reveal themselves physically to some people during the day or during the night and in the bush or in the village, or they can be inhabited in the form of a spirit in the mind of a human being who will immediately enter into a trance. This individual, possessed by the spirit called *kézou* or *benjinè*, can identify the sites inhabited by invisible beings. He/she is also endowed with the power to see them and communicate with them. The following interviews between Marcello Monteleone and Kai and then Daifourou illustrate the occasional encounters of humans with spirits:

> There are men who used to disappear into the bush and come into contact with the *yèbèn.* One day, a man from the village of Nacombo met some *yèben* and ate with them. Subsequently, with his family, he ate only beer [*dolo*], because he had eaten elsewhere, with the *yèben.* He had acquired the gift of talking about strange things. He knew the *Sigui-sò*[5] without anyone

[2] The Dogon are a Mandingo sub-group, a farming people who currently live in the Mopti region of central Mali, an area known as Dogon Country. They are one of the few Mandingo peoples who have best preserved their ancestral culture and values before the Arab-Muslim and Judeo-Christian invasions (their pre-colonial cultures).

[3] *Ougourou* is a cyclical sacrificial ceremony of purification and renewal of the pact between the Dogon people and the invisible world inhabited by the spirits and the spirits of the ancestors within a given territory.

[4] Antoumbouloum are small bipedal creatures with large heads and feet pointing away from the face, endowed with speech and extraordinary magical powers from which some hunters are said to derive their knowledge of animals and plants, which some people occasionally encounter in the forests or around villages but which disappear immediately.

[5] Sigui-ò: is the language of the masks that only those initiated into the mask brotherhood understand. It is totally different from the everyday language of the ordinary Dogon.

> having taught him, as well as the Koran in Arabic [even though he was not a Muslim]. He never ate at the homes of people who received him (Kai, interview.3, 2013, pp. 63-64).

It is clear from this account that the encounter of the villager of Nacombo, a village about ten kilometres from Bandiagara in the Mopti region in the centre of mali, with the spirits, completely transformed his life by granting him the power to speak two new languages that were previously totally foreign to him. It is important to point out that the person who received extraordinary powers from invisible beings (who are no longer invisible to him) is obliged to keep secret everything he has seen and heard. This is clearly expressed in the following interview with Daifourou:

> In the Ouolo village lives a boy called Daifourou Maba; he is the son of a teacher. His whole family is still alive. In the past, he used to eat outside the house, near a group of trees where he claimed that a friend offered him food. The boy was not even ten years old. His father questioned him every day and he replied that he had eaten well, as his oily mouth attested. In fact, he would enter a small wood and... basta! No one knew anything more about him and no one went there. The boy was not allowed to add anything, because if he did, the *yèben* [his friend] would kill him (Interview 23, 2013, pp. 64–65).

The *kézou* or *benjinè*, now endowed with extraordinary divinatory powers, are the mediators between the visible and invisible worlds. Knowing the places of 'residence' of the *yèben* spirits in the waters, in the forests and in the caves, they are the guarantors of the purity and purification of the land, the waters and the forest in case of defilement. They are therefore the only ones who have the meaning and scope of the *ougourou*. This sacrificial ceremony symbolises the very first pact to preserve the equilibrium of the ecosystem in Dogon society, as this account by Kai eloquently proves:

> The first people who occupied our territory developed a ceremony for its preservation. This ritual concerned the earth, the rocks, the water, the trees, the animals, in short everything that moves [biodiversity], including the *yèben*. These are neither devils nor angels, but invisible creatures that inhabit the caves, as well as the heavens and the earth. Our ancestors tried to understand their environment in order to better respect it. It should be noted that the same *yayye* can include several types of environment (Interview, n.3, ibid., p. 65).

The ceremony that enshrines the preservation of the territory requires two attitudes towards the occupants: an understanding and respect for the environment. Understanding means an awareness on the part of the human beings that they are not the only occupants of the surrounding universe and that they must therefore act with due regard for the existence of other beings, even if they are invisible. By respect for the environment, we must not only understand the visible concreteness of the fauna and flora, but also the sites inhabited by the spirits of nature, sites identified by the divination of the *kézou* or *benjinè*. These 'residences' of the *yèben* also house the altars for the *ougourou* ceremony, and access to them is only allowed to the initiated. This is why the notion of respect, which is a fundamental ethical value among the Mandingo in general and the Dogon in particular, has a metaphysical origin in reference to the first sacred pact between the spirits of nature and the founding ancestors of the clans and villages. At the same time, it has implications for environmental and human ethics. In clearer terms, the notion of respect is only understood in its fullness when it is understood in its three intrinsically related dimensions.

Respect for the *ougourou* shrine has several implications for environmental ethics. For example, several moral norms for the protection of flora and fauna are dependent on the existence of the *ougourou* altars. For example, it is forbidden to cut down trees within a perimeter of one hundred metres around an *ougourou* site, as this could threaten the tranquillity of the *yèben*. It is also forbidden to hunt in the area. This means that to this day in all Dogon villages where the *kézou* or *benjinè* are alive, the altar sites are surrounded by forests which are the last refuges of wild animals whose shelters have been destroyed by humans. These sacred sites cannot be inhabited by humans otherwise the consequences would be disastrous. This was the case in Doucombo, a village located 5 km from Bandiagara.

> In Doucombo, people had converted to Islam and, ignoring the teachings of their tradition, had built a house on a sacred site. Disgrace followed and the millet no longer produced. The villagers then went to Sokolo [to the Karembé] and tried to resolve the issue by making reparation for the offence committed. They had to go to Sokolo three times in a row and on the fourth time their wives joined them. Finally, a rite was celebrated near the sacred place and a second one near a site where the ougourou is still performed. This is how everything came back to normal. (Ankemé Albert Karembé, interview n. 9).

The ethical rules that regulate relations between humans are, in the Dogon society, essentially rooted in benevolence and the fear of reprisals from the spirits of nature. The *ougourou*, which is the cyclical sacrificial ceremony in respect of the sacred pact that establishes harmonious relations between the spirits of nature and the ancestors, is also the guardian of purity, purification and social cohesion within the community and between communities. Thus, crimes, lies, betrayals, sexual relations in the bush are considered impurities whose consequences can be individual reprisals from the living and nourishing earth, as well as a series of mysterious deaths in the wrongdoer's family, or a drought or an epidemic that would strike the whole community. For example, in a land dispute between two families or two clans in the Dogon country, to resolve the conflict, the wise men of the community will ask the two belligerents to offer a sacrifice to the *ougurou* and the dispute will be quickly resolved. Why would the problem be so quickly and simply resolved? Because the person who is not the owner (the intruder) of the *yayyé* (agricultural field) and who sacrifices to the *ougourou* will die during the said ceremony, struck down by the spirits and founding ancestors of the clan or community. Thus, telling the truth is a sacred and vital duty for peace and social cohesion among the Dogon of Mali. The abandonment of this practice by a large part of the Dogon people who have converted to Islam or Christianity is probably one of the fundamental reasons for the persistence and multiplication of land disputes in the Dogon community today.

What we need to recall from this is that human beings are not the only inhabitants of the earth, let alone the cosmos. Human beings are in constant interaction, consciously or unconsciously, with the forces of nature. They are not the masters of nature, as evidenced by natural disasters such as tsunamis, increasingly dangerous and uncontrollable pandemics such as COVID-19, as well as as new scientific discoveries of living things.

Moreover, although the *ougourou*, the most important agrarian rite of the Dogon community, may seem mythical and mysterious, it is nonetheless the principal foundation that determines all Dogon moral norms and values. This is what Monteleone points out:

> The *ougourou* sacrifice does not teach a fetishistic technique for thanking or negotiating with the invisible forces of nature, but educates in the principles of adherence to morality, specific to the circular system of giving. In doing so, the *ougourou* refers to the mythical precepts related to the sacrifice par excellence, the *nommo*, which guaranteed the emergence of a new purifying force, against the infection that had tainted the earth, for the benefit of all humanity. (2013: 67).

The permanent interactive relationship between human beings and the invisible world, whose knowledge and respect are the necessary conditions for the harmony of the ecosystem, the guarantor of the survival of humanity; the moral precepts emanating from prayers, rituals, in short, from the sacrifice of the *ougourou,* require us to consider ethics not simply from the point of view of correct behaviour between humans, but also how to better harmonise the relationship between human beings on the one hand, and between humans and other living beings in the universe. This calls for a cosmo-anthropocentric ethics.

9.4 Cosmo-Anthropocentric Ethics

Because of the consubstantiality of living things and the permanent interaction with nature, which is the result of the sharing of the cosmos as a unique existential space, human beings must rethink their relationship with nature, which has been exclusively anthropocentric. This conception of ethics is called anthropocentric because its ultimate goal is to achieve the well-being of humans at the detriment of animals and plants, and the spiritual world. The consequences of such a conception of ethics are well known: global warming, climatic disruptions, existential threat to forests and consequently to related flora and fauna, the appearance of natural disasters such as tsunamis and increasingly uncontrollable epidemics, etc. Since the status quo is unsustainable for humanity, how can we get out of this apocalyptic picture? This is the challenge that cosmo-anthropocentric ethics will attempt to address.

The cosmo-anthropocentric ethic perceives of the living world (human, animal, plant and spirits) as an organic world whose preservation of harmony is the *conditio sine qua non* for the survival of each of its part as well as the whole. This intrinsic existential relationship maintains each part in such an interdependent manner that any attempt to dominate an organic element causes a destabilising or even destructive dysfunction. Cosmo-anthropocentric ethics is based on certain principles: the equal moral value of all life, the universal connection of living things, the need to preserve the whole as a condition for the protection of each part.

9.5 Equality of the Moral Value of All Living Things

This principle states that consubstantiality (all life is generated by mud, *gnoron gnoron*) renders arbitrary any temptation to create a hierarchy between plant, animal, spirits, and human life. This is the first statement of the Mandinka Hunters' Oath[6], which stipulates:

Every (human) life is a life. It is true that one life comes into existence before another life, but one life is not more 'ancient' (older), more respectable than another. Just as one life is not superior to another (Cissé 1991, Appendix I). If the author translates 'all life' as human life, the great hunter, master, Fodé Moussa Sidibé, considers that the notion of 'all life' refers to all living things and not only to humans, because this is the first teaching that is given to each young candidate for initiation into the hunters' brotherhood. He says:

> In the vision of the *donso* (hunter in Mandinka), 'every soul is a soul; no soul is more important than another soul.' This is the first lesson taught to candidates during their initiation into the *dankun* to learn to respect life in all its manifestations, and especially the life of the game that the hunter destroys. To support this idea, the masters add: "No destruction of a soul goes unpunished; any harm inflicted on a soul deserves reparation (2020: 144–145).

The sacredness of life is not unique to human life. Cosmo-anthropocentric ethics is holistic and extends to all living things, which is the only guarantee for the efficient and effective preservation and protection of human beings and their surrounding environment. Respect for the sacredness of life in all its manifestations is an absolute necessity for the full realisation of the sacredness of human life as part of the whole. This is called the principle of universal connection.

9.6 The Principle of the Universal Connection of All Living Things

This principle is based on the original uniqueness of their primary essence, which is the muddy mud, the *gnoron gnoron*. Although concrete life manifests itself in many forms, the fact remains that the balance of each constituent element of the cosmos can only be maintained in the long run by preserving the whole ecosystem. The universal connection is comparable to the maternal bond between the different children of a woman. Each of the children pursues his/her own individual career and thus seems to follow a personal path independent of the others, but in the African reality this is an illusion because no matter how individualistic you are you are always caught up in your family or communal network. Hence the existence of a much stronger bond of solidarity between children of the same mother.

[6] The Mandinka hunters' oath or Manden kalikan is a collection of legal texts proclaimed in 1222 by Soundjata, founder of the Mali Empire, and his peers to regulate relations between the Mandinkas and also to regulate hunting in the empire. It was the first legal text in the world to formally abolish slavery.

The world has become a global village thanks to the new communication and information technologies. This is because human beings did not previously realise that all people live and share the same earth in different places. This isolationist vision had led, for centuries, to the abusive exploitation of the fauna and flora as well as non-European races and peoples, ignoring the universal connection of life and erroneously thinking that the disastrous consequences would be suffered exclusively by local populations. It is this same partial and biased vision of the universe that led Europeans and the Western world in general to bury their nuclear waste on other territories and to relocate their environmentally polluting factories, hoping that the harmful effects would be confined to them. However, cosmo-anthropocentric ethics teaches us that it is not the mere performance of new communication and information technologies that makes the world a global village, but rather the principle of the universal connection of living things that makes technological connection possible. This is also what the Kenyan philosopher, Henry O. Oruka, advocates in his "Parental Earth Ethics", in which he contends that the earth or the world is a Family Unit, a global village in which members have kith and kin relationship with one another (1993). There is, therefore, the need to consider global solutions to the challenges of the contemporary world.

9.7 The Principle of the Necessity for the Preservation of All for the Protection of Each Other

The excessive humanisation of relations between living things in the cosmos, based on a hegemonic conception of the human being, has led the latter to ignore or even despise the crucial role of the intrinsic and constitutive interaction of all living things. This desire for domination has led humans to place other living things at their service, relegating to the background the interdependence of each constituent element with the other, which is the only guarantee of the preservation of the ecosystem's balance. This conception of the relationship between human beings and other beings is supported by a mechanical and experimental rationality that excludes spiritual rituals of contact with the invisible world. Yet, the sacrificial ceremonies of the *ougourou* lay emphasis on the fact that we are not the only living beings on earth and in the cosmos and that a happy and sustainable human life cannot be envisaged in disregard of the existence and interaction with others. If the Dogon *ougurou* may seem mystical to a Cartesian, it is the Dogon people who discovered Cyrus centuries before modern science and organised rituals of its cycle every 60 years. It is also the same people who had discovered since the 16th century that a woman could conceive a child without seeing her menstrual period, something that modern biological sciences only confirmed at the end of the 20th century. In the same vein, the Kurukan Fuga Charter[7] proclaimed prohibitions that protect nature:

> In the Mande the law will be: the bush, the trees, the springs are our most precious goods. They provide us with food. Everyone has a duty to protect and preserve this heritage. Delimit the area to be burnt; do not destroy the bush by fire"; "We must also protect the forests where the spirits are housed. Respect the spring heads, make offerings to the deities of the waters, springs and forests.

The proclamation of the Mandinka Hunter's Oath and the Charter of Kurukan Fuga, in the 13th century, definitively institutionalised a cosmo-anthropocentric ethic that extended ethics beyond the sphere of humans to include the natural world and the world of spirits with an obligation for all to respect each entity in its specificity, but also and above all, in its original uniqueness. It is only at this price that humanity may be able to face the challenges posed by global warming and climate change, the disappearance of several plant and animal species, in short, a generalised imbalance of the ecosystem.

9.8 Conclusion

In conclusion, we would say that the current challenges of the contemporary world have shown the limits of anthropocentric ethics with omnipotent attempts to control evil and to advocate human welfare. The violence exerted by humans on the plant and animal kingdoms as well as the destruction of the "residences" of the spirits of nature, has unfortunate repercussions on human life, such as natural disasters and uncontrollable epidemics that are increasingly destructive. Cosmo-anthropocentric ethics, based on absolute respect for all life, the universal connection of living things and the need to envisage global holistic solutions to the challenges of climate change and global warming, are the only relevant ways to lead us out of a disaster programmed by an exclusively anthropocentric morality.

References

Bujo, B. 2009. Ecology and Ethical Responsibility from an African Perspective. in *African Ethics: An Anthology of Comparative and Applied Ethics*, ed. Murove, F.M. Scottsville: University of KwaZulu-Natal.

Ciceron, 2004. *Tusculanes*, V.4, 10 (trad. Humbert) in Dorion, L-A. 2 *Socrate*, Paris: Presses Universitaires de France.

[7] The Charter of Kurukan Fuga: it was proclaimed in 1236 by Soundjata Keita, the founder of the Mali Empire after having defeated and reunited with his allies the whole Mandingo world. Kurukan Fuga is a vast sentence located at Kangaba in the present rural commune of Mandé, about fifty kilometres from Bamako. It was the second legal text in the history of the Mandingo that guaranteed the peaceful and harmonious living together of peoples of different cultures and beliefs.

Cisse, Y.T. 1991. *Soundjata, la Gloire du Mali* (Soundjata, the Glory of Mali), éd Karthala, ARSAN.
Dorion, L.-A. 2004. *Socrate*. Paris: Presses Universitaires de France.
Mbiti, J.S. 1969. *African Religions and Philosophy*. London: Heinemann.
Monteleone, M. 2013. *Le Culte de la Terre au Pays Dogon (Mali) : Entre coutumes foncières et décentralisation avec le témoignage de Ambaéré André Tembely (The Land Cult in the Dogon Country (Mali): Between Land customs and Decentralisation with the Testimony of Ambaéré André Tembely)*. Paris: L'Harmattan.
Niane, D.T. 2010. Kouroukan Fouga : *Soundjata et l'Assemblée des Peuples* (Kouroukan Fouga: Soundjata and the Assembly of Peoples), Abidjan, NEI/CEDA.
Oruka, H.O. 1993. Parental Earth Ethics. in *Quest*, vol. VII. no. 1 (June 1993), 20–27; Rejoinders. in *Quest*, vol. VII. no. I, 106–109.
Sidibe, F.M. 2020. *De donsoya à donsologie « La confrérie des chasseurs traditionnels – donso ton Matériaux – Concepts – Notions – Croyances et spiritualité »* (From donsoya to donsology: "The brotherhood of traditional hunters - donso ton Materials - Concepts - Notions - Beliefs and spirituality", Bamako, EDIS.
Tangwa, G.B. 1996. Bioethics: An African Perspective. *Bioethics* 10 (3).

Belko Ouologuem holds a PhD in Philosophy from the University of Fudan in Shanghai, China (2008). He is currently professor of ancient philosophy in the Department of Philosophy at the University of Letters and Social Sciences, Bamako, Mali. Ouologuem teaches a module on bioethics at the *Ecole Normale Supérieure* of Bamako and another on the philosophy of Aristotle at the Catholic University of West Africa in Samaya. He is the author of several articles on comparative philosophy, moral and political philosophy, secularism, and has published a book entitled *La philosophie de Confucius*, L'Harmattan, Paris, 2019.

Chapter 10
Ɨwu-ɨ-Kom-ɨ-Twal: Kom Agrarian-Environmental Ethics

Mbih Jerome Tosam

Abstract In this chapter, I show that Kom agrarian-environmental thought is encapsulated in the Kom triadic worldview commonly referred to as *ɨwu-ɨ-kom-ɨ-twal* (the Kom three hands) which includes "*wayn*" (a child), "*afo-aghina*" (food), and "*nyamngvin*" (communal flourishing). The child denotes perpetuation of culture; food signifies human subsistence; and "*nyamngvin*" symbolizes communal flourishing, which includes not only human, but also ecologic, spiritual, and cosmic flourishing and good health. According to this triadic worldview, agriculture and the use of natural resources must take into consideration the interest of the present, past and future generations. In this chapter, I examine some beliefs, values, and practices connected with agriculture and environmental preservation in indigenous Kom culture. In Kom ontology, all of creation, natural and supernatural, stands in intimate relation with each other in perpetual quest for harmony. It is an outlook which discourages the inconsiderate pollution (moral and spiritual) of nature. Here, emphasis is placed on maintaining fertility, ecological protection, and communal flourishing. Humans do not have a special mandate to dominate and/or exploit nature. As co-occupant and interdependent members of an extended community, moral worth is not restricted to humans or rational agents; it is also extended to include nature. Hence, caring for nature means caring for each other. The right attitude to adopt is inter-species reciprocity and solidarity, with humans having the moral responsibility to ensure cosmic balance.

Keywords *ɨwu-ɨ-kom-ɨ-twal* · Kom agrarianism · African agrarian philosophy · Kom triadic worldview · Environmental ethics · Future generations

---

M. J. Tosam (✉)
Department of Philosophy, Faculty of Arts, The University of Bamenda, Bamenda, Cameroon
e-mail: mtosam2002@yahoo.com

M. J. Tosam and E. Masitera (eds.), *African Agrarian Philosophy*, The International Library of Environmental, Agricultural and Food Ethics 35,
https://doi.org/10.1007/978-3-031-43040-4_10

10.1 Introduction

Agriculture and the use of the natural environment in most indigenous African cultures have greatly influenced religious beliefs, moral norms, values, and the emergence of social organizations (Kugedera et al. 2021), to regulate not only inter-human, but also inter-species relations. In pre-colonial Kom society it was difficult to separate agriculture from culture. Agriculture as a dominant lifestyle of the people shaped indigenous beliefs, customs, and practices that transcended a wide range of issues related to environmental protection, food security, religion, medicine, biological reproduction, indigenous knowledge, and social organization. Freyfogle has emphasized the role of nature in forming norms and customs when he says: "nature helps shape norms and social structures..." (Freyfogle 2008: 547). How can these traditional values and customs help to ensure food security and environmental protection today with the growing pace of environmental degradation, pollution, and disappearance of flora and fauna caused by modern technology, population explosion, and especially the mutations that Kom culture has experienced as a result of the influences of colonization, globalization and capitalism?

Kom agrarian-environmentalism is relational. It is a continuum that includes humans, the ecosystem, spirits, the living dead, and the unborn in the moral community. Humans depend on nature for survival and in order to show gratitude to preceding generations, they ensure that they do not deplete nature and the available resources for the sake of future generations (Wiredu 1994: 46; Murove 2004:184). According to Fern, "human life cannot be sustained, let alone achieve well-being, apart from the manifold goods of ... nature" (2004: 11). For instance, in most cultures of the world, "trees have been important to humans as 'natural symbols' of the central values of communal life, as sources of food and medicine, and as signs of spiritual realities" (Roothaan 2018:135). However, it is not only for their instrumental ends that humans protect nature; it is also for nature's sake—for its intrinsic value.

Kom agrarian thought is encapsulated in the Kom triadic ontology commonly referred to as *ɨwu-ɨ-kom-ɨ-twal* and roughly translated as the Kom three hands, which includes "*wayn*" (a child), "*afo-aghina*", (food), and "*nyamngvin*", variously translated as welfare, prosperity, communal flourishing, and good health. *Ɨwu-ɨ-kom-ɨ-twal* provides insights into Kom agrarian philosophy because it provides norms, values and guidance on how to relate with other members of the community and nature and how manage the environment to ensure the sustenance and welfare of the community. Telleen argues that agrarianism functions "in such a way that it honors and maintains the earth, sustains and perpetuates the community, shelters and benefits the citizens thereof, and respects the commonwealth for what it is: the common wealth. It is neither a warrior society nor an industrial society" (Telleen 2013: 29). From the same angle, Thompson maintains that, "agrarian philosophies ...offer formulations on the way that norms, values and social institutions emerge from human beings' interaction with nature in the form of material subsistence practices such as obtaining food, clothing and shelter" (Thompson 2008: 528). In the Kom triad, the child denotes continuity and perpetuation of culture; food symbolizes human subsistence; and communal flourishing signifies common wealth, the reverence for life and the natural environment. According to the Kom triad, agriculture and the use of natural resources must be carried out taking into consideration the interest of past, present, and future generations. Humans must procreate to ensure continuity; produce food for their subsistence, health, and the common good of both humans and nature.

It may be argued that the quest for development based primarily on Western modern materialistic and individualistic ethos has resulted in the general neglect of some indigenous Kom values and practices connected with farming, food ethics, and environmental conservation. This has happened as if to say development and modernization are incompatible with indigenous ontologies and values. However, although Kom agrarian values have been changing, as it is the case with most African societies today, as a result of the epistemologically and culturally disruptive influences of colonialism, Christianity, capitalism, and globalization, some traditional Kom agrarian values have endured.

In this chapter, using Kom relational ontology, I show that *ɨwu-ɨ-kom-ɨ-twal* forms the kernel of Kom agrarian-environmental philosophy. The Kom believe that humans are integral members of a physical and supernatural community in which they entertain a reciprocal and interdependent relationship. Hence, to count, morally, a being must not necessarily be real, rational or be able to defend his/her interest. This view has been underscored by many African thinkers. For example, Mbiti says "the spiritual universe is a unit with the physical, and… these two intermingle and dovetail into each other so much so that it is not easy, or even necessary, at times to draw the distinction or separate them" (Mbiti, 1969: 74). The notion of the universe as a continuum (Bujo 1998: 20; Ikuenobe 2006: 63, 2014: 2), triangular (Udeani 2008: 67), and interdependent (Wiredu 1994: 46; Tangwa 1996: 191; Murove 2004: 185; Behrens 2014: 62; Chemhuru 2014: 75; Tosam 2019: 177), where there is no strong separation between subject/object. These beliefs and values ensured that traditional farming methods and land use did not pollute or degrade the environment, and that the living did not exhaust natural resources for the good of past and future generations.

It is a also from the same perspective that Ramose argues that "the principle of wholeness applies also to the relation between human beings and …nature. To care for one another, therefore, implies caring for physical nature as well. Without such care, the interdependence between human beings and physical nature would be undermined" (Ramose 2009: 309). Without care, cosmic equilibrium will also be threatened. This view is fundamental not only to Kom agrarianism, but also to the agrarian ideal as a whole. As Wirzba contends:

> Agrarianism is about learning to take up the responsibilities that protect, preserve, and celebrate life. The first requirement of such responsibility is that we give up the delusion that we live in a purely human world of our own making, give up the arrogant and naive belief that human ambition should be the sole measure of cultural success or failure (Wirzba 2013: 7).

According to agrarian ethos, since we are interdependent beings, the interest/survival of all beings, natural and spiritual, must count in our moral consideration.

This chapter is divided into four sections: In section one, which is largely descriptive, I present the triad as a heuristic for understanding Kom agrarian-environmental thought. Section two examines Kom agrarianism-environmental ethics. In section three, I discuss ways in which traditional authorities and institutions in Kom used to regulate farming and the use of natural resources to ensure responsible exploitation of nature. Finally, in section four, I focus on the moral considerations of future generations in Kom agrarian-environmental philosophy.

10.2 The Triadic Worldview as the Basis of Kom Agrarian-Environmental Thought

There are three foundational values in Kom ontology. These include: *a child, food, and communal flourishing.* Apart from capturing Kom agrarian and environmental thought, *ɨwu kom ɨtwal* also highlights some beliefs, customs, and norms which are relevant to indigenous Kom epistemology, metaphysics, and moral and social philosophy.

The Kom people belong to the Tikar group who migrated from northern Cameroon around the 18th Century fleeing from Islamic Jihadists (Nkwi 2017: 36–37). They are located in the Bamenda Grassfields in the North West region of Cameroon. Kom is largely an agricultural community comprising mainly of subsistent farmers with a few exceptions who engage in fishing, and palm wine tapping in riverine areas, hunting and gathering of non-timber forest resources. Before exploring the Kom triad and its significance to agrarianism, it will be necessary to briefly examine the Kom myth of origin, which also has ecological and agrarian implications.

Myth-creation is a fundamental part of human existence, and myths arise from human epistemic limitation. Hence, "all human cultures engage in myth-making" (Tangwa 2010: 201). Without myths, some ultimate questions about the origin and meaning of human life cannot be answered (Tangwa 2010: 201). Across sub-Saharan Africa, creation myths abound.

According to Kom orature, the people migrated from Ndobo in northern Cameroon with other Tikar groups escaping the 19th century jihads of Uthman dan Fodio (Nkwi 2017: 36). Oral sources maintain that the Kom first moved to Babessi, in the present-day Ndop plain, where they settled briefly. Here, the king of Babessi hatched a sinister plan to eliminate them (Nkwi 1976). The myth is recounted that one day the *foyn* (king) of Babessi told the *foyn* of Kom that some of their people were becoming stubborn and might trigger a war between the two groups. He therefore suggested that they should each build a house in which the trouble-makers would be burnt. The *foyn* of Kom, Muni, agreed to the plan and the houses were constructed. However, while the king of Babessi constructed his house with two doors, the *foyn* of Kom built his, unsuspectingly, with a single door, as agreed. After locking the front doors, the houses were set ablaze. The Babessi people escaped through the second doors while the Kom people were burnt to death (Nkwi 2017: 37). This scheme reduced the size of the Kom population in Babessi and exasperated the *foyn* of Kom. In anger, Muni decided to avenge the death of his people. He told his wives and sisters that he would commit suicide on a tree in a nearby forest and on that spot a lake would emerge and all the maggots from his putrefied body will turn into fish, but he warned the Kom people against fishing in the lake. Shortly after hanging himself, a lake was discovered by a Babessi hunter who immediately reported to the palace and a royal fishing mission was organized. At the peak of the fishing the lake mysteriously closed up and all the Babessi people who were in it got drowned. Following Muni's instructions, a mythical python's trail, believed to be the reincarnated *foyn* (Nkwi 2017: 36), led the surviving Kom people from Babessi to Nkar and Idien, and then to Laikom, their current location and the traditional capital of the Komland.

According to oral sources, the sacred forest is located in the spot where the mythical python is believed to have disappeared. It is also where the dead *foyns* are buried (Nkwi 2017). "In the sacred forest, no farming [is] carried out; there was no fetching of wood or felling down of trees. It is in this forest that religious rituals are usually performed to mark the start of the planting season" (Nkwi 2017: 40). Although this is not a specifically African practice, the practice of the sacred forest helps in the promotion of ecological conservation.

In most African societies different animal and plant species were granted moral status based on their symbolic powers and cultural influence on the people or on the basis of the belief of the people about the contribution of these animals, plants, and natural features to the founding of their clan. For example, in some communities, when certain types of snakes are seen at home they are fed because it is believed that the snakes are ancestors (Gumo et al. 2012; Nkwi 2017). In Kom, the green mamba (*ɨssɨ-ɨ-ɨjva*) is still revered because it is believed that children, especially twin, transform into them. Another example of a revered snake is the *ɨmueyn* (*amphisbaena fuliginosa*) which is considered as ancestral manifestation. "The belief in animal affinities, or spiritual beings and that dead kings were residing in waterfalls and deep pools… are widespread across the grassfields" (Chilver and Kaberry 1967: 47) of the West and North West regions of Cameroon. Beliefs and customs about nature play a fundamental role in African ontology. As Imafidon argues,

> Deified divinities… are heroes and founding fathers of African communities who contributed…to the founding of a people and are believed, in death, to be in a position to influence the community positively by relating their problems to the Supreme Being. A man who sacrificed himself to save his village, a man who started a settlement or a woman who performed great feats for her village was, at her death, deified and venerated and became the primary deity for the respective community. Shrines and statues are erected for them and either once, or some instances more than once, in a year festivals are held to commemorate their lives and feats (Imafidon 2014: 41).

It is for the above reason that in most indigenous African cultures, animals were not hunted for fun. "They were hunted for food and for religious reasons…. The African people know, like the Native American people knew, that if you destroy the environment, you will ultimately destroy the human race" (Mutwa 1996: 19). This shows the very close and interdependent link between humans and nature. It is also for this reason that humans did not regard themselves as superior to, but as part of nature (Mutwa 1996: 9–10) in traditional African culture. They considered themselves as part of nature. Hence, when we protect nature, we are also protecting humans. From this outlook, it is evident that indigenous African agrarian thought is neither anthropocentric or biocentric, it is "eco-bio-communitarian" (Tangwa 2004: 387) where there exists a "slim and flexible line between plants, animals, and inanimate things, between the sacred and the profane, matter and spirit, the communal and the individual" (Tangwa 2004: 387). Let me now return to the Kom triad.

10.3 Wayn (Child (ren))

A child (*wayn*) stands for the future and cyclical continuity. Children are the living continuum of their forebears; a means of cultural reproduction; the only way through which their beliefs, values, history, and customs can endure. In Kom culture, procreation and child bearing are important aspects of life in the community. Although this is true for all human cultures, because it is the only "means by which each society can perpetuate itself, its customs and institutions" (Cook et al. 2003: 3), in Kom society, there is a special premium attached to procreation such that when a couple does not have children or is unable to have them, it may result in the loss of dignity. One's personhood and social recognition is determined by his/her ability to have children. Reproduction is so important that it is the primary reason why people get married. According to Mbiti, "children in an African marriage are an absolute necessity. Without procreation, marriage is a waste and incomplete. A person who has no children, in effect quenches the fire of life, and becomes dead, since his line of physical continuation is blocked" (Mbiti 1977: 192). In Kom thought, *abzɨ* (reproduction) children are perceived as "wealth for the family and community as a whole" (Nsom 2015: 45). In some cases, failure to have children is considered as sufficient justification for divorce or for a man to marry another wife.

According to the above outlook, reproduction ensures that the welfare of the living will be taken care of when they die. In this interconnected and interdependent cosmos, "the dead depend on the actions, especially ritual sacrifices, of the living for their well-being and the living, in turn, depend on the solicitations and intermediary of the dead for their health, progress and well-being" (Tangwa 2010: 17).

There are a number of Kom proverbs which highlight the importance of procreation and child bearing in Kom culture:

Wayn nɨ ghɨ nka' mɨ dviyn (A child is one's firewood at old age).

Njɨ nɨ dvɨna yi nyuŋ wàyn (When a sheep grows old it is suckled by its offspring).

These two proverbs show that for the Kom, children are one's security or life assurance at old age; they are not only one's security in the physical, but also in the supernatural world. It is believed that without children, the health and well-being of the living dead cannot be assured and vice versa; there will be no one to offer ritual sacrifices to the ancestors and no one to solicit on their behalf in the ancestral world. Hence, children guarantee that cultural and spiritual continuity is not blocked. Bujo contends that "…the living dead can "enjoy" their being ancestors only through the living clan community. In this way, a kind of "interaction"—hierarchically organised from top to bottom and vice versa—is created … The goal of this interaction is the increase of vitality within the clan" (Bujo 1998:16). Without children, no culture is certain of its survival. Commenting on the importance of adherents and followers of a culture for its continuous survival, Fern opines that "every tradition has as a point-of-view, an *ethos*, that defines its identity and shapes its unfolding over time. Adoption of this *ethos* is part of what makes one an adherent of a given tradition. No tradition would exist apart from its affirmation by its adherents…." (Fern 2004: 179). For the Kom, and certainly for all human societies, it is only through children that the continuity of culture and vitality can be guaranteed. This outlook is morally important because it engenders a culture of responsible use of natural resources taking into consideration the welfare of the past and future generations. The *fɨchuo* ritual which is usually performed by Kom traditional authorities is aimed at improving fertility of the land (for all species), as well as vitality in the universe.

10.4 *Afo-aghina* (Food)

The second arm of the Kom triad is *afo-aghina*. Food, agriculture and human survival are intimately interwoven (Nath 2012: 1). Food is not only a means of survival, but also an essential part of a people's culture (Wirzba 2013: 12). "…indigenous food-ways have reflected a worldview grounded in principles of reciprocity that actively nourish health, culture, and nature" (Global Alliance for the Future of Food 2021: 5). Food can "communicate information in terms not only of occasion but also social status, ethnicity and wealth" (Murcott 1982: 203). In Kom society, the question of food security and human subsistence is preponderant; it has always been the concern of all and traditional authorities have the obligation to ensure that the land is fertile and that there are no natural disasters which may affect the abundant supply of food crops. There are beliefs, norms, taboos linked with farming, hunting, and fishing. There are rituals to enhance fertility, sustainable use of natural resources, and to ensure food security in the land. The aim of these was to maximize the productivity of the land, prevent natural disasters like floods, strong winds, droughts, and to improve the well-being (nyamngvin) of people by limiting damage to natural resources.

Food also has a symbolic meaning. In Kom culture, food is a way of expressing love, solidarity and friendship. Kola nuts and palm wine, for example, are used not only for entertainment, but also as a symbol of life, love, and peace. Moreso, there are some common foods used to perform rituals namely, *egusi* (pumpkin seeds), *corn fufu*, tadpoles, palm wine, and kola nuts. Food is therefore a medium through which Kom communal culture is expressed and transmitted.

To ensure sustainable use of natural resources and food in particular, there was subsistence farming which involved rotation farming. Hunting was also strictly regulated. There were particular periods of the year dedicated for hunting and there were taboos against killing young and pregnant animals to ensure that plant and animal species were not extinct. As in most of the world's religions, food plays a very significant role. For example, Jewish tradition is oriented towards seeing in the act of nourishment a meaning that teaches choice and continuous certification, defines the relationship of humans with nature and is deeply concerned with sacredness (Sibal 2018).

Moreover, food is a symbol of love and social communion. "Food and food habits as a basic part of culture serve as a focus of emotional association, a channel of love, discrimination and disapproval and usually have symbolic references. The sharing of food symbolizes a high degree of social intimacy and acceptance" (Reddy and Anitha 2015: 613).

10.5 *Nyamngvin* (Communal flourishing/Welfare)

The third and last arm of the triad is *nyamngvin*. In *itanghɨ kom* (the Kom language), *nyamngvin* does not have a straightforward and simple meaning. It symbolizes welfare, prosperity, the common good or communal flourishing, and good health. Etymologically, the word *nyamngvin* is derived from two Kom words: *nyam*, which invariably stands for meat, animal, property, and all the good things that nature and human ingenuity has to offer; and *ngving*, which is commonly translated as arable land, fertility (human, animal, and crop fertility), good harvest, abundance, success, etc. In Kom social philosophy, *nyamngvin* is the ultimate good, the highest good, the *summum bonum*, for which all members of the community must strive to achieve for the common good of all. The idea of the common good or communal well-being as the ultimate value of African communitarian ethics has been emphasised by many African thinkers (Gyekye 1995, 2010; Wiredu 1996, Bewaji 2004; Metz 2022). *Nyamngvin* is sought for to provide for or to satisfy *wayn* (the child) which includes all members of the community as well as the young and the unborn—future generations. Because in Kom moral outlook the community is a continuum which encompasses humans, animals, plants, inanimate beings, the unborn and ancestors, the notion of *nyamngvin* also includes ecological and cosmic well-being. Agriculture always involves the interaction with and between all these spheres and forces of nature.

According to Kom moral thought, each member of the community at their individual level is expected to strive towards achieving *nyamngvin* and when they achieve success in their individual pursuits, they ought to use it to promote the common good since, the individual is the product of his/her community, and as Mbiti argues, "the community …make [s], create[s], and produce[s] the individual, for the individual depends on the corporate group. Physical birth is not enough…." (Mbiti 1969: 141). A Kom proverb which underscores this fundamental role of the community in the nurturing of a person is the following: *wayn wul imo' nɨn go' ɨlv-a,* which is translated as 'a personal child is only in the womb.' Once a child is born, she becomes the 'property' of the community and not simply the 'possession' of his/her biological parents/

family. Here, a special weight is placed on communal belonging. According the Kom communitarian ethos, an action is right if it promotes the interest of the community, and is wrong if it does not. Another proverb which accentuates the importance of communion in Kom culture is the following: *Nɨnyɨŋ nɨn jofɨ kɨsɨ avɨŋa,* which is translated as "aloofness is good only for a witch/wizard." The proverb highlights the importance of solidarity, reciprocity, community in Kom moral thought. From the same perspective, Nyerere outlines three fundamental values which define communal living in traditional African community. These include: "mutual involvement in one another"; "property", that is, "all basic goods were held in common and shared" (Nyerere 1968a: 107) equally among members of the community, "no one could go hungry while others hoarded food, and no one could be denied shelter if others had space to spare" (Nyerere 1968a: 107). The third principle is that "everyone had an obligation to work" (Nyerere 1968a: 108). *Nyamngvin* can only be achieved when the community as a whole flourishes.

It is only in the community that an individual can attain genuine personhood. This idea of the interdependence and shared humanity of persons is further underscored in the following proverb: *awu ami'a ka'kɨ bû læ kul ibu,* which means "one hand cannot tie a bundle." This means that no individual person is self-sufficient and no one can succeed without the support of others, which is further summed-up in the following proverb: *wul nɨn ghɨ wul bom wul,* commonly translated as "a person is a person through/because of other persons" (Tosam 2014: 12). In this moral outlook, no one is considered to be truly wealthy if his/her wealth does not benefit others. Communal flourishing is the basis of Julius Nyerere's *Ujamaa* (familyhood, or brotherhood) philosophy. According to Nyerere, the aim of African socialism is to build a society in which all members have equal rights and opportunities; a society in which all members can live in peace with each other without suffering or injustice; and in which all members have a basic level of material welfare before any individual lives in luxury (Nyerere 1968b: 340). The significance of this communal ethos is that it promotes cooperation and, unlike capitalism, it does not encourage excessive individualism, unnecessary competition, the tendency to exploit the weak, and does not promote inequality in the society. However, African communitarianism also has its pitfalls. It is liable "to encourage complacency and individual laziness" and "a generally stagnant and unprogressive society" (Tangwa 2002: 2019).

In Aristotelian ethics, *nyamngvin* would denote *eudaimonia*, which is a Greek word also commonly translated as human flourishing, happiness, well-being. For Aristotle, *eudaimonia* is the highest human good, the only good that is desirable for its own sake rather than for the sake of something else or as a means toward some other end. In Kom moral worldview *nyamngvin* refers to mutual caring, happiness, peace, health, and prosperity and communal well-being. When there is prosperity and good health, where there is general fertility (humans, animals and plants), the Kom talk of *nyamngvin.* Hence, when the community is flourishing economically, culturally, politically, ecologically, and spiritually, there is *nyamngvin.* However, unlike *eudaimonia, nyamngvin* does not merely denote human good or "activity of soul in accordance with virtue," (Aristotle, 2004, Bk I, ch. 7, 1097b) or rational activity performed virtuously or excellently, but rather denotes actions that lead to the well-being of all members of the community. When a new *foyn* is enthroned, he is usually empowered, through installation rites, with the responsibility to ensure that nothing, humanly or spiritually, stands on the way of the attainment of the above-mentioned three values (Nkwi 1976: 50).

10.6 Kom Agrarian-Environmental Ethics

Kom agrarian and environmental philosophy can be easily gleaned through traditional Kom ecological beliefs, values, and practices as stated above. It is a wholistic worldview in which we share our world with nature, with the earth, its soils, plants, and animals (Fern 2004: 202). According to the Kom worldview, because humans share the cosmos with nature, they cannot cut the earth or use its resources without consulting or obtaining permission from nature. This outlook is in line with one of the core values of agrarianism. As Wirzba opines, "agrarianism builds on the acknowledgment that we are biological and social beings that depend on healthy habitats and communities" (Wirzba 2013: 4). In African ontology, this interdependence goes beyond the physical universe. For instance, Mbiti argues that "the spiritual universe is a unit with the physical, and… these two intermingle and dovetail into each other so much so that it is not easy, or even necessary, at times to draw the distinction or separate them" (Mbiti 1969: 74). For Tangwa, "within the Nso' worldview the distinction between plants, animals, and inanimate things, between the sacred and the profane, matter and spirit, the communal and the individual, is a slim and flexible one" (Tangwa 2004: 389). Murove contends that "…the distinction between humanity and nature, the living and the dead, the divine and the human is blurred to such an extent that human existence becomes continuous with the natural world" (Murove 2004: 185) such that there is "…an indissoluble solidarity between humanity and the natural environment" (Murove 2004: 202). On his part, Kelbessa talks of a "…positive relation between individuals, humans and the natural environment" (Kelbessa 2005: 25). In traditional Kom society, there were sacred zones where animals and plants were thought to inhabit spirits. For example, there were pools and areas in some rivers where fishing was proscribed because the rivers were considered to inhabit water spirits. The green mamba or the *ɨmueyn* are snakes which were offered food or spilled with camwood instead of being killed because of the human spirits they were (are) thought to incarnate. When the *ɨmueyn* was killed inadvertently, purification was done.

H. Odera Oruka and Calestus Juma have called for a 'Parental Earth Ethics' in which the notion of the community is extended to include the natural environment. "We hope it is clear that the earth or the world is a kind of family unit in which the members have kith and kin relationship with one another" (Oruka and Juma 1994: 125–126). From the same relational viewpoint, Bujo contends that in traditional African thought, "all beings, organic and inorganic, living and inanimate, personal and impersonal, visible and invisible, act together to manifest the universal solidarity of creation" (1998: 209; Ouologuem, chapter 9 in this book). Ikuenobe also contends that "reality is seen as a composite, unity and harmony of natural forces. Reality is a holistic community of mutually reinforcing natural life forces consisting of human communities (families, villages, nations, and humanity), spirits, gods, deities, stones, sand, mountains, rivers, plants, and animals (2014: 2). According to Ikuenobe, "everything in reality has a vital force or energy such that the harmonious interactions among them strengthen reality (2014: 2).

What the above perspectives from African philosophers imply is that African agrarian-environmentalism is neither anthropocentric nor biocentric; it encompasses both and goes beyond. Tangwa thinks that it would be inaccurate to characterize African environmental thought as "either eco-centred, bio-centred or anthropocentred" (1996: 188–189); because it is an all-embracing moral worldview. Tangwa broadly describes the African outlook as "eco-bio-communitarian" (2004: 395); which involves the "recognition and acceptance of inter-dependence and peaceful coexistence between earth, plants, animals and humans" (Tangwa 2004: 387–395). In this interdependent cosmos, human beings are "more humble and more cautious, more mistrustful and unsure of human knowledge and capabilities, more reconciliatory and respectful of other people, plants, animals, inanimate things, as well as sundry invisible/intangible forces, more timorous of wantonly tampering with nature, in short, more disposed towards an attitude of live and let live ...." (Tangwa 1996: 191). Emphasizing this interdependent interconnectedness in the African universe, Bujo avers that: "The relationship between those living on earth and the ancestors is very close, since the living owe their existence to the ancestors from whom they receive everything necessary for life... (Bujo 1998: 16). Hence, both members of the extended community need each other for their mutual subsistence.

In Kom culture, it is believed that some spirits incarnate animals, plants, forests, rivers, and surrounding mountains and if injured through inconsiderate exploitation of nature, this could result in different forms of natural disasters, birth malformations, and misfortunes. For example, it is forbidden to cut any plant at night with a machete or to through any objects. However, if one must do so, he/she must seek the permission of the spirits (through rituals or some other means) as commonly requested in the saying: *"yi leu's ɨ itu ɨyvɨ"*, translated as "take off/shift your heads." According to Kom cosmology, nature can be harmed, and since the health of the environment is intrinsically linked to ecological and spiritual health, everything must be done to ensure a safe and healthy environment for both past and future generations. Each generation is bestowed with the responsibility to protect nature, for the sake of the unborn and for the sake of the death—in a sort of trans-generational solidarity. In this sense, therefore, any irresponsible exploitation of nature was considered a sacrilege. According this view of the cosmos broadly defended by a number of African scholars, "the necessity of having an attitude of respect toward nature as a good of its own, not just preserving nature for the ultimate good of persons.... The notion of the interrelatedness of all natural things is taken as providing grounds for a moral obligation to treat nature with respect" (Behrens 2014: 59).

Moreover, in Kom culture, it was believed that humans with extraordinary powers can transform into different animals or creatures. The Kom people believe that spirits co-habit their natural environment. Therefore, during the sacred days of the week—*ɨtuh ɨbolɨ and ɨtuh ɨyvɨ nɨ kom*, no farming is allowed since it is believed other spirits do farm on these days. Among the Kom people, it is held that the *foyn* can transform into different types of animals. This idea is widespread among most African communities. Kanu underscores this idea when he contends that "sacred animals occupy a fundamental place in African traditional religion. If they are not considered the symbol of the ancestors and deities, they are considered more suitable for the offering of sacrifice, etc." (2021: 2). The people were expected to show respect for totems, and the totems in turn were supposed to protect them in times of despair. In communities where there are totems, there are taboos, laws, proscriptions against the killing of such animals except in self-defence. There were taboos against the killing of pregnant animals, young animals, sick animals, and against the depletion of species. Special rites and prayers were offered before some trees were felled down. For example, before a kola nut tree is cut down in Kom society, prayers and rituals are performed. This practice is informed by the metaphysical idea that there is no clear distinction between the natural and spiritual world. Tempels avers that in African ontology, "a human group and an animal species can occupy in their respective classes a rank relatively equal or relatively different. Their vital rank can be parallel or different. A chief in the class of humans shows his royal rank by wearing the skin of a royal animal" (Tempels 1959: 61). In Kom society royalty is expressed by putting of feet on a leopard skin and the *foyn* is referred to as the Leopard. In traditional Nso' society, the people clapped or bowed down when they saw a lion, because it symbolizes royalty and power. In the Nso' land, "the King … is … believed to be capable of transformations as an aspect of his kingly attributes" (Tangwa 1996: 191). In this interdependent universe, to ensure good health, one must live in harmony, not in conflict, with nature. People have rights and duties not only towards one another, but also towards nature. The following Kom proverb highlights this attitude towards: *Wa k æ ajva wa fayti tɨnteŋ*, which translates as: "when you harvest garden eggs, mend its branches." The idea invoked in this maxim is that if you benefit from something then use it responsibly because other members of the community will also need it for their survival. It was therefore morally wrong to completely expend a species.

In traditional Kom society, nature was considered as the commonwealth of the community and people owned property in common. You could go to anybody's farm and harvest any crop provided you harvested only the quantity necessary to mitigate your hunger (Tosam 2019). If you harvested too much it was considered stealing because your action would be inconsiderate. In some African communities there are moral codes and practices which regulate inter-human and inter- species relationship. For example, in the Oromo society in Ethiopia, *Saffu* is used to guide the proper, fair, and sustainable management of the environment (Kelbessa 2005: 24). For Kelbessa, African environmental thought ". . . does not allow irresponsible and unlimited exploitation" of natural resources (2005: 24). Moreover, in African ethics, "justice, integrity and *respect* as human virtues are not only applicable to human beings but they extend them to nonhuman species and mother Earth" (Kelbessa 2005: 24).

10.7 Traditional Authorities and the Enforcement of Agrarian-Environmental Values

Kom agrarianism is not only expressed theoretically, agrarian values and customs are enforced through rituals, taboos, injunctions, and sanctions exacted against environmental transgressors. In the Kom society, the *foyn* exerts great, but not absolute, political, moral and spiritual power over the community he/she is charged to administer, because he could also be dispossessed of these powers by the *kwi'foyn* (the traditional council), if he/she abuses them. "The [*foyn*] is the spiritual leader, the chief priest… of his people" (Nkwi 2017: 34) and is also the "…protector of fertility of the land, promoter of good harvest, prosperity...." (Nkwi 1976: 60). In the domain of agriculture and environmental protection, the influence of the *foyn* is most evident. The *foyn* provides guidance on land distribution and use, how farming is to be carried out, religious rituals connected with farming, hunting, and conflict resolutions. This fundamental role of ensuring peace, justice and prosperity of his/her people is emphasized at the enthronement of the *foyn*. During the rite of purification that precedes the enthronement of the *foyn*, the presiding priest makes the following or similar prayer to the ancestors and the gods:

We have brought the new *foyn* to show you,
Make him to rule Kom with peace and justice
Let no son of Kom cry
Let him rule with one voice,
Let him rule the child, food, and animals [ensure communal flourishing/*nyamngvin*]
We have come to feed you,
So that you might teach him the right ways to rule.

In Kom society, rituals meant to ensure good yield, to protect crops from violent weather and natural disasters, and from destruction by evil forces (de Wit 2009: 126–127), were presided over by the *foyn*, since the weather and climate are associated to the supernatural. Under the authority of the *foyn* is a powerful traditional council, the *Kwi'foyn*, which enforces these regulations. They also mediate between the natural and supernatural world (de Wit 2009: 125–126). Moreover, traditional authorities have the responsibility to organize communal labour (construction of roads, bridges, and community buildings, etc.), dealing with crimes and governing the country (de Wit 2009: 135).

The fields are communally owned and are placed under the custody of traditional rulers. The later, in consultation with other traditional governance structures, decides the fields that are to be cultivated and those to be allowed to fallow. The traditional rulers controlled the programing of the various periods of the farming season—farming, planting, weeding, and harvest. Moreover, regulatory authorities are responsible for maintaining cosmic, moral, and social harmony caused by human misconduct. Traditional authorities also ensure that taboos aimed at protecting the natural environment are respected and defaulters are sanctioned. According to the Kom worldview natural disasters occur as a result of human misbehaviour. Hence, "fighting and assaulting people, was considered as… 'polluting' and, hence, the cause of misfortune. Accidental death by drowning or lightening, fire destroying property, or destructive natural events in general, are also considered as polluting the land" (Wanier and Nkwi 1982: 57; de Wit 2009: 135). Moreover, traditional authorities were charged with restoring lost cosmic harmony by eliminating the evil forces that defiled and polluted the land. Rituals and expiatory sacrifices were usually carried out to eliminate misfortunes caused by human pollution.

10.7.1 Agrarian Rituals and Environmental Protection in Kom Culture

Rituals are an important part of indigenous African agrarian regulatory and remedial system. Fayemi contends that "…ritual is the principal tool used to approach that world of felt but unseen forces in a way that will rearrange the structure of the physical world and bring about ecological equilibrium" (Fayemi 2016:). The *fɨchuo, ntul, azhea, aghoi, and mukain* rituals are performed to promote "fertility, health, prosperity, peace and justice" (Nkwi 1976: 53) throughout the Kom fondom. The *Fɨchuo* is carried out by the *achaff* and *ekwi* priests (founding lineages of Kom) to inaugurate the planting season, to promote fertility and ensure food security (Nkwi 1976: 55). In the week following the *fɨchuo* ritual, social gatherings and any occasions where there is loud noise are prohibited (Nkwi 2017: 41). Although the *ntul* ritual was performed to promote "harmony, cohesion… reconciliation, and appeasement among rival groups or individuals…" (Nkwi 1976: 53), it was also performed for those found guilty of environmental crimes like destroying food "crops without any reasonable justification, or cutting down trees forbidden by law, were usually fined and they paid what was called *bzi-ntul* (the *ntul* goat)" (Nkwi 1976: 55). The *azhea* ritual was performed "to secure regular rainfall, sunshine and harvest" (Nkwi 1976: 56). The *aghoi* ritual was performed to "prevent excessive rainfall, hailstones, and thunderstorms" (Nkwi 1976: 56). Finally, the *mukain* ritual was performed "to protect people from harm and misfortunes at home and abroad" (Nkwi 1976: 56).

These rituals are performed not only to ensure regular rainfall and sufficient sunshine to enable good harvest (Chilver and Kaberry 1967; Nkwi 2017), they are also aimed to "…enforce directives with regards to community's use of its environment …" (Schoffeleers 1978: 145). The aim of the rituals, therefore, is to achieve ecological and cosmic purity. These agrarian rituals are informed by the belief that "management of nature depends on the correct management and control of society" (Nkwi 2017: 41). The *fɨchuo* and *azhea* rituals demonstrate the importance of the natural environment to the life and well-being of the Kom community. However, in recent years, the moral, spiritual, and political authority vested on the *foyns* have become a source of conflict and communal disharmony as some traditional leaders abuse their powers by grabbing and selling lands at the expense of their poor subjects.

In a large part of the Bamenda grassfields of the North West region of Cameroon, land disputes involving the *foyns* and their communities abound.

The *fɨchuo* ritual usually takes a whole week of abstention, ritual prayers and sacrifices to the gods of fertility. During this week seeds are blessed by the village chief priest. The *fɨchuo* also purifies the Earth and the people who defiled it. Farming and hunting was prohibited in some specially preserved and protected forests. According to the Kom people, the universe is viewed as God's creation and He is considered to be found in all of His creatures. Hence, humans and the rest of nature are all infused with spirits and forces. In this regard, there are forests and trees which are specially preserved for spiritual as well as material good including initiation rites, religious ceremonies, medicine, food, and human well-being. For example, there were forests specially preserved for medicinal purposes commonly called *akua mɨfɨ* (forest for medicines).

It may be argued that what I have discussed above are about spiritual beings, which are non-material, outside of time, and therefore not susceptible to performing a variety of rational acts which may be 'good' or 'bad' as modern science has taught us. However, science does not provide us with complete knowledge of reality. As a result of the extraordinary success of the new sciences in the 17th and 18th centuries, scientific "methods became the norm to understand nature" and, therefore, "the paradigm for trustworthy knowledge" (Roothaan 2019: 44). This endorsement led to the exclusion of spiritual ontologies, like the one I discuss in this chapter, from the sphere of valid knowledge. However, as Feyerabend and other critics of the methodology of modern science like Goodman (1990), Latour (2013), and Roothaan (2019), have argued, modern science limits its gaze merely to the material and empirical realm thereby narrowing the scope of reality and valid knowledge. The world is not only made of humans, animal, plants, minerals, and the material things we see; there is a spiritual realm which is beyond the reach of scientific experience. Criticizing this unfortunate turn of events, Feyerabend in *Conquest of Abundance* (1999) avers that:

> The search for reality that accompanied the growth of Western civilization played an important role in the process of simplifying the world…this search has a strong negative component. It does not accept the phenomena as they are, it changes them, either in thought (abstraction) or by actively interfering with them (experiment)… In both cases, things are being taken away or 'blocked off' from the totality that surrounds us. Interestingly the remains are called 'real', which means they are regarded as more important than the totality itself (Feyerabend, cited by Roothaan 2019: 2).

In the same light, Roothaan contends, further that:

> Scientific discourse doesn't include earth, or non-human others in the palaver, presupposing they don't speak. When secular conservationism for instance makes a case that 'the great five' should be protected, it will not look into the interconnectedness with everything that would make survival of those animals possible—as they may need preserving the entire network which they are a part (Roothaan 2019: 138).

According to Roothaan, in order to bring back the spiritual into the realm of valid knowledge, it will be necessary to "take spiritual ontologies and those who hold them seriously, as negotiating partners …. This means trying to take their voices seriously"

(Roothaan 2019: 5). For her, it should begin by recognizing that the colonial actions of the modernists have been responsible for the suppression of indigenous people's rights and their ways of being in the world. The colonial imposition of naturalism as the only surest, reliable and valid source of knowledge has deprived humanity of an enormous source of knowledge about the world stemming from indigenous ontologies.

10.8 Kom Agrarian-Environmental Ethics and Obligations to Future Generations

Deeply implicated in Kom relational agrarian environmental ethics, as seen in the triadic ontology, is not only a strong commitment to future generations, but also to present and past generations. The moral community is a continuum of past, present, and future members. The environment is the habitat of the ancestors and spirits (Chemhuru 2016); and the living have moral obligation to manage the environment sustainably to ensure that they bequeath to future generations a clean and healthy environment just as their ancestors bequeathed to them—a sort of intergenerational justice. The question of environmental ethics and intergenerational obligation has been largely explored within contemporary Western moral and political thought (Partridge 2003; Westra 2006: 136); there has not been any serious scholarly attention given to the question in African environmental philosophy. However, some African philosophers have advanced ideas which may be considered as indigenous African perspectives or contributions to the question of agrarian-environmental ethics and obligation to future generations (Behrens 2011; 2014).

Partridge has highlighted some theoretical difficulties connected with having moral obligations towards future generations (Partridge 2003: 429–432). Before I delve into the question of moral obligations to future generations in Kom agrarian thought, it will be imperative to examine some of these theoretical difficulties.

The first concern is about having responsibilities towards future beings; people who do not (yet) exist. Traditional Western moral philosophy has mainly focused on moral obligations towards real persons. The second theoretical difficulty is that the very existence of future persons will depend on the actions of the living. Hence, how can we, therefore be held responsible for future persons who will owe their existence to our decisions in the present? The third concern is that most of our ethical obligations are expressed to self-determining agents. We do not know who these future people will become. The fourth difficulty is that moral obligations are usually grounded on reciprocity. Future or past generations cannot reciprocate. The fifth challenge is that future persons cannot negotiate a contract, since they do not really exist. We can only negotiate a contract with real people (Behrens 2011:123). The sixth concern is that not only do we not know who future persons will be; we also do not know what their actual desires and interests will be. They may discover new sources of renewable energy, making all our efforts to conserve nature for their sake pointless. The seventh difficulty is that even if it is clear that we have obligations towards future generations, the fact that they do not exist, are dependent on the living, cannot reciprocate, and we do not know their interests, makes it difficult for us to be motivated to preserve nature on their behalf (Partridge 2003: 429–432).

In Kom/African moral ontology, on the contrary, most of these difficulties, if not all, may not apply because of the interdependent connection between the living, past, and future members of the community in Kom thought. In the African moral universe, the community is a continuum involving all the above-mentioned members; they are treated as real persons and are consulted when there is a major decision to be taken concerning the community. Imafidon avers that "in an African community, the ancestor is intrinsically intertwined with the living. Their consent is sought regarding family decisions and they are appeased when a family member errs" (Imafidon 2014: 41). For Bujo, "the living dead can "enjoy" their being ancestors only through the living clan or community" (Bujo 1998: 16). In such an interdependent moral community, it is not only possible to reciprocate between real and future persons as well as past generations, but, above all, to be held accountable by future and past generations by reprimanding them when they get astray. However, this does not imply that indigenous African moral outlook is not dynamic; moral values change as human knowledge and experience improves. Values which violate human dignity, pollute the environment and negatively affect the vital force or harmony that binds and sustains being in the cosmos are generally subject to change as new and better ones are discovered.

The nature of indigenous Kom/African ethics makes it more inclined to take into account the welfare of future and past generations in their reflections about nature than other ethical traditions. For example, Udeani proffers that "traditional African spirituality and ethics are, in a way, a triangular relationship which incorporates, firstly, the natural beings in their relationship with other natural beings. Secondly it pertains to the relationship between natural and spiritual beings, and then that of existing spiritual beings in their relation to one another. Ethical and moral issues go beyond the issues that arise within the context of interactions among natural beings" (Udeani 2008: 67). In the same light, Behrens argues that obligations to future generations in African environmental philosophy can be gleaned from the widespread "belief that the ancestors continue to exert an influence over the lives of their living descendants, guiding their behaviour, and, on some accounts, even punishing and rewarding them." (Behrens 2011: 125).

There are Kom proverbs which highlight this moral commitment to future generations, for example, and as stated earlier, the proverb: *Wa kæ ajva wa fayti tɨnteŋ*, which means, "if you harvest garden eggs, mend its branches." The proverb invokes the idea of sustainable use of natural resources for the benefit of others (Tosam 2014: 14). The moral rule here is not to take from nature more than you need and even when you take, leave it in a state that it can still sprout for others to benefit. It was common practice in Kom subsistence farming to allow some farms, or portions of it, for a year or two, for the soil to heal or regain its fertility before it could be cultivated. According to Bujo, because of this human interconnectedness with the cosmos "personal salvation" is perceived "as being connected to the cosmos: one can only save oneself by saving the cosmos too. It is not surprising then that many liturgical African rites call for the participation of all of nature" (Bujo 2009: 282). On his part, Wiredu maintains:

> Of all the duties owed to the ancestors none is more imperious than that of husbanding the resources of the land so as to leave it in good shape for posterity. In this moral scheme the rights of the unborn play such a cardinal role that any traditional African would be nonplussed by the debate in Western philosophy as to the existence of such rights. In upshot there is a two-sided concept of stewardship in the management of the environment involving obligations to both ancestors and descendants which motivates environmental carefulness, all things being equal (Wiredu 1994: 46).

According to Wiredu, the living have double moral responsibility towards past and future generations. Towards their ancestors, the living owe them respect and gratitude, since it is thanks to their "stewardship" and sustainable management that they are enjoying the natural environment and all its resources. An important way of showing gratitude towards past generations is by taking good care of the natural environment to ensure future generations benefit from their stewardship. Behrens considers this as a "sense of anthropocentric inter-generational justice" (Behrens 2011: 125).

The idea of inter-generational obligation is also echoed in the environmental philosophy of Murove. He posits that: "The ethical aspiration of doing good beyond the grave is an explicit expression of immortality of values.... Those who are still living re-enact their relatedness with the past through rituals of remembrance, the living (the present) enter into communion or fellowship with their ancestors" (Murove 2004: 184). In Kom culture, agricultural rituals are carried out with the idea of inter-generational justice in mind. As seen in the triad, each generation has a moral responsibility to ensure that the natural resources bequeathed them by their ancestors are not depleted. It is also in line with the idea of moral responsibility of the living towards future generations that there are taboos against killing young animals, pregnant animals or destroying young plants in Kom culture (Tosam 2014: 15). According to this moral outlook, to count morally, a being must not necessarily be real, physically, or have force to preserve things as they are, or be rational to be able to engage in a contract. According to Murove, "...the present will contribute to its own existence into the future when it has become the past. Through anamnestic solidarity, the communal life of the living (the present) and that of the ancestors (the past) is re-enacted as a gift to be shared and passed on to posterity" (Murove 2004: 184). Murove refers to this moral obligation towards past and future generations as the "immortality of values" (Murove 2004: 18). This immortality of values involves an inter-generational educational process where "...cultural moral values are passed on from each generation to the next. One of these inherited values is that the environment should be protected for the sake of posterity. So, succeeding generations simply learn to take future generations into account" (Behrens 2011: 126).

10.9 Conclusion

In this chapter I have shown that the *iwu-i-kom-i-twal* is the foundation for understanding Kom agrarian-environmental thought. Kom agrarian customs, beliefs, and values ensured that food production and farming methods did not cause pollution and that the living did not exhaust the available resources for the common good of all members of the community including past and future generations. Indigenous Kom relational agrarian-environmental thought seem to have a more plausible justification for the moral consideration of future persons than some leading Western moral theories like utilitarianism and Kantianism. This is because Kom agrarianism does not restrict the moral community to real and self-determining persons, who are capable of entering into mutual agreements with other (real) members of the community. In Kom worldview, plants, animals, humans, the earth, and supernatural beings are included in the moral community. To ensure responsible use of natural resources, taboos and rituals have been instituted to govern everyday life and to safeguard the environment. Indigenous Kom beliefs and values may be a valuable resource for the mitigation of the multiple agrarian and environmental crises that humanity is currently confronted with such as pollution, desertification, and the cumulative disappearance of flora and fauna.

References

Aristotle 2004. Nicomachean Ethics, tans. and edited by Roger Crisp, Cambridge, Cambridge University Press.

Behrens, K. 2011. African Philosophy, Thought and Practice, and their Contribution to Environmental Ethics. PhD Thesis, Faculty of Humanities at the University of Johannesburg.

Behrens, K. 2014. Toward an African Relational Environmentalism. In *Ontologized Ethics*: *New Essays in African Meta-Ethics,* ed. Imafidon, E. and Bewaji, J.A., 55–72. Lanham: Lexington Books.

Bewaji, John Ayotunde Isola 2004. Ethics and Morality in Yoruba Culture. In *A Companion to African Philosophy*, ed. Wiredu, Kwasi, 396–403. Oxford: Blackwell.

Bujo, B. 1998. *The Ethical Dimension of Community: The African Model and the Dialogue between North and South.* Nairobi: Paulines Publications.

Bujo, B. 2009. Ecology and Ethical Responsibility from an African Perspective. In *African Ethics: An Anthology of Comparative and Applied Ethics*, ed. Murove, F.M., 281–297. Scottsville: University of KwaZulu-Natal.

Chemhuru, M. 2014. The Ethical Import in African Metaphysics: A Critical Discourse in Shona Environmental Ethics. In *Ontologized Ethics*: *New Essays in African Meta-Ethics*, ed. Imafidon, E. and Bewaji, J.A., 73–88. Lanham: Lexington Books.

Chemhuru, M. 2016. The Import of African Ontology for Environmental Ethics. PhD Thesis, University of Johannesburg.

Chilver, E.M., and Kaberry P.M. 1967. *Traditional Bamenda: The Pre-colonial History and ethnography of* the *Bamenda Grassfields*. Buea, Cameroon: Government Printers. Ministry of Primary Education and Social Welfare, and West Cameroon, Antiquities Commission.

Cook, R.J., B.M. Dickens, and M.F. Fathalla, eds. 2003. *Reproductive Health and Human Rights: Integrating Medicine, Ethics, and Law*. Oxford: Oxford University Press.

De Wit, S. (2009). Climate Warning: An ethnography of the encounter between global and local climate-change discourses in the Bamenda Grassfields, Cameroon, Bamenda, Langaa.

Fayemi, A.K. 2016. African Environmental Ethics and the Poverty of Eco-Activism in Nigeria: A Hermeneutico-Reconstructionist Appraisal. *Matatu* 48 (2016): 363–88.

Fern, R.L. 2004. *Nature, God and Humanity: Envisioning an Ethic of Nature.* Cambridge: Cambridge University Press.

Frede, D. 2003. Plato's Ethics: An Overview. *Stanford Encyclopedia of Philosophy*, 2016. https://plato.stanford.edu/archives/win2016/entries/plato-ethics/.

Freyfogle, E.T. 2008. Fostering a Culture of Land Commentary on "'Agrarian Philosophy and Ecological Ethics."' *Science and Engineering Ethics* 14: 545–549.

Global Alliance for the Future of Food 2021. *The Politics of Knowledge: Understanding the Evidence for Agroecology, Regenerative Approaches, and Indigenous Foodways.* https://futureoffood.org/.

Goodman, F.D. 1990. *Where the Spirits Rule the Wind: Trance Journeys and other Ecstatic Experiences.* Bloomington and Indianapolis: Indiana University Press.

Gumo, S., S.O. Gisege, E. Raballah, and C. Ouma. 2012. Communicating African spirituality through ecology: Challenges and prospects for the 21st century, *Religions*, MDPI AG Journals 3(2): 523–543. https://doi.org/10.3390/rel3020523

Gyekye, Kwame. 1995. *An Essay on African Philosophical Thought.* Philadelphia, PA: Temple University Press.

Gyekye, Kwame 2010. African Ethics. In *The Stanford Encyclopedia of Philosophy*, ed. Zalta, Edward. http://plato.stanford.edu/archives/fall2010/entries/african-ethics.

Ikuenobe, P. 2006. *Philosophical Perspective on Communalism and Morality in African Traditions.* London: Lexington Books.

Ikuenobe, P. 2014. Traditional African Environmental Ethics and Colonial Legacy. *International Journal of Philosophy and Theology* 2: 1–21.

Imafidon, E. 2014. On the Ontological Foundation of a Social Ethics in African Traditions. In *Ontologized Ethics: New Essays in African Meta-Ethics,* ed. Imafidon, E. and Bewaji, J.A., 37–54. Lanham: Lexington Books.

Kanu, I.K. 2021. Sacred Animals and Igbo-African Ecological Knowledge System. In *African Indigenous Ecological Knowledge Systems : Religion, Philosophy and the Environment*, ed. Ikechukwu Anthony Kanu, 1–17. Maryland, The Association for the Promotion of African Studies.

Kelbessa, W. 2005. The Rehabilitation of Indigenous Environmental Ethics in Africa. *Diogenes* 52: 17–34.

Kugedera, A.T., N. Sakadzo, T. Muiseva, E. Chivehenge, S.L. Kugara, F. Chimbwanda, and M.T. Decide. 2021. The Link Between Indigenous Knowledge Systems and Sustainable Agricultural Production in Zimbabwe. *African Journal of Religion, Philosophy and Culture* 2 (2): 19–39.

Latour, B. 2013. *An Inquiry into Modes of Experience: An Anthology of the Moderns.* Cambridge: Harvard University Press.

LeVasseur, T., Pramod Parajuli, and Norman Wirzba, eds. 2016. *Religion and Sustainable Agriculture: World Spiritual Traditions and Food Ethics.* Lexingtone: University Press of Kentucky.

Mbiti, J.S. 1969. *African Religions and Philosophy.* London: Heinemann.

Mbiti, J.S. 1977. *Love and Marriage in Africa.* London: Longmans Group LtD.

Metz, T. 2022. *A Relational Moral Theory: African Ethics in and Beyond the Continent.* Oxford: Oxford University Press.

Murcott, A. 1982. The Cultural Significance of Food and Eating," *in Proceedings of the Nutrition Society.* https://www.semanticscholar.org/paper/The-cultural-significance-of-food-and-eating.-Murcott/

Murove, F.M. 2004. An African Commitment to Ecological Conservation: The Shona Concepts of Ukama and Ubuntu. *Mankind Quarterly* XLV: 195–215.

Mutwa, C. 1996. *Isilwane: The Animal*, Cape Town, Struik.

Nath, P., ed. 2012. *The Basics of Human Civilization: Food, Agriculture, and Humanity*, vol. 1. New Delhi: New India Publishing Agency.

Nkwi, P.N. 1976. *Traditional Government and Social Change: A Study of the Political Institutions among the Kom of the Cameroon Grassfields*. Fribourg: Fribourg University Press.

Nkwi, W.G. 2017. The Sacred Forest and the Mythical Python: Ecology, Conservation, and Sustainability in Kom, Cameroon, c. 1700–2000. *Journal of Global Initiatives: Policy, Pedagogy, Perspective* 11 (2): 31–47.

Nsom, J. 2015. The modern Kom Society: Culture, Customs and Traditions, Yaoundé: Nyaa Publishers.

Nyerere, J.K. 1968. *Freedom and Socialism*. Dar es Salaam and New York: Oxford University Press.

Nyerere, J.K. 1968. *Ujama'a: Essays on Socialism*. London: Oxford University Press.

Oruka, H. and Juma, C. 1994. Ecophilosophy and Parental Earth Ethics. In *Philosophy, Humanity and Ecology*, ed. Oruka, H., 115–129. Nairobi: ACTS Press.

Partridge, E. 2003. Future Generations. In *The Environmental Ethics and Policy Book*, ed. Van DeVeer, D. and Pierce, C. Belmont: Wadsworth.

Ramose, M.B. 2009. Ecology through *Ubuntu*. In *African Ethics: An Anthology of Comparative and Applied Ethics*, ed. M.F. Murove, 308–314. Scottsville: University of KwaZulu-Natal Press.

Reddy, S., and M. Anitha. 2015. Culture and its Influence on Nutrition and Oral Health. *Biomedical & Pharmacology Journal* 8: 613–620.

Roothaan, A. 2019. *Indigenous, Modern and Postcolonial Relations to Nature: Negotiating the Environment*. London and New York: Routledge.

Roothaan, A. 2018. Hermeneutics of trees in an African Context: Enriching the Understanding of the Environment 'For the Common Heritage of Humankind.' In *African Philosophy and Environmental Conservation*, ed. Jonathan Chimakonam, 135–145. London: Routledge.

Schoffeleers, J.M. 1978. *Guardians of the Land; Essays on Central African Territorial Cults*. Harare, Zimbabwe: Mambo Press.

Sibal, V. 2018. Food: Identity of Culture and Religion. https://www.researchgate.net/publication/327621871_FOOD_IDENTITY_OF_CULTURE_AND_RELIGION.

Tangwa, G.B. 1996. Bioethics: An African Perspective. *Bioethics* 10, Number 3.

Tangwa, G.B. 2002. The HIV/AIDS Pandemic, African Traditional Values and the Search for a Vaccine in Africa. *Journal of Medicine and Philosophy* 27 (2): 217–230.

Tangwa, G. 2004. Some African Reflections on Biomedical and Environmental Ethics. In *A Companion to African Philosophy*, ed. Wiredus, K. Malden, Blackwell.

Tangwa, G.B. 2010. *Elements of African Bioethics in a Western Frame*. Mankon Bamenda: Langaa Research and Publishing.

Telleen, M. 2013. Mind-set of Agrarianism... New and Old. In *The essential agrarian reader: The future of culture, community, and the land*, ed. N. Wirzba, 2013, pp. 28–35. Lexington, Kentucky: The University Press of Kentucky.

Tempels, P. 1959. *Bantu Philosophy*. Paris: Présence Africaine Éditions.

Thompson, B.P. 2008. Agrarian Philosophy and Ecological Ethics. *Science and Engineering Ethics* 14: 527–544.

Tosam, M.J. 2014. The Philosophical Foundation of Kom Proverbs. *Journal on African Philosophy* Issue 9: 1–27

Tosam, M.J. 2019. African Environmental Ethics and Sustainable Development. *Open Journal of Philosophy* 172–192.

Udeani, C.C. 2008. Traditional African Spirituality and Ethics—A Panacea to Leadership Crisis and Corruption in Africa? *Phronimon* 9: 65–72.

Wanier, J.P., and P.N. Nkwi. 1982. *Elements for a History of the Western Grassfields*. Yaounde: Publication of the Department of Sociology University of Yaoundé.

Westra, L. 2006. *Environmental Justice and the Rights of Unborn and Future Generations: Law, Environmental Harm and the Right to Health*. London: Earthscan.

Wiredu, K. 1994. Philosophy, Humankind and the Environment. In *Philosophy, Humanity, and Ecology*, ed. Henry Odera Oruka, 30–48. Nairobi, Kenya: ACTS Press.

Wiredu, Kwasi. 1996. *Cultural Universals and Particulars: An African Perspective*. Bloomington: Indiana University Press.
Wirzba, N. ed. (2013). The Essential Agrarian Reader: The Future of Culture, Community, and the Land, Lexington, Kentucky, The University Press of Kentucky.

Mbih Jerome Tosam is Associate Professor of Philosophy at the University of Bamenda, Cameroon. He obtained his PhD in Philosophy from the University of Yaoundé I, Cameroon in 2011. He is former Chair of Philosophy at the Higher Teacher Training College (HTTC) Bambili (2011–2017) and at the Faculty of Arts of the University of Bamenda, Cameroon (2017–2021). His research interests are in the areas of Bioethics, African Philosophy, and Intercultural Philosophy. Some of his publications have appeared in the folowing joiurnals: *South African Journal of Philosophy, Annali di studi religiosi, Medicine, Health Care and Philosophy: A European Journal, Deveolping World Bioethics, Journal of World Philosophies*, and *Polylog: Forum for Intercultural Philosophy.*

Chapter 11
Land Ethics Among the Traditional Annangs of Southern Nigeria: Traditional Environmental Ethics, Challenging Contemporary Hostilities Towards Our planet

Dominic Umoh

Abstract Ethics as a normative science concerns what is proper or improper in human conduct. When this is applied to what ought to be people's relationship with "The Land" in Annangland, a whole boundless and infinite spectrum of reality is unfolded, because of the socio-religious connotations of the "Land" in the Annang Nation. There is something of the divine in the land; she is a Deity; hence the common designation THE MOTHER EARTH. She cares, sustains, nourishes, grooms, upholds, pampers, caresses and shelters her children in the warmth and tenderness of her bosom. In the past this lavishing tenderness called for total submission to and respect for environment whose greatest representative the Land is. This paper is a sensitization in an area of environmental ethics, seriously menaced in today's conscienceless civilization. The paper advocates a return to the African root; to the ancestral wisdom. This attitudinal return to the root should become a mark of identity wherever Africans are found. The method employed in this research paper is critically exploratory.

Keywords Mother earth · Annang culture · Religion · Human development

11.1 Introduction

One of the greatest contemporary menaces to healthy human existence is the environmental crisis. The devastating negative human impacts on the environment turn out to pose the greatest threat to man's wholesome subsistence. In a bit remote past,

D. Umoh (✉)
Philosophy of Religion and Religious Ethics, Akwa Ibom State University, Obio Akpa Campus, Mkpat-Enin, Nigeria
e-mail: labbedominic2009@yahoo.co.uk

M. J. Tosam and E. Masitera (eds.), *African Agrarian Philosophy*, The International Library of Environmental, Agricultural and Food Ethics 35,
https://doi.org/10.1007/978-3-031-43040-4_11

we were hearing of nuclear tests sometimes in open seas, destroying aquatic lives, yet with terrible negative impacts on human health. One cannot forget the age-long horrible experience of Chernobyl in Russia. Today, however, we are threatened by a drastic climate change, the threat of ozone layers and pollution from industrial wastes which calls for international treaties that some superpowers have refused to sign.

Throughout this chapter, our stance on Annang Religion or Culture is purely phenomenological. We are not here to promote Annang traditional Religion and Culture or any other religion and culture for that matter, indicating such as better than others. Rather, our intention is simply to highlight succinctly how a particular traditional religion has infiltrated all dimensions of a certain culture thereby impacting on them tremendously. Within our context, these impacts are felt in the attitude of typical Annang nationals towards the environment, which they refer to as "Our Closest Neighbour." The Annangs of yesteryears were in good, healthy and very cordial relationship with the surroundings. Traditionally, in Annangland the natural environment was recognized as a living entity of its own right. From this angle already, nature was considered worthy of respect on the same par with humans. It has rights and privileges and was allowed to flourish, mature and fructify—that is generate, regenerate, recycle or reproduce. This would save it from extinction. Communities were very much aware of the negative consequences of the reverse: namely that the environment, just as the human organism, can deteriorate in health, be impoverished and can be eliminated from existence. Such a possibility was already recognized as tragic and disastrous not only according to the framework of the holistic, social, cum cultural belief flowing from the Annang worldview, but more so because of the natural conscious belief of Annangs that the environment is man's closest neighbour and that to the largest extent, human existence depends on it. Any injustice done to nature, here taken synonymously with environment, was considered a provocation of divine wrath, a curse and a doom on humans and their society. This would have been the peak of insult to the ancestors and the deities of the land, who were considered very much present in their communities as custodians.

Through the lenses of this traditional Annang community one can now see the aberration arising from contemporary inimical attitudes of humans towards the natural environment and how humans are already paying for such crimes. These crimes range from man's insensitivity and positive hostility to the surroundings to becoming actually nature's major destroyer. The modern man does this in multifarious ways: by the pollution of land, water and air; the distortion of the eco-system, deforestation, over-exploitation of nature's resources, overpopulation, indiscriminate dumping of toxic wastes, oil spillage, and insensitivity to sustainable development. Granted that some of these injustices are accidental, the majority are premeditated and very purposeful sometimes for pure economic reasons. In plain language Taiwo Osemwegie affirms:

> The phenomenon of global warming and climate change continue to cast threat on the face of all mankind on planet earth. Most of the activities leading to the perpetual rising of global

> warming are unarguably caused by man, either while trying to explore nature or in the course of making provision for the survival of his well-being. These activities elicit moral questions, such as ought the natural environment be degraded in a bid to satisfy man's quest or need? How should man relate with his natural ecology in order to maximize the happiness of both? (2011: 15–16)

11.2 African Environmental Ethics

Ethics is a branch of philosophy which considers moral values, moral obligations and other factors related to human conduct in terms of its moral rectitude and turpitude. Environmental ethics applies this style of philosophical evaluation to the ways humans interact with the earth, natural resources and non-human animals; in short with the environment—land, air and water. This is the real stereo-typed etymological implication of the term "Eco" meaning house or home; describing lives in the natural environment. In our present context, eco means the natural relationship between plants, animals and people, and the places in which they live. Put otherwise, ecology means the study of the connection that exists between sentient and non-sentient beings and the places in which they live. (Ekwealo 2011: 4) would affirm this about the environment within the African context:

> (In Africa) The fundamental principles that govern the relationship between man and the environment (are) based on African worldview… As a field of research, it believes that the natural environment and man are composed of invisible energies and their relationships result to one form of manifestation or the other. In other words, unlike in the West where there is discord and a separation from nature, African Environmental Philosophy has a background belief in the linkage of nature, community and man, from which an ethical relationship is defined.

According to this worldview, the heavens hang so low as to touch the earth to the point of becoming one with it. The line separating the spiritual from the physical is very slim and very often none-existent. This rallying point between the physical and the spiritual is found in the community, which is overwhelmingly religious-oriented. The community provides the locus for all. This is the subject of this remark by Nwachukwu-Udaku:

> The community is the meeting point between the living and dead, between the spiritual and the physical world, between the divine and the human, between the invisible and the visible, between the One and the many. The community further provides a locus for interaction between human and other animate and inanimate beings. It is in the community that the individual is born, reared, and given basic identity through different rites of passage. (2011: 9)

That is why there is no dimension of life or aspect of culture in Africa that is untouched by religion. Mbiti (1980) affirms:

> Chapters of African religions are written everywhere in the life of the community and in traditional society there are no irreligious people. To be human is to belong to the whole community, and to do so means participating in the beliefs, ceremonies, rituals and festivals of that community. A person cannot detach himself from the religion of his group, for to do

> so is to be severed from his roots, his foundation, his context of security, his kinships and the entire group of those who make him aware of his own existence. To be without one of these corporate elements of life is to be out of the whole picture. Therefore to be without religion amounts to a self-excommunication from the entire life of society and African peoples do not know how to exist without religion. (2)

Elsewhere Mbiti (1980) has it that in their traditional life, African peoples are deeply religious. In fact it is religion, more than anything else, which colours their understanding of the universe and their empirical participation in that universe, making life a profoundly religious phenomenon. To be is to be religious in a religious universe. That is the philosophical understanding behind African myths, customs, traditions, beliefs, morals, actions and social relationship. Up to a point in history this traditional religious attitude maintained an almost absolute monopoly over African concepts and experiences of life (262).

To understand any sector of life in Africa therefore, the impact of religion must first and foremost be sought for and fully appreciated. This is exactly where many non-African writers are lost off by trying in vain to study various aspects of African community and cultural life in isolation. This "colouring of their understanding of the universe and their empirical participation in that universe" by religion in the final analysis amounts to utter filial respect, adoration and full submission to the environment in which the Africans find themselves. This is the backbone of the African environmental ethics which is religious in nature. Ethically, existence means relating friendlily with one's surroundings. This flows from the awareness that nobody can subsist without the universe, for the individual is part and parcel of his surroundings. This awareness extends to the belief that one is subject to the whole, for it is the whole, in terms of society that gives him his identity. There we are with the concept of "holism" strongly advocated by Osemwegie (2011: 16–17) as "the doctrine that emphasizes the priority of a whole over its parts. It is an understanding of a particular phenomenon/entity in relation to the whole, i.e. part-whole relation." "That is why the interdependence in the African socio-cultural world is a peculiarity per distinction. The entire social-cum-cultural setting is prized (more) than any individual entity. The essence of individual or particular entity is rendered crystal-clear in relation to the whole. This reinforces the doctrine of communalism (as opposed to individualism) in the African socio-cultural setting."

The above structure is a pointer to the doctrine of Communitarian Logic very much propagated in the traditional African social system. This doctrine, in the words of Mbiti is expressed thus: "I am because we are, and since we are therefore I am" (1980: 113). Further, Umoh (2014a: 48) explains: "It is the community that makes me who and what I am, who I ought to be and who/what I am capable of becoming. The society covers the entire spectrum of the individual personhood: social, economic, religious, political etc." And elsewhere he has the following remarks to support the same view: "The Annang person is not simply an individual. He is a corporate society not just because he is a symbol of or ought to represent his community, but because *per se* he is a miniature society" (2012a: 107).

Ethically speaking, there was a culture of respect, dignity, mutuality, collaboration and accommodation for all organisms, such that before one invades or ventures into another creature's space, it must be absolutely based on necessity and very often with due permission. This mutual respect did not exclude the environment. The case of the mystical "Nkwa-Uwia" tree used for the manufacture of ancestral masks in the past drives home this point. Permission had to be sought from the tree long before it was mowed down for carving:

> Because of these deeply religious beliefs about the 'nkwa-uwia' tree, sculptors had to consult traditional priests and would go with them with various items to sacrifice at the foot of the 'nkwa-uwia' tree before felling it…. The sacrifice was to appease the displaced spirits that used to employ the tree as their habitat and thereby preventing their anger raging against the lumbermen and eventually the artists. (Umoh 2015: 76)

Human is subjected to the environment in many regards and this in spite of any anthropocentric insinuations, for more often than not the environment is regarded as a divine entity. For example, among the Umueri people in Anambra State Nigeria, before one fetches the bark of a particular tree, maybe for medicinal purposes, one has to first seek permission by incantations to the 'spirit' of the tree. Although the spirit is always willing to yield its curative energy, humans must first make connection through supplication. This is part of the respect for the dignity of nature, a form of recognition of the 'rights' and 'values' that all beings possessed…. Philosophically, therefore, African environmental philosophy/ethics is basically ecocentric where recognition is accorded to other lives and creatures that exist (Ekwealo, 4–5). Among the Annangs, during hunting expedition, the hunters do not call each other by name. This is because of the fear for the spirits inhabiting the bush, especially the forces in the animals they are hunting. This is because in Annang culture, anyone who knows your name has you at his beck. Before they even step into the bush, they must perform incantations and throw a handful quantity of sand into the bush requesting the spirits to withdraw for awhile as they entered. This symbolizes full respect for the environment.

11.3 The Annangs and the Environment

The philosophy of the Annangs concerning the environment is very spectacularly unique. Its central nexus hinges on the spiritual powers within nature; calling upon nature's life forces in their respective interventions in human endeavours. All adults in Annangland, whether male or female, were farmers. Even those engaged in other trades like commerce, were still in agriculture. People's wealth was determined by the number of farm plots, palm plots and yams barns they possessed. This very fact made them very close allies of the land. Everybody, even the new born child, was brought up to be environment friendly. The traditional Annang society was agrarian and all their farming was meticulously religious. Plots for cultivation could not be cleared until the chief priest had offered sacrifices to the land goddess for fertility; neither could people harvest the first products of the land without sacrifice. This was the origin of new vegetable and new yams festivals. The Annangs were quite aware that they are part of the environment or simply that they were inseparable from its grips. After all, they depend entirely on the environment. They took cognizance of the liveliness, agitations, potentialities, principalities and the whole animated life forces evident in the environment. In respect of the deities of the land, on certain market days, farm works were forbidden. All these sentiments of awe before nature generate in them formidable feelings of fear, obedience, dependence and reference. Nature was deified. This gave vent to the peculiarity of their worldview which was prodigally religious.

11.4 The Annang Worldview

As with the entire African nation, the Annang worldview is a completely religious one. This portrays what Christopher Bryant would call in depth psychology "the self and the experience of God". He argues that:

> …for (such a) believer, God is present as the ruling agent in everything that happens: in the structure of the atom, in the evolution of animal species, in the lives of men and women and of groups, classes and nations, and in the history of the human race. He is therefore an invisible and perhaps unknown actor in the life of every person without exception. (1987: 31)

This is very true of the Annang nation and people. Referring to Africa as a whole, (Obi 2002) would argue that: "It is this strong sense of God that gives the African worldview its characteristic Theo-centric and religious consciousness. Thus the traditional African person is God-conscious. He respects God, devotes himself to him and seeks to do his will in his daily living. It is God that gives meaning to his being" (45). Annang cosmology unfolds a tripartite conception of the universe. The three domains are separate and distinct yet very much interrelated and intertwined. And everything is religiously couched and evaluated. There is 'ikpa-anyọñ'- the upper ether, 'Ikpa isọñ'- the lower ether and 'Awio-ekpo'- the domain of the dead.

11.5 The Upper Ether

'Ikpa-anyọñ' refers to the sky, the heavens, heavenly bodies and whatever of the planetary systems that was known. Included in the term Upper Ether are the sun, moon, stars and all heavenly powers which constitute a sort of entourage around the Sky God who is the overall Ruler of everyone and everything. The term can as well designate the domain of this Head of all deities and His entire administration. Since the sky is so far removed from the earth, human beings claim total ignorance of what are going on there and the form of governance operational over there. Yet they are aware that their very existence and that of every other organism as well as the welfare of beings in the great beyond eventually depend on that domain. Hence issues regarding the Sky God and His authority over the world must be acknowledged with fear and trembling, for if provoked the consequences could be disastrous and only diviners and traditional priests have a glimpse of his character. The Sky God is never taken for granted, especially as nobody can reach Him directly. This Head of deities does not exercise direct authority over the earth and humans. He has departmentally distributed his power among his vassals—the deities. Like Moses vis-à-vis the people of Israel at Mount Sinai, human beings would plead with the Sky God not to deal with them directly or else they die, for they are mere creatures. Some communities consider Him so pure and so sublime to mingle directly in profane human affairs. There is no shrine assigned to Him. What therefore obtains in reality is a delegation of power over the earth to the minor gods (the deities), spirits and the ancestors. In line with this, there is 'Awasi ukod'—controlling affairs between in-laws; 'Awasi ajejen' governing grand-parents/grand-children relationship; 'Awasi ụtụk'—the god of justice and fair play. 'Eka Awasi' takes care of issues about maternity and procreation; while "Ekpo-nka-agwo" penalizes violators of marital vows. In this sense the relationship between the Sky God and humans is quasi-deistic in nature. For deists:

> ...accepts the existence of God, but only as an impersonal creator of the world. It denies all divine interventions or 'interferences' in the course of its development or progress; that means that God has no hands in the (day to day) running of the universe after creation. God only traces the route for his creatures but does not accompany them on their journey". (Umoh 2012b: 45)

Also Feaver et al. (1967) opines that the transition from theism to deism is what obtains in the tendency of modern science which tries to do away with God as a principle of explanation. God has been pushed by science into the farthest distance, as an ailing, useless grandfather placed in an old people's home by a more vigorous younger generation. Natural law has effectively displaced divine intervention and purpose in explaining events.

However, the explanation above does not capture wholly the type of aloofness that obtains with regards to the Supreme Being in Annang indigenous religion. Rather, as we have seen above, the divine being is considered so pure and so non-human to meddle in the affairs of mere terrestrial beings. Here their deistic inclination is as a result of the absolute transcendence of God, who after creation handed over the management of the universe to the deities and the ancestors while He retired to his celestial domain. This sense of dedicated authority over the universe is the object of this poignant observation by Obi (2002):

> Africans believe in the existence of an all-powerful supreme and eternal God who created the world of both visible and invisible beings. But being transcendent in nature, he rules the visible and invisible universe through the intermediary deities, spirit forces, guardian spirit, ancestors and other agents in the ontological hierarchy. This faith permeates every African culture (45).

In his turn, Frans Vansina has these comments with regards to the deistic tendencies of some Eastern and African religions: "Here God is professed to be so high above man and the world that He takes no interest in the course of human life. God is almost apathetic with regard to man's life and the world's events." Borrowing from

Mircea Eliade's *Myth and Reality*, in connection with the Fangs of Equatorial Africa, Vansina reiterates: "God is above, man is below; God is God, man is man, each in his own place; each in his house" (21). Through the intermediation of the deities, the Sky God is all the same very much present and powerful in the world. This is the dimension of his all-important and indispensable immanence in the universe, without which the world would disappear into nothingness. We should note that even in Christian theology, a world without God's immanent presence would be non-existing. This immanence does not in any way jeopardize his transcendence. It is God who is the crux of the very existence of the world. Hence He cannot completely withdraw, even for a moment, from the universe otherwise all existents would be annihilated. It is of this same phenomenon of divine transcendence in Annang religion that Udondata and Ekanem (2013: 183) are making allusion to in the following lines: "*Awasi Ibom* (Sky God) was believed to dwell above and to have created human beings, the world and everything in it. Apart from this work of creation, he often had no direct link with man but through the lesser gods." And Essien and Umotong (2013: 168) paint the picture of Annang reality showing that the two worlds of nature and spirit are but one:

> The Annang have a firm belief that there are two worlds: the physical world and the spiritual world. The physical world is the world of human beings, plants, animals and inorganic beings. The spiritual world is the realm of the Supreme Being (Awasi–Ibom), the gods (Nnem), Ancestors (Mme Ete-Ete), (and) Spirits (Ekpo). It is believed that there is a very close link between these two worlds. While Awasi-Ibom is believed to be high up beyond the sky, his influence is only felt through the gods, who act on his behalf and who are closer to the people. Because of his unlimitedness, Awasi-Ibom has no shrine or temple, since this would confine him and so, contradict his nature.

The transcendence of the Supreme Deity is at the center of the lengthy citation below by Nwanaju. According to him, before the arrival of Islam and Christianity to the African continent, there was already a huge heritage of African myths and legends, although they may have been mere sparse account about the creation of the world by a Supreme Deity. In the context of their development and civilization, such myths and legends expressing their belief in God could be said to be adequate.

> Such myths and legends contain stories about lesser divinities and spirits who represent what is sometimes erroneously referred to as 'the withdrawn high god', **deus otiosus**. Such myths "record the founding charters and accounts of the primordial activities of a society's gods. They also contain information about the people's beliefs regarding the nature and purpose of human existence and the principles underlying relationships among humans, the divinities, and other creatures. It was a popular belief that after creating the world of the living, the high god withdrew to the outer reaches of the universe, handing over to lesser divinities the responsibility for ordering the daily lives of the people, and supernaturally validating the legitimacy of the founding ancestors' authority.
>
> (www.academicexcellencesociety.com/challenge_of_african_traditional_religion_and_culture to_christianity_and_islam.html) assessed August 22, 2019.

11.6 The Lower Ether

As opposed to the heavenly realm, the upper ether, 'Ikpa-isọñ' stands for the lower ether, the earth and all in the terrestrial order. This is the whole of the contingent order. The management of the lower ether, 'ikpa-isọñ' is entrusted to deities and the ancestors, whose visible representatives are the community leaders and chiefs. There was a sort of division of labour among the gods, assigned by the Sky God according to their respective departments. With their anthropomorphic conception of God, the Annangs considered the management of the entire universe too much for 'one person', no matter how omnipotent he might be. Hence the sky God had to hand over the administration of the affairs in the lower ether to the lower deities, who nonetheless were still accountable to him. These deities were bilingual; speaking human as well as divine languages. This character was advantageous in view of their assignment. They shared the authority and powers of the 'Sky-God' sufficient enough for their task. Since these numerous deities were very close to humans in the universe, overseeing their ethical conduct in their various domains posed no problem. The lower ether had then its own jurisdictional areas, corresponding to the various deities that took charge of them. There were shrines (Iso Ikpa-isọñ) dedicated to such deities in each locality. At such shrines, evergreen trees such as 'akọrọ', 'mkpafere' and 'nnụñ-nụñ' were planted. Such evergreen trees signified immortality. It was at such shrines that those deities were sacrificed to and entertained especially during festivals. Families, villages and clans had their own shrines—**'Iso ikpa-isọñ'**. (Umoh 2014b)

The literal analytic meaning of this term "Iso ikpa isọñ" unfolds a whole wealth of implications, significance and meanings for the Mother Earth. **"ISO"** means face, the front part, the *Recto*—the first side say of a card as opposed to the *Verso* which is the reversed side, the opposite side of something. "Iso" is the most exterior, the outermost part and therefore the most visible, prominent and pronounced. In case of war, "iso" is the most frontal or the most exposed part—that is the frontline—of the battle field and so the battle line of the war or the war front. **"Ikpa"** can mean the skin of something; say a leopard, used as red carpet for receptions of monarchs or other high ranking traditional dignitaries. With this comes an idea of being spread out; a layout, an expanse or an open landscape or landmass. Another technical literal interpretation of "ikpa" is a cane, traditionally used to discipline wayward children or even adult law-breakers like thieves. From this perspective already, when it is associated with the Land, it is portrayed as the custodian of morality, capable of meting out punishments to law breakers. The word **"Isoñ"** itself has multifarious meanings. It can mean the soil, the ground, or the land in the sense that it is the very first point of contact for both animate and inanimate organisms beginning with humans. At birth, new born babies of Annang origin, fresh from the mother's womb, or as it is believed, from the land of the spirits, are first and foremost initiated into the community by being seated on the ground and being scrubbed with the sand of his/her Fatherland. The land is what is cultivated and what provides nourishment and sustenance for its inhabitants—her sons and daughters. Here lies the origin of the powerful expressions: The sons or the daughters of the soil. The same provides resting point for our feet when we walk, stand or our seats when we sit. It is the layer on which we lie, whether on a bed, on a mat or on bare floor. It provides the basis for our structures: such as buildings or other constructions and a platform where our communities stand. "Isọñ" refers to a territory or a nation.

> It is also the lower ether where shrines and worship places are erected as opposed to the upper ether, the abode of the celestial God. In another sense, it is the womb that collects back millions of her children at the end of their earthly sojourn. "Isong" is historical, anthropological, cosmological, sociological, physical, spiritual, ethical and political. It is mythical, cultural, religious and divine. (Umoh 2016: 298)

"Isọñ" is the same word used for the village cabinet or the council of chiefs. Thus it is the visible authority of the ancestral forbears, who are the custodians of moral or norms of the community. "Isọñ" is the first contact point where the spiritual and the terrestrial meet. It is the rallying point of all realities. Eventually at death it is the Mother Earth that swallows up her child, receiving him back into her womb. There is therefore something mysterious, something sacred, and something awesome in the name 'Land'. It is deified. Before bush clearing for farming and before the actual planting of farm crops, sacrifices had to be offered to the deities of the land, especially the gods of agriculture, to ensure fertility of soil. Before harvesting those crops, sacrifices were again offered at the "iso ikpa isọñ". From the above, the weighty implications of the name earth shrines cannot be overemphasized. It is the primary, the most original and the foremost contact point between the community and any external agent; be it terrestrial or spiritual. In these regards its protectively safeguarding authority and sheltering capacity for the community is most assured. No external influence, good or bad can reach the community without passing by the scrutiny of the "iso ikpa isọñ". One can see now why during boundary disputes or tribal wars, indigenes would fight tooth and nail to defend their fatherland from external aggression. The sacrifices offered at the "iso-ikpa-isọñ were in thanksgiving to the deities not only for the abundant harvests of the current year but also as solicitation for land fertility during the next planting season. Sacrifices were equally in appreciation of other benevolences enjoyed by the community during the period in time. These were the moments for the seasonal appearance of masquerades which were symbolic representatives from the world of the spirits to appreciate the human sacrificial offerings. Sacrifices were offered also before every war either to solicit or appreciate divine intervention in that course of protecting the motherland. From all these, it is evident that nature (environment) was sacred and could not be treated anyhow.

11.7 The Land of the Dead

The third domain of reality was 'Awio-Ekpo'—the land of the dead. If heaven is where the just are rewarded and hell where the wicked are punished according to Christian belief, for the Annangs these domains are not located in the sky, where nobody knows what is going on between the Sky God and his vassals, but in the netherworld. Since one must die and be buried in the ground in order to be rewarded or damned, then the reward of the dead can only be offered in the land of the dead—'Awio-Ekpo.' This netherworld is located under the earth, because the dead are buried in the ground or better still are accepted back into the womb of the "Mother-Earth". This beautiful bit of poetry of Shorter (1975: 56) portrays it well: "Earth and dust, the dependable one, I lean upon you. Earth, when I am about to die, I lean upon you, when I am alive, I depend on you". As quoted in (Ekwealo 2011: 5) "…beneath the material world that we can see and touch and feel is an energetic world, the world of spirit, whose vitality enlivens not only all living things, but the very geography of the world that holds life. In the indigenous views, the world of spirit and the material world co-exist: each needs the other because each feeds the other." The "Awio Ekpo" is thickly populated with all the dead since the beginning of time. The departed who were morally upright members of families are regrouped along the lineage of their earthly families. This reunion with ancestors "of old" and never to separate again is the Annang concept of eternal life. Though of the same stock, wicked members of the families are not granted admission into this marvelous company of the saintly members of the extended families. The wicked dead members of families are frustrated and unsettled in the land of the dead, the very reason why they return to the world as "Ekpo-Ulọk-Isọñ" (the underground ghosts) to torment and frighten the living. Such become visible mostly at night or in lonely places during the day. With their horrifying appearance, they would cast spell on beholders, provoking untold spellbound and instant temporary paralysis in them. Because of this, the term "Ekpo" is ambivalent and always needs specification. If somebody, say a stroller at night, runs back home screaming that he has encountered "Ekpo", everyone's mind would turn immediately to the underground ghost. However, if he explains that it is "Abụbụb Ekpo" (the masked 'Ekpo': Masquerade) then all would know that he is referring to Ekpo masquerade—the personified ancestral spirit. At any rate both the underground ghost and Ekpo masquerades are cousins. Both must be frightening and fright-provoking in appearance. Those good dead have no reason to return to roam the world menacing and terrorizing the living, for they have their beautiful mansions, beatific vision and restful homeland in the abode of the dead. If the good dead (the Ancestors) must appear at all, such appearance must be very rare; it may be to resolve pending family intrigues or to warn their families against impending dangers. Or to show an approval for a decision well taken on what affects the family. But whatever the case both the ghosts of the evil dead as well as the upright ones come from 'Awio-Ekpo' because they both had died and were buried. They all must cast spell on beholders. In line with this Annang belief, it was from this underworld that the representatives of the spirits of the dead, the Ekpo masquerades, came to monitor, inspect and sanction or appreciate what the living was doing on earth. Hence their outings were regarded by every Annang as the appearance of the dead among the living and so very significant.

11.8 Nature

Nature, according to the Annang worldview is considered a living creature. It has the same rights, privileges and legitimacy as any living human being. This is panpsychism, a doctrine that the world, or nature, which gives birth to living creatures, must itself be thought of as alive and animated. It was "literally describable as possessing reason, emotion, and a 'world-soul.'" (Blackburn 2005: 170). This particular viewpoint is thus approved by Taiwo.

> A distinctive feature of the African world-view is the understanding that every existent possesses a spiritual element like monads. However, these monads are known as forces in the African ontology. Force pervades everything in the African environment. These forces are active and conscious. This is what makes the environment to be perceived as animated organism. (Cf. Ekwealo 17)

This view stresses that matter is either a conscious substance or that the entire universe is a shroud of an interminable realm of mental existent. This is similar to the views of Leibniz, Schopenhauer and Schelling, and to an extent Hegel. In Hegel's metaphysical interpretation of the universe, everything is a microcosm and a manifestation of the invisible Absolute Spirit. The universe is not an unconscious or blind entity; it is a rational universe with a definite purpose (Ekwealo, 17). It is this African world view that was christened animism by Westerners; a term that was considered derogatory by early African scholars. However, I personally think that these Westerners were somehow right and logical in their nomenclature because the whole of nature according to Africans, was considered animated, 'ensouled' or invaded and inhabited by powerful, lively forces and the Latin word "anima" from where 'animism' takes its name means soul.

If one suffering from severe ailment is abruptly relieved of his sickness through a simple use of a balk, some leaves or a root of a plant in the name of medication, then one would not be fully reasonable to conjecture that there is no power of whatever sort in such an element of nature. The surging of the sea, tornados and tempest, cyclones, thunderstorm, earthquake, flood or tsunamis with their turbulence solicit such basic thoughts that nature is potent and is imbued with forces. The world-view of the Annangs is built around reflections on such natural phenomena. Such generate beliefs, fears, respects and taboos and they embellish myths relating the community to its primordial origin. Human environment was considered as harbouring live souls and forces. Certain meteorological phenomena were interpreted as expressions of God's emotions and sentiments. Lightning for instance was considered the Creator's lantern, the stars his flashlight, while thunderstorm was his groaning in annoyance and the release of His canon shots—an expression of his severe anger. In other places the sound of thunderstorm was interpreted as God's stomach upset and so the abdominal noise. During such moments, people were urged not to provoke the Creator as perhaps he would be of bad temperament.

11.9 Nature is Sacred

Nature, which in Annang dispensation has many nuances, including the natural environment and much more, is sacred. In some sense the word Nature refers directly to God the Creature or even human destiny. As Umotong and Essien and Umotong (2013: 168) opine:

> In Annang thought, there is no clear-cut distinction between God and nature. Though there is a distinctive name for God in Annang, Awasi, yet when it comes to causation, God shares the same name with nature, Abot. This gives a clue to God as the causal agent of the Annang Universe. In terms of this rapprochement in the Annang conception of causation, the Annang are not far from pantheism. This is quite understandable due to their deep sense of religion, where they see God everywhere, yet without a temple, which they believe would confine him.

Abod is the Originator of all things; the One who creates and then assigns each and every one of its creatures its appropriate nature—that is its essence—and its legitimate end. Therefore the name *Abod* is synonymous with God, especially with reference to creation. The Annang proverb: "*Abod adedeme ubọk-utom, mfine ademe eku.*"—Nature, the Creator is the distributor of people's means of livelihood, just as the trap is the selector of the type of animal it catches for its owner. Ufied Keed K'Usen. Viewed against the backdrop of the above, the consequential filial relationship with Nature is inevitable. The Annangs maintained a cordial referential relationship with the environment represented principally by the Mother Earth. In all regards this filial respect to the earth is visible. This sentiment is highly promoted and protected by African religion as a whole. As Sarpong vividly asserts, the nature of African Religion is such that gives serious significance to the physical cosmos. The religion itself evolved via the conception of the African peoples of their environment. To the people, God is seen, felt, and experienced in everything. The African holds in high esteem all of the creatures of God, both animate and inanimate. Everything, no matter how little, everyone no matter how seemingly insignificant, has its importance in the order of the created world.

Sarpong Groups.google.com/d/topic/usaafricadialogue/kuJ0

This religious reference to the environment is most prominent in the area of agriculture, an area in which the assistance of the Mother Earth is solicited for the nourishment of humanity. As no indiscriminate style of farming is authorized in Annang land, before preparing the plot for planting right away from bush clearing, the chief priest of the community must first sacrifice to the land, soliciting a fruitful yield. Bush clearing, bush burning and cultivation must strictly abide by the prescribed norms and regulations of the community in honour of the Mother Earth. The harvesting of crops right from the first vegetables through maize harvesting and new yam, are ceremoniously celebrated as agricultural cycles. This is the origin of the ecological rites of passage that witnessed the outings of masquerades representing the spiritual world.

Expiatory sacrifices as well, harbour a great deal of this cordial relationship and reconciliation with one's environment in Annang culture. There are certain types of crime considered abominable and that are said to contaminate the entire land and which must be atoned for in order that the land may be sanitized and rededicated so that it may continue to yield its products for human sustenance. Its inhabitants should therefore be renewed in their covenant with God through the deities by means of expiation. For instance, the murder of an innocent person by a member or members of the community or harming an in-law or grandchild is an abomination which imposes a curse on the community which has forced an innocent blood down the throat of the Mother Earth. This rhymes with this biblical text in the case of the murder of Abel by his brother:

> What have you done? The voice of your brother's blood is crying to me from the ground. And now you are cursed from the ground, which has opened its mouth to receive your brother's blood from your hand. When you till the ground it shall no longer yield to you its strength; you shall be a fugitive and a wanderer on the earth. (Gen. 4: 9–12)

In this way the environment exerted an enormous influence on the daily behaviour of the people and their interpersonal relationships which extend to their relationship with the closest neighbour, the environment.

11.10 The Mother Earth

As seen above, among the Annangs the land has something of the mysterious, even in nomenclature. The term denotes the land, the Mother Earth on which we walk, stand, lie and pour libation; which we cultivate, from which we harvest our crops and which therefore is our sustainer, whose sand was used to rub our little bodies immediately we arrived from our mother's wombs. This was a means of identifying the new born with its homeland. It is on the land that human habitations are built, that forests providing lumbers and cooking woods thrive. It is on the ground that water, the source of our lives, flows. It is to the womb of the Mother Earth that our ancestors returned. It is through its gate that the righteous gain access to the land of the spirits. Therefore the land has something of the divine. Another connotation of the term "isong" is our human domain as distinct from the sky (the heavens). This is the world of human beings as opposed to the domain of spirits (Umoh 2014a: 65). With these intermediaries, man must live religiously in a religious environment. No wonder then why the same word "isong" is used to designate the governing council of the community which by extension and delegation plays the intermediaries to the Sky God in the governance of the community. The land is therefore sacred in its own right and therefore has to be respected not only as the abode of the ancestors but as man's closest neighbour.

11.11 Critical Evaluation, Proposals and Conclusion

We are just rounding up our itinerary of the Earth-based spirituality of the Annangs, which showcased how a particular culture has impacted on the attitude of its people rendering it very submissive to the promptings of nature and the rhythm of their environment. This is obtained thanks to its religion which is the hallmark of the people's way of life. With all the above in mind one could see the outrageousness of the hostilities of the contemporary societies against the environment. Harm done to the surroundings today both individually or collectively leaves much to be desired. And already man has started paying heavily for it.

It is very obvious that what people do about their ecology depends on what they think about themselves in relation to their surroundings. Human ecology is deeply conditioned by religion and dependent on beliefs about human nature and its destiny. Among Africans, religion infiltrates into all nooks and crannies of life to sanctify and to sanitize them. Religion flows freely into and out of Annang worldview to the point of making them inseparable from and inextricably bound to one another. Hence, there is a harmonious and unbroken bond between religion and the remaining sectors of human existence. (Umoh 2014a: 65).

Imeh Umoh strongly opines that:

> Some cultures have the reverential concepts to protect the environment. They worship natural powers and animals as a means to champion the cause of ecology. They have traditional values to regard the rivers as mothers, land as mother, cows as mothers and trees as divine, protecting all from harm or destruction. These notions combine eco-friendliness with culture contrary to how the environment is being treated by the modern science and technology. (Umoh 2011: 30)

In traditional African societies, people lived in a religious universe where humans and nature were partners. This explains why 'environment' as a term, has no single direct translation in most African languages and dialects. To Africans, environment means life in its totality. Among many ethnic societies in Africa, natural objects and phenomena are regarded as manifestation of God's authority. (Gitau 2000: 4).

11.12 The Hazardous Divorce from Nature

In Western Environmental Philosophy, the material and physical world is separate and distinct from the spiritual. This leads to dualism and competition between the contingent and the supernatural orders. According to this setting, nature is devoid of sacredness and this is what accounts for the rape and abuse of the environment (Ekwealo: 5–6). This is how the emergence of western science and technology is grossly responsible for the wanton degradation of the eco-system. This is not to say that science and technology are absolutely awful but the truth is that whatever has advantages also has disadvantages. The continuous depletion of the ozone layer is largely due to western technology. The aftermath of science and technology constitutes bulk of the reason for environmental degradation. This is seen in the emission of poisonous gases in the atmosphere from mobile and industrial sources such as vehicles, air-crafts, marine vessels, train, and electronic gadgets, nuclear weapons, toxic gases, germ warfare and rocketry to mention but few (Ekwealo: 30).

The contemporary world, as we have it today, has lost its grips on the environment. Very often leaders of today's world opt for mere economic benefits as opposed to the promotion of healthy surroundings. And that is a great error, especially on the part of the world's Super Powers with their obnoxious stance against eco conservation. Humanity is currently experiencing harsh natural disasters and a threat to its survival as a result of human inconsiderateness. These hazards and their undesirable consequences include overgrazing, deforestation, desertification, torrential rainfall, depletion of scarce resources such as fossil fuels, topsoil and clean air, changing weather patterns caused by the greenhouse effect, severe overpopulation, overcrowding, et cetera. These are in addition to the social effects of pollution, at times by corporations, all of which lead to frustrations, stress and confusion resulting in violence, disease and a bleak future (Ekwealo 2011: 1–2). Awkward decisions of some international bodies jeopardizing biodiversity do not promote the conservation of the environment. All these are the results of the distorted anthropocentrism of our contemporary world and its reckless dominance of and encroachment upon the environment.

> The environmental crisis is rooted in an anthropocentric Western Christian worldview, inherited from Judaism through the biblical cosmology, in which nature is transformed from **a subject to be revered to an object to be used** (Our Emphasis). This Western Christian worldview destroyed pagan animism, separated humans from nature, enabling humans, in part, to share in God's transcendence and to exploit the natural world with indifference to the feelings of natural objects. (Simkins 2018: 166)

In connection with environment, the contemporary world, especially in the West, represents a fallen humanity divorced from the divinities. Environmental issues have been politicized and monetized, giving birth to various obnoxious ideologies. That is why in some Western nations, there are political parties pro and con environment. The Green Party or Ecolo symbolizes political ideologies that are environment friendly; whereas opponents criticize such stance as options for financial bankruptcy in the name of pollution eradication. Here is what Alokwu has to say about the contribution of foreign cultures to environmental degradation: "Before the introduction of Christianity in Africa, Africans were ecologically responsible, but the introduction of Christianity brought with it a mechanistic view of the world which has hugely contributed to the present state of the environment." (Alokwu 2011: 34). In some instances some developed nations in collaboration with some money hungry citizens export dangerous and toxic wastes to some third world countries designated as dumping grounds on payment of little tokens. Such endangering of the life of the poor masses sold over by some unscrupulous and greedy citizens is to say the least immoral in social and environmental ethics.

11.13 Proposals

Our stance in this paper is simple: As the world has completely turned into a global village, global communities should adopt an unequivocal common strategy of moral principles in favour of the environment. We know that some attempts have already been made in this regards in the past; for instance the popular annual events of The World Earth Day inaugurated on the 22nd April 1970 in the United States. But we must equally observe with great dismay that some world super power nations have bluntly refused to be parties to such endeavours due to their selfish motives which are inimical to the fostering of environmental protection. These remarks of Krech and Ballachey (1962) come true in their regards opining that we often think that our perspective of the world is the only one; that other people must see the world in the same way that we do. This egocentric assumption hinders our attempts to understand the behaviour of others. It hampers the adoption of a common strategy that would foster the conservation of our environment. Turning deaf ears to the echo of the pleadings of past generations (which were always environment-friendly) for the benefits of the present in view of the future, is to say the least, self-destructive because our well being as individuals and community depends to a great extent on the survival and the welfare of the planet earth. Environment is life.

In all sincerity, we would not, and no one should, advocate a total and an uncritical return to the past with its pre- and unscientific overtones and its pantheistic worship of nature. Yet we cannot ignore the fact that we are already paying gravely for our misuse and degradation of the Mother Earth. And we should look back with nostalgic admiration and yearning at the bounteous gains of our forebears through their environment friendly style of existence and how they handed on the same treasurable attitude to us their posterities only for it to be downtrodden and exchanged for a self-destructive attitude towards the environment in vogue today. Professor Imeh Umoh has it right when he says: "In an age of advanced civilization, mass education, incredible technology, computerized life, the rejection of the past way of life of a society and the substitution for fashionable but destructive trends leaves present generation without the standards, values, and knowledge to judge the present madness and corrupt way of life in the society" (Umoh 2011: 1). The rejection of one's past, is the uprooting of oneself from one's source of existence—that is from one's culture, for the real proof of culture is in the lives people lead, which proves to be a trip back to one's primeval past; culture being, as it were, a journey through history; a way of life of a society. Hence our contemporary societies should adopt, purify, modify, modernize and sanitize our ancestors' attitude towards our planet.

11.14 Conclusion

> This recommended wholesome, qualitative environmental ethics and values for the 21st century would therefore be summed up as right attitude to animals, non-animals (plants and other things in reality) embodied in a correct, respectful relation and association. Through this attitude, natural balance would be maintained and the life of Man and Environment ensured. These can only be achieved through a new consciousness anchored from a neo-metaphysics like the African already shown (sic) in which there is a coherently ordered system of complementary dualism, a necessary psychology of association that would promote healthy and productive human–environmental relationship. (Ekwealo 6)

The contemporary society is invited to briefly return to the ancestral past, to embrace and borrow a leave from their primordial wisdom. There should be no wastage, efficient utilization and management of resources in the universe. This is a new civilization of 'ecological humanism'. The forefathers were, in our terminology, illiterate, but they were not fools. They were not in affluence as we are today, but they were a million times happier, more satisfied, more contented and felt more fulfilled than the present generation. They respected the environment and were thereby recognized by their surroundings as friends, offspring and collaborators, whose interest the environment respected to the fullest. That friendship with the nearest neighbour paid off. In full confidence, nature did not withhold from them its secrets. It unfolded to them the bountiful treasures of its wealth and wisdom in terms of its forces, healing potentials and more so its predictability.

References

Alokwu C.O. 2011. The Synthesis of Oikotheology and African Ecological Ethics as a Model for Environmental Protection in Africa. *Journal of African Environmental Ethics and Values* 1: 34–54.

Blackburn, S. 2005. *Oxford Dictionary of Philosophy*. New York: Oxford University Press.

Bryant, C. 1987. *Depth Psychology and Religious Belief*. London: Darton, Longman and Todd.

Ekwealo, J. 2011. Environmental Ethics and Values in the 21st Century: An Africanist Philosophical Analysis. *Journal of African Environmental Ethics and Values* 1: 1–13.

Essien, E.S., and Umotong I.D. 2013. Annang Philosophy: Foundations and Outline. *British Journal of Arts and Social Sciences* 13 (11): 167–187. https://www.bjournal.co.uk/BJSSaspx.

Feaver, C., and W. Horosz. eds. 1967. Religion in philosophical Perspective (New Jersey: Van Nostrand).

Gitau, S.K. 2000. *The Environmental Crisis: A Challenge for African Christianity*. Nairobi: Acton Publishers.

Krech, D., and E. Ballachey. 1962. *Individual in Society*. Tokyo: McGraw Hill.

Mbiti, J. 1980. *African Religions and Philosophy*. London: Heinemann.

Nwachukwu-Udaku, B. 2011. Remembering Who We Are: The Role of Memory in the African Moral Life and Praxis. *The Reach-Out* 10: 9–10.

Obi, I. 2002. African Traditional Values in Dialogue with Christianity. *Nacath Journal of African Theology* 12: 27–35.

Osemwegie, T. 2011. Eco-Bio-Holism: An African Environmental Value and Approach to Environmental Justice and Sustainability. *Journal of African Environmental Ethics and Values* 1: 15–33.

Oyeshola, D. 1998. *Politics of International Environmental Regulations*. Ibadan: Daily Graphics.

Sarpong Groups.google.com/d/topic/usaafricadialogue/kuJ0.

Simkins, R.A. 2018. Religion, Environment, and Economy Living in a Limited World. *Journal of Religion and Society, Supplement* 16 (2018): 165–178.

Udondata, J., and J.B. Ekanem. 2013. The annang people of Nigeria: History and culture. (Texas: Altrubooks).

Umoh, I.S. 2011. *Annang Culture: Way of Life in Annang Nation and Moments in Time*. Lagos: Newswatch Books.

Umoh, D. 2012. Annang Expiatory Rituals Enriching Christian Sacrament of Reconciliation. *Abuja Journal of Philosophy and Theology* 2: 106–119.

Umoh, D. 2012. *A Philosopher looks at Religion: A Concise Introduction to Philosophy of Religion*. Port Harcourt: University of Port Harcourt Press.

Umoh, D. 2014. Annang Religious Culture. In *A compendium of Annang Culture*, ed. Etete Ineme, P., and Udondata, J., 58–71. Ikot Ekpene: Ritman Press.

Umoh, D. 2014. *Traditional Rites of Passage Enriching Christian Rites and Rituals*. Saarbrücken: Lap Lambert Academic Publishing.

Umoh, D. 2015. *The Beautiful Ugly Masquerade: "Ekpo Ikpa-Isong Annang" in Nigeria*. Saarbrücken: Lap Lambert Academic Publishing.

Umoh, D. 2016. Annang Political Consciousness as Contribution to Nigeria's National Consciousness. In *Annang National Consciousness*, ed. Ekanem, D.S. Ikot Ekpene: Ritman University Press.

Vansina, Frans. 1985. *Absence and Presence of God*. Leuven: Acco.

Dominic Umoh is associate professor of Philosophy of Religion and Religious Ethics in Akwa Ibom State University, Obio Akpa Campus, Nigeria.

Chapter 12
Shangwe Environmental Ethics: A Panacea for Agrarian Problems in Gokwe

Dorcas Hwati

Abstract The chapter examines Shangwe agrarian environmental ethics in the Gokwe area of Zimbabwe. The Shangwe, like any other Shona societies in Zimbabwe, have a rich corpus of environmental values which are enshrined in their traditions cultures and customs. In pre-colonial Africa, different communities used different moral norms, beliefs, and customs to protect their environment, a factor which enhanced sustainability in Agriculture. These moral beliefs and customs which are attributed to sages were attributed to sages were applied by the Shangwe people to solve agrarian problems including agrarian conflicts, environmental degradation, land pressure and scarcity of water resources. The Shangwe have a rich reservoir of environmental values to deal with such problems. For instance, Shangwe taboos foster acceptable human relations and good relations with the environment. The Shangwe regard nature as sacred. As such people were prohibited from abusing or destroying nature thereby preserving the ecosystem and promoting a sane and healthy environment. The essence of this chapter is to explore the various Shangwe ethical principles in preserving the environment for the sake of boosting productivity in agriculture. These ethical codes could be employed in tandem with some contemporary environmental legislations in order to achieve sustainability in agriculture. I contend that government policies should not undermine traditional ethical values in environmental management.

Keywords Environmental ethics · Agrarian · Shangwe · Philosophy · Sacred · Taboos

D. Hwati (✉)
Nemangwe Secondary School, Gokwe, Zimbabwe
e-mail: dorcashwati33@gmail.com

M. J. Tosam and E. Masitera (eds.), *African Agrarian Philosophy*, The International Library of Environmental, Agricultural and Food Ethics 35,
https://doi.org/10.1007/978-3-031-43040-4_12

12.1 Introduction

Gokwe is primarily the home of the Shangwe ethnic group. The Shangwe like any other Shona societies in Zimbabwe have a rich array of environmental values which are embedded in their beliefs and cultural practices concerning the environment. From time immemorial African communities have used these values to preserve their environments, a factor which enhanced sustainability in African Agriculture prior to colonialism. The Shangwe, likewise, have sophisticated indigenous environmental management values and customs which were used in agriculture in the Gokwe area. The entire Gokwe area used to be one of the major sources of white gold (cotton). Against this background proponents of European agrarian laws started denigrating indigenous environmental customs in favour of European Agrarian laws and practices.

The Shangwe have a plethora of literary genres like proverbs, idioms, formulae, songs, legends, myths and folktales which define their environmental and agrarian management systems. This study therefore endeavours to analyse the relevance of Shangwe environmental management ethics in solving agrarian problems that sometimes rock the Gokwe communities. Prohibitions and restrictions through taboos, proverbs, idioms, formulae, songs, legends, myths and folktales on unsustainable utilisation of some plant species, forests, mountains, rivers, pools, animals and all other ecological species is not a new phenomenon among the Shangwe people. These moral codes play a pivotal role in fostering sustainability in agriculture and environmental protection.

As such indigenous Shangwe environmental values continue to shape the agricultural system of the region although they have been relegated to the background as a result of modern Western agricultural approaches. Kelbessa (2010: 2) defines environmental ethics as the philosophical enquiry into the nature and justification of general claims relating to the environment. It is theory about appropriate concern for, values in and duties to the natural environment and about their application. It is concerned with what the people are committed to doing concerning the natural environment. Environmental ethics examines the moral basis of environmental responsibility (Ojomo 2011: 573). According to Chemhuru, human beings have tended to look at themselves as the only morally superior creatures on the planet earth. As such human communities design ethical principles that enable them to dominate the natural ecosystems including land resources. Environmental ethics provides the principles for managing the environment in a sustainable manner. Some Shangwe people have remained accustomed to their traditional environmental management ideologies and they have always prospered in Agriculture.

12.2 Agrarian History and Land Reform in Zimbabwe

Prior to the 1950s, the Gokwe area of north-western Zimbabwe was perceived as the wild, remote, and culturally backward domain of the Shangwe ethnic group (Nyambara 2002: 75). Between the 1950s and the 1960s the Gokwe region saw an influx of the Madheruka people who were drifting from the southern part of Zimbabwe due to discriminatory land policies which dispossessed Africans of their lands through unfair land policies such as the Land Apportionment Act of 1930 and the Native Land Husbandry Act of 1951, a factor that created land shortages in other regions (Moyana 1984: 26–27). The indigenous people were in turn called the Shangwe by the Madheruka.

Since the introduction of small-holder cotton production in the 1960s, and the influx of Madheruka from the south which has been represented as a miracle of agrarian transformation, a frontier of commoditization, and more broadly, as an example of the transition to modernity (Nyambara 2002). Before the coming of Madheruka, the indigenous Shangwe people were not familiar with the growing of cotton seed. The advent of cotton production was a product of the migration of the Madheruka into the region. Prior to cotton production in Gokwe, the Shangwe were growing maize, sorghum, millet and rapoko. At first the Shangwe people were reluctant to adopt cotton production but they eventually ventured into the industry and Gokwe became one of the major cotton growing regions in Zimbabwe.

After Zimbabwe attained independence in 1980, the wave of migration into Gokwe increased rapidly largely due to severe land shortages in other areas, the post-independence cotton boom, and the effects of the Economic Structural Adjustment Programme (ESAP) initiated in the early 1990s, which resulted in massive retrenchment of state workers (Siziva 2022). The majority of the retrenchees found their way into Gokwe villages. By the 1990s, there were clear signs that the frontier was closing and land pressure was manifested in ubiquitous land disputes among various land claimants (Nyambara 2001). As land shortages became a reality in Gokwe villages, landless households resorted to various forms of sharecropping arrangements with land rich households (Nyambara 2002: 73).

The main goal of the white settler farmers from the 1890s to the 1960s was to gain control over prime agricultural land and depress the earnings and profits of small-scale farmers. The achievement of control over prime land guaranteed white economic dominance and black poverty during the 90-year colonial period, from 1890 to 1980. This strategy of depressing the profits of small-scale farmers and tenants and the wages of farm workers was also pursued historically by large-scale farmers, in collaboration with the state, in Namibia, South Africa, Kenya, Algeria and many countries in Latin America (Ibid). Such policies by the white settlers created a barrage of agrarian problems in African designated areas in Zimbabwe.

12.3 Agrarian Problems in Gokwe Region: A Menace in the Shangwe Communities

This part of the chapter focuses on agrarian problems faced by both the Shangwe and the Madheruka in the Gokwe region. These problems include droughts, environmental degradation, poor farming methods and agrarian conflicts. Here I attempt to explore the causes and effects of agrarian problems in the area in question.

People in the Gokwe region encounter agrarian conflicts. These agrarian conflicts include land inheritance, land ownership, stock and crop theft and also conflicts in relation with land degradation. Agrarian conflicts are a common feature in Gokwe region and Zimbabwe as a whole having untold social and economic consequences on the people. These agrarian conflicts call for appropriate management and resolutions.

Villagers in the Gokwe region are continuously facing conflicts associated with land boundaries. One of the informants from Tiyeri Village pointed out that land boundaries and ownership conflicts are leading to quarrels, allegations of witchcraft, physical confrontation and even death. Conflicts associated with land inheritance are also a common feature among the Shangwe communities in the Gokwe region like in most Shona communities in Zimbabwe. Usually, conflicts over resources arise when people are competing for the same resources, when there is a dispute over who has the right to certain resources and when people want to take someone else's resources or prevent someone from getting needed resources (Mertz and Liber 2004). Research shows that in most cases after the death of the head of the family, who usually is the father, the land is inherited by the husband's male relatives at the expense of the wife and children of the deceased and this has resulted in dire conflicts within extended families.

A number of factors seem to influence the management of water in Gokwe just as in other rural areas. Ethnicity, gender, ethnic affiliation, political affiliation and social status influence water management, and water scarcity is widely recognized as one of the causes of significant violence and conflict within nations (Ahmad 2010: 9). Many disputes erupt in water stressed areas especially in canal end points. In rural communities there are various factors that are directly linked with irrigation water management conflict like political faction struggling for power, lack of basic resources, corruption involving access to opportunities and funds, gender discrimination, age group, ethnicity, tribal status and lack of common vision (Gehrig and Rogers 2009). It is important to note that despite these problems the Shangwe communities continue to employ their traditional customs to prevent and resolve conflicts associated with agriculture.

Environmental degradation is another serious agrarian problem experienced in the Gokwe region. Due to land pressure people are resorting to stream bank cultivation. Writing on the effects of the fast-track land reform, Zembe et al. (2014) opined that increased population caused by escalated migration has resulted in land pressure and scramble for the remaining arable farmlands which are the stream banks both in rural and urban areas. Mhiribidi (2010) also postulates that population pressure, disparities in access to productive lands and civil strife are pushing farmers into cultivating ever-steeper slopes like in mountains and river banks for small-scale food crop production. Stream bank cultivation is leading to poor management of resources since people are striving to secure their livelihood. Most of the farmers are cultivating along river banks without conserving the soils resulting in soil erosion which leads to flooding, siltation, landslides and loss of arable land (Derman 2016). To avoid predicaments associated with stream bank cultivation, the Shangwe local leaders have declared riparian zones of major rivers such as Sasame River, Svisvi River and Sengwa River sacred places (Interview with a Headman from Zarova community).

Over clearing of stream bank vegetation is a poor environmental management practice which is resulting in accelerated rates of bank erosion which has degenerated into environmental degradation in the Gokwe region. River bank cultivation is also leading to siltation of rivers leading to reduced reservoir storage and there is increased land overflow. Stream bank cultivation chokes rivers with silt which leads to impervious surfaces resulting in increased overland flow, shorter lag time between a storm and surge in discharge and increased incidences of flow downstream leading to floods in low lying areas (Bola et al. 2014). Since water absorption in the river is reduced, runoff is increased leading to floods in the area and also destruction of crops.

Nyambara (2001) postulates that the post-independence cotton booms which attracted more immigrants; the effects of the Economic Structural adjustment Programme (ESAP) initiated in the early 1990s which resulted in massive retrenchments of workers and the majority of the retrenchees found their way into Gokwe villages. One villager opined that multitudes of unemployed young generation in Gokwe today are scrambling for pieces of land resulting in severe deforestation leading to land degradation which is also affecting the hydrological cycle. As such, environmental degradation is a severe agrarian problem in the area. However, the Shangwe people always try to avert the precarious conditions that emanate from land degradation through their diverse environmental ethical codes which are enshrined in traditions and cultural practices such as taboos which prohibit people from cultivating in sacred places.

Drought is one of the most complex and least understood of all the natural disasters affecting more people than any other hazard in socioeconomic terms. Drought is a normal feature of the climate and occurs in both high and low rainfall areas. This implies that all areas in Zimbabwe are liable to experience droughts at some points (Ndlovu 2011). Drought has been a common feature in most parts of Gokwe and Zimbabwe in general. The 1991–1992 drought which is considered the "worst in living memory," placed more than 20 million people at serious risk. In recent years, concern has grown worldwide that droughts may be increasing in frequency, severity, and duration given changing climatic conditions and documented increases in extreme climate events (Peterson et al. 2013).

Responses to drought by governments throughout the world are generally reactive—poorly coordinated and untimely—and are typically characterized as "crisis management." Drought differs from other natural hazards in several ways. Drought is a slow-on set natural hazard often referred to as a creeping phenomenon (Sivakumar

2012). Gokwe region is experiencing a dramatic increase in disasters related to precipitation. Lack of precipitation as well as heavy downpours are causing disasters ranging from extreme droughts to unprecedented floods, climate change and environmental degradation (Peterson et al. 2013). Droughts have also culminated into another agrarian problem in Gokwe which is crop and stock theft. As a result of droughts some members in different communities end up engaging in stealing crops and livestock. During drought seasons, crop and stock theft become common features in Gokwe. Research has shown that Gokwe is a drought prone region and many farmers are losing their crops and livestock as a result of theft. Droughts and water scarcity in Gokwe region are particularly visible in agriculture.

12.4 Environmental Ethics as a Prescription for Agrarian Problems in Gokwe

The Shangwe people who occupy most parts of the Gokwe region have mechanisms to manage and preserve their environment. Their beliefs and practices are environmentally friendly. The Shangwe like other Shona societies believe that environmental protection and management is sanctioned by the creator God and the ancestors of the land and their religion is centred on relationships with living people, spirits of the dead, animals, land and plant life (Makamure and Chimininge 2015: 7). The Shangwe Environmental Taboos results in a sustainable use of the environment (Chemhuru and Masaka 2010: 121). Shangwe taboos are ethical considerations which foster acceptable human relations and also promote good relations between human beings and the environment. Chemhuru talks of communitarian environmentalism, the relationship between human community and non-human community such as soil, water, air and animals. These two communities are integral aspects in agriculture. Therefore, such ecosystems need to be properly managed. Behrens (2014: 63–82) argues that African communitarian environmentalism is capable of informing environmental ethics. This implies that the environment in indigenous communities is managed by communities not by individuals. As such members of the community are obliged to abide by the community's environmental ethical codes. This promotes the sustainability of the environment which is an important factor of agricultural development.

The Shangwe believe that the environment belongs to the spirits. The land is sacred because it belongs to their ancestors. In their belief systems the Shangwe attitude towards the environment are primarily about their relationship with the spirits. The Shangwe believe that the spirits have the responsibility to look after their property. The Shangwes have strong beliefs in their ancestral spirits, taboos, totems and sacred places which bear witness to the contention that their practices are meant to conserve and manage the environment (Makamure and Chimininge 2015: 7). For their environmental management and protection they base on fear and respect of ancestors. According to the Shangwe, the land, water bodies, animals and plant life are the sources of life therefore they need to be preserved and well managed. Their relationship with nature is environmentally conservative. The Shangwe manage and preserve their local environment through their environmental ethics. Their beliefs and practices are environmentally friendly.

Makamure and Chimininge (2015: 8) define taboos as avoidance rules that forbid members of the human community from performing certain actions, such as eating some kinds of food, walking on or visiting some sites that are regarded as sacred, cruelty to nonhuman animals, and using environmental resources in an unsustainable manner. According to the Shangwe people, taboos are known as specific rules that guide human behaviour. They believe that violating the moral code will lead to misfortunes such as death, diseases, bad luck and drought to individuals and the community.

According to the Shangwe, taboos (zviera) form an important part of their environmental ethics. Environmental taboos play a pivotal ethical role towards sustainable management of the environment. Taboos prohibit unsustainable utilisation of certain plant species, forests, mountains, rivers, pools and nonhuman animals, among other ecological species in the ecosystem (Makamure and Chimininge 2015: 8). Despite the tide of modernity and cultural imports, contemporary Shangwe like any other Shona communities have continued to observe their deep-seated taboos (Gelfand 1973). The Shangwe environmental prohibitions teach people to be mindful of various animal species and the natural environment. Shangwe environmental ethics take the leading role in preventing and mitigating agrarian problems in Gokwe. Therefore, the role of indigenous environmental ethics in resolving agrarian problems cannot be overemphasised.

The Shangwe regard different objects, sites, birds, animals and even land as sacred. Land itself is believed to be sacred. It is sacred in so far as it belongs and bears the remains of the ancestors in the form of graves and it is where the people's umbilical codes are buried (Ibid). According to Shangwe environmental ethics, land belongs to the living, the unborn and the ancestors, a factor which makes land a sacred resource.

Forests are also sacred among the Shangwe. It is for this reason that chiefs, headmen and kraal heads authorise through rituals the gathering of fruits in forests regarded as sacred. Chiefs prohibit the cutting down of certain trees like *muhacha (mobola plum), muonde (fig tree), mushozhowa (psudalachnostylismaprouniflolia)* and other big trees (Makamure and Chimininge 2015: 10). The Shangwe are prohibited from cutting down certain trees especially fruit trees. The Shangwe sacred sites include certain mountains and other places that members of the Shangwe communities are prohibited from visiting, cutting down trees, and hunting wildlife in them.

Gokwe South has a sacred Mountain in Mufungo area known as Gomoremhembwe where people are not allowed to hunt animals or cut down trees in it. According to Siziva (2019: 39) at one point the Ndebele wanted to attack a certain Shangwe Chief called Mafunga and his people. Mafunga and his people climbed up the mountain within their domain. When they were excited a buck appeared in their midst and mysteriously died. It is believed the buck was mystical in nature. Through this they were alerted against the Ndebele who were already closer to the vicinity (Ibid). Mafunga and his people were able to hide and escape from Ndebele raid. Since the buck saved them from Ndebele attack, it became a taboo to kill the animal. They changed their totem to Mpunzi in honour of the buck and named their mountain stronghold Mhembwe, hence, Gomo Remhembwe (Ibid). The Gomo remhembwe Mountain automatically became a sacred place. As such the mountain has been saved from unsustainable land use practices such as deforestation. This has a positive bearing on the hydrological cycle and ecological management systems of the Shangwe.

Moreover, the Mafunga chiefdom which is present day Mufungo area has another sacred place known as Matehenya, which means place of skulls. It is alleged that the skulls were left behind by the Tonga people under Nyamakwena before the arrival of the Karanga Shangwe (Siziva 2019: 37). The Mufungo people are prohibited from practising any form of agriculture at this place. Since the place is adjacent to the Gwave River, it has been saved from land degradation resulting from cultivation of riparian zone of the river in question. Thus the river is also saved from siltation and eutrophication.

Siziva (2019: 37) also posits that when Chimera and his people arrived at Mupare's domain in Gokwe North in the 1870s they were obsessed with the desire to destroy Mupare and his people. Mupare realised their intentions and begged them not to destroy his territory. Instead Chimera and his people were given boiled maize (*mangai*). While the strangers were busy eating the meal Mupare gathered his people and vanished (Ibid). Mupare used his magical powers and disappeared into the ground together with their animals and crops. At this place cries of people and animal sounds could be heard. The place became known as Gombaremhere (a pit of cries) (Ibid). The place is one of the notable sacred places in the Gokwe region. The Gombaremhere has developed into a sacred dam serving animals and people within the vicinity. It is also believed that the utilisation of the water in the Gombaremhere dam is regulated by Nevana spirit medium. As such the water is used sustainably by the local members to avoid violating the ancestors of the region.

It is also believed that in Gokwe North there is a tree called *musumha waVahombe* which must not be cut because a certain old lady called *Vahombe* was killed by a lion at that tree. The place is now respected as sacred and people are discouraged to cut down that *Musumha waVahombe* tree. This shows that these myths are helping in encouraging and ensuring a harmonious relationship between nature and human communities. Shangwe environmental ethics through taboos prohibits human beings from misusing natural resources. It is believed that sacred sites, such as mountains, have symbolic importance and these sacred mountains developed some natural fires as a way of informing people of the advent of the rainy season (Makamure and Chimininge 2015: 11).

When approaching the rainy season mysterious fires appear on the summit of this mountain. It is believed that these natural fires alerted people about the looming rainy season and them to start preparing their arable land for the farming season. These beliefs help villagers to get prepared for the farming season and to perform their farming activities. This also discourages people from visiting these sites thereby aiding the cause for harmonious living between human beings and nature (Makamure and Chimininge 2015).

The Shangwe people, also perform rituals under *muhacha* (mobola plum) *muonde (fig tree), mumvee* (sausage tree) *and mushozhowa* (psudalachnostylismaproanifolia) trees and this shows that rituals have served as environmental conservation ways (Ibid). Big and tall trees like *muuyu* (baobab tree), *mumvee* (sausage tree) and *muhacha* (mobola plum) are believed to be sacred and are left in the fields thereby preserving nature and promoting the hydrological cycle as well as providing medication for people and animals. Maintaining human and animal health is fundamental in ensuring sustainability in agriculture.

Shangwe environmental ethic which takes into account the interests of nature and human beings. Makamure and Chimininge (2015: 12) reiterated that people do not disapprove of a sustainable use of natural resources, including other living creatures for draught power and food; they are against wanton destruction of fauna and flora without justification. They also take great exception to the cruelty of animals because for them, all animals are sentient and therefore deserve to be given moral consideration. For them, a person who exhibits violent surges through cutting down trees without any need for them, and cruelty to other living creatures lacks morality (*unhu*).

In Gokwe, sacred places play a significant role in mitigating drought. Residents of the Nemangwe community in Gokwe perform the Mukwerera ceremony in a sacred place called Nevana shrine as a way of preventing droughts. The major indigenous method of preventing droughts is by performing the Mukwerera ceremony. One clan member of the Nevana clan opined that traditional leaders take an active role in the Mukwerera ceremony which is a fundamental phenomenon in terms of averting droughts. According to this respondent the village heads and headmen mobilise people to go to the Nevana shrine and perform the ceremony in order to appease the ancestors to bring sanity to the community. The same informant went on to say that in times of drought village heads and headmen in the community approach thet chief for the way forward in terms of managing the calamity. The chief will then advise the village heads and headmen to mobilise their people to come and perform the Jichi dance at his court. The chief sends messengers (vanyai) to spirit mediums of the community. There are two spirit mediums for the Nemangwe community that are Chinemakwati or Garamukanwa, the female medium and Nevana, the male medium. It is believed that Chinemakwati was Nevana's wife. This female accompanies the vanyai to Nevana, the senior medium of the Nemangwe chiefdom and say, "*Ndiniamaindauyanevana. Mhuriyanguyoparara here murume.*" The female medium pleads with Nevana spirit to protect her children from the wrath of the drought. Eventually the Mukwererais performed to plead for rains from the ancestors. The chief leads the proceedings. Usually, soon after the ceremony is done there is rain fall as a sign that the plea has been heard by the ancestors.

Another respondent observed that the mukwerera though a critical part of the traditional drought management systems is slowly losing its validity because of Christianity and the influence of Western secular culture. The Shangwe people believe that some droughts are caused by violation of traditional customs. A village head of the Nevana clan believes that violation of traditional customs has culminated into severe droughts in Nemangwe area in the period between 2008 and 2020. He argued that in Nemangwe area many people are Christians and as such they are reluctant to participate in the rainmaking ceremonies, *mukwerera.* Some are even disregarding sacred days like *chisi.* People are expected to perform the Mukwerera ceremony at the Nevana shrine before the onset of the rain season (Ngara 2019: 121). According to the chisi tradition of the Shangwe people, it is a taboo to work the fields on Thursday. The above informant sees Christianity as a threat to the traditional customs, a factor which brings about poor seasons in Gokwe.

The Shangwe believe that land is sacred. They beliefs in the sacrality of the land and also believe that land is the back (*musana*) of the ancestors on which nature and humanity are carried (Makamure and Chimininge 2015: 8). Through their religious beliefs, taboos, totems and sacred places, no one can freely gather wild fruits, cut down trees, hunt certain animals or pollute certain water bodies in areas or places regarded as sacred. The Shangwe religious beliefs play an important role in determining positive values and attitudes towards the environment and are a crucial component of any efficacious environmental policy among their communities (Makamure and Chimininge 2015: 8). Among the Shangwe communities' religious beliefs, taboos, totems and sacred places plays a vital role in environmental degradation prevention. In Shangwe communities, people must live in peace with all the sacred things and the environment. The Shangwe consider rocks, water bodies and mountains as living beings and this is why before climbing a particular mountain or entering a particular forest one must ritually ask for permission (Ibid).These practices help in environmental protection and preservation.

12.5 Shangwe Environmental Ethics and Water Conservation

Water is considered as one of the most important resources in human life. Water can be used for domestic purposes, such as drinking, cooking, washing, bathing, and irrigation and also contributes significantly towards sustaining the lives of other living things, such as nonhuman animals and plants (Chemhuru and Masaka 2010: 125). The Shangwe like other Shona societies, appreciate the value of water to the lives of human beings and other living things. Shona people are conscious of a moral code that promotes the well-being of not only human beings, but also the environment and they have some taboos that prohibit abuse of water sources, such as wetlands, rivers, and wells (Chemhuru and Masaka 2010). Shangwe communities manage to promote sustainable use of water resources through taboos. Chemhuru and Masaka (2010: 126) point out that the Shona environmental management and conservation taboos validate the claim that Shona people had, and still have, an environmental consciousness that seeks to protect water sources like rivers, pools, dams, wetlands, wells, and springs. Traditional Shona communities have a heavy dependence on open wells, rivers and springs for drinking, cooking, bathing, washing, and agricultural and industrial uses. One of the environmental taboos that is used by the Shangwe people in Gokwe to protect their water sources is *Ukachera mvura nechinhu chitema, mvura inopwa* (if you fetch water with a black container the water source will dry up).

Taboos are used to foster environmental awareness among the Shangwe people. As human beings, we carry the whole weight of moral responsibility and obligations for the whole world on our shoulders and this responsibility is not human centred, but takes into account the interests of all there is in the world (Tangwa 2006). In Gokwe South there is a water body called *Gawaramakanda* (Makanda's pool) *and* it is believed that the place is sacred and no one is allowed to practice any agricultural activity around the area. The place is densely vegetated and is a source of water for domestic animals. Thus these Shangwe beliefs help to conserve the water sources and vegetation thereby promoting the hydrological cycle and providing grazing land and drinking water for humans and domestic animals. Chemhuru and Masaka (2010: 128) are of the view that environmental awareness takes a spiritual dimension in that ancestral spirits (*midzimu*) are said to be the custodians of nature and they have a conscious interest in the way the living interacts with the environment. They also pointed out that the environment, including water sources, should be treated with respect since their misuse may provoke ancestral spirits who may in turn punish the human community with droughts and floods.

12.6 Conclusion

Shangwe environmental ethics remain an integral part of Gokwe agrarianism. The Shangwe, like any other African society, continue to employ indigenous values when dealing with problems associated with agriculture. However, some of these indigenous values and customs have not been given voice because of the domineering influence of colonialism and globalisation on indigenous ways of being and practices. It is the author's appeal that the local leaders like village heads, headmen, chiefs, traditional healers and spirit mediums and indigenous sages should continue to promote the preservation of these environmental ethical codes. The government should support these traditional leaders and sages in their endeavour to inculcate Indigenous Knowledge Systems (IKS) that help to manage agrarian predicaments that sometimes rock rural communities in Zimbabwe. Some government departments like the Forestry Commission and Environmental Management Agency (EMA) should work closely with traditional institutions to achieve their development goals. Cooperation of indigenous institutions and contemporary institutions is paramount if agrarian problems are to be diligently addressed.

References

Ahmad, W. 2010.*The Causes of Conflicts Among Rural Communities of District Archi,Afghanistan;* A Reference to Irrigation Water Management. Wageningen, Van Hall Larenstein University of Applied Sciences.

Behrens, K.G. 2014. Environmental Ethics. *An Interdisciplinary Journal Dedicated to the Philosophical Aspects of Environmental Problems* 36 (1): 63–82. doi:https://doi.org/10.5840/enviroethics20143615

Bola, G., C. Mabiza, J. Goldin, K. Kujinga, I. Nhapi, H. Makurira, and D. Mashauri. 2014. *Coping with droughts and floods*: A Case study of Kanyemba, Mbire District, Zimbabwe. *Physics and Chemistry of the Earth, Parts a/b/c* 67: 180–186.

Chemhuru, M., and Masaka, D. 2010. *Taboos as Sources of Shona People's Environmental Ethics.* Department of Philosophy and Religious Studies, Great Zimbabwe University. *Journal of Sustainable Development in Africa* 12 (7). ISSN: 1520-5509.

Derman, B. 2016. Nature, Development, and Culture in the Zambezi Valley. In *Life and Death Matters*, 101–124. Routledge.

Gehrig, J., and Rogers, M.M. 2009. Water and conflict incorporating peace building in to water development. Catholic Relief Services.

Gelfand, M. 1973. *The Genuine Shona: Survival Values of an African Culture.* Gwelo: Mambo Press.

Kelbessa, W. 2010. Indigenous and Modern Environmental Ethics: A Study of the Indigenous Oromo Environmental Ethics and Modern issues of Environment and Development, *Cultural Heritage and Contemporary Change* Series 11, Africa volume 13.

Makamure, C., and Chimininge, V. 2015. Totem, Taboos and Sacred Places: An Analysis of Karanga People's Environmental Conservation and Management Practices. Zimbabwe Open University. *International Journal of Humanities and Social Science Invention* 4 (11): 07–12. ISSN (Online): 2319-7722, ISSN (Print): 2319-7714. www.ijhssi.org.

Mertz, G., and Lieber, C.M. 2004. Five Dimensions of Conflict. ESR, Educator for Responsibility. http://www.sd70.bc.ca/_SocialResp/resources/Misc.%20SR%20Lessons%20-%20high%20school/ESR_Five_Dimensions.pdf. Accessed on: July 6, 2010.

Mhiribidi, S.T. 2010. Promoting the Developmental Social Welfare Approach in Zimbabwe: Challenges and Prospects. *Journal of Social Development in Africa* 25 (2): 121.

Moyana, H.V. 1984. *The Political Economy of Land in Zimbabwe.* Gweru: Mambo Press.

Ndlovu, B. 2011. *Drought Copying Strategies at Mutasa District in Zimbabwe*, University of the Free State South Africa.

Ngara, R. 2019. *Kayanda Musical Arts for the Installation of Shangwe Chiefs: An Epistemological, Gendered, Symbolic, Interpretive, Community—State Model for Sustaining Tangible and Intangible Heritage in Zimbabwe.* Pretoria: University of Pretoria.

Nyambara, P.S. 2001. The Politics of Land Acquisition and Struggles Over Land in the Communal Areas of Zimbabwe. The Gokwe Region in the 1980s and 1990s. Africa: *Journal of the International African Institute* 71 (2): 253–285.

Nyambara, P.S. 2002. Madheruka and Shangwe: Ethnic Identities and the Culture of Modernity in Gokwe, Northwestern Zimbabwe, 1963–79. *The Journal of African History* 43 (2): 287–306.

Ojomo, P.A. 2011. *Environmental Ethics: An African Understanding.* Ojo: Lagos State University.

Peterson, T.C., M.P. Hoerling, P.A. Stott, and S.C. Herring. 2013. Explaining Extreme Events of 2012 from a Climate Perspective. *Bulletin of the American Meteorological Society* 94: S1–S74.

Siziva, E.T. 2019. Shangwe Chieftainship: Succession Disputes in the Nemangwe Chiefdom and the Quest for Legitimacy cc 19th–21st, Great Zimbabwe University, Available at: http//ir.gzu.acc.zw.8080/xmlui/handle/123456789/337

Siziva, E.T. 2022. From Resilience to Capitulation? Problematising Morgan Tsvangirai's Political Career in Zimbabwe. In *Morgan Richard Tsvangirai's Legacy: Opposition Politics, The Struggle for Human Rights, Democracy and Gender Sensitivities*, ed. N. Marongwe, M Mawere, and F.T.P. Duri, 243–256. Laanga Publishers. https://muse.jhu.edu//pub/358.

Tangwa, G.B. 2006. Some African Reflections on Biomedical and Environmental Ethics. In *A Companion to African Philosophy*, ed. K. Wiredu, 387–395. Oxford: Blackwell Publishing.
Zembe, N., E. Mbokochena, F.H. Mudzengerere, and E. Chikwiri. 2014. An Assessment of the Impact of the Fast Track Land Reform Programme on the Environment: The Case of Eastdale Farm in Gutu District, Masvingo. *Journal of Geography and Regional Planning* 7 (8): 160–175.

Dorcas Hwati holds a Master's Degree in Development Studies from the Great Zimbabwe University (2020). Her research interests are Development, Ethnicity and Politics. She is currently a Secondary School Teacher at Nemangwe Secondary School.

Chapter 13
Agrarian Rituals, Food Security and Environmental Conservation in the Bamenda Grassfields of Cameroon

Michael Kpughe Lang

Abstract Agriculture has been crucial in enabling the people of the Bamenda Grassfields of Cameroon to satisfy their subsistence. They have consistently observed ritual practices aimed at sustaining agricultural production and protecting the environment. These agrarian rituals constituted an integral part of indigenous agriculture and food systems that hinged on ethics, creativity, and innovation. Communities across the region had values and constantly changing practices accruing from their ingenuity which had a bearing on the way they interacted with nature to obtain food. This chapter uses African agrarian ritual practices that were deeply embedded in African worldview to demonstrate the connection between religion and agriculture. It argues with empirical evidence that these rituals, informed by indigenous agrarian values, contributed to local food security and environmental protection. The chapter opens with a contextualization of Bamenda Grassfields' indigenous belief systems and agrarian thought as a means of interpreting the origins of agrarian rituals. This is followed by an analysis of rainmaking and planting rituals which are usually performed at the beginning of the farming season. The chapter further discusses the celebration of bountiful harvest through rituals as indicative of food security. Finally, the crucial role of these rituals in ensuring conservation of the environment is examined. This study suggests that a new emphasis be put on agrarian rituals as a defining feature of African agrarian philosophy because of their unending bearing on the success of agricultural endeavours and their capacity to control ecological conditions in ways that contribute to environmental protection.

Keywords Grassfields · Rituals · Agriculture · Food security · Environmental conservation

M. K. Lang (✉)
The University of Bamenda, Bamenda, Cameroon
e-mail: mickpughe@yahoo.com

M. J. Tosam and E. Masitera (eds.), *African Agrarian Philosophy*, The International Library of Environmental, Agricultural and Food Ethics 35,
https://doi.org/10.1007/978-3-031-43040-4_13

13.1 Introduction

The peopling of the Bamenda Grassfields of Cameroon was largely conditioned by the natural environment, most notably the richness of the soil in the area. The reliability of food production led to the emergence of settled communities. This bearing of agricultural production on the growth and resilience of these communities has long been recognized (Kaberry 1952), and this is evidenced by the people's reliance on a combination of cereal cultivation and livestock production. The centrality of food to the welfare of people across the Bamenda Grassfields made the people to place agriculture under the direct control of traditional rulers and associated governance institutions. The fields were communally owned and were placed under the custody of traditional rulers. The latter, in consultation with other traditional governance structures, decided the fields that were to be cultivated and those that were to be allowed to fallow (Goheen 1996). Decisions about the timing of the various phases of the farming season—cultivation, planting, weeding, and harvest—were taken by traditional rulers. The farming season was characterised by a chain of agricultural rituals and ceremonies performed at the start of each phase.

In fact, indigenous religious beliefs and practices shaped economic activities in the Bamenda Grassfields including agricultural production. But any talk of traditional religious influence on agricultural development in this region has to consider the central importance of land. This is not only because many people depended on agriculture for a livelihood, but also because peoples' ideas about the proper use and ownership of land were often expressed in terms of religion, which was informed by local philosophical knowledge. One of the common characteristics of the grassfields culture was the traditional religious belief that land (which was used for many purposes including farming) was the ownership of the Supreme Being. This explains why traditional functionaries such as *fons* (chiefs) and ritual priests exercised spiritual control over the land. Seen this way, the land that was used for agricultural production was therefore sanctified by its possession by God and ancestral spirits. In all societies in the area, land had a value linked to a community, its *fon* and the spirits of ancestors. No wonder the fon in every grassfields chiefdom was called 'owner of the land'. This ownership of the land by the *fon* anchored on his/her supposed connections with mythological founder-ancestors of his/her chiefdom. Ancestors were believed to have chosen him and given him power and authority over land and his subjects. Hence, the *fons* were sacred and had divine authority linked to the farming land and the supernatural spirits that owned it. Consequently decisions on the communal farmlands to be cultivated were taken by the *fon* of each chiefdom in consultation with earth priests. Sadly, some fons abused the tenets of communal land ownership despite the existence of governance institutions whose role was to check their powers. In some cases, *fons* engaged in land grabbing, which resulted in numerous cases of land conflicts in the area (Tosam, in this volume).

The availability of fertile land and its close association with the worldview of the people yielded a thriving agricultural economy upon which the sustenance of communities depended. Recognising the contribution of agriculture to the well-being of society as well as the urgency of finding ways to ensure sustainable agricultural production, the people developed agricultural values and practices which aligned with what some scholars label as African agrarian philosophy (Oruka and Juma 1994; Niekerk 2005; Tosam 2019). It was a creative and explicit agrarian thought whose development was necessitated by human and natural risks, such as droughts and land depletion, which threatened the thriving agricultural economy. Faced with these challenges, communities across the Bamenda Grassfields adopted a plethora of strategies in the hope of ensuring food security and environmental protection. Prominent among these strategies was recourse to agrarian rituals that were part of their indigenous agrarian philosophy built upon their worldview, moral norms, creativity and innovation. It was an agrarian philosophy that recognized agriculture's centrality in enhancing the welfare of society. Indigenous religious beliefs and practices were at the heart of this agrarian philosophy as communities leaned on religion to address problems that threatened food security. This concurs with Jules Janick's observation that "the close affinity of agriculture and religion has made some types of agricultural practice take on the sheen of religious practices themselves." (Janick 2013: 14) The Supreme Being was considered to be responsible for the fertility of the soil, the prosperity of the land, and the livelihood of its people. This can be likened to Ancient Egypt where agricultural productivity was understood as the product of a theology that was based on a pantheon of gods and goddesses (Janick 2013: 20).

Just like in many other societies, agricultural production in the Bamenda Grassfields developed a spiritual ethos that was largely faith based. Little wonder these communities observed farming rituals that involved cosmological ideas of the Supreme Being, the ancestral spirit world, and the role of ritual experts. Traditional food production in the region was perceived and understood in cultural and spiritual terms (Chilver and Kaberry 1968). With insights from their worldview, the people developed agricultural technologies capable of ensuring food production that even exceeded required quantities and varieties. Apart from ensuring bountiful harvest and food security, agricultural rituals in the Cameroon Grassfields were also intertwined with the preservation of the natural environment. Derived from supernaturally supported worldviews and customs, agrarian rituals were marked by prayers and sacrifices to God and the ancestors during times of extreme weather like prolonged drought. Through rites and ceremonies at various sacred sites, ritual experts controlled the environment and weather and pronounced on directives aimed at ensuring a sustainable use of the environment. Clearly, agrarian rituals were agricultural and environmental in both purpose and outcome (Ikeke 2018; Nkwi 2017).

An excursion into the growing literature on African agrarian philosophy reveals the centrality of an indigenous agricultural ethos in explaining the success of a society. There is a consensus among agrarian philosophers that beliefs, values, practices and institutions have a combined influence on how communities guarantee sustained food supply. Nyerere (1968) who pioneered scholarly discourse on African agrarian philosophy espoused philosophical views on African agrarianism at the core of which was a communal agrarian ethos anchored on the *Ujama'a* philosophy. To him, communal ownership of land and its collective exploitation for food production were informed by this communal orientation of life and the desired collective welfare. This communality of life and the necessity of sustenance yielded what Oruka and Juma (1994) describes as an ethic of duty towards humans and nature. Because agriculture was intertwined with the natural environment, farmers, informed by local beliefs, values, norms, and practices, could not overlook the sustainable exploitation of land. Farmers, going by the thoughts of Oruka, had the moral responsibility to sustain the fertility of land and to restore it whenever it was threatened by excessive exploitation. Little wonder Van Niekerk (2005) stresses that constraints to African agriculture can only be addressed through ethically acceptable solutions. The thoughts of Nyerere, Oruka and Van Niekerk are consistent with the agrarian philosophy of communities across the Bamenda Grassfields, which was hinged on beliefs, values, practices and institutions that sanctioned the way land was exploited for agricultural purposes. This chapter emphasizes the nexus between indigenous religious worldview and agricultural production as a lens through which African agrarian philosophy can be appreciated. I concur with Mbiti (1969: 1) that Africans are notoriously religious to the point that every sphere of life, most notably agriculture, has a spiritual dimension.

This chapter therefore explores how indigenous agrarian philosophical knowledge, especially cultural traits such as values, attitudes, beliefs, and practices that were prevalent among people in the Bamenda Grassfields shaped agrarian rituals on agricultural production and environmental protection. How did agrarian religious rituals that were embedded in indigenous agricultural philosophy shape agricultural production and human well-being? To address this question effectively, the chapter argues that indigenous African agrarian philosophical perspectives, viewed through the lens of religion, determined the success of agricultural production, resulting in food security and sustainable exploitation of the environment. Existing scholarship on agriculture in the Bamenda Grassfields reveals too much concentration on the economic significance of farming, with little focus on the imprint of local philosophical knowledge and religion on food production and the protection of the environment. In *Women of the Grassfields*, Kaberry (1952) explore the economic position of women in the Bamenda Grassfields, focusing on how agriculture has contributed in shaping their status. In their study, Chilver and Kaberry (1968) discuss the traditional agricultural economy of the peoples of Bamenda Grassfields, noting that crop cultivation was their major occupation. Without any details, these scholars emphasize the importance of agricultural rituals in communities across the area. Banadzem's "Catholicism and Nso' Traditional Beliefs" examines Nso' beliefs and practices and has scratched from the surface the imprint of these beliefs on agricultural production (Banadzem 1996: 129). Kah (2016) explored the Laimbwe *Ih'neem* ritual and concluded that its performance ensured food sufficiency. In a previous paper, I examined the intertwined relationship between religion and agriculture in the Cameroon Grassfields. The paper established how Traditional Religion and Christianity shaped agriculture in this region. It also emphasized the need to focus more on indigenous rituals for agriculture to better understand their bearing on food production and landscape conservation (Lang 2017).

Evidently, existing scholarship has not paid any serious attention on how agrarian philosophy, particularly agricultural rituals, influence agriculture and food security in the Bamenda Grassfields. Based largely on conversations and interviews with elders in some Tikar, Chamba and Widikum polities, this paper rescues the agrarian ritual perspective of African agricultural philosophy from academic neglect by showing how farming ritual practices ensured food security and environmental protection in the Bamenda Grassfields of Cameroon. I talked to older people in some Grassfields communities to obtain data to explore the agricultural and ecological dimensions of rituals. To demonstrate this, I present the environmental and human setting of the Bamenda Grassfields and the complex nature of indigenous agricultural rituals that marked the farming season. The contribution of agricultural rituals to food security and environmental preservation is handled in the final two sections in view of according relevance to indigenous African agrarian philosophy.

13.2 The Setting: Landscape, People and Worldview

Densely settled, the Bamenda Grassfields constitutes part of the Cameroon Grassfields with a diversified natural environment. Its ecosystem ranges from extensive mountain areas, savannah and dry land areas, to low lands, coastal plains and tropical forests. The Bamenda High Plateau, characterised by mountains, hills, valleys, and numerous rivers and lakes, is the most outstanding geographical feature in the area. The rainy and dry seasons together with other climatic conditions made possible the presence of rich alluvial soils, raffia swamp forests, and extensive grassland vegetation. This diversified natural environment attracted people who settled and organized themselves into ethnic communities. The Bamenda Grassfields therefore supported large farming communities, with agriculture constituting a major source of livelihood. Forming a fairly distinct geographical unit, the area covered the former Bamenda Division during the first two decades of British colonial administration. Presently, the area corresponds to the North West Region of the modern Cameroon State.

The area is littered with numerous ethnic communities that are broadly categorized under the Tikar (Nso, Kom, Bafut, Oku, Yamba, Mbum, Nkwen, Weh, Esu, Bambili, Bambui, Kejom-Keku, etc.), the Widikum (Ngemba, Moghamo, Meta, Oshie, Ngie, Ngwaw, and Mundum), the Chamba (Bali-Nyonga, Bali-Kumbat, Bali-Gham, and Bali-Gashu), and the Tiv (principally the Aghem). The crafting and growth of these communities hinged largely on the natural environment as it determined their traditional economy which was a blend of subsistence agriculture, animal breeding and handicraft (Tangie 2007: 23). Agriculture was the major source of living and various food crops were cultivated: guinea corn, maize, cassava, beans, coco-yams, plantains, bananas, groundnuts, and potatoes. The history of this traditional agricultural civilization began with the crafting of these polities and was anchored on the worldview of the people, at the centre of which were culturally-defined agricultural practices. Traditional agricultural practices ensured high agricultural production while guaranteeing sustainable exploitation of land. In all these ethnic communities, both agriculture and religion were intertwined, and this took the form of ritual practices and ceremonies that were aimed at boosting production and ensuring environmental protection.

There were many common features of African Traditional Religion such as the belief in a Supreme Being whose appellation was as varied as the ethnic groups. For example, among the Weh, God is *Keze* (Great Spirit and Creator). Among the Nso, God is *Nyuy* while the Bali-Nyonga call Him *Nyikob.* Other common features included the belief in spirits/divinities, belief in life after death, the existence of religious personnel and sacred places and numerous magic practices. Generally, traditional religion affected all aspects of life, from farming to hunting, from travel to courtship, and from birth to death. In this dimension, John Mbiti (1975) underlines the religious notoriety of Africans. Prior to the introduction of other faith traditions, traditional religion had permeated all aspects of societal life: political, economic and social. As Moyo (1996: 12) writes, traditional religion was "a way of life in which the whole community is involved, and as such it is identical with life." Undeniably, therefore, traditional religion influenced the lives of many people in the Bamenda Grassfields through its absolute and communal ritual practices. The religious authorities exercised jurisdiction over many aspects of spiritual life. Communally, the people involved in traditional religious practices to win the favour of God and ancestral spirits on issues like marriage, journeys, war, farming, and infertility (Mbaku 2005).

In most of the ethnic groups in this region, traditional rulers and ritual priests conducted rituals, organized ceremonies and offered sacrifices and prayers with the intent of ensuring the dual fertility of land and people (Lang 2015). The highly religious nature of these people resulted in an unending entanglement between religion and agriculture. The peopling of the area by these people placed the region on the path to becoming the centre of traditional religion. As the various people migrated into the area and sedentarised thanks to their adoption of an agricultural culture, their indigenous religion took shape. As the people interacted with each other over time and circumstances, their indigenous religion intertwined, spread and absorbed new ingredients. Consequently, this traditional religion that has survived until today has always contained similar insights, such as has been expressed in belief in the Supreme Being, ritual practices and sacrifices. It was not limited to these practices and affected all aspects of life. As I have argued elsewhere, the observance of traditional religious practices within their ethnic boundaries was communal and absolute prior to the introduction of Islam and Christianity (Lang 2017). Considering that these people were dependent on agriculture, they developed a plethora of agrarian rituals, most notably those that were intended to control ecological and whether conditions.

13.3 Rituals and the Control of Ecological and Weather Conditions

The importance of food and the natural and human risks that threatened the agricultural economy in communities across the Bamenda Grassfields explains why sacred farming rituals were performed. The bearing of ecological and weather conditions on the fertility of farmlands and agricultural production necessitated ritual practices that were aimed at influencing rainfall in view of addressing harsh weather conditions such as droughts, strong winds and floods. These rituals, which were embedded in the people's indigenous agrarian philosophy, preceded the preparation of farmlands and were performed after the traditional rulers had decided on the farmlands that were to be cultivated. With the farmlands selected, ritual ceremonies were organized in all chiefdoms to guarantee the reproductive capacity of the earth at the chosen sites. Anchored on the human-earth relationship and determined by local moral norms, creativity, innovation, and belief systems, rainmaking was an indigenous response to changing environmental conditions which were intended to alter the pattern and quantity of rainfall. It was an adaptation to an ecological challenge that negatively affected agricultural production. Ikeke (2018: 131) observes that "African people have always related with nature and have cultural practices and beliefs about the environment that have enabled them to live in a sustainable manner from ancient times."

Prevalent across the Bamenda Grassfields, rainmaking rituals were aimed at boosting agricultural production in a context of extreme climatic conditions that were compromising the survival of entire communities. These rituals were communal, absolute and indispensable because of the people's dependence on rain-fed agricultural production. In all communities, the patterns of rainfall determined the agricultural calendar of activities. The start of the farming season was based on rainfall, whose pattern was quite often disrupted by environmental shifts. Each community had specialists bestowed with the power and knowledge to perform rainmaking rituals. The rituals rested on the belief that through sacrifice, prayer and incantations, God's rain could be summoned. Traditional leaders and ritual priests ensured sufficient rains through annual rituals and ceremonies. In Nso for example, the people believed in the Supreme Being (*Nyuy*), divinities and ancestors from whom they solicited rain. Banadzem (1996: 126) recounts how Nso rainmakers (*Fon*, *Taawon*, and *Yeewon)* journeyed to Kovvifem royal sacred site every year to observe the *Cu* ritual. The ritual was directed to *Nyuy*, and it involved objects such as palm wine, camwood, a double bell, and a ram. The ritual took place in the month of March at the beginning of the farming season when rain was in great need. At Kovvifem, the *Fon* and his two sacred deputies (*Taawon* and *Yeewon*) approached *Nyuy* in prayer, ritual and sacrifice (Banadzem 1996: 127). The Fon slaughtered the ram on the royal graveyard (*fem*) and allowed its blood to drip on the stone grave marker. He then prayed for rain and other blessings for the entire community. In most cases, the *Fon* recited the following prayer as translated by Joseph Lukong Banadzem:

> God of Way (market) Taa, we are all present here, no one is left behind. As it is the beginning of the planting season, we have then come to speak, just as our fathers used to speak in the past. For our sake, we implore as the planting season has come. We appeal for abundant rain, food and offspring. We also appeal that our land should progress. (Banadzem 1996: 128)

While the rainmaking ritual was an annual ceremony in the Nso Chiefdom, other communities observed these rituals only when it did not rain at the expected time as well as when rain was not enough to support the growth of crops. Among the Bali-Nyonga, the science of rainmaking was the preserve of members of the *Voma* society, most notably the *ndaghanbira* title-holders and those who had acquired the title of *gwan*. Members of *Voma* society, who may qualify as indigenous philosophers, performed a ritual called *vom naba* at the beginning of the farming season. A key aspect of this ritual was the making of rain when it did not come in abundance at the start of the farming season. The Bali-Nyonga expected rain to fall naturally, but they believed that God (*Nyikob*) and ancestors usually interrupted rainfall for a plethora of reasons. This explains why the *ndaghanbira* met in a lodge called *dolu* with the medicine required for rainmaking. The *dolu* lodge was an enclosed sacred site at the centre of which was a clay pot used for rain-medicine. The seven rainmakers gathered at the *dolu* and sat on the stone seats that surrounded the medicine pot upon the request of the officiating *ndaghanbira*. With the medicine in the pot, rainmaking prayers were said for rain to fall abundantly. These rainmakers are described by Chilver and Kaberry (1968: 67) as persons who controlled rain and lightning.

Similarly, the Mbot people performed rainmaking rituals at the start of the farming season. In this Tikar polity, rainmaking was the responsibility of the *ndamgong* lodge (the house of the country). The chief priests of this lodge (*Fai Njila-Nkur, Ndi Taramfo, the tawong and yewong and the nkfu*) made regular sacrifices over the *mfu-nggong* (hole of the country) whenever drought threatened agricultural productivity (Chilver and Kaberry 1968: 108). Performed on behalf of the whole of Mbot, the sacrifice involved an offering of food items, a ram and palm wine. Fai Njila-Nkur led the officiants to the sacred site and poured libations for rainfall. On their part, the Weh had *Ndau Keze* (house of God) whose task was the procuring of rain during the growing season. The members performed rainmaking rituals only during this season. When the rituals commenced, women stayed in the village for three days and were not allowed to work the fields (Geary 1979: 67). Generally, the correlation between water and agriculture caused rainmaking through traditional religious means to play a central role in the Bamenda Grassfields agricultural system. In all communities in the region, there existed cults that were engaged in the art of rainmaking. This is because the dry and rain seasons determined the farming cycle, and in many societies the change of the seasons and its bearing on agriculture was a topic that preoccupied indigenous philosophers.

The emergence of rainmaking as part of the people's agricultural ethos was thanks to ritual priests and leaders of cult societies who reflected on problems threatening agriculture. Their philosophizing yielded the view that drought had divine roots and necessitated the observance of rainmaking rituals in communities across the region. Clearly, harsh climatic conditions such as drought caused rainmaking rituals to be associated with agricultural production. Indigenous philosophizing had yielded the consensus that properly performed rainmaking rituals enabled the people to mitigate drought and to ensure that there was sufficient rainfall throughout the farming season. The success of rainmaking rituals among the Nkwen was confirmed by the rain priest, George Chefor Tanwie (Interview 6 December 2021). He stressed that "God and our ancestors always answer our prayers by bringing us rain at the beginning of the farming season." This success of rainmaking rituals is also supported by field data from interviews conducted with informants in many other communities in the Bamenda Grassfields and some of whom can be likened to the sage philosophers described by Oruka and Juma (1994). This indigenous science of controlling weather and climate was, as demonstrated already, ingrained in religion; it was proof of the peoples' relationship with the supernatural world. Indigenous thinkers therefore provided their wisdom on drought as a challenge to agriculture which led to the recognition of rainmaking rituals as an effective way of responding to extreme weather conditions in view of ensuring food security.

13.4 Annual Planting Rituals

In the Bamenda Grassfields, land was treated as a sacred natural resource entrusted to the indigenous people by the Supreme Being. Even though the people naturally expected land to be fertile, they believed that God and ancestors, for various reasons, could interfere with soil fertility. As such, the people performed planting and harvesting rituals with the intent of maintaining the fertility of the soil and guaranteeing high agricultural production. These rituals were absolute and communal because of the general belief that the sustainability of their communities was not possible without soil and farming. Kah (2016: 58) is therefore right when he says that "many Cameroonian ethnic groups believe that a good harvest is the handiwork of the Supreme Being through the intercession of the ancestors." Throughout the pre-colonial period, the principal method of farming in the Bamenda Grassfields was shifting cultivation. The people moved within a demarcated zone, clearing the grassland, farming for some years and moving on. The abundance of land was such that competition for farmland was almost absent. The agricultural cycle which was shaped by the seasons and religious observances covered an entire year. The dry season was devoted to the preparation of farm plots (Kaberry 1952: 61). Crops were planted following the onset of the rains in late March. From May to August, farmers were chiefly involved in weeding; and from August to November with harvest. As noted already, this agricultural cycle was underpinned by religious observances since farming was considered as a religious act. The entire farming cycle was marked by ritual practices which included sacrifice to, and appeasement of, the spirits or God; prayer and requests for communal intercession. In every community, there existed traditional religious specialists whose roles were connected with agriculture. They carried out religious observances throughout the year in an annual cycle of rituals intended to promote agriculture.

The farming season in communities across the Bamenda Grassfields was launched with the performance of planting rituals. Among the Kom, for instance, the inauguration of the planting season was marked by the observance of a ritual ceremony called *fuchuo*. The ritual was performed at a sacred site in Laikom by a sacred team led by the traditional leader. Performed on the royal graveyard, the ritual began with the offering of food and palm wine to royal ancestors. With the food and wine placed on the royal grave, the *fon* poured the palm wine in libation and prayed to his royal ancestors appealing for the fertility of the farmlands (Kah 2016: 58). In Awing, the Council of the Nine (*angoeroe*) performed agricultural rituals at the Awing Lake owing to the belief that their ancestors lived in the lake. The ritual was an offering of a white ram, salt and camwood to the ancestors. The ram was slaughtered and its pieces were smeared with salt and camwood and thrown into the lake. The Awing Fon said a prayer requesting the ancestors to bless the farmlands with abundant crops (Nkwi and Warnier 1982: 120).

In Weh, the *Fon's* ritual functions enabled him to perform farming rituals which were necessary for the protection of the fields and the unperturbed growth of crops (Geary 1973). The Weh also had the *Ndau Kenyi* (*kenyi* is the name of the medicine that the experts produce) which was a cult specialized in performing farming rituals. The *Ndau Kenyi* was responsible for the growth of the plants, and on each Weh Saturday (*Tsu-u-kpeghe*) its medicine was distributed in the fields (Geary 1979: 67). *Ndau Kenyi* was in full session during the three months growing period (April through June/July) and met after then at regular intervals once a week. Finally, the Weh have the *Ndau Asang* (house of guinea corn) whose sessions start at the beginning of the planting season. For three months, the members meet once a week and offer sacrifices to ancestral spirits. The cult also holds emergency sessions on the *Fon's* request in times of severe famine. Because of the importance of maize in the livelihood and culture of Weh people, they had a secret society, *ndau kedzuule*, which specialized in the production of medicine for plentiful maize harvest. As explained by Geary (1979), *ndau kedzuule* began its rituals when maize had just been planted. Its medicine was buried in the cultivated fields, especially in the communal farm that was called *Ibeghe*. The Laimbwe people observed similar rituals that were intended for soil fertility and growth of crops. Kah studied one of these rituals (the *ngaang)* and summarized its sequence:

> The ritual of *ngaang* involves the erection of shrines at road crossings leading to the farms and also at the entrance and exit of the villages. The ritual is meant to protect maize from the destructive effects of tornadoes and other evil spirits brought to the land by wizards and witches. It also prevents people having evil intentions of causing harm to crops. (Kah 2016: 62)

This ritual of *ngaang* had the dual function of ensuring soil fertility and protecting crops throughout the farming season. This explains why experts of this ritual were active from when crops were planted to when they were harvested.

In the Wimbum communities of War, Wiya and Tang-Mbo, farming rituals were observed in the *ndamgang* (house of the country). Nkwi and Warnier record that ritual priests in these communities carry out farming rituals in the *ndamgang* in

March and December every year. According to these Anthropologists, these ritual priests slaughter a fowl when the maize in the farms is about a foot high before the first weeding. Besides, fons in this area blessed women's hoes by placing them in the *ndap-ngon* (house of stones). In this ritual lodge, the leaves of special plants were macerated in palm wine and the hoes were sprinkled with it (Nkwi and Warnier 1982: 164). The religious tradition of blessing crops in the farmlands before harvest offers a glimpse into the longstanding relationship between agriculture and faith in Nso land. According to Kaberry (1952: 33) and Banadzem (1996: 132), the Fon of Nso usually performed farming rituals at the Kovifem sacred site and other alters spread across the fondom. At Kovifem, the Fon and his ritual associates performed the major sacrifice to his ancestors and to *Nyuy* (God) to ensure the fertility of all Nso land. Among the Nso, the link between *Nyuy*, the earth, and the people who live on and cultivate the earth, was a close one and was expressed in moral and ritual terms. Evidently the rituals carried out at Kovifem and other alters in Nso by lineage heads constitute an expression of dependence on the supernatural and of appreciation for good harvest. In the Nso palace, there was a lodge called *fai shishwaa* in which apotropaic medicines were prepared at irregular intervals during the growth period. The medicines were often distributed at cross-roads and on some farms. At the beginning of the farming season, women take their hoes to the *fai shishwaa* for blessing (Chilver and Kaberry 1968: 103).

The Nkwen observed an annual agricultural ritual to ensure high agricultural productivity. The ritual, *nggang*, was an important public ceremony observed at the beginning of the farming season in March. The ritual team consisted of owners of herbal medicine who assembled at the palace on the day of the ritual. The *nggang* ritual consisted of a sequence of actions. First seeds of various food crops like maize, groundnut, beans, and cocoyam were taken to the palace of the Sub-Chief (*Atanche*) of Menteh by several women led by females called *Kikem*. The *Atanche* of Menteh, acting as the chief officiant, then used palm wine and a variety of medicinal plants to approach the Supreme Being and ancestors. In a sustained prayer, he called upon the ancestors to bless the seeds for a plentiful harvest. The blessed seeds were given to the *Kikem* and some were placed at major road junctions (Ngufor, Interview 5 December 2021). With the seeds ritually blessed at the *mbitum* lodge in Menteh, herbal medicines were harvested from the natural environment by traditional doctors from all over Nkwen in preparation for another ritual ceremony intended to bless the farmlands. This ritual began at the palace when chief priests from the Menteh and Mbelewa quarters who went by the name *betabenggang* collected, mixed and ground herbal medicine in the *kwifon* lodge. The *betabenggang* added their own medicine and slaughtered a ram over it. As the blood dripped on the medicine, they prayed that the farmlands be blessed and protected by the Supreme Being (Waji, Interview 6 December 2021). This was a community-wide ritual which brought together medicine men from all over the chiefdom. The *betabenggang* chanted incantations asking for the farmlands and crops to be protected. They also blessed the seeds with the herbal medicine in order to protect them from evil spirits. In pre-colonial Nkwen, there were various communal farmlands such as Alahloung, Ntebejong, Mifie, and

Atiela. The *betanekuru* (quarter heads) took the mixture to their respective quarters where it was laid across road junctions (Chilver and Kaberry 1968: 61).

Overall, communities across the Bamenda Grassfields observed a plethora of farming rituals in the course of the farming season. At the start of the farming season, the farmlands were ritually selected and seeds were blessed by chief priests who presided over ritual ceremonies. Some of the rituals were primarily concerned with the protection of the crops from witches and wizards who were believed to be at the origin of destructive winds and poor harvest. The people also believed that embittered spirits and ancestors were responsible for destructive winds and poor harvest. While powerful medicines were prepared and spread at major road junctions to contain witches, embittered ancestors were appeased with sacrifices. Summarizing these rituals, Kaberry notes that:

> The rulers or village heads in many tribes perform sacrifices to the gods of the earth and to the ancestors for the fertility of land and women, and for the general welfare of the people. In addition to these rites there are others associated with particular crops – maize, finger millet, and guinea corn – which may be carried out by lineage heads or individual farmers prior to planting and again at harvest. (1952: 70)

Through these rituals, which came into existence thanks to local sages' insights into particularly troubling agricultural issues, God and the ancestors revealed themselves to the living through the fertility and protection of farmlands which guaranteed plentiful harvest? Little wonder the present generation of rural elders (sages) still accord relevance to farming rituals, insisting that they represent a successful indigenous response to food insufficiency. As will be discussed in the next section, ethnic communities in the Bamenda Grassfields observed harvest celebration rituals to thank God and the ancestors for blessing and protecting their farmlands.

13.5 Harvest Celebration Rituals: Evidence of Food Security

The emergence and expansion of ethnic communities across the Bamenda Grassfields was evidence of successful agricultural production which was a major economic activity. Plentiful harvest was interpreted in all these communities as divine blessings accruing from rainmaking and planting rituals. The consistent abundance of food, as noted by Asongwa Peter Mufora (Interview 8 December 2021), was considered to be proof of the intervention of God and the ancestors in agriculture. In this section, it will become clear that the success of agrarian rituals was evidenced in the celebration of plentiful harvest through rituals, and this was strongly reflected in the centrality of food in the livelihood and culture of the people, which I interpret as evidence of the connection between rituals and food security. What follows is an exploration of some harvest rituals, with a particular focus on food security as a product of agrarian rituals.

In the Bamenda Grassfields, sedentary life and the emergence of chiefdoms were anchored on agriculture. Chilver and Kaberry (1968: 39) are right in their observation that Grassfields polities were sufficient in food. Agrarian rituals began with rainmaking and ended with the harvest celebration rituals, which in most communities was interpreted as food sustainability and sufficiency. Grassfields people developed agricultural technologies and accompanying rituals capable of ensuring food production that even exceeded required quantities and varieties, with some communities specializing on particular crops. These people recognized the intervention of God and the ancestors in agriculture and observed thanksgiving ritual ceremonies. Throughout the region, the traditional farmers saw the crops from their farmlands not as mere material for food but as blessings accruing from God. Among the Weh, it was believed that plentiful harvest was the handiwork of God (*Keze*) through the intercession of the ancestors. This explains why the Weh performed a ritual to launch the harvesting of crops as a sign of gratitude to *Keze*. The *Ndau Asang* (House of Guinea Corn) held weekly rituals for three months during the farming season. Its final rituals were meant to thank God and the ancestors for a bountiful harvest. It was only performed when maize was ready for harvest in the farmlands. Harvest was preceded by a thanksgiving ritual performed by members of *Ndau Asang*. The ritual involved an offering of guinea corn and palm wine. After this, women were then permitted to reap the first crops on a day set by the most senior female notable (*Nahtum*). On the agreed day, Weh women and their female children dressed in traditional attire and went to the farmland in the morning carrying baskets. Led by the *Nahtum*, these women harvested the first crops (mostly maize and vegetables). The journey home was usually a procession marked by the signing of joyous songs in celebration of the plentiful harvest. They thanked the chief priests, especially members of *Ndau Asang* for performing the rituals which attracted blessings from the Supreme Being.

Similar harvest celebration rituals and ceremonies were observed in neighboring chiefdoms such as Aghem and Esu. The Aghem celebrated the successful prosecution of agriculture through rituals that were performed by members of *Kwifo* and *etshuidigha*. The Fon of Aghem presided over the ritual in *etshuidigha* by offering a ram and palm wine to God and the ancestors. Members of *Kwifo* spent an entire night celebrating bountiful harvest in prayer and music before officially launching the harvesting of crops the next morning. This permitted women to participate in a harvest ceremony led by the Queen Mother (*Nahtum*). They went to the communal farm (*Ibagha*) and reaped the first crops before returning home singing thanksgiving songs. While the Nkwen lacked such a communal harvest celebration ritual, special rituals were performed to thank God for plentiful harvest. One of my informants, Francis Ngongeh (Interview 6 December 2021), explained that harvesting was preceded by a simple ritual performed in the farms by women by reaping a crop of their choice and throwing it into any of the shrines erected by *betabenggang* and members of *kwifo* at the beginning of the planting season. This was done in gratitude to the Supreme Being and ancestors for blessing them with good harvest. Quite often, the crops were thrown into the shrine Kwa-kwa because it was found on the main road to many farmlands.

The observance of harvest rituals in Bamenda Grassfields communities was evidence of adequate average food availability which was the result of sufficient agricultural production made possible by agrarian rituals. The striking reality is that food was usually grown far more than required even though there were always imbalances in crop types owing to geographical differences. The majority of the communities relied on guinea corn and later maize which were always produced in excess. This is supported by Nkwi and Warnier (1982: 43) in their observation that "The most productive regions were probably those that produce large quantities of maize nowadays, that is the Ndop plain, Wum, Kom, and the Bamenda Plateau." While these areas specialized in the production of maize, others specialized in the production of oil palm, cow-peas, beans, coco yams, and groundnuts. By necessity therefore, there was local agricultural specialization which generated imbalances in food resources. The imbalances that were caused by geographical differences were compensated through inter-chiefdom trade over short and long distances. For example, Bamenda Plateau chiefdoms such as Bafut, Mankon, Nkwen, and Bambili specialized in the production of cow-peas that were traded in exchange for other crops (Nkwi and Warnier 1982). Trade provided deficit chiefdoms with the volume of food resources they could not produce. So, there was food security as all chiefdoms had access to sufficient food which enabled the people to live healthy and active lives.

This abundance of food was at the origin of demographic expansion across the Bamenda Grassfields. Nkwi and Warnier who studied the Western Grassfields as a whole concluded on the basis of ethnographic evidence that the increase in population was a product of the successful execution of agriculture. They declared:

> The population density of the Western Grassfields is unusually high by African standards, reaching perhaps 30 or 40 people per square km as an average in the nineteenth century, with localized areas of much higher density. This may have been the case for a very long time before the nineteenth century, for without modern medical facilities, the growth rate of the population could not have reached the staggering figures of the twentieth century. It is directly connected to food production. (1982: 39)

It is therefore clear that agrarian rituals played an important role in ensuring that there was adequate food availability which was responsible for population growth and the sustainability of chiefdoms across the region. This was manifested in the use of food by men and women for initiation into various regulatory societies. Among the Laimbwe for instance, men were required to provide varieties of food crops to qualify for initiation into *kuiifuai* and *tschong* societies (Kah 2016: 64). In the Aghem, Weh and Esu chiefdoms, initiation into various associations required the provision of considerable amounts of foodstuffs. The Weh had associations for men such as *djitisem, okum, kweifo, ndau ifa, and ndau mbaa.* Associations for women included *fumbwi* and *kefab.* There were similar associations for men and women in Aghem and Esu. Membership dues into these associations were high and consisted of various foodstuffs such as oil pam, maize, egusi, cocoyam, and beans (Geary 1979). It was the same situation in other Grassfields chiefdoms as various foodstuffs were required for initiation into secret societies such as *ngwerong, kwifon, nggiri, and njong.* The abundance of food was also evident during annual festivals such as *Lela* of the Bali-Nyonga and *Ngonso* of the Nso.

Taken together, agrarian rituals ensured that there was steady food security across the Bamenda Grassfields even though this was in rare moments disrupted by locust invasions. This abundance of food was responsible for demographic growth and the sustainability of ethnic communities across the region. But the people equally possessed a traditional ecological knowledge that was also reflected in their agrarian rituals. In the next section, I explore the bearing of agricultural rituals on the preservation of the environment.

13.6 Environmental Preservation

The observance of agrarian rituals permitted Bamenda Grassfields people to control ecological and weather conditions, which contributed to the preservation of the environment. In fact, the emergence and expansion of ethnic communities in the Bamenda Grassfields was thanks to the peoples' interaction with the natural environment in a sustainable manner. The people's worldview about land, agriculture and the aura of rituals contributed to respectful and sustainable environmental practices. These rituals played a major role in shaping the manner in which the people interacted with the natural environment. Considering the bearing of the natural environment on agriculture, agrarian rituals could not be performed without an ecological component. Rituals of agriculture impacted on the natural environment in ways that contributed to a friendly treatment of landscape. Hence, this section explores the link between agrarian rituals and environmental sustainability.

Rainmaking rituals, it should be emphasized, were intended to affirm the fertility of land and the protection of flora and fauna. These rituals for example helped indigenous communities to overcome drought which led to the preservation of water supply sources, groves, and sacred forests. I am therefore in agreement with Ombati's observation that "the ethnoscience of rainmaking rituals is a prototype of African indigenous knowledge on climate change." (Ombati 2017: 74) As explained by Tamasang Clement (Interview 4 December 2021), the making of rain did not only ensure that there was enough water to support plant growth, but also provided the various animal species with drinking water. The Oku believed that abundant rains that were yielded by rituals were beneficial to the environment by ensuring the growth of various plant species (Talla et al. 2019: 381). The checking of drought through rainmaking rituals helped in preserving the environment. In this regard, rainmaking was part of traditional ecological knowledge, and it was aimed at controlling weather. Ombati (2017: 88) studied rainmaking rituals in Africa and concluded that "The rites indicate the African peoples' pragmatic adjustment to nature."

Apart from ensuring agricultural production, planting and first fruit rituals were indicative of the peoples' understanding of their environment. These rituals played a role in enforcing directives intended for the sustainable exploitation of the environment. In all Bamenda Grassfields communities, shifting cultivation was practised to enable the soil regain its fertility after years of cultivation. It was through rituals that decisions were made about when to suspend farming on a given field and the new one to be cultivated. During the fallow period, it was believed that the land was ritually returned to God and the ancestors for them to restore its fertility. This explains why each community always had fields that were not cultivated. In communities like Weh, Aghem, Esu, and Bafut where agricultural land was abundant, the fallow period for a particular field exceeded ten years. This period was enough for the soil to regain its fertility and for various species of animals and plants to flourish. This indicates that agricultural rituals formed part of what Ikeke (2018: 229) describes as "African traditional ecological knowledge."

Besides, the people of the Bamenda Grassfields saw ritual sites as sacred and ensured the protection of such sites, most of which were located in the natural environment. Most chiefdoms in the area had sacred forests hosting various agrarian ritual sites. Among the Kom for instance, *fuchuo* and *azhea* farming rituals were performed at ritual sites in the sacred forest. The Kom were conscious of the environmental importance of this sacred forest and took measures to conserve it. Walter Nkwi studied the Kom sacred forest and found that the people had taboos surrounding the forest. He insists that the rituals that were performed in the forest necessitated its conservation:

> The sacred forest played a role in Kom agriculture and helped to sustain the fertility of the land and its people. No farming is carried out in the sacred forest. There was no careless fetching of wood of felling of trees. The Kom believe that the forest has to be preserved for societal sustenance. (Nkwi 2017: 40)

Like the Kom, the Oku had ritual sites in the Kilum-Ijim forest which they considered as sacred. They believed that the sacred groves in this forest were inhabited by God and the ancestors and preserved them through local taboos and sanctions. According to Talla et al. (2019: 380), ordinary citizens were restricted from entering this forest; it was accessed only by ritual experts and members of secret societies for ritual purposes. Clearly, the transformation of forests into sacred sites for agrarian rituals and their preservation through taboos contributed towards the preservation of various species of plants, animals, birds, and water sources.

Evidently, Bamenda Grassfields people had indigenous environmental knowledge that was reflected in their agricultural rituals. They were conscious about the relevance of the natural environment to human welfare and fashioned beliefs and ritual practices aimed, *inter alia*, at ecological preservation. The rituals ensured the preservation of sacred forests and other ritual sites that were found in the natural environment. It was also thanks to rituals that ecological and weather conditions were controlled for the good of fauna and flora. Clearly, agrarian rituals were agricultural and environmental in both purpose and outcome; they guaranteed plentiful agricultural production and sustainable conservation of the environment.

13.7 Conclusion

This chapter has examined how indigenous agrarian philosophical knowledge, especially cultural traits such as values, attitudes, beliefs, practices, and ingenuity that were prevalent among people in the Bamenda Grassfields shaped the bearing of agrarian rituals on agricultural production and environmental protection. After discussing agricultural work and rituals as key elements of an indigenous farming system in the Bamenda Grassfields, the chapter submits that indigenous African agrarian philosophical perspectives, viewed through the lens of religion, determined the success of agricultural production, resulting in food security and sustainable exploitation of the environment.

The influence of agricultural rituals on food security and environmental protection necessitates the reconsideration of African agrarian philosophical knowledge, especially agrarian rituals and ecological practices, in the battle against food insufficiency and climate change. The rituals were therefore agricultural and environmental in both name and outcome by ensuring the successful execution of agriculture and the sustainable conservation of the environment for the collective welfare of society, both present and future. These conclusions concur with the thoughts of Nyerere (1968), Oruka (1991) and Presbey (2014) whose works link African agrarian philosophy to food security. Indigenous agrarian rituals and their impact on agricultural practices and environmental conservation were anchored on the thoughts and ideas of traditional thinkers or philosophers, described by Oruka as sages.

Surprisingly, current global efforts at mitigating food insecurity and climate change ignore traditional African ways of ensuring food security and environmental conservation. It would clearly be a mistake to think that the war against food insecurity and environmental degradation can be won without recourse to an inclusive approach that takes into consideration all existing agrarian philosophical and ecological beliefs and practices. This mistake is already on course and some scholars have begun attacking it, stressing that Western ecological and agricultural values are best applicable in the Western setting and cannot be successfully implemented in Africa (Ikeke 2018; Nkwi 2017). It is therefore obvious that Western agricultural and landscape values are being enforced in African societies in problematic ways, with no possibility of rolling away food and environmental crises that were initially occasioned and sustained by the exploitative Western colonial enterprise. But if African agrarian philosophy is important as this study has demonstrated, why is it still overlooked by those who make and manage agricultural policy and farmers? In fact, agrarian philosophical ritual practices are an important and ignored factor in agricultural development. The fact remains that practitioners are uncomfortable dealing with culture and specifically with the traditional values, practices, and innovations that have shaped agriculture. To surmount this chief problem, policymakers and farmers ought to pay attention to local agricultural philosophical perspectives, especially those rooted in religion, in order to enhance agricultural production and human progress. If agricultural policymakers were to spend more time examining indigenous cultures, especially religious practices, they would be able to fashion agricultural policies that are congenital with African agrarian philosophy. Simply put, giving voice to indigenous African agrarian and religious values and practices in agricultural development is the chief recommendation of this study.

References

Banadzem, J.L. 1996. Catholicism and Nso' Traditional Beliefs. In *African Crossroads: Intersections Between History and Anthropology in Cameroon*, ed. Ian Fowla and David Zeitlyn, 125–140. Oxford: Berghahn Books.

Chilver, E.M., and P.M. Kaberry. 1968. *Traditional Bamenda: The Pre-colonial History and Ethnography of the Bamenda Grassfields*. Buea: Government Printing Office.

Geary, C. 1979. Traditional Societies and Associations in We (North West Province), Cameroon. *Paideuma* 25.

Geary, C. 1973. The Weh Chiefdom in Menchum Division. In *Paper Presented at CNRS's Symposium on the Contribution of Ethnographic Research to the History of Civilizations in Cameroon, Paris, 24–28 September*.

Goheen, M. 1996. *Men Own the Fields, Women Own the Crops: Gender and Power in the Cameroon Grassfields*. Madison: University of Wisconsin Press.

Ikeke, M.O. 2018. Eco-philosophy and African Traditional Ecological Knowledge. *IDEA* XXX (1): 228–240.

Interview with Asongwa Peter Mufora, 79 Years, Bujong Quarter, Nkwen, 8 December 2021.

Interview with Ngongeh Francis, 74 Years, Nkwen, 6 December 2021.

Interview with Ngufor Sebastien Waji, 90 Years, Notable, Alalieh, 5 December 2021.

Interview with Tamasang Clement, 90 Years, Nkwen, 4 December 2021.

Interview with Tanwie George Chefor, 70 Years, Rain Ritual Priest, Menteh, Nkwen, 6 December 2021.

Interview with Tita Waji, 72 Years, Member of Nkwen Kwifo, 6 December 2021.

Janick, J. 2013. The Intersection of Religion and Agriculture. In *The Basics of Human Civilization: Food, Agriculture and Humanity*, vol. 1, ed. Prem Nath. New Delhi: New India Publishing Agency.

Kaberry, P. 1952. *Women of the Grassfields: A Study of the Economic Position of Women in Bamenda, British Cameroons*. London: Routledge.

Kah, H.K. 2016. The Laimbwe Ih'neem Ritual/Ceremony, Food Crisis, and Sustainability in Cameroon. *Journal of Global Initiatives* 10 (2): 53–70.

Lang, M.K. 2015. The Fight against Infertility through Indigenous Religious Rituals in the Bamenda Grassfields: The Case of the Kengbeum in Weh Fondom since Pre-colonial Times. In *Indigenous Political Hierarchy and Sustainable Collective Meaning in the Changing Cameroon Grassfields*, ed. Funteh Mark Bolak, 43–65. London: Dignity Publishing.

Lang, M.K. 2017. The Role of Religion in Agriculture: Reflections from the Bamenda Grassfields of Cameroon Since Pre-colonial Times. *Afro-Asian Journal of Social Sciences* 4 (8): 1–19.

Mbaku, J.M. 2005. *Culture and Customs of Cameroon*. London: Greenwood Press.

Mbiti, J.S. 1969. *African Religions and Philosophy*. London: Heinemann.

Mbiti, J.S. 1975. *Introduction to African Religions*. London: Heinemann.

Moyo, A. 1996. Religion in Africa. In *Understanding Contemporary Africa*, 2nd ed, ed. April A. Gordon, and Donald L. Gordon. London: Lynne Rienner Publishers.

Nkwi, W.G. 2017. The Sacred Forest and the Mythical Python: Ecology, Conservation, and Sustainability in Kom, Cameroon, c. 1700–2000. *Journal of Global Initiatives: Policy, Pedagogy and Perspective* 11 (2).

Nkwi, P.N., and J.-P. Warnier. 1982. *Elements for a History of the Western Grassfields*. Yaounde: SOPECAM.

Nyerere, J.K. 1968. *Ujama'a: Essays on Socialism*. London: Oxford University Press.
Ombati, M. 2017. Rainmaking Rituals: Song and Dance for Climate Change in the Making of Livelihoods in Africa. *International Journal of Modern Anthropology* 10.
Oruka, H.O. 1991. *Sage Philosophy: Indigenous Thinkers and Modern Debate on African Philosophy*. Nairobi: ACTS Press.
Oruka, H.O., and C. Juma. 1994. Ecophilosophy and Parental Earth Ethics. In *Philosophy, Humanity and Ecology*, ed. Henry Odera Oruka, 115–129. Nairobi: ACTS Press.
Presbey, G.M. 2014. African Sage Philosophy. *The Internet Encyclopedia of Philosophy*. ISSN 2161-0002. http://www.iep.utm.edu/. Accessed May 2023.
Talla, R.T., C.A. Ngwa, and D.B. Mbain. 2019. Assessing Indigenous and Colonial Forest Conservation Policies on the Kilum-Ijim Forest, Precolonial to 1961. *International Journal of Research and Innovation in Social Science* 3 (11).
Tangie, E.N. 2007. *From Friends to Enemies: Inter-ethnic Conflicts Amongst the Tikars of the Bamenda Grassfields (North West Province of Cameroon) c. 1950–1998*. PhD thesis in Peace Studies, University of Tromso, Norway.
Tosam, M.J. 2019. African Environmental Ethics and Sustainable Development. *Open Journal of Philosophy* 09(02): 172–192. https://doi.org/10.4236/ojpp.2019.92012
Van Niekerk, A., ed. 2005. *Ethics in Agriculture—An African Perspective*. Dordrecht: Springer.

Michael Kpughe Lang is an Associate Professor of Religious History and Chair of the Department of History in the Higher Teacher Training College at the University of Bamenda. He was a SUSI Fellow at the University of California, Santa Barbara in 2013. His research interests are in religion and society with a focus on ecumenism, gender, conflict, medical ministry, and development. He has two books and over fifty peer-reviewed articles to his credit.

Chapter 14
Indigenous African Eco-communitarian Agrarian Philosophy: Lessons on Environmental Conservation and Sustainability from the Nsoq Culture of North West Cameroon

Peter Takov

Abstract Humankind's activities on the planet earth have evolved to unprecedented levels, thus putting the ecosystem on the verge of the abyss. The exigency of caring for the environment has ushered in an urgent need for philosophies, theories, advocacies and actions to preserve the ecosystem and to ensure environmental sustainability. There is a reawakening awareness that humans need to rethink their use of the environment and to enliven their concerns for its care. In addressing this need, Euro-centric models have been the dominant perspectives advanced; little appeal has been made to indigenous African environmental philosophies and their potential to contribute to the desired change is largely ignored. There is, therefore, a need to explore alternative approaches in indigenous thought systems that may appeal easily and rapidly to the consciences of peoples. African agrarian philosophy of the environment seems to be an area with global implications. The Nsoq' of the Grass fields of Cameroon subscribe to what some have described as an "eco-bio-communitarian". Tangwa observes that within the traditional African metaphysical worldview, the dichotomy between plants, animals, and inanimate things; between the sacred and the profane, matter and spirit, the communal and the individual, is a slim and flexible one. He postulates that such an outlook has very significant implications on the way nature is approached and treated by traditional Africans. (Tangwa in Bioethics: An African Perspective, 1996). The "reverence" for mother earth, which is prevalent in

We prefer to use the name "Nsoq" among other alternatives. According to Mzeka P. in *The Core Culture of Nso'*, the evolution of the people is global cutting across multiple facets. One of such facets is the epistemic or linguistic strophe. The name of the Nsoq tribe has evolved greatly from both external and internal influences. The Germans adopted the "Nsaw" version, the missionaries used "Nso", others use Nso' and the most recent acceptable name of the people is "Nsoq." These appellations do not in any way constitute a problem. They can be used interchangeably. (cf. P. N. Mzeka, *The Core Culture of Nso'*, Jerome Radin Co, Agawan (Mass.), 1980 (c. 1978), p. 6.)

P. Takov (✉)
Catholic University of Cameroon (CATUC), Bamenda, Cameroon
e-mail: ptakov@catuc.org

M. J. Tosam and E. Masitera (eds.), *African Agrarian Philosophy*, The International Library of Environmental, Agricultural and Food Ethics 35,
https://doi.org/10.1007/978-3-031-43040-4_14

the Nsoq approach, can serve as a contributing voice in saving mankind from an impending environmental catastrophe.

Keywords Nsoq Agrarianism · Environmental sustainability · Worldview · Ecosystem · Biodiversity · Eco-communitarianism

14.1 Introduction

The universe of things is generally understood to be the composite of two realms of life: the supernatural and celestial on the one hand and the natural and or the 'immediate' on the other; that is the spiritual and the material. Both realms co-exist within certain binding principles. The nexus between the two is governed by certain in-built values and forces. These values usher the resilience, stability and sustainability of the spiritual as well as the natural operational forces *sine qua non* for life's subsistence therein. Any abuse of the rubrics' basic principles of governance here necessitates a threat to the very fabrics of existence in each realm. For instance, the Judeo-Christian religion mythically narrates the war between the angels where Lucifer was overthrown in order to restore peace (Percer 1999). Orderliness in celestial or spiritual world presupposes that of the natural world. Each cultural philosophy approaches nature from a particular or peculiar bend. Cultural worldviews vary from place to place and from people to people. Cultural philosophies share more in common than in diversity. They have certain social, moral, religious, and cultural values common to them than the few experiences that separate them like language, customs, among others that may separate them. One arena of diversity is that of the maximization or minimization of the divine mandate: "... conquer the earth and subdue it..." (The New Jerusalem Bible 1990). Varied tendencies and philosophies in an effort to exact this divine mandate tend to subscribe or not to the binding underlying maxims of preservation, conservation and propagation. It is on this count that we have to understand that the natural environment or eco-system is being threatened by human approaches to the principles: *conquer and subdue*—sometimes construed and sometimes misconstrued by many cultures.

The Western approach to the exploitation of the natural environment has to a greater extent been seen as an indirect threat tantamount to its eventual extinction. The Western agrarian worldview is more mechanical, scientific, anthropocentric and individualistic. Western radical agrarianism holds that the earth having been given to mankind in a common occupancy, each individual has by nature a right to possess and cultivate. Henry George further argues that individuals should purchase or confiscate private property in land; they can retain it if they want to; buy and sell, bequeath and devise it according to their means. (Montmarquet 1985: 10). The massive abuse of land according to this spirit and urge for technological progress keeps threatening the nature on daily basis. Many perceive these as the causes of the destruction of the ecosystem, environmental degradation, destruction of the ozone-layer, etc. Western

Baconian and Cartesian approaches to the environment via the use of science, technology and today's techno-progressions ideology has illusively and cosmetically solved many global problems. But to what extent can Western science and technology go in solving all human problems as they claim? Tosam argues that Western epistemological arrogance has arrogated to itself the powers of mankind's prophet and master of everything; all aspects of life and even human development depends on the West; a type of self-imposed master over all cultures and peoples; the *illuminaté* of other parts of the world (Tosam 2019: 172–179). Exploiting Francis Bacon's slogan that 'Knowledge is Power', Mbessa depicts the Western thought thus:

> The industrial revolution drew its impetus from Francis Bacon's slogan…convertible into commercial value, from the idea that knowledge is unqualifiedly good, from the belief that nature is, in principle at least, completely knowable and controllable, from perception of the universe as something which ought to be explored, subdued, dominated and exploited…. (Mbessa 2020: 446)

Western cultural agrarianism hinges on this view of Bacon. Insofar as knowledge solves human plight, it can be explored and exploited irrespective of the means as the euro-centric philosophy is based on the maxim: the end justifies the means. Therefore, "bridle the horse to doom" aptly describes the misconstrued erroneous understanding of the human intercourse with the environment and or ecosystem by the West. It is in this note that Tangwa describes the Western omnivorous spirit of research's (bio) destructive consequences that extend to bo1996th the environment and human nature. For in spite of the goodies of Western techno-progressism, it all ends up stripping the universe of its biological footprint and humans of their spiritual fabric (Tangwa : 76).

On the other hand is the African worldview vis-à-vis the divine mandate concerning human relation with the environment or ecosystem. The African cultural perception comes to terms with the understanding that the link between man and nature is inseparable and divinely ordained. The African, and to be precise, the Indigenous African Agrarian cultural philosophy, is pregnant with salient values that clearly and succinctly express the African's perception of the universe of things. The Indigenous African cultural perception of nature or the world is one of four actors, equal in strength and value, ranked in importance and not in terms of the one exploiting the other, but rather a blend of the eco-communitarian spirited element of co-existence in the universe of things or beings at large. These four actors are: the realm of spiritual beings and divinities, the Abode of the Ancestors, the human realm and other beings. This link clearly spells out each one's role; all of which are governed by the interplay of certain restrictions, prescriptions and taboos geared at enhancing a flourishing state of harmony between these realms of life. Thompson argues that:

> The relational ontology existent between entities is so rich that humans are inseparable from nature or the world around them. Indigenous African Agrarians' cultivation of the soil provides a special direct contact with nature and the other beings around and above all is blessed with a closer nexus to the divine spiritual realm. (Thompson 2008: 538)

A popular Nsoq saying holds that: *alim nsaiy, a nsaiy lim wo*; translated as: 'we cultivate the land while the land cultivates us.' There is a traditional distributive justice between the occupants of the universe. Its effects or consequences are dangerous when the environment is threatened by humans or other occupants: it harms the weaker as well as the stronger party (Montmarquet 1985: 10). This justifies the close connection between the human species and the world around, affirming the inseparable relationship between the agrarian values of: land, beauty, food, work and the communities in the universe of things (Wirzba 2003: 50). This argument is based on the notion that man by nature is agrarian as one of the fundamental human needs from the genesis of her/his existence is 'feeding' or food, besides shelter and clothing. In cognizance to this fact and need, the Indigenous African has as her prime preoccupation an agrarian philosophy. Ipso facto, the perception of the world around or nature at large by a defined people largely influences the way it manages it. This can either be positively or negatively, improving or degrading the health of the ecosystem. Mbessa argues that when humans perceive that they are an integral part of nature or other beings, and equal partners, with whom they share a common habitat, they can treat nature with respect. But if on the contrary, they consider it inferior, then they treat it with contempt and such is an anthropocentric Western view which is unsustainable and abusively exploits the natural environment (Mbessa 2020: 452). One such Indigenous African agrarian commonwealth of humans is the *Nsoq* Man of the North-West Region of the Cameroons. We cannot delve into a better understanding of the *Nsoq* Man's worldview and agrarian cultural philosophy without situating the *Nsoq* people historically and geographically.

In this chapter, we provide a brief geo-historical location of the Nsoq people which shall lead us to their world-view of nature as a cultural value inseparable from human beings or the Nsoq man's way of life. We equally have a look at its agrarian philosophy which is communitarian *cum* sacrosanct. The Nsoq Man's perception is existential considering the various communities: spiritual, ancestral, human, animal, vegetative, animate and inanimate others. We consider the ethical values existent between these realities and equally exploit the logic of Nsoq Man's eco-communitarian agrarian dimension of farming and some ethical overtones and a conclusion.

14.2 A Brief History of the *Nsoq* People

The historical roots and or origins of the *Nsoq* People are traced back to the *Tikari* world. The *Tikari* world is a cluster of clans of *Bankim* in the Adamawa Region of the Republic of Cameroun. Oral sources maintain that the chief or traditional ruler of the people of *Rifem*[1] or *Kimi* of the *Tikari* clans in the fourteenth century (1387)

[1] However, the exact place may not be easily located today. There are doubts as to whether this was really on the other side of the River Mbam. There is something in the story about the lack of an ability to swim across the river. It is not clear if it were the Mbam people who could not swim or the other two.

was said "to have been missing" (Fomine: 69–92).[2] And there was a power tussle for succession to the throne between his children. *Mveng* emerged victorious and when he was enthroned, some of his brothers and a sister vamoosed on self-exile: *Nchari Yen, Moruntar* and their eldest sister, *Ngoun. Yen* after crossing the River *Mbam* is said to have settled in present day *Noun* where he founded the *Bamoum* dynasty. *Noun* is in the Western Region of the Republic of Cameroun with her seat at *Foumban. Moruntar* continued towards the northeast of the River *Mbam* and finally settled in present day *Bafia* in the Centre Region of the Republic of Cameroun. *Ngoun*, their elder sister continued westward with her followers. She earlier settled at *Mbaw-Nso* where she established the *Nsoq* dynasty. The name *Nsoq* came to be used because of the inter-nuptial intercourse between the indigenous *Nsaw-Mntār*[3] and the *Ngoun* followers. *Ngoun* later on moved to *Ndzən-Nsaw*. Due to the nuisance from the Fulani raiders, she migrated with her followers to *Kov-Vifəm*. Later on she moved to *Tâ-Visa*. For security reasons, she later relocated again to *Kov-Vifəm*. Due to the interplay of the forces of nature, she moved to present site of *Kimbo*. Her dynasty (*Nsoq* Fondom) remains one of the largest in the Grass-Field Area of the people of the Cameroons (Fomine 2009: 69–92).

14.3 The Geographical Location

Geographically, the *Nsoq* Fondom is located in the Bamenda Grass-Field Area in the Northwest Region of the Cameroons. She is situated approximately 111 km away from the capital city of the region, Bamenda. To the West, she is bordered by the *Kom* Fondom and fondly described by the *Nsoq* Man as the *Vijem* people. Eastwards she shares boundaries with her brother—the *Bamoum* dynasty. To the South, she shares boundaries with the *Ndop* Fondoms and northwards she is bordered by the *Mbuum* villages commonly called the *Viyaa* people of the Donga and Mantung Division. The seat of the Fondom is in *Kimbo* (administratively called *Kumbo*). The language spoken by the *Nsoq* Man is called *Lam-Nsoq*.

The *Nsoq* Man is notoriously a traditional and a cultural person. Their traditional values are inseparable from their *modus vivendi* and *modus operandi*. The *Nsoq* culture is like a religion to the people. Traditional, cultural, religious values are interrelated in a way that one cannot distinguish what is cultural from what is religious and traditional and vice versa. Such values as beliefs, customs, mannerisms, institutions etc. cut across their way of life and their way of doing things. The

[2] It's a taboo to say the Fon or Chief is 'dead'. Culturally or traditionally whenever a leader or ruler passes on, the common parlance is to say that s/he is 'missing'. This is based on the fact that they do not die but move on to the world here-after where they commune with the other ancestors and from the abode of the ancestors continuously influence life in the here-now.

[3] The Invincible Group called "*The Thirty*". They are not members of the royal family but their place in the enthronement of the Fon, carrying out certain rites, rituals and duties in the land is indispensable.

Nsoq Man's source of livelihood is agriculture. Farm production is his main preoccupation. Goheen captioned her work on the Nsoq people: "*Men Own the Fields, Women Own the Crops.*" (Goheen 1996). Goheen aptly explains the agrarian nature and life of the *Nsoq* Man. Her agrarian nature is inseparable from her cultural life and traditional beliefs. That is why there is the Nsoq Man's expression: "*Ndzə dzə-wir: a wir-dzə ndzə! A'yi ghansho'a dzə-kibam! Boyo' dzə n'kem*". This is translated as: 'Nature is humanly: to be human is to be in nature. It must be sacredly and secretly lived and never x-rayed.' This is a synopsis of the *Nsoq* Man's underlying belief and guiding principle of human ontological nexus with nature at large. It is generally argued that each cultural community's perception of the universe of things or nature profoundly influences the way it manages and treats nature. The *Nsoq* Man's perception of nature is so rich that it is necessary for us to delve into as this paper seeks to add to the discussion on African Agrarianism. To do this we need to investigate into the indigenous *Nsoq* agrarian cultural worldview. An in-depth perception of the *Nsoq* worldview, her eco-communitarian, agrarian spirited element and praxis will in one way or the other add to the world of Indigenous African Agrarian philosophy.

14.4 The Nsoq Worldview

As a notoriously traditional and cultural being, the *Nsoq* Man subscribes to her traditional and cultural values as a religion. Her worldview is centered on certain values, ways of being, customs and beliefs preserved and transmitted from one generation to the other. Living as a family; living "in-unism" as a commonwealth is an essential element of life. The spirited element of oneness and or togetherness is an indispensable cultural value. It is this communitarian philosophic element that generates and activates the spirit of co-existence. "*A'dzə bōm-dzə! A'sidzə bo'a sidzə*" is a typical *Nsoq* Man's sagacious saying which stands to describe the spirited element of belonging and oneness. This is translated as: "if you are, therefore I am; and living together, is life…" The *Nsoq* philosophy of gross tribal happiness and a flourishing life for all takes pride of place and governs the principle that: *ngha mbuŋ yo'dze*—no one is in need and want. There was no stranger; slaves bought were integrated (assimilated) into the family without strings or traces. For instance, oral history has it that the father of Bernard Fonlon was not from the *Nsoq* land. When he was rescued from the river in the neighbouring village and brought to Nsoq, he was considered to be equal to every other *Nsoq* man and it was a taboo to call him stranger-*kingwiy* or *wir-kitum*. Many such individuals and families existed in *Nsoq*.

Family and tribe store-houses *(vitav ve'lah* and *kitav ke woŋ* respectively) were often stocked with food stuffs for the needy, visitors and strangers. There was a general family house and the tribe assembly house (*ngaiy ye lah* and *ngaiy ye woŋ* respectively) for the needy, strangers and visitors in transit. Food and drink was always available in these houses especially on the two exceptional days of obligation *(vishiy vebam ve-woŋ)* from private and routine activity-*Kiloviiy* and *Ngoiylum* (scornfully called *Kontry Sundays*). There are eight traditional days of the Nsoq Man's week: *Kaavi, Rhəəviy, Kilōviy, Nzəəri, Geegee, Ngoiylum, Waiylun and Ntaŋrin.* Even though some community activities were carried out on the other six days of the week, the two days of obligation were reserved for general communal activities.

The *Nsoq* eco-bio-communitarian, agrarian cultural philosophy of life is tailored at enhancing the wellbeing of all, a flourishing state of life and the general good or happiness of individuals, the families and the community, respectively. This worldview is pregnant with a chapel of values and virtues that go a long way to kindle these goals. It is characterized by customs and beliefs; taboos, injunctions and prescriptions handed down from generation to generation. The *Nsoq* worldview therefore places emphasis on individual roles and obligations to the self, the family and then the community; admonishing laziness as irrelevant thus: *Vinjoh vi-dze suum virim*; which stand to mean, 'laziness is a fertile ground for witch-hunting and/or witchcraft.'

Life is considered to be sacred. Everything was sacrosanct. There is a special obligation towards the spiritual realm of life. Respect for things sacred, the tradition, elders, the land, the ancestors, spirits, divinities, the gods and the creator. Tangwa and M'Dzeka argue that such a mindset has very significant implications in the way nature is approached and treated. The Nsoq attitude toward nature is that of respectful co-existence, conciliation, and containment (Tangwa 2004: 37). These are the beings or entities that constitute the *Nsoq* world. This eco-communitarian element is seen in the interplay of the spiritual, ancestral, human and other environmental forces within the *Nsoq* universe of things or beings. F. D. Goodman describes this relationship existent in nature thus:

> A rediscovery of a new relationship of participating in a mutual, playful, dialogical way with the beings around us – opposed to the 'normal' modern relationship to nature, which filters our conscious and intentional agency of any being but the rational ones, making the world around us, even though the beings in it move and interact, in a sense inert and alien…such that the tree I want to fell to build a shelter will not strike back at me…but address the spirit of the tree to conjure it or atone it to fence off retaliation. (Goodman 1990: 4–5)

Goodman's analysis and description clearly explain the reality of the spiritual relationship existing between the natural and the spiritual realms as perceived by the *Nsoq* man. For there exists an Indigenous African (*Nsoq*) language that humans, animals and plants understand what cannot be understood and spoken by Western technology. Hierarchically, there is the God of creation (*Nyuy-Mbom*), the gods (*anyuy*), good and evil spirits (*vibaiy vejuŋ wuna vebivi*) and divinities who occupy the zenith of the spiritual world and who are said to greatly influence the abode of the Ancestors (*mbov-akfəə*) and humans. The ancestors in turn influence or affect the world of the humans and human co-existence with the nature around. Human nexus with the immediate environment is sanctioned by the prescriptions, prohibitions and taboos. This chain clearly explains the rubrics of the eco-communitarian spirited element in the *Nsoq* cultural philosophy. James Sire argues that it is simply the collection of attitudes, values, stories and expectations about the world around in relation to human thought and action. It explains how a culture works in individual praxis (Sire 2015: 366) and visualized in the commonwealth. In the same light Segun Ogungbemi argues that humans are stakeholders in the definition of the African environment and their existence find expression in their association with non-human beings/animals.

This means that without this association, human beings are incomplete. He aptly calls this "the ethics of nature-relatedness." (1997: 23–24).

14.5 The Nsoq Concept of Nature

The word nature in *Nsoq* is understood to be *ndzə*. *Ndzə* is a word used to explain the *Nsoq* Man's experience of the world around; the world of the phenomena. For nature cannot be thought of in isolation of the supernatural element and touch. The *Nsoq* Man's perception of the cosmos or nature consists of multiple strophes running from the supernatural to the natural realms. W. James argues that there exists around us a spiritual universe, and that universe is in actual relation with the material. From the spiritual universe comes the energy which maintains; that which makes the life of each individual spirit. Our spirits supported by a perpetual 'indrawal' of this energy… (Sire 2015: 366). The Nsoq cosmology considers at the apex of nature the being called *Nyuymbōm* (God the creator) said to be the architect of every reality existent. That is the God who fashioned out of clay everything. Besides this God are other gods (*anyuy*) having attributes tied to their specific duties and functions and closely related to realities and things in the universe called *Nsoq*. Such gods as:

- *Nyuy*: This is an appellation reserved to the 'gods'. The gods do not work in isolation but in synergy, except the God considered to have been the creator or the designer of all beings, animate and inanimate. All other gods work in communion. That is where the notion of communal spirit of the *Nsoq* Man has its roots. So the concept "*Anyuy*" is the plural form of '*nyuy*', and defines their unitive way of action in the universe of *Nsoq*. Such gods would include.
- '*Anyuy-woŋ*': These are the gods of the land or clan. They are called by their names and in relation to their function and common action if there is need for any rituals or sacrifices to be offered to them and not to a particular one. The gods of the land are responsible for bringing about peace and order, protection, wellbeing, happiness, a flourishing life to all among other duties.
- '*Anyuy-Kwa*': These are the gods of the particular farming-land(s). They are responsible for keeping the soil fertile, good yield and harvest and keeping at bay intruders. Penalizing whoever flouts the laws or injunctions of the farm-land. Closely linked to these gods are the *anyuy suum*—farm gods. They too have the same functions as the '*Anyuy-kwa*' but at a lower level only to particular farms.
- '*Anyuy-Lah*': These are the gods of the compound or common settlement of a family or an extended family. Just like the tribe's gods, they play a similar role but at the level of the various families.
- '*Anyuy-Ngvən*': These are the gods of a defined farming area allotted to a particular family or settlement for agricultural purposes. The gods preserve and protect the land for farming and other usages or activities.

- '*Anyuy-Shiv*': These are the gods of traditional medicine. They are responsible for directing medicine men and women in the healing, curing, protecting of individuals and families and the entire tribe. These gods direct traditional medicine men on what to do in health matters.
- '*Anyuy-Kikwiy*': These are the gods who direct seers and medicine men on what type of herbs and or 'grass' to use in healing, curing and protecting individuals and families from evil spirits and abnormal forces.
- '*Anyuy-Kibahri*': These are construction gods. They play a major role in the construction of private, family or community houses and other community and tribe's projects. For instance, in the construction of bridges, they are often consulted as rituals and sacrifices have to be carried out at night to chop up particular trees and or materials necessary for this activity. In road construction, these gods are also consulted.
- '*Nyuy-Arhim*': This is the god of evil spirits. For instance, children are not beaten-out-of the house at night and it is forbidden to sweep dirt out of the house at night because of these gods. It is generally maintained that night is the hour of these gods and the evil spirits.

We have to note here that the Nsoq man understands the role of these beings not only from a religious dimension of relatedness but most importantly from a socio-moral perspective. They stand to regulate sound co-existence between humans, human beings and other stakeholders in the universe of things. They make for peaceful co-existence and the abuse of it is often sanctioned without any recourse except by atonement, appeasement, sacrifice, as Tangwa notes here: in Nsoq, there are no frequent offering of sacrifices to God, to the divine spirits, both benevolent and malevolent, to the departed ancestors and to the sundry invisible and inscrutable forces of nature. The simple reason being that the taboos, injunctions, and prescriptions are there to regulate life; being-with-others (Tangwa 2004: 37). For instance: there are certain days meant for the divinities; such as *Kilōviy* and *Ngoiylum*. There is no farming on such days. The gods and spirits hover around to bless the lands and farms. Social, moral and cultural prescriptions are respected to the letter.

Thus, there is a litany of gods and divinities in the Nsoq cosmological understanding of nature. Equally, there are spirits of all sorts. Some are the good spirits and others are the evil/negative spirits. The abode of the ancestors constitutes only the good-living-dead as there are no evil or bad ancestors. Only the good reputable family members are invoked from the world here-after for one reason or the other. For instance: a dead child, youth, unmarried, childless persons, individuals who have committed suicide, persons bought over as slaves and strangers assimilated into a family or tribe can never be considered in the lineage of family or tribe's ancestors. This is the composition of the spiritual realm of the Nsoq cosmological order. The third realm is the human world. It is composed of all the members of the living family and the community. Bought-over slaves and permanent strangers are not left out. The last of the realms is the animal, the plant or vegetable and the inanimate world orders of reality. Godfrey Tangwa argues that the lowest (animal, plant, inanimate) world is very significant as humans coexist with them for good or for evil.

There is the belief that humans can transform into animals, plants, such as trees, or forces, such as the wind. This has very significant implications to nature as a whole (Tangwa 2010: 76). Humans use the animal and plant kingdoms to interpret the signs of nature. That explains why palaces and family headship houses are constructed in areas surrounded by the forest in the Nsoq world. Plants, birds and animals are believed to have a language. For instance: the presence and sounds of birds are interpreted in multiple forms. The swallow/*shigha* is considered the bird of a potent and carrier of a good will; the ugly-bird/*shigha'rin* is a harbinger of good news. The language of the birds is understood by those who flock with them. Thus the Nsoq saying: *Lam minnuəni yu'akwo'*. For instance, swallows do not just construct their habitat because they have found a house or compound. Weaver birds do not build their nests because they have found a beautiful tree. Wherever they go, they carry with them a tiding or an omen. They will not abandon their nests just for the sake of an intruder but signal some pending good or evil tiding. Imani Sanga argues that in the highlands of Tanzania, the Wawanji people speak with the birds and understand their language. There is a cultural aesthetic value attributed to the natural sounds of the birds and animals and they concern human experiences such as fear, work, joy, and hope (Sanga 2006: 96–102).

14.6 The Relationship Between Man and Nature

Man in *Lam-Nsoq* is *Wir*. The concept '*Wir*' ontologically is understood to be the composite of the body (*wuun*), the intellect (*doŋ*) and the spirit (*kiyoiy*). This is considered to be the vivifying element in a human being. *Kiyoiy* can be divided into two classes. One keeps and sustains the body and the other can subsist after the other ceases to vivify the body. *Kiyoiy kebamki* would simply mean 'the last breath'. At this moment, the other spirit subsists and carries with it the spiritual body to the world-hereafter. If there is then 'the last breath' and the spirit thrives thereafter, there is still another force or element that subsists after the cessation of the bodily vivifying force. The reality of the abode of the ancestors from which they actively influence physically and spiritually the lives and activities of the living justifies the existence of the second spirited element in man.

On the other hand is the concept 'Nature', the *Nsoq* Man calls *Ndzə*. *Ndzə* is not only the material reality or phenomena around the universe, but equally the immaterial or the spiritual reality which influences the natural or material. When the *Nsoq* Man says: *ndzə ma'ni mo*; it means that 'the forces of nature have turned against him/her'; 'nature and the things of nature are not fair towards him/her.' Sometimes it is said, *ndzə binkirni*-which simply means that 'the tides of nature in ones favour have been upset by its forces.' Proverbially it stands to mean that 'life presented its ugly face to someone.'

The inter-relatedness between humans and nature as perceived by the *Nsoq* cultural philosophy from the analysis above is that of coexistence of the two. It is not a parallel nexus as such but one in which Man acts as a custodian and guide to nature. Humans merely share of the good and ugly faces of nature. The famous *Nsoq* sage justifies this argument: *Ataŋ woŋ a 'woŋgi wiy binkir suumi wo'a…*? 'Why should one strive to possess the earth when the earth shall soon possess one?' This maxim clearly spells out and stipulates the relationship between man and nature. There is an African proverb which states, "the restive grasshopper rests in the end in the gizzard of a bird". Augustine of Hippo justifies this when he says that "our souls are restless until they rest in God." And if the *Nsoq* Man were to borrow from the Judeo-Christian theology, then the relation between man and nature can be described thus: everything will pass away, but the invincible word (nature) will not pass away. Man's duty, therefore, according to the *Nsoq* cultural philosophy, is to guard, protect, and preserve nature; keep it in tact as one found it for it is eternal. It is not an egalitarian relationship as misconstrued by some cultures but that of man's dependence on nature for sustenance and subsistence. Besides man's relationship with nature is that of man and other beings, a tenet worth noting in the *Nsoq* cultural philosophy.

14.7 The *Nsoq* Concept of Man as a Being-With-Others

The human species relates in multiple ways and fashions. The most notable are the social, moral, political, religious and cultural. As social beings in relation to their ways of living; as moral beings in relation to their ways of doing things; as political beings in the ways they manage and govern their affairs; as religious beings in their connection and operation with the supernatural world; and as cultural beings in sense of belonging to a commonwealth. These are the various strophes among others which explain human relation with others. According to Ramose "the principle of wholeness applies also to the relation between human beings and …nature. To care for one another, therefore, implies caring for physical nature as well. Without such care, the interdependence between human beings and physical nature would be undermined" (Ramose 2009: 309). The *Nsoq* Man shares these values as other people but salient in the *Nsoq* cultural philosophy of 'being-with-others' is the famous maxim or principle: *wirdzə-wir bi'a-mo'oh* which means: the person is the person because of others known and unknown. Being-with others is indispensable. Alone, one cannot survive. This best explains social cohabitation without strings than as can be expressed in: *wirdzə-wir bi-wir.* These mean that ones success depends on those known and close. The first explains the fact that human success and failure is seen in the way others appreciate him or her and the second gives the impression that man's success depends on man. "*Bi'a-mo'oh*" explains and emphasizes the cooperative and communitarian value in living with others. "*Bi-wir*" imagines and assumes the fact that by necessity humans are what they are by the machination of others irrespective of their cooperation. That is, man's being is the creation of the other(s). In *Nsoq* cultural philosophy, a man is what he is in the way and manner in which he relates with others and vice versa. Being a man is not in the successes and or failures but in the achievements. So the saying: *Wir-dzə wir bi' keŋrhə! Wir yo' dzə kav! Wir-dzə keŋ!* "A man is a man because of his wealth or achievements; a person is not riches, but wealth." This can be seen in occasions as: *kintchu'um* (the unforeseen) in the likes of sudden human or natural hazards; *ki'nkah* (programmed) in the likes of memorials, marriages, ceremonies etc. These and many others are a situation where the worth of man's relation-with-others is rated. The legendary Fon of *Nsoq, Sehm Mbinglo* had opted for a more apt maxim on the count that the development of a people comes more from without than from within when he said: *Nsoq dzə Nsoq bi'a mo'oh*...the riches, wealth and development of *Nsoq* comes from the strangers and outsiders.

In sensu stricto, the notion of being-with-others according to the *Nsoq* Man is communitarian *par excellence*. Living together as a family has no exclusions. Everyone in the commonwealth has a place, role and duty. There is no foreigner or stranger. There is no slave or freed man or woman within the communitarian activity. There was no visitor except that the person was specially taken care to the admiration of others on the count that it could be one of the gods or ancestor in transit or visiting. Everyone ate and drank from the common public pot and calabash. In relation to farming, the planting, harvesting, transportation and feasting thereafter is always communitarian. Family and tribe stock-houses are never empty as they provided for the needy, visitors and strangers.

14.8 The Eco-bio-communitarian Dimension of Farming in Nsoq

Nature, soil, land and in particular farming land, is considered sacred and revered by all. Sacred pieces of land, forests and vegetation are reserved for the gods and spirits within the farming land surface. These are the habitat of the gods and spirits of the defined farm-land owned by the tribe and managed by a particular family settlement. *Ngvən* is the *Lam-Nsoq* word for the farm-land. The gods of the land are often not far-off the farming land. The *Anyuy-Ngvən* (gods of the farming land surface), *Anyuy-kwa'* (gods of the farms), *Anyuy-suumsi* (gods of the particular farms) were often seen majestically basking-off themselves in the early morning or late evening sun. These are considered the totems of the land. Most often these totems are animals of the snakes and cats kingdoms. For instance, in the *Kov-Kinka* (Mbiame Forest), a headless snake is often seen especially on market days *(Waiylun)* crossing the road carrying with it cowries on its supposed forehead. The same scenario is witnessed below the Nsoq palace, a snake with cowries on its forehead is often seen basking itself at a particular hour and day. Sometimes it majestically leaves the palace to the central market at Squares and back to the palace. This takes place once in a while at a particular hour and day. This is a sign that something mysterious is going to happen or wrong and there is an urgent need to appease, atone or sacrifice to the gods.

Annual rituals and sacrifices are offered to them for appeasement, atonement and or in appreciation. It is always a family activity accompanied by dinning and winning. This is a uniting factor of the people, family members and outsiders who farm in this land; an activity carried out by the *Ta'Ngven* (custodian of a defined farming area). The gods of the farm-land have their own share of raffia palm-wine and oil; blood from the skinned cock or a goat if it warrants it; specially prepared *egussi* and pounded coco-yams. It was a family feast and celebration. It had a communitarian tone. Those who could not answer present sent representatives and defaulters had a fine to pay. Failure to do this could warrant an injunction on one's farm-land not to farm the set piece of land until further notice and this could eventually lead to forfeiting the farm-land. There were many pieces of farm-land with such injunctions for one reason or the other. An injunction of '*mba-ngwerong*' (A staff made out of a bamboo with three black marks at the topper part and no person uses a bamboo as a walking stick) from the *ngwerong* secret society and or '*kirang*' (fibered injunction made out of palm front) from *Ta'ngven* deterred anyone from daring around the set farm-land. Farming is more of a religion considering the communitarian touch, the sacredness of farming activity and the land. The land or soil was said to have some human and spiritual qualities—talks, eats, drinks from the sacrifices to the gods. There was no one above the farming taboos. A taboo flouted was flouted and retributions from the gods awaited any defaulters or a sacrifice offered to avert the wrath of the gods. Wirzba (2003: 64) complements this view thus:

> We can see a deliberate way of life in which the integrity and wholesomeness of the peoples and neighbourhoods, and the natural sources they depend upon are maintained and celebrated … an indispensable tool for cultural flourishing. Agrarianism is a cultural contract. Its job description is to function in such a way that it honors and maintains the earth, sustains and perpetuates the community … respects the commonwealth for what it is … a preservationist than a conqueror.

The eco-communitarian strength is drawn from the sacredness of the land and nature. That is why agricultural activity is sacred. It was inseparable from the world-beyond and the supernatural forces omnipresent in the farming world. The invisible and active presence of the gods, spirits and ancestors was indisputable and indubitable. They were appeased for good yields and harvest for everyone. Wirzba further argues that it is a quasi-religion and needs some commandments which he calls The Ten Commandments of Agrarianism. The first being that the Earth is our mother, not always forgiving and not a treasure house to plunder (Ibid: 69). They ensured gross tribal flourishing, wellbeing and happiness. There was no one in lack or in need. The presence of intruders today put the gods to the test and the consequences are visible; ipso facto, the urgent need to revisit, rethink, and or go back to the roots as doom threatens all on every side. It is on this count that we can talk of the visible advantages and or lessons of the eco-communitarian farming system in *Nsoq*. These advantages include:

- The fruits of community life were plentiful. The unitive purpose and goals of community life left no one outside the good yields and harvests, thanks to the reverence accorded nature via the respectability of the rubrics and laws that governed farming in *Nsoq*. There was nothing like a poor harvest. Harvest was generally rich as each one played his or her role from the gods, spirits, the ancestors, landlords *(Ata'ngvən)*, traditional chief priest and priestess (*Ta'woŋ* and *Yeewoŋ*) and the farmers.

- The pasture land was exploited by all. It was not owned by a particular person. Pasturing of the animals was a communal activity. It is said that to engage in animal production or farming, one needed a grazing permit only from the *Ta'ngvən*. For instance in tethering the animals and the birds (chicken), everyone cared for them as if they all belonged to him or her. No one complained of a missing animal or chicken. Even the strayed ones were caught and brought to their respective owners. They could even remain with those of the neighbours for a period of time without any problem. They were well catered for indiscriminately. Hens often laid, incubate and hatch eggs in a neighbour's house without any problem. They were even fed by the neighbours during this period. The owners did not go in search of them. They knew they were in a neighbour's safe protection. Back-yard poultry farming flourished in the eco-communitarian *Nsoq* world.
- Land was never sold out. Agrarian-land was shared or given out. A symbolic gift of *kitem ke shilu' wuna gwəv (kiyuu)*. 'That is a calabash of raffia palm-wine and a healthy cock' offered to the *Ta'ngvən* suffices for one to beget a piece of farm-land. Subsequently one is expected to give to the *Ta'ngvən* at the end of every harvest season as payment of royalty*: nkem ngwasang or nka'tu'*. That is a basket/kenja of corn or a bucket of Irish potatoes. The *Ta'ngven* in turn takes from his collect, the share of the Fon to the palace. This was called *Nsuu' Ngvən*. The boundaries of the farming land given to a person were defined by one's strength or neighbour or a tree or a raffia palm-bush, a foot-path or a stream; but in cases of open agrarian farm-land, the *Ta'ngvən* determines the size—length and width by throwing a stone from where he stood to where it fell. One's strength indirectly defined the borders of one's farm-land.
- Seasonal agrarian activities were communal. Every farmer had to '*kwir*' or 'fallow' the farm-land once every three years. It was compulsory or obligatory. At the end of which the *Ta'ngvən* invited everyone working in his jurisdiction to inaugurate the new farming season from his own farm and thereafter individuals could start farming. To inaugurate the new planting season, he will plant the first corn-seed, '*sov-ngwasaŋ*'. Then everyone planted in *Ta'Ngvən's* farm-land to usher the start of the planting season. One could plant or harvest for consumption anything on a 'kontry sunday' with the exception of corn or *ngwasaŋ*. It was the same rite carried out for the harvest season to start. This was real Communitarianism in praxis.
- One advantage was that a variety of food production for local consumption was carried out on the same piece of farm-land. Agrarian produce was not sold. In fact everyone had enough. The surplus was used in exchange for other necessaries a household could not provide for, because household sizes were not the same. In effect, trade by barter was carried out. For instance, huckleberry was never sold. When in need one visits a neighbour's farm-land/garden and got the quantity necessary for the household. Eco-communitarian farming system in Nsoq was at its best.

- Land was properly conserved and sustained as the fallowing system of farming was prescribed to everyone in a particular land surface area by the *Ta'ngvən*. Some special farm-lands were not farmed. They were reserved for unforeseen exigencies. All the farm-land was not shared out. There could be a visitor or a stranger in need.
- Community mobilization and development was ensured via the eco-communitarian, agrarian system of life.
- A piece of farm-land could be shared by two or more persons. A neighbour or friend could be given part of the farm-land to work for instance huckleberry or other vegetables. They could plant fruit trees like cola-nuts, oranges, mangoes, etc., raffia palm-bush, plantains and bananas in their farm-lands. This was real eco-Communitarianism in praxis.

The eco-communitarian, agrarian system of life among the *Nsoq* people did not exist without some disadvantages. The absence of flaws or weaknesses maybe an oversight as sometimes the advantages outweighed the disadvantages.

- One disadvantage of this cultural philosophy is the absence of the sense of economic growth. This system lacked the philosophy of expanding her economic prowess. The economy was stagnant. The economic progress was dormant or sterile. This is because of the absence of the interplay of the market forces of demand and supply. Even where there was marketing, it was at a small scale and not for any investment but to acquire some basic necessities for the household. Large scale production was absent. Empty-belly philosophy stifles economic growth.
- A sound economic growth is the fundamental pillar of development. A porous economy is tantamount to poor development. The *Nsoq* eco-communitarian, agrarian cultural philosophy has as one of its weaknesses the absence of progress and development. The developmental elements are too narrow. That is why *Sehm Mbinglo* and *Ngah Bifon II* advocated an open-door philosophy within the tenet and maxim: *wir Nsoq dzə-wir bi' vitum*. That is, the *Nsoq* man can only advance because of the active presence of the outsider. On their own, they cannot advance, progress and develop.
- The communitarian system kills a lot of initiative. One such area is that of the degradation of farm and pasture-land. The absence of the spirit of ownership tends to neglect the land. Land preservation, conservation and sustainability maybe abandoned leading to destruction of the ecosystem. The result is often that of putting the blame on the wrong shoulders. That is, attributing the problem to the negligence on the part of the gods. Masaka writing about the Shona people like the Nsoq people's shares the same experience as the members of the community were forbidden from certain actions, visiting sites regarded as sacred; using sources in an unsustainable manner as religion was central to a Shona and Shona spiritual beings were both feared and respected thereby hindering better agrarian philosophical development (Masaka and Chemhuru 2010: 41).
- Poor yields and harvest maybe as a result of human or natural forces; but this is often not the case in a situation where agriculture is inseparable from religion.

Poor yields and harvest is often attributed to the gods for failing to protect the land. In this situation, the propensity to hang on superstition is very high.

We cannot pretend to exhaust the possible weaknesses and or advantages of the indigenous *Nsoq* eco-communitarian, agrarian philosophy. Other human, natural and spiritual lessons abound vis-à-vis the *Nsoq* cultural, agrarian philosophy.

14.9 The *Nsoq* Farming Ethics

With the understanding that ethics deals with the *modus vivendi cum modus operandi* of man at large, the norms and guiding principles of human acts therefore have no boundaries as man-as-man is an ethical being. The meaning of 'ethics' is hard to pin down as the views many people have about ethics are shaky. Ethics is not limited to one's feelings, nor can it be confined to religion and or merely following the law or laws of the land or state as many persons would think, it is more than just these tendencies. The Markula Center for Applied Ethics succinctly puts 'ethics' thus:

> Ethics is two things. First, ethics refers to well-founded standards of right and wrong that prescribe what humans ought to do, usually in terms of rights, obligations, benefits to society, fairness, or specific virtues…secondly, ethics refers to the study and development of one's ethical standards…constantly examine one's standards to ensure that they are reasonable and well-founded (Velasquez et al. 2010).

This approach to an understanding of ethics clearly falls within our scope here. It entails the continuous effort of studying our own moral beliefs, conduct, and striving to ensure that humans and the institutions live up to standards that are reasonable and solidly-based. Here we can understand that ethics is not morality as often misconstrued. Ethics is imposed and morality is personal. One is personal opinion on values and the other is a set of impositions to force people into supporting values and standards. We can see why Garret Hardin says that one of the Ten Commandments of Agrarianism is: living with limits. He argues that the agrarian must beware of nature-free equations (Wirzba 2003: 70). The Nsoq man like many other African people recognizes the fact the African environmental or agrarian ethics is defined in terms of relatedness. What Charlton calls the Sacrament of Co-existence (Ibid: 70). From this standpoint, we can comfortably talk of the *Nsoq* agrarian moral beliefs, moral conduct and the institutions that set the parameter for this eco-communitarian community. The *Nsoq* farming ethos is both moral and social and at the same time embracing certain religious undertones. That is why from the point of view of morality, it is a cultural value. From the viewpoint of ethics, there are sacred taboos and prescriptions as well as institutional or cultural principles and norms governing the ethics and aesthetics of farming. Agriculture is a cultural, religious and ethical activity; with each moral body and institution playing an indispensable role in the flourishing of the agrarian life in the Nsoq cultural society.

- The gods and the spirits have a pride of place in the institutionalization of the ethical values in the *Nsoq* agrarian cultural setup. The role and or duty of the gods and spirits is that of a supervisory moral body who sanctions human agrarian acts as good by reward and as bad by punishment. For certain ugly agrarian acts unseen and unknown by others are publicly sanctioned by the gods and only a seer could decipher the source and the issue at stake when an agrarian calamity befalls an individual, family or the tribe as a whole. Taboos such as: destroying the crops of your neighbour knowingly (an act which calls for retributions from the gods), harvesting from a neighbour's farm without her consent, and farming on a day of obligation as *Kilōviy* and or *Ngoiylum* abound. There is nothing like secretly breaking or flouting a taboo. The gods are the heart-beat of *Nsoq* ethical agrarian philosophy.

- The ancestors come in second place. These are the living-dead of the land. They too had farmed these lands and had bequeathed it to others. The ancestors play an intermediary role between the gods and the living. They sanction agrarian activity directly or indirectly. In the appeasement and atonement of the gods and sacrificing to the gods, the ancestors supplicate the gods on behalf of the living. Ancestors assist the gods in the implementation of agrarian ethical values.
- In the third place is the *Ata'ngvən*. They too play an indispensable role in the success of the *Nsoq* agrarian ethics. They appease the ancestors and the gods of the different farm-lands they control. The *Ata'ngvən* handle and manage petty-petty litigations for the smooth running of the farming ethics. They are the eye of the tribe, the Fon, the ancestors and the gods as far as the *Nsoq* agrarian ethics is concerned.
- The Fon is the direct custodian of the land. The *Ngwerong*, *Ta'wong* and *Yeewong* are the immediate collaborators of the Fon in the overseeing of the affairs of the land. The *Ngwerong* is traditionally known as *Ta'nsaiy*. This is a special secret group in the palace that foresees and ensures the smooth-running of the traditional and cultural values. It is the direct eye of the entire tribe. When described as "*Ta'nsaiy*", it means that as a moral institution, it is the 'owner of the soil or land.' All the sacrifices in the land are carried out by the Fon, *Ngwerong*, *Ta'wong* and *Yeewong*. Keeping at bay intruders and any persons from desecrating the land is the sole prerogative of these moral bodies and institutions.
- We cannot minimize the place of the commoner or the farmer. The farmer markets or sells out the cultural agrarian ethics to the other tribes. If they keep the moral and ethical laws and principles, then there is peace, order and tranquility in the land. The reverse will be true. They are the direct sustenance of the people through their agrarian activity. That is why Wirzba argues that the traditional agrarian in many African tribes or communities has as his/her natural home: the field, the garden, the stable, the prairie, the forest, the tribe, or village and these serve as the cradle of agriculture. They know a lot about local soils, local weather, local crops, animal behaviour and each other (Ibid: 65). The mutual existence between them makes for the African environmental ethics.

The Nsoq farming ethics thrives within the framework of certain moral and social values-rights, obligations, observances, taboos and prescriptions. These values and virtues are put in place to enable the smooth running of the agricultural sphere of community life without stress and strings. The sacredness of farming necessitates the preservation, conservation and sustainability of the eco-system or environment. Where each individual or institution exacts his or her duty and functions meticulously, the entire farming season is guaranteed and there is a positive yield and harvest. The farm-lands do not suffer or undergo wear and tear as is the case with the exaggerated westernized notion of "subdue and conquer". With the division of work and role, the implementation of ethical norms, taboos and injunctions, there is the enhancement of the conservative and sustainable spirited elements of the ecosystem. What are some of the lessons one can find in *Nsoq* agrarian ethics?

14.10 Conservative and Sustainable Elements of Nsoq Agrarian Ethics

- Bush-fallowing is obligatory. It is mandated by each *Ta'Ngven* after three farming seasons. The Nsoq farming philosophy of '*kwir suum e'bam viya' vitar*' was fashionable. That is the praxis of fallowing the cultivated land after three farming seasons. Alternatively the farmers farm from one *ta'ngven* to the other. This remains one of the advantages of conserving and sustaining the ecosystem. Farm-land degradation is equally avoided.
- Bush-fallowing is twofold. There is the first model as described above and there is the second which is called the *kinntchari*. It is an extended or long-term fallowing model which can take place for more than two years. While to *kwir-suum* is mostly for a year or two, the *kinntchari* takes more than two years of fallowing the farmed land. This system was rich as it entails multiple advantages in the *Nsoq* agrarian world. The *kinntchari* became the best system of the conservation and sustainability of the farm-land or environment. For the farm-land surface area gained optimal fertility; temporarily used as mini-hunting arena where villagers set hunting traps for animals to spice their meals as these were considered a seasonal delicacy. Therefore, the *kinntchari* model remained one of the best systems for the *Nsoq* agrarian land conservation and sustainability.
- The praxis of not sharing out all the farmable land was a healthy method of conserving, sustaining and evading farm-land degradation. Some of such farm-lands were reserved for unforeseen exigencies, family visitors or strangers in need.
- Some specific land surface was preserved and revered as the abode or habitat of the gods, spirits and ancestors. It could be a forest or farmable land. Others were preserved for the growth of special grass for the roofing of houses. Some of the preserved forest was for medicinal herbs (trees, backs of trees, roots, leaves). Such forest reserves (kovsi) included: *Ngongba', Kinka* of *Mbiame, Diyri* of *Banten, Kintosong (Nsa')* of *Tatum, Buh-kpu* of *Ta'du* area, *Oku* etc. These forest reserved areas are often exploited minimally in acquisition of medicinal plants and herbs, hunting, kitchen wood and special wood for the construction of houses and bridges

etc. But above all, this was an epitome of the conservation and sustainability of the ecosystem.

- The ecosystem was conserved and sustained via the conventional 'fire-line paths'. Farm-lands, pasture-lands, plantations (of fruits and nuts) and tubal crop farm-lands (cassava, sweet and Irish potatoes, local carrots and onions) were conserved via the use of fire-line paths called *kintini.* This was a system of protecting these agricultural areas from fire disasters. Bush-fire was one of the handicaps and a threat to the *Nsoq* agrarian philosophy. That is why agrarian areas were annually preserved during the dry seasons especially in the months of January and February. This agrarian system was to conserve and sustain the ecosystem.

These were some of the advantages of *Nsoq* agrarian ethos vis-à-vis the conservative and sustainable perspective of the environment. Wendell distinguishes between the notions of 'nurturing' and 'exploiting' in agrarian life found among the Nsoq people. He argues that the standard exploiter is efficiency and the standard nurturer is care. The exploiter's goal is money, profit; the nurturer's goal is health—his land's health, his own, his community's, hiscountry's (1977: 7–8). The Nsoq Man subscribes to the logic of nurturing the land. Farm land and produce were not sold. This was a plus in the Nsoq agrarian philosophy. On the other hand there were some disadvantages or weaknesses.

- The tendency of reserving some farm-lands for the habitat of the gods, spirits and abode of the ancestors reduced the farm size for production. This equally applies to the conservation of some farm-lands for the unforeseen exigencies. Families or households were not always static in size. They were increasing not in proportion to the food production and *ipso facto* an unhealthy precedence.
- The tendency of bush-fallowing reduces and affects the quantity of production. Concentration of energy towards the cultivation of a small land surface was not healthy.
- Bushfires in some limited cases were another way of preparing farmland though this would appear to be against environmental preservation. However, it is important to note that it was not really a recommended method and that is why it was common among people who were lazy and could not clear the grass or bushes before farming.

The Nsoq agrarian farming ethos remains an area of special interest considering the lessons we have explored in this work. A revisit of this system of the indigenous African cultural, agrarian philosophy remains paramount if we have to fight the westernized mechanical, scientific and technological approach towards the environment.

14.11 Conclusion

Agrarian philosophy is simply that of Man's nexus to the land or environment considered to be the source of agricultural products. This relationship is communal and sacred. That is why Ogungbemi succinctly defines the African environmental ethics as the ethics that leads human beings to seek to co-exist peacefully with nature and treat it with some reasonable concern for its worth, survival and sustainability. The potential crisis threatening the agrarian life in the African world needs to be visited. Agrarian values should constitute a system of ethics that can restrain the self-interest syndrome that threatens the gross communal happiness. Considering the advantages and lessons explored in the logic of the Nsoq agrarian philosophy, we can still upset the apple cart that keeps threatening the very existence of the agrarian values: land, beauty, food, work and the community. These values, if preserved, can enhance a much more sustainable agrarian life suitable for the generations now-after.

References

Berry, W. 1977. *The Universe of America: Culture and Agriculture*. New York: Avon Books.

Fomine, F.L.M. 2009. *A Concise Historical Survey of the Bamoum Dynasty and the Influence of Islam in Foumban, Cameroon, 1930-Present,* vol. 16, no. 1 & 2. https://www.ajol.info/index.php/aa/article/view/87547.

Goheen, M. 1996. *Men Own the Fields, Women Own the Crops*, 1st ed. London: University of Wisconsin Press.

Goodman, F.D. 1990. *Where the Spirit Ride the Wind: Trance Journeys and Other Ecstatic Experiences*. Bloomington and Indianapolis: Indiana University Press.

Masaka, D., and M. Chemhuru. 2010. Taboos as Sources of Shona People's Environmental Ethics. *Journal of Sustainable Development in Africa* 12 (7).

Mbessa, D.G. 2020. African Bioconservatism and the Challenge of the Transhumanist Technoprogressism. *Open Journal of Philosophy* 10: 443–459. 10pite.4236/ojpp.2020.104031.

Montmarquet, J.A. 1985. Philosophical Foundations for Agrarianism. In *Agriculture and Human Values*. Springler.

New Jerusalem Bible (The). (1990). *Reader's Edition*. New York: Doubleday.

Ogungbemi, S. 1997. *An African Perspective on the Environmental Crisis in PA. 2011. Environmental Ethics: An African Understanding*. https://www.readperiodicals.com./201103/2323094941.html. Accessed August 20, 2022.

Percer, L.R. 1999. *The War in Heaven: Michael and Messiah in Revelation 12*. Texas: Baylor University.

Ramose, M.B. 2009. Ecology Through *Ubuntu*. In *African Ethics: An Anthology of Comparative and Applied Ethics*, ed. M.F. Murove, 308–314. Scottsville: University of KwaZulu-Natal Press.

Sanga, Kumpolo I. 2006. Aesthetic Appreciation and Cultural Appropriation of Bird Sounds in Tanzania. In *Folklore*, vol. 117, no. 1, 96–102. Taylor & Francis Ltd. https://www.jstor.org/stable/30035324. September 23, 2021.

Sire, J.W. 2015. *Naming the Elephant: Worldview as a Concept*, 2nd ed. Intervarsity Press. https://www.amazon.com>Naming. September 22, 201 21.

Tangwa, G.B. 1996. *Bioethics: An African Perspective*. Bamenda: Langaa Research & Publishing and Common Initiative Group.

Tangwa, G.B. 2004. Some African Reflections on Bio-medical and Environmental Ethics. In *Ojomo, PA 2011. Environmental Ethics: An African Understanding*. https://www.readperiodicals.com./201103/2323094941.html.

Tangwa, G.B. 2010. *Elements of African Bioethics in a Western Frame*. Bamenda Langaa Research & Publishing C.I.G.

Thompson, P.B. 2008. Agrarian Philosophy and Ecological Ethics. In *Science and Engineering Ethics*. https://www.researchgate.net>2340.

Tosam, M.J. 2019. African Environmental Ethics and Sustainable Development. *Open Journal of Philosophy* 9: 172–192.

Velasquez, C. Andre, et al. (eds.). 2010. *Issues in Ethics, "What is Ethics", 11E V1 N1*. Markula Center for Applied Ethics, Santa Clara University Press. https://www.scu.edu/ethics/ethics-resources/ethica-decision-making/what-is-ethics/. September 26, 2021.

Wirzba, N., ed. 2003. *The Essential Agrarian Reader: The Future of Culture, Community, and the Land*. New York: The University Press of Kentucky.

Peter Takov holds a Ph.D. in Philosophy from the Pontifical Urban University in Rome. He is currently chair of Philosophy at the Catholic University of Cameroon, Bamenda and a visiting lecturer in the Departments of Philosophy of the Saint John Paul II Major Seminary, Mamfe and the University of Bamenda. He is a researcher in the Ethics of Responsibility with a bias on Environmental Philosophy.

Part IV
Indigenous Knowledge Systems, Agrarianism, and Higher Education

Chapter 15
The Emergence of a Re-humanizing Pedagogy for African Agrarian Philosophy

Birgit Boogaard, Bernard Yangmaadome Guri, Daniel Banuoku, David Ludwig, and David Fletcher

Abstract Until today, an externally imposed epistemological paradigm is dominant in most educational curricula at universities in Africa. Despite ongoing Eurocentrism and Western hegemony in mainstream agricultural trainings in Africa, Indigenous knowledge on agriculture still exists: it has been preserved for generations by farmers and wise elders in rural communities who often are knowledge authorities on African agrarian Indigenous knowledge, values and practices. An imposed epistemological paradigm on the African continent reinforces epistemic injustice by dominating and ignoring Indigenous African ways of doing and knowing, which is deeply dehumanizing. Inspired by Paulo Freire's 'humanising pedagogy' and Mogobe Ramose's call for *mothofatso* in human relations, we explore a re-humanising pedagogy for African agrarian philosophy, which addresses the following questions: Can African agrarian philosophy contribute to re-humanization by reviving and restoring Indigenous agricultural knowledge, values and practices? If so, what pedagogy and educational methods are appropriate? In search for answers to these questions, we reflect on a two-week educational program on an endogenous approach to community resilience in Ghana, as a pedagogy that is emerging. In doing so, the chapter is firmly rooted in African soil, both practically—through trainings in Ghana—and theoretically—through African agrarian philosophy. By connecting educational experiences with insights from theory, seven themes for a pedagogy for African agrarian philosophy come to the front: an African agrarian philosophy *with* memory; a dialogical

B. Boogaard (✉) · D. Ludwig
Knowledge Technology and Innovation (KTI) Group, Wageningen University and Research, Wageningen, The Netherlands
e-mail: birgit.boogaard@wur.nl

D. Ludwig
e-mail: david.ludwig@wur.nl

B. Y. Guri · D. Banuoku
Centre for Indigenous Knowledge and Organizational Development (CIKOD), Accra, Ghana

D. Fletcher
People Development Ltd, Antigonish, Canada

M. J. Tosam and E. Masitera (eds.), *African Agrarian Philosophy*, The International Library of Environmental, Agricultural and Food Ethics 35,
https://doi.org/10.1007/978-3-031-43040-4_15

student–teacher relation; the value of lived experiences; intergenerational and spiritual methods of education; relationality of human beings and Mother Earth; unity between theory and practice; critical consciousness about people's rights. The thesis defended is that African agrarian philosophy is without relevance if it remains a theoretical exercise that is not developed and put to use with farmers, wise elders and youth in the communities. Instead, when combined with a re-humanising pedagogy, African agrarian philosophy can contribute to reconnecting and reviving African agricultural knowledge, values and practices, which subsequently contributes to transforming African food systems that heals the environment and produces enough food to feed communities.

Keywords African philosophy · Humanising pedagogy · Agricultural education · Indigenous knowledge · *Mothofatso*

15.1 Introduction

African agrarian philosophy is a relatively new field, in the sense that limited attention has been paid so far to agriculture within African philosophy (Boogaard 2019). The current handbook makes an important and urgently needed contribution to give shape and content to this emerging field that is characterised by "pre-colonial African agrarian beliefs, values, and practices as well as contemporary issues" (current volume). Despite the relative absence of this field in academic discourses agrarian philosophy has always been part of the lifeways and cosmovision of African people. For thousands of years, African Indigenous knowledge on agriculture has survived and been passed on from generation to generation by farmers and wise elders in rural communities. Women and men farmers and wise elders are knowledge authorities on African Indigenous knowledges, values and practices, who have been constantly innovating and adapting practices to harmonize with their changing environment. Yet while African agrarian philosophy existed before universities did, African philosophy as an academic discipline at universities tends to focus mainly on academic theory, while being less engaged with agricultural practices and people's livelihoods in rural communities. The thesis defended in this article is that African agrarian philosophy will be without relevance if it remains a theoretical exercise that is not put into practice to revive and reconnect with African agrarian knowledge, practices and values among farmers, wise elders and youth in rural communities across the African continent. This chapter aims to bring theory and practice on African agrarian philosophy together by starting from grounded practice with educational experiences from the African Learning Institute (ALI). ALI is a pan-African initiative that emerged out of the need among practitioners in Ghana to revive and reconnect people with Indigenous agricultural knowledge, practices and values. While ALI is not a formal educational institute (yet), the shared need resulted in a two-week educational program that explored Indigenous knowledge for community resilience and searched for ways to support learning processes through mutual learning. The training was a collective process and exploration by the organisers and participants from 10 African countries. In this article we reflect on this novel initiative as a way to explore a pedagogy for African agrarian philosophy.

We first elaborate on pedagogical theories, and then describe the two-week training that was provided in Ghana. Thereafter, we connect these experiences from ALI with insights from theory which reveals seven themes for a pedagogy for African agrarian philosophy. In doing so, the chapter is firmly rooted in African soil, both practically—through trainings in Ghana—and theoretically—through African agrarian philosophy. As such, insights from this chapter are relevant for the design of agricultural trainings and education on African agrarian philosophy in Africa and beyond.

15.2 The Need for a Re-humanising Pedagogy

Until today, an externally imposed epistemological paradigm is dominant in most educational curricula at universities in Africa. As Ramose describes: "the epistemicide commited at colonisation was left virtually intact by ensuring that the coloniser's epistemological paradigm shall remain dominant" (Ramose 2016b: 548). As a result of this dominant epistemological paradigm, African philosophy has been actively marginalized, delegitimized and discriminated against in academic curricula. Active and systemic exclusion and oppression of African ways of knowing and doing reinforces epistemic injustice (Mungwini 2017), which can lead to inferiority and superiority complexes. As Oelofsen (2015) writes: "the values of the coloniser or oppressor, which judge the colonised to be inferior and backward, are internalised by the colonised, and thus she suffers a loss of self-esteem and starts to hate herself and what she represents in the worldview of the oppressor." (Oelofsen 2015). On the oppressor side is a superiority-complex, which can also be described as Eurocentrism: the view that the West understands itself as superior to other cultures, and as such defines what philosophy and science is (Kimmerle 2016).

Agricultural curricula and trainings neither escape inferiority-superiority thinking. Eurocentrism is for example reflected in the idea that Africa's agriculture should resemble Western, modern, commercial agriculture (Boogaard 2019). Likewise, through ongoing Eurocentrism, mainstream agricultural development projects and trainings reinforce epistemic injustice (Boogaard 2021).[1] Despite ongoing Eurocentrism and Western hegemony in mainstream agricultural trainings in Africa, Indigenous knowledge on agriculture still exists: it has been preserved for generations by farmers and wise elders in rural communities who often are knowledge authorities on African agrarian Indigenous knowledge, values and practices. Such knowledge has been taught, handed down through apprenticeship relationships, cultural festivals and initiation rites, and dynamically reinvented in various ways within the lifeways

[1] As mentioned elsewhere (Boogaard 2021), there are organisations and participatory trainings that actively try to move away from Eurocentrism and take farmers' knowledge as a starting point. Prolinnova is such an example (Promoting Local Innovation, https://www.prolinnova.net/).

of African peoples. To challenge Eurocentrism and epistemic injustice in agricultural trainings, education in African philosophy can play a crucial role (Boogaard 2019), specifically education in African agrarian philosophy.

In search for a pedagogy of African agrarian philosophy, we can learn from debates in African philosophy about education at universities and other knowledge institutes, (see e.g. a special issue[2] in South African Journal of Philosophy 2016). A pedagogy can be described as a complex philosophy, politics, and practice of education (Del Carmen Salazar 2013), and includes questions about *what* should be taught as well as *how* this should be taught (Oelofsen 2015). With regard to a pedagogy for African philosophy, Oelofsen (2015) pleads for a humanising pedagogy as a way how "the African mind can start to escape the belief that it is inferior". Similarly, work will need to be done for the non-African mind to shed the belief in its superiority. The origin of a humanising pedagogy goes back to Paulo Freire (2005), who described it in 'Pedagogy of the oppressed'. While Paulo Freire (2005) developed this idea of a humanising pedagogy in the context of Latin America, his pedagogy seems highly relevant for other contexts as well. In this line, Del Carmen Salazar (2013) emphasize that in Freire's view it is "not transferrable across contexts, but rather should be adapted to the unique context of teaching and learning" (Del Carmen Salazar 2013: 127). As such, Freire invited others to adapt and refine these ideas to other contexts in other parts of the world. Freire's humanising pedagogy seems still highly relevant in present-day Africa and debates on 'decolonizing' minds, for the reason that he proposed a humanizing pedagogy to counter dehumanization:

> Dehumanization, which marks not only those whose humanity has been stolen, but also (though in a different way) those who have stolen it, is a *distortion* of the vocation of becoming more fully human. This distortion occurs within history; but it is not an historical vocation. Indeed, to admit of dehumanization as an historical vocation would lead either to cynicism or total despair. The struggle for humanization, for the emancipation of labor, for the overcoming of alienation, for the affirmation of men and women as persons would be meaningless. This struggle is possible only because dehumanization, although a concrete historical fact, is *not* a given destiny but the result of an unjust order that engenders violence in the oppressors, which in turn dehumanizes the oppressed. Because it is a distortion of being more fully human, sooner or later being less human leads the oppressed to struggle against those who made them so. (Freire 2005: 44)

An imposed epistemological paradigm on the African continent reinforces epistemic injustice by dominating and ignoring Indigenous African ways of doing and knowing. This is deeply dehumanizing in the sense that it is an ongoing "distortion of being more fully human". In this line, Ramose (2020) builds on Freire's dehumanization argumentation as follows:

> The ethical-historical counter to dehumanisation is learning to be human. It is indeed a life-long learning, because each historical moment can have its peculiar manifestations of dehumanisation. Because of this re-humanisation presents itself as the fundamental ethical counter to dehumanisation. On the basis of this reasoning, it is re-humanisation and not 'decolonial' that speaks to the basic issue of the struggle for truth, justice, and peace in the world. (Ramose 2020: 304)

[2] South African Journal of Philosophy, Volume 35, 2016—Issue 4: Special Issue: Africanising the philosophy curriculum in universities in Africa.

Ramose thus proposes *mothofatso*—a re-humanisation of human relations—rather than the term 'decolonization'. The loss, destruction, and disconnection of African people with Indigenous agricultural knowledge, values and practices due to the unjust wars of colonization can be seen as an example of dehumanization. Thus—inspired by Freire's humanising pedagogy and Ramose's call for *mothofatso*—it becomes clear that there is a need for a re-humanising pedagogy in agricultural trainings and African agricultural philosophy. It should be noted though that it is not new to connect Freire's pedagogy with agricultural trainings. In fact, Freire's pedagogy was one of the main sources of inspiration for Robert Chamber's Participatory Rural Approach (PRA) that has been widely applied in agricultural training and extension in Africa over the past decades (Chambers 1994). However, Freire's pedagogy has not been used to explore a pedagogy for African agrarian philosophy.

Based on the above, we aim to explore a pedagogy for African agrarian philosophy that addresses the following questions: Can African agrarian philosophy contribute to re-humanization by reviving and restoring Indigenous agricultural knowledge, values and practices? If so, what pedagogy and educational methods are appropriate? In search for answers to these questions we reflect on a two-week training in Ghana, as a pedagogy that is emerging.

15.3 In Practice: A Two-Week Training by the African Learning Institute

In July 2016, an intensive two-week educational program to explore local and Indigenous knowledge for community resilience took place in Techiman, Ghana. Community resilience was chosen as the theme for the training of the African Learning Institute (ALI) because of its relevance to the participants, and their work with African agrarian communities. Resilience and community resilience are current in various disciplinary discourses (Béné et al. 2012), yet they are often blind to, or marginalize, indigenous African conceptualizations (Apusigah 2008; Crane 2010; Fletcher 2017). A stated intent of the ALI was to explore community resilience from the perspectives and world views of the participants, transcending disciplinary boundaries. The training was designed by the Centre for Indigenous Knowledge and Organisational Development (CIKOD) (Ghana) and the Coady Institute (Canada). Three authors of the current article were co-designers and facilitators of the training (Bern Guri, Dan Banouku, David Fletcher). The origin of the ALI training finds itself in the desire to revalorize and revitalize the local Indigenous wisdom of African peoples and feelings of frustration of CIKOD staff with dominant Western educational systems in Ghana. CIKOD staff wanted to further their work on promoting Indigenous knowledges and institutions and to 'decolonize African minds'. They found a willing collaborator at the Coady Institute in Canada that was beginning their own journey of valuing and appreciating Indigenous knowledges in their own pedagogical practices. The Coady

Institute has much experience on designing and providing transformational educational programs in a multicultural environment in which diverse world views and knowledge paradigms are respected (Bean 2019). Sources of inspiration to design the two-week ALI training in Ghana were among others Linda Tuhiwai Smith's book 'Decolonizing methodologies' (2001), David Millar's 'Our sciences: indigenous knowledge systems of northern Ghana' (2012) and George Sefa Dei's work on Indigenous knowledge (2000) and his book 'Teaching Africa' (2010).

The main aims of the training were to gain a deeper understanding on endogenous development and to promote Indigenous knowledge for community resilience among practitioners across the African continent (Fletcher et al. 2017). The focus on endogenous development led to a clear 'how to'-question as the main objective of the training: how to facilitate processes so that communities can develop their own processes and knowledge? In this line, the training aimed to re-connect people with their knowledge—e.g. passed on through their parents and elders in rural communities—and to re-gain a sense of pride in one's community and Indigenous identity. Moreover, a strong ambition of the ALI training was to build capacity and connect people across the African continent. Hence, the training was no individual learning exercise, but it was designed to create a collective learning experience in which participants would learn mutually and help each other. It aimed to create a sense of community between participants—as a pan-African learning community. In this line, African-based organisations were selected on the basis of the following criteria: (i) their work is community-based and they are dedicated to a different way of working with communities, (ii) they show commitment to respect Indigenous knowledge and the rights of Indigenous people, (iii) they have been working on these issues for several years, (iv) they expressed their interest to use an endogenous development approach and to build resilience in their communities.

In total, 37 people (13 women and 24 men) participated, who represented 15 different organisations across 10 African countries, including Ghana, Benin, The Gambia, Ethiopia, Kenya, Uganda, Zimbabwe, Zambia, South Africa and Lesotho (see Appendix) (Fletcher et al. 2017). Most participants were practitioners on biocultural issues, while their educational and cultural backgrounds and roles were rather diverse, including e.g. people from NGO's, farmers, government officials, students, a university lecturer, and community members. The main assumption of the ALI training was that all participants came to the course with a wealth of knowledge already and that this would be a major opportunity to learn from each other and co-create. Therefore, the training aimed to share knowledge among participants, as a fundamental principle of the endogenous approach: as well as bringing in external resource persons to share their own experiences, participants of the course were also the teachers and vice-versa—teachers were participants as well. We will revert to this point when describing the educational methods of the training.

In terms of content, the main focus of the training was on agriculture and Indigenous knowledge. However, agriculture should not be seen in isolation: instead, it is part of life, which involves other topics as well, such as natural resource management, wildlife management, human health, and care of the environment. The training started from an endogenous development approach, meaning to start from within—thus to start from people's lifeworld and experiences, rather than an exogenous approach, which comes from outside (Millar 2014). As such, endogenous development builds on an African cosmology that combines three interwoven worlds: the spiritual, the human and the natural world (Millar and Haverkort 2006). This approach also means that the exact content of the training was evolving throughout the course, depending on the topics that were addressed by the participants and communities. In doing so, the training offered space to respond to priority challenges, such as the documentation and recording processes on African biocultural contexts and Indigenous people's rights. Some of the training topics included: community resilience, Ecological Organic Farming (EOF), biodiversity, Indigenous people's rights, Biocultural Community Protocols (BCP), and African cosmology and spirituality (Fletcher et al. 2017).

15.3.1 Story Telling: Starting from People's Own Stories

An essential part of endogenous development is that participants bring in their knowledge, in this case about community resilience. So rather than having "external experts" explaining what community resilience is and what it should—or should not—include, the course started with participants' own stories of community resilience, based on their experiences, Indigenous knowledge and worldviews. Since different countries and people have their own stories on agriculture, a central guiding question was: what issues are oppressing or uplifting people? This resulted in stories about, for example, Indigenous healing in Uganda, large community-based music and cultural celebrations in Ethiopia, and protection of Indigenous people's rights in Kenya. Bern Guri told the story of how a Biocultural Community Protocol (BCP) was instrumental in halting gold mining in his community Tanchara, Ghana. People were struck by this story in the sense that a small community challenged a big international mining company, as well as by the time it took (multiple years) to develop a Biocultural Community Protocol (BCP), and the tensions the procedure caused within the community (Fletcher et al. 2017). Subsequently, the stories were deconstructed to identify characteristics of resilient communities and related to theory. The BCP-example shows that story-telling is a powerful way to raise awareness: it raised awareness about the fundamental issue of Indigenous people's rights and the challenges to fight for these rights.

15.3.2 Elders Panel: Reviving Indigenous Knowledge

The ALI training combined theory with practice, in the sense that it moved between real life experiences and stories from participants on the one hand, and general theoretical concepts—such as colonialism, globalization, Indigenous knowledge, endogenous development, and community resilience—on the other. The theoretical concepts were not brought from an "exogenous frame" and imposed on people as another foreign-popular concept, but instead these were brought by elders and holders of Indigenous knowledge, who were also participants in the program. The elders panel was inspired by the Ghanaian symbol of *Sankofa* (going back to our roots): a number of wise elders shared their experiences and expertise about Indigenous wisdom. For example, there was a community elder who spoke about the leadership system; there was a Ghanaian professor—also a community elder—who spoke about endogenous development and emphasized that elders, African women and men farmers are also scientists by systematically observing and experimenting in their own environment; and there was a woman leader who was active in community development. The latter—Queen Mother from Wenchi Community in Ghana—spoke about the value of Indigenous knowledge by drawing on her own experiences, including some of the challenges she faced when using Indigenous wisdom and spirituality.

15.3.3 Spiritual Practices: Burning Ritual and Connecting with the Ancestral-World of the Living-Dead

A powerful effect of story-telling is that people connect to other people's stories, which allows them to feel interconnected with each other's life (Chilisa 2020). Connecting to someone's story often goes together with emotional attachment, such as joy, anger, fear and sadness. Often these stories related to colonialism and other—historical and ongoing—forms of oppression of African people. The deep pain and sadness in these stories resonated with many participants. Before going ahead with training after these stories, it was therefore important to give space to release such deep emotions which sometimes have been kept inside for years, a life-time or even generations. The ALI training therefore developed a burning ritual, in which participants were invited to write their frustrations, pain, anger and feelings they wanted to let go of on a paper. Subsequently, these were burned in a ceremonial fire outside the classroom, facilitated by an elder from Southern Africa. In doing so, the burning ritual gave space to feelings, emotions, pain, anger and frustration of how colonialism and oppression influenced people's life, now and in the past. After this process, participants could move forward together for the rest of the training.

The legacy of colonization exists in various ways until today, including a devaluation of African spirituality (Kimmerle 2006). Ongoing imposed Westernization and 'modernization' disconnect African people from the spiritual voices of Africa speaking to the soul. The ALI training therefore included meditative moments—named 'mystica'—in which participants were invited to slow-down, to listen, and to connect with themselves and the ancestral world. Sometimes such meditative moments were used at the beginning of a session to set a sacred intention and build positive energy for the day. At other times, these moments were used to close a session or day, in which participants processed and reflected on what they had learned. Another meditative exercise was the collective creation of a seed mandala: at the end of each day, participants added different varieties of seeds in a circle. At the end of the course, the seed mandala was shared among participants who took part of it home. These meditative moments made spirituality practical and contributed to creating a conducive and healing environment for the participants.

15.3.4 Community Visits and Residential Program

The main assumption in the design of the ALI training was that learning about Indigenous knowledge does not only occur inside the classroom through stories and theory, but one should also experience the value of Indigenous knowledge in practice. Therefore, the major part of the training occurred 'outside the classroom', in communities in the area of Techiman—such as Forikrom—which exposed participants to lives of people in the communities. They were also close to the knowledge of the communities. This opened the space for the participants to gain some understanding of the knowledge of the communities. During these community visits, participants engaged with agroecological practices, visited sacred natural sites, shared in community celebrations, and joined an Indigenous food exhibition. The residential program was a central method: participants stayed for one or multiple night(s) with a host family in a community, where they experienced community members' daily life routines (see Text Box 15.1). Participants stayed for example with queen mothers, community elders, farmers, healers and traders. The aim of these overnight stays was that participants learned from community members, about their way of life, their knowledge, and tried to understand why they do the things they do.

Text Box 15.1: Residential Program: Overnight Stay with a Host Family in a Community

"Participants were taken to stay with queen mothers, chiefs, farmers, healers and traders in local villagers. They were introduced by Nana Kwa Adams a traditional leader in the area and a fellow course participant. The hospitality experienced and the solidarity developed was an overwhelming experience for some. Their hosts became the experts on Indigenous knowledges for community resilience. Participants took part in their host families daily routine lives to understand their culture and their well-being. They learnt about traditional protocols, traditional health and medicine, agroecology and traditional trading practices depending on who was their host. Participants were unanimous with their praise for this component as it provided exposure to real lives of people in their biocultural context and demonstrated their local knowledge and resilience. Many people had a-ha moments during this experience! What was being discussed in the hotel training room became real. Building community resilience is not about some theory or practices from outside communities. Building community resilience is about what community people do themselves in their own environment, from their own world view perspective, to create the world they want to see and respond to the risks and challenges they encounter along the way. The everyday challenges and complexities of rural life are real, but at the same time the pride and strength people have in solving African problems with African wisdom is the path to the future." (Fletcher et al. 2017: 24).

15.3.5 Evaluation, Post-training Activities and Challenges

The official course evaluation showed a high score of 4.61/5 on 'overall satisfaction'. Moreover, participants were unanimous in their high appreciation of the residential program as it provided exposure to real lives of people in their area, through which participants experienced how Indigenous knowledge contributed in practice to biodiversity, nature conservation, agroecology, and community resilience (Fletcher et al. 2017). In addition to the evaluation score, participants were asked to give written feedback on the training. This feedback showed that participants gained a deepened understanding of Indigenous knowledge and endogenous development—as one participant wrote:

> The most significant change [of the training] for me was a more deepened understanding on the role of Indigenous knowledge systems, culture and cosmovision in the domain of biocultural diversity conservation and management. This is for the benefit of the present generation and posterity. In actual fact it is very encouraging to realize that the entirety of the rural communities respect the interdependence of the natural, human and spiritual worlds. They put a lot of reverence on natural resources conservation using local knowledge systems experiences…..the attendance at the ALI was very empowering and inspirational. It strengthened the aspects with regards to culture, endogenous development, social and economic endeavors. As we left Techiman I know all of us applied aspects and new development approaches and processes we learnt. (Fletcher et al. 2017: 18)

Despite these high evaluations, the designers and participants of the training were also confronted with some recurring challenges throughout the training. The endogenous development approach starts 'from within', which means that the training tried to move away from knowledge hegemony by 'external experts', and instead make use of the wisdom of the participants through story-telling and an elders panel. However, since many participants have had more 'conventional' education by 'experts' at the front of the classroom, it was sometimes challenging for participants to take note of and recognize the wisdom being shared by their peers. In addition, the training tried to contribute to epistemic justice by revalorizing Indigenous knowledge and recognizing that Indigenous knowledge is also valid in terms of its ontology and epistemology. At the same time, however, it was important that participants did not romanticize Indigenous knowledge but developed a critical attitude. This means that input from academic knowledge, new technologies, and other exogenous knowledge were not rejected, instead different forms of knowledge were combined to create alternative perspectives. In practice the integration of different forms of knowledges turned out to be quite challenging. For example, some participants from typically Christian family backgrounds could not come to terms with the idea and practice of rain making. A final concern was that while participants in general much appreciated the endogenous development approach, they also expressed that it would be challenging to apply these skills in the dynamic context of communities where they work. However, this was an important ambition of the ALI training: while the training ended after two weeks, the endogenous development process should not. Hence, participants were encouraged to take these lessons and practices into their communities and apply their learnings in their ongoing work. After nine months, participants from nine organizations in eight countries reported back on their activities (see Text Box 15.2) (Fletcher and Guri 2017).

After the course, participants experienced several challenges, threats and risks when applying their skills and knowledge in other places. To start with, in some places—such as South Africa and Ghana—young people became increasingly immersed in their culture and belief systems (see Text Box 15.2), while in other places—such as Ethiopia—it remained particularly challenging to involve the youth (Fletcher and Guri 2017). In this line, several participants reported back that the strong influence of Western education and lifestyles on young people and community members is hard to break through, in the sense that ideas of 'modernization', 'Westernization', and 'commercialization' often go together with devaluing Indigenous knowledge, values and practices (Fletcher and Guri 2017). By giving trainings on endogenous development and Indigenous knowledge in communities, some communities re-connected to Indigenous food and agro-ecological practices (e.g. in Gambia, Uganda, Lesotho, South Africa, Zimbabwe), which changed the way they look at their own communities. At the same time, however, the role of spirituality and cultural taboos was sometimes not fully understood nor accepted by some communities e.g. in the Gambia (Fletcher and Guri 2017). It shows that it is difficult for these organisations to break through the hegemony of Western knowledge—which is still seen as the 'norm' or 'standard'—and move towards a re-appreciation of spirituality. The ongoing lack of reference and written material on Indigenous and spiritual knowledge in education and at national level, is reinforcing this hegemony rather than going against it.

Text Box 15.2 Post-training Activities by the Participants

Some of the post-training activities, reported by participants (Fletcher and Guri 2017):

- People's rights were articulated and protected by developing Biocultural Community Protocols (BCPs) (Zimbabwe, Ghana, Kenya)

- People visited a community that successful defended their sacred site, and learned to document sacred sites, consider legal action, and involve many stakeholders (Ethiopia)
- The youth of a community engaged their elders and the entire community to document their customary laws guiding the management of their sacred groves (Ghana)
- The youth became deeply immersed in their culture and belief systems (Lesotho, South Africa, Zimbabwe)
- Community members affirmed the value of the diversification of Indigenous food crops to stabilize food security. (Lesotho, South Africa, Zimbabwe)
- A press conference was held to promote local foods (Gambia)
- Communities reaffirmed their deeply entrenched spiritual values including the significance enshrined in small grains like sorghum, finger and pearl millets. (Lesotho, South Africa, Zimbabwe)
- Ecological organic farming was revalued in a community (Uganda)
- Communities felt motivated to continue using organic manure in their agricultural practices rather than synthetic fertilizers in order to preserve the soil texture, water quality, and human health. (Lesotho, South Africa, Zimbabwe)
- A conference was organized to review the traditional medicine policy (Uganda)
- A cultural spiritual retreat was organized for spiritual mediums from across the country (Uganda)
- An Elders Council was established to promote Indigenous knowledge system and the role of elders as the source of wisdom to ensure peace and security (Ethiopia).

Another major challenge is related to the struggle for Indigenous people's rights. The ALI training started from the point that raising awareness about one's rights is fundamental. However, only raising awareness (consciousness) and learning about one's rights is not sufficient: it should go together with preventing those rights from being violated. Often communities find themselves in situations where their rights have already been systematically ignored or abused. Communities then need to consider the kinds of actions they will take to seek compensation, create alternatives, and advocate for or against policies or practices that contribute to the realization and protection of those rights, as well as the risks that potential actions may entail (Fletcher and Guri 2017). To protect community's rights several organizations—e.g. in Ghana, Kenya and Zimbabwe—were involved in developing Biocultural Community Protocols (BCPs). However, often this did not go without challenges or even threats. For example, in Zimbabwe mining companies were not keeping their promises after taking away peoples' lands. Thus, reclaiming people's rights may lead to conflicts between communities and external companies, as well as within communities, e.g. between elders, youth, or government officials. For example, in Ethiopia elders went to court to reclaim their sacred natural site, which had been taken away by youth for agriculture. In Kenya, there was initial tension between community elders and local government officials: elders in pastoralists communities felt that they were not involved in part of the decision-making process on issues affecting the community, such as natural resource management. After discussing this with community and government officials, quarterly multi-stakeholder meetings were organised with local government and Indigenous leaders (Fletcher and Guri 2017). These examples show that differences in interests and power as well as views on community development can potentially lead to serious conflicts and tensions within communities and beyond. Moreover, it reflects how government officials—at local as well as national level—can support or withhold endogenous development in communities. In a way, this also affects the African Learning Institute itself: the importance of these trainings is not widely recognized and supported among government officials (in Ghana) and (international) funding organisations, which makes it difficult to find structural financial support for the training institute. However, endogenous development processes are long-term processes for organisations and communities, which require sustained resource allocations beyond one two-week educational program.

15.4 Exploring a Re-humanising Pedagogy for African Agrarian Philosophy

When we connect the above described ALI training on endogenous development and Indigenous knowledge with insights from Freire's humanising pedagogy and Ramose's call for *mothofatso* in human relations, seven themes come to the front that seem relevant for a re-humanising pedagogy on African agrarian philosophy. It should be noted that the list of themes is not exhaustive but should be seen as a start to explore a re-humanising pedagogy for African agrarian philosophy. Moreover, themes are not mutually exclusive, but are interrelated and may overlap.

15.4.1 An African Agrarian Philosophy with Memory

Through stories, rituals and ceremonies, such as the burning ritual, the ALI training acknowledged, heard and gave space to the deep and ongoing effects and pain of the colonial experience. These methods are important because they revealed that "the actual living experience of the continual interaction between the conquered and the conqueror, oppressor and oppressed, exploited and exploiter" (Ramose 2019: 61) is still present in our lives. This also counts for agriculture and agricultural education: while agriculture has been practiced from time immemorial on the African continent, the West primarily—though not only—continues to impose a 'modernization' paradigm on Africa's agriculture, thereby ignoring indigenous ways of knowing and doing (Boogaard 2021). When such epistemic injustices are internalized it may result in an inferiority complex or 'colonised mind' (Oelofson 2015), in which Western agricultural knowledge is considered superior over indigenous agriculture knowledge and practices. This is also reflected in one of the main challenges that organisations in the ALI training experienced when trying to implement insights from the training in communities: it is difficult to break through the hegemony of Western knowledge—which is still seen as the 'norm' or 'standard'. In this line, we concur with Ramose (2019: 71) that "a philosophy without memory cannot abolish epistemic and social injustice". This means that it is essential for a re-humanising pedagogy on African agrarian philosophy to pay specific attention to the history of agriculture in the African continent and the effects it has until today, including the colonial legacy of industrial agriculture (e.g. GAFF 2021) with unequal epistemic relations and prevailing Eurocentrism. We thus plead for an African agrarian philosophy *with* memory in order "to pursue an effective struggle for epistemic and social justice" (Ramose 2019: 70). This also means that teachers, philosophers, agricultural scientists and practitioners with a Western cultural and educational background—who want to work on a re-humanising pedagogy of African agrarian philosophy—should engage in critical self-reflection and actively work against prevailing Eurocentrism and epistemic superiority.

15.4.2 A Dialogical Student–Teacher Relation

The ALI training started from participant's knowledge, rather than bringing in external teachers: through the elders panel and storytelling participants were teachers and vice versa. This can be described as a dialogical student–teacher relation, which is essential for a humanising pedagogy (Del Carmen Salazar 2013; Oelofsen 2015). As Freire wrote: "Through dialogue, the teacher-of-the-students and the students-of-the-teacher cease to exist and a new term emerges: teacher-student with students-teachers. The teacher is no longer merely the-one-who-teaches, but one who is himself taught in dialogue with the students, who in turn while being taught also teach. They become jointly responsible for a process in which all grow." (Freire 2005: 80). Freire adds that in such a joint learning process epistemic curiosity—i.e. curiosity about the object of knowledge (Freire 2005)—is fundamental. According to Freire, a humanising pedagogy requires epistemic curiosity as well as a clear ethical and political commitment by teachers to transform oppressive social conditions and work towards the humanization for all people (Del Carmen Salazer 2013). In this line we see a parallel with Ramose's plea for a critical pan-epistemic orientation of teachers and students in which "the teacher and the student ought to engage in the pursuit of justice and peace in national and international relations." (Ramose 2016b: 547). For a pedagogy on African agrarian philosophy this means that a dialogical student–teacher relation—rather than traditional top-down teacher-student relations—is required, with a pan-epistemic orientation that includes an ethical and political commitment by teachers and students to work towards re-humanisation in the domain of agriculture and food.

15.4.3 The Value of Lived Experiences

Educational methods in the ALI training did not only take place in-class, but also outside the classroom. An important outside-of-the-classroom method was a ceremonial cultural and historical field visit, and an overnight stay in the communities—which had a deep impact on the participants as reflected in the evaluations: by working together on the land, preparing and sharing food, being part of people's daily routines, participants experienced how Indigenous knowledge contributed to biodiversity, nature conservation, agroecology, and community resilience. Freire emphasized the importance of lived experiences in a humanizing pedagogy, in the sense that learning and dialogues should be grounded in people's lived experiences and a curriculum should reflect realities of people's lives (Del Carmen Salazar 2013). For a pedagogy of African agrarian philosophy this means that the learning space is critical for the learning processes: what a person learns is informed by the space where it takes place. African agrarian philosophy has existed and been preserved for generations by knowledge holders like farmers and wise elders in communities. It has been taught, handed down, and dynamically reinvented in various ways within the lifeways of African peoples. So, a central question is what happens when Indigenous knowledge is taught in an academic setting, for example inside the building or a classroom of a university. From the above follows that trainings organized in ways that people can really connect with Indigenous knowledge holders, such as with community visits will be much more effective, because connecting to a territorial space creates deeper meaning and understanding to re-connect, re-appreciate, and re-vive Indigenous agricultural knowledge, practices and values.

15.4.4 Relationality with Human Beings and Mother Earth

The ALI training was based on endogenous development—i.e. development from within—as a way to re-connect with Indigenous knowledge and agro-ecological practices. In doing so, the ALI training started from an African cosmology that consists of the human, natural and spiritual worlds. These worlds are intertwined with different emphases at different times and in different situations. In an indigenous African worldview, "land, water, animals and plants are not just a production factor with economic significance", but "they have their place within the sanctity of nature" (Millar and Haverkort 2006: 23). In other words, Mother Earth—as the preserver of life—is sacred, which deserves ontological and normative commitment to respect and non-injure (Ramose 1999, 2004). Sacredness of Mother Earth does not mean that land, water, plants and animals cannot be used by humans, but it means there should be harmony between ethics of preservation and ethics of production (Kelbessa 2015). Humans are thus integral parts of nature, in which relationality is the main underlying principle (Ramose 1999; Chivaura 2007). The maxim "*umuntu ngumuntu nga bantu*" can be translated as "to be a human be-ing is to affirm one's humanity by recognizing the humanity of others and, on that basis, establish humane relations with them" (Ramose 2003a: 272). This maxim refers to the relationality of human beings with other human beings, while also encompassing relations with non-humans and nature (Eze 2017; Ramose 1999; Kelbessa 2015). From this *ubuntu*-maxim we learn that through our relationship with other human beings we become human—which requires a humane, respectful, and polite attitude towards other human beings (Ramose 2003a). With regard to a re-humanising pedagogy, this means the following: "Humanisation in the sense of *ubuntu* is, at core, about having a healthy collective self, and that involves having good relationships with others in the society" (Oelofsen 2015: 143). Such relationality between human beings also seems to be an important part of Freire's humanising pedagogy, in the sense that the aim of education should be—according to Freire (2005)—to become more human. Summarizing, when we bring experiences from the ALI training together with insights from Ramose and Freire, it becomes clear that relationality of human beings and Mother Earth is a central theme for a re-humanising pedagogy on African agrarian philosophy.

15.4.5 Intergenerational and Spiritual Methods of Education

The ALI training used a variety of educational methods, such as storytelling, a burning ceremony, visits to sacred natural sites, participation in community celebrations, and an Indigenous food exhibition. Informed by endogenous development, most of these educational methods were not new, but go back to traditional education methods that have been—and continue to be—used in communities. For example, Indigenous knowledge, values and practices have been transferred from generation to generation through rites of passage, festivals, ceremonies, community rituals, funerals, and totemic taboos. These intergenerational methods of education often are based on oral traditions, in which stories are central—as a way "to collect, deposit, analyze, store, and disseminate information" (Chilisa 2020: 193)—and which can take the form of a song, dance or poem. Through these methods, indigenous knowledge, values and practices on agriculture have been transferred intergenerationally.

While the importance of oral traditions in African philosophy has been widely recognized—in the sense that much wisdom and logical reasoning can be found in stories and proverbs (Gyekye 1987; Kimmerle 1997; Olúwolé 1997)—, the ALI training shows that it is also important to recognize oral traditions as specific educational methods. In addition to oral traditions, several educational methods in the ALI training had a strong emphasis on spirituality. For example, many of the "mystica" activities involved speaking of relationships with other living things and nature—such as discussing the nutritional, medicinal and spiritual value of plants—, which all helped in connecting to Mother Earth and the ancestral world. In doing so, the

ALI training put African spirituality into practice. African cosmology is a spiritual worldview through interdependent relations between the living, the living dead, and the yet to be born—together forming the visible and invisible worlds (Ramose 1999, 2004; Chilisa 2020). Despite the important relation between the visible and invisible worlds in African cosmology, the post-activities of the ALI training showed that exercises relating to the spiritual dimension were sometimes not fully understood nor accepted in communities. This was largely due to fact that Western knowledge was seen as the 'norm' or 'standard'. However, the imposition of a Western epistemology, ontology and religion on the African continent through Christian missionaries as well as colonial science and technology (Mavhunga 2017), is precisely the central cause that devalued African spirituality as something primitive and superstitious (Kimmerle 2006). Hence, the intellectual legacy of colonization is clearly present through ongoing epistemic suppression of indigenous and spiritual ways of knowing. For a pedagogy of African agrarian philosophy this means that it is important to recognize the relevance of African spirituality as an integral part of shaping relations with environments, especially in agriculture (Boogaard et al. 2023).

In this line, we follow Kimmerle who pleads for a re-appreciation of African spirituality (Kimmerle 2006). This means that it is important to include spiritual traditions as educational methods, such as meditative activities, as well as oral traditions, e.g. invite participants and/or wise elders to tell their stories. This also means that African agrarian philosophy has to be taught inter- and intra-generationally, through deeply rooted cultural practices and apprenticeship systems. At the same time, there is currently a scarcity of documentation of spiritual practices and indigenous intergenerational educational methods. Although it may seem contradictory to put oral and spiritual traditions in written form, such written documentation would help to make these indigenous intergenerational educational methods more widely available and as such contribute to further develop and implement a re-humanising pedagogy of African agrarian philosophy.

15.4.6 Unity Between Theory and Practice

The ALI training brought theory and practice together through an ongoing process of learning and experimentation by farmers in the fields. It combined theory with practical experiences. In doing so, the training content aimed to be relevant to present-day situations and challenges in communities. The combination of theory and practice is much in line with Freire's pedagogy, who wrote: "I never advocate either a theoretic elitism or a practice ungrounded in theory, but the unity between theory and practice." (Freire 2005: 19). At the same time, the unity between theory and practice brings at least two points of attention to the front for a re-humanising pedagogy of African agrarian philosophy. To start with, it raises a question about the relevance of theories on African agrarian philosophy: what is the relevance of such theories if these are not put into practice with knowledge holders such as farmers, wise elders and youth in communities?

In order to revive, reconnect and restore Indigenous knowledge, values and practices, there needs to be cooperation between these knowledge holders and academic philosophers, as well as other (agricultural) scholars, NGO's, governments, students, and community members. Secondly, Indigenous knowledge is not static, but adapting over time. So, when trying to unite theory and practice, it is important to recognize that this is a creative and dynamic learning process. This means that an endogenous development approach is not about going backwards and rejecting new ideas or innovations. Instead, it is about critically reflecting what values are maintained and adapted to deal with present-day challenges. While the ALI training puts significance on starting from within, external knowledge has not been rejected entirely. On the contrary, exogenous knowledge types—such as academic knowledge or new Western technologies—were considered appropriate for creating alternatives in perspectives. Yet, a main challenge is to try to relate and possibly integrate different perspectives and to recognize overlapping and diverging epistemologies (Millar 1996; Haverkort 2006; Ludwig and El-Hani 2020). While a re-humanising pedagogy for African agrarian philosophy aims explicitly to move away from Western hegemony and Eurocentrism, this does not mean that Western knowledge is rejected by definition. As Oelofsen (2015) explains: "Humanising pedagogy will therefore also encourage engagement with Western material, as long as it is with the understanding that Africa, and African students, have a valuable perspective to contribute alongside what is found in the Western canon" (Oelofsen 2015: 144).

15.4.7 Critical Consciousness About People's Rights

Through the ALI trainings, participants learned to appreciate the history and culture in their respective areas, which made them realize that this way of life and livelihood is valuable, and that they may want to continue living this way. However, they also realized that to maintain this livelihood they had to find out about laws and their rights, including the indigenous right to living one's lifeways with free, prior and informed consent (see e.g. UN Declaration on the Rights of Indigenous Peoples, United Nations 2007).

While the ALI training recognized from the start that raising critical consciousness about people's rights is fundamental, it was not a human-rights training imposed from outside: the determination to work for people's rights emerged through grounded experiences of wanting to continue living in away close to the land and the realization that these rights were threatened—or already violated—in various ways. The right to food (Ramose 2003b) and the right to land are intrinsically bounded with agriculture, and these rights may be under severe threat—if not violated—when people's livelihoods in communities are threatened. Thus, through the ALI trainings, participants developed a critical consciousness about their rights. Here, we see a parallel with '*conscientização*' in Freire's humanising pedagogy, that is: "the deepening of the attitude of awareness characteristic of all emergence" (Freire 2005: 109). However, the consequence of becoming aware that one's rights have been violated is to take action. As Freire describes: "Obviously, *conscientização* does not stop at the level of mere subjective perception of a situation, but through action prepares men for the struggle against the obstacles to their humanization." (Freire 2005: 119). In this line, many of the ALI post-training activities in communities aimed to support people in their struggles to protect or reclaim people's rights.

These struggles for people's rights do not go without challenges and risks. It is therefore helpful to support and learn from each other across the African continent. Hence, the ALI institute aims to be a pan-African community of mutual help. One of the recurring struggles lie in the fact that rights—such as the right to land—may be defined differently and function differently in a community compared to rights at the level of national or international law. The discrepancy between land rights and (inter)national laws goes back to the unjust wars of colonisation and deserves a separate assessment that goes beyond the scope of the current chapter (see e.g. Ramose 2016a). For a re-humanising pedagogy of African agrarian philosophy it is essential that people develop a grounded critical consciousness about their right to food, land and lifeways, while for educators it is important to realize that there is a consequence to such a pedagogy: to act upon critical consciousness.

15.5 Concluding Remarks

Based on insights from literature as well as practical experiences from the African Learning Institute, we have identified the following seven themes for a pedagogy of African agrarian philosophy: an African agrarian philosophy *with* memory; a dialogical student–teacher relation; the value of lived experiences; intergenerational and spiritual methods of education; relationality of human beings and Mother Earth; unity between theory and practice; critical consciousness about people's rights. These themes show how ALI provides a radically different way of doing agricultural education. One of the things that makes ALI trainings innovative is that—by using an endogenous development approach—it starts from African ontology and epistemology in which relationality of human beings and Mother Nature are central. Moreover, it puts a *re-humanising pedagogy* into *practice* in an African context by bringing together theoretical and practical elements that are essential in African agrarian philosophy, through a dialogical student–teacher approach. The importance of such agricultural training centres in present-day Africa cannot be underestimated.

African agrarian philosophy is without relevance if it remains a theoretical exercise that is not developed and put to use with farmers, wise elders and youth in the communities. Instead, when combined with a re-humanising pedagogy, African agrarian philosophy can contribute to reconnecting and reviving African agricultural knowledge, values and practices, which subsequently contributes to transforming African food system that heals the environment and produces enough food to feed communities.

Acknowledgements We are grateful to the Christensen Fund for their financial contribution to the project and to the participants of the COMPAS Africa Network for contributing to the ALI training in 2016. We also like to thank Mogobe Ramose for providing valuable feedback on this chapter.

Appendix

Participating organisations in the two-week ALI training in Techiman, Ghana (2016):

1. National Coalition of Farmers Associations (NACOFAG), The Gambia.
2. Jinukun, Benin.
3. Centre for Indigenous Knowledge and Organizational Development (CIKOD), Ghana.
4. WaterAid, Ghana.
5. Rural Women's Farmers Association (RUWFAG), Ghana.
6. Sirigu Women Potters Association (SWOPA), Ghana.
7. Widows and Orphans Movement (WOM), Ghana.
8. Abrono Organic Farmers Project (ABOFAP), Ghana.
9. MELCA, Ethiopia.
10. Institute for Culture and Ecology, Kenya.
11. Kivulini Trust, Kenya.
12. PROMETRA, Uganda.
13. Southern Africa Endogenous Development Program (SAEDP), South Africa/Lesotho/Zimbabwe.
14. Muonde Trust, Zimbabwe.
15. Women for Change, Zambia.

References

Apusigah, A.A. 2008. *Tullum: A Gendered African Wisdom with Possibilities for Development*. Ghana: University of Development Studies.

Bean, W. 2019. Towards an Abundant Life for All Through Transformative Adult Education in Castle. In *Seeds of Radical Education at the Coady Institute*, ed. D. Fletcher, and O. Gladkikh. Antigonish: People Development.

Béné, C., R.G. Wood, A. Newsham, and M. Davies. 2012. *Resilience: New Utopia or New Tyranny? Reflection About the Potentials and Limits of the Concept of Resilience in Relation to Vulnerability Reduction Programmes*, 1–61. IDS Working Papers. https://doi.org/10.1111/j.2040-0209.2012.00405.x.

Boogaard, B.K. 2019. The Relevance of Connecting Sustainable Agricultural Development with African Philosophy. *South African Journal of Philosophy* 273–286. https://doi.org/10.1080/02580136.2019.1648124.

Boogaard, B.K. 2021. Epistemic Injustice in Agricultural Development: Critical Reflections on a Livestock Development Project in Rural Mozambique. *Knowledge Management for*

Development Journal 16 (1): 28–54. https://www.km4djournal.org/index.php/km4dj/article/view/475.

Boogaard, B.K., D. Ludwig, B.Y. Guri, D. Banuoku. 2023. A Reconsideration of African Spirituality in Agricultural Development Projects: Traditional Ecological Knowledge from Dagara Elders in Koro, Ghana. In *Beauty in African Thought—Critique of the Western Idea of Development*, ed. Roothaan, Angela, Bolaji Bateye, Mahmoud Masaeli, and Louise Müller. Lexington Books. p175-196.

Chambers, R. 1994. The Origins and Practice of Participatory Rural Appraisal. *World Development* 22 (7): 953–969.

Chilisa, B. 2020. *Indigenous Research Methodologies*, 2nd ed. SAGE Publications.

Chivaura, Vimbai Gukwe. 2007. Hunhu/Ubuntu: A Sustainable Approach to Endogenous Development, Biocultural Diversity and Protection of the Environment in Africa. In *Endogenous Development and Bio-cultural Diversity*, ed. B. Haverkort, and S. Rist, 229–239. Leusden: ETC/Compas.

Crane, T.A. 2010. Of Models and Meanings: Cultural Resilience in Socio-ecological Systems. *Ecology and Society* 15 (4): 19. https://www.ecologyandsociety.org/vol15/iss4/art19/.

Dei, G.J.S. 2000. Rethinking the role of Indigenous knowledges in the academy. *International Journal of Inclusive Education* 4 (2): 111–132.

Dei, G.J.S. 2010. *Teaching Africa. Towards a Transgressive Pedagogy*, 9. Springer, Explorations of Educational Purpose 9.

Del Carmen Salazar, M. 2013. A Humanizing Pedagogy: Reinventing the Principles and Practice of Education as a Journey Toward Liberation. *Review of Research in Education* 37 (1): 121–148.

Eze, Michael Onyebuchi. 2017. Humanitatis-Eco (Eco-Humanism): An African Environmental Theory. In *The Palgrave Handbook of African Philosophy*, ed. Adeshina Afolayan, and Toyin Falola, 621–632. Springer.

Fletcher, D.G. 2017. *Illuminating Community Resilience from an Indigenous Perspective: Insights from a Qualitative Study with Dagara Communities of Lawra District, Ghana*. Ph.D. dissertation, University for Development Studies, Tamale, Ghana.

Fletcher, D., B. Guri. 2017. *Activities and Impacts Following the Techiman Course August 2016 to May 2017*. African Learning Institute (ALI) on Local and Indigenous Knowledges for Community Resilience, Ghana.

Fletcher, D., B. Guri, and P. Bansa. 2017. *Final Narrative Report April 2016 to August 2017*. African Learning Institute (ALI) on Local and Indigenous Knowledges for Community Resilience, Ghana.

Freire, P. 2005. *Pedagogy of the Oppressed*, 30th Anniversary Edition. Translated by Myra Bergman Ramos. New York: Continuum.

Global Alliance for the Future of Food. 2021. *The Politics of Knowledge: Understanding the Evidence for Agroecology, Regenerative Approaches, and Indigenous Foodways*. https://futureoffood.org/.

Gyekye, K. 1987. *An essay on African Philosophical Thought. The Akan Conceptual Scheme*. Cambridge: Cambridge University Press.

Haverkort, B. 2006. Discourses Within and Between Different Sciences. In *African Knowledges and Sciences: Understanding and Supporting the Ways of Knowing in Sub-Saharan Africa*, ed. D. Millar, S.B. Kendie, A.A. Apusigah, and B. Haverkort, 38–52. Barneveld: COMPAS, UDS Ghana, UCC Ghana.

Kelbessa, W. 2015. African Environmental Ethics, Indigenous Knowledge, and Environmental Challenges. *Environmental Ethics* 37 (4): 387–410.

Kimmerle, H. 1997. The Philosophical Text in the African Oral Tradition. The Opposition of Oral and Literate and the Politics of Difference. In *Philosophy and Democracy in Intercultural Perspective. Studies in Intercultural Philosophy*, 3rd ed, ed. H. Kimmerle, and F. M. Wimmer, 43–56. Amsterdam: Rodopi B.V.

Kimmerle, H. 2006. The World of Spirits and the Respect for Nature: Towards a New Appreciation of Animism. *The Journal for Transdisciplinary Research in Southern Africa* 2 (2): 249–263.

Kimmerle, H. 2016. Hegel's Eurocentric Concept of Philosophy. *Confluence: Journal of World Philosophies* 1: 99–117.

Ludwig, D., and C.N. El-Hani. 2020. Philosophy of Ethnobiology: Understanding Knowledge Integration and Its Limitations. *Journal of Ethnobiology* 40 (1): 3–20.

Mavhunga, Clapperton. 2017. *What Do Science, Technology, and Innovation Mean from Africa?* Cambridge, Massachusetts: The MIT Press.

Millar, D. 1996. *Footprints in the Mud. Reconstructing the Diversities in Rural. People's Learning Process*. Ph.D. thesis, Wageningen.

Millar, D. 2012. *"Our Sciences": Indigenous Knowledge Systems of Northern Ghana*. Indigenous Studies Department Collection.

Millar, D. 2014. Endogenous Development: Some Issues of Concern. *Development in Practice* 24 (5–6): 637–647.

Millar, D., and B. Haverkort. 2006. African Knowledges and Sciences: Exploring the Ways of Knowing of Sub-Saharan Africa. In *African Knowledges and Sciences: Understanding and Supporting the Ways of Knowing in Sub-Saharan Africa*, ed. D. Millar, S.B. Kendie, A.A. Apusigah, and B. Haverkort, 11–37. Barneveld: COMPAS, UDS Ghana, UCC Ghana.

Mungwini, P. 2017. 'African Know Thyself': Epistemic Injustice and the Quest for Liberative Knowledge. *International Journal of African Renaissance Studies—Multi-, Inter- and Transdisciplinarity* 12 (2): 5–18.

Oelofsen, R. 2015. Decolonisation of the African Mind and Intellectual Landscape. *Phronimon* 16 (2): 130–146.

Olúwolé, S.B. 1997. *Philosophy and Oral Tradition*. Lagos: African Research Konsultancy (ARK).

Ramose, M.B. 1999. *African Philosophy Through Ubuntu*. Harare: Mond Books Publishers.

Ramose, M.B. 2003a. The philosophy of Ubuntu and Ubuntu as a Philosophy. In *The African Philosophy Reader*, 2nd ed, ed. P.H. Coetzee, and A.P.J. Roux, 270–280. London: Routledge.

Ramose, M.B. 2003b. Justice and restitution in African political thought. In *The African Philosophy Reader*, 2nd ed, ed. P.H. Coetzee, and A.P.J. Roux, 541–588. London: Routledge.

Ramose, M.B. 2004. The Earth 'Mother' Metaphor: An African Perspective. In *Visions of Nature. Studies on the Theory of Gaia and Culture in Ancient and Modern Times*, ed. F. Elders, 203–206. Brussels: VU University Press.

Ramose, M.B. 2016a. "To Whom Does the Land Belong?" Mogobe Bernard Ramose Talks to Derek Hook. *Social Psychology* 50: 86–98. https://doi.org/10.17159/2309-8708/2016/n50a5.

Ramose, M.B. 2016b. Teacher and Student with a Critical Pan-Epistemic Orientation: An Ethical Necessity for Africanising the Educational Curriculum in Africa. *South African Journal of Philosophy* 35 (4): 546–555.

Ramose, M.B. 2019. A Philosophy Without Memory Cannot Abolish Slavery: On Epistemic Justice in South Africa. In *Debating African Philosophy. Perspectives on Identity, Decolonial Ethics and Comparative Philosophy*, ed. G. Hull, 60–72. New York: Routledge.

Ramose, M.B. 2020. Critique of Ramon Grosfoguel's 'The Epistemic Decolonial Turn'. *Alternation* 27 (1): 271–307. http://alternation.ukzn.ac.za/Files/articles/volume-27/14-Ramose-Fin.pdf.

Smith, L.T. 2001. *Decolonizing Methodologies: Research and Indigenous Peoples*. New York: Zed Books.

United Nations. 2007. *United Nations Declaration on the Rights of Indigenous Peoples*. Resolution Adopted by the General Assembly, 107th Plenary Meeting, 13 September 2007.

Birgit Boogaard (Ph.D.) works as lecturer and researcher at the Knowledge Technology and Innovation Group at Wageningen University, where she teaches courses on African philosophy and social justice. Her research interests include African philosophy, agriculture, international development, and epistemic justice. She has an interdisciplinary Ph.D. in Rural Sociology and Animal Science from Wageningen University. She was post-doctoral researcher at the International Livestock Research Institute, for which she lived two years in Mozambique.

Bernard Yangmaadome Guri is the founder and executive director of the Center for Indigenous Knowledge and Organizational Development (CIKOD) in Ghana. He has over 30 years of experience in community development. He holds an M.A. in Development Studies (Politics of Alternative development strategies) and a Post Graduate Diploma in Development Studies (Rural Policy and Project Planning) both from the Institute of Social Science, The Hague, and a B.Sc. (Agriculture) from the University of Cape Coast, Ghana. He is founding member of the Alliance for Food Sovereignty in Africa (AFSA), of which he was the chair.

Daniel Banuoku is a Deputy Executive Director of the Center for Indigenous Knowledge and Organizational Development (CIKOD) in Ghana. He has been directly responsible for the implementation of and coordination of the programs in the Northern sector of Ghana. He holds a Bachelor's degree in Integrated Development Studies with specialization in Environment and Natural Resources Management from the University for Development Studies. He obtained a certificate in Community-driven Health Impact Assessment from the Coady international Institute. He is currently a post graduate student of the University for Development Studies.

David Ludwig (Ph.D.) is an associate professor at the Knowledge Technology and Innovation Group at Wageningen University, where he focuses on global dimensions of science and on local knowledge about biocultural diversity. He coordinates the research program "Global Epistemologies and Ontologies" (GEOS) (https://www.geos-project.org/) at Wageningen University. His work has been published in journals such as Philosophy of Science, Public Understanding of Science, and Current Anthropology.

David Fletcher (Ph.D.) is an educator, researcher and consultant with Guyanese and Canadian ancestry. He works in the areas of community resilience, food sovereignty, racial-cultural justice, adult education and personal and social transformation. Over the past 35 years he has lived and worked in Nigeria, Ghana, Ethiopia, The Gambia and Canada and completed assignments in other African and Asian countries. His work has always been guided by an appreciation for local knowledges and indigenous wisdoms. David holds a Ph.D. (Endogenous Development) from the University for Development Studies, Ghana, and a Masters in Adult Education, STFX University, Canada. He is currently with the consulting network People Development (www.pdltd.net).

Chapter 16
African Endogenous Knowledge and Sustainable Development: Evolving an African Agrarian Philosophy

Alloy S. Ihuah

Abstract In Africa, the human person is the supreme force, the most powerful and dominant among all created beings. While this decreed power makes the lower beings subservient to humanity, it is only intended to be a source of harmony in the advancement of the hospitality and the joy of the human species. Today, however, the traditional lifestyles of Africans are threatened with virtual extinction by insensitive development over which the indigenous peoples have no participation. Africa has not only acquiesced a scientific technology that has developed its own laws far removed from the known system of nature, it has no self-limiting principle in terms of size, speed and violence to nature and humanity. Consequent upon this incongruity, the machine has replaced the human person thus degrading the ecology and resources necessary for harmonious living. This chapter argues that, African endogenous knowledge is strategic heritage resource to salvage the technologically abused environment for sustainable agro-ecological living. The chapter argues further that African endogenous knowledge system encodes environmental management and policies for resource development in Africa's complex and ageing ecological system. In doing this, we engage in conversation about current and potential land sovereignty work in African Indigenous communities. This will achieve food security and improved nutrition, promote sustainable agriculture and improved economic status of the African indigenous farmer. We conclude that, in seeking the whole from the units of the earth resources through cultivation of the soil, the African derives high credit balance of spiritual good, virtuous life, self-reliance, courage and moral integrity.

Keywords Africa Tiv Agriculture · Heritage resource · African Agrarianism · Endogenous knowledge · Environmental ethics · Sustainable development

A. S. Ihuah (✉)
Benue State University, Makurdi, Nigeria
e-mail: aihuah@bsum.edu.ng

M. J. Tosam and E. Masitera (eds.), *African Agrarian Philosophy*, The International Library of Environmental, Agricultural and Food Ethics 35,
https://doi.org/10.1007/978-3-031-43040-4_16

16.1 Introduction

African cultural traditions acknowledge the relevance of science and technology in sustainable agrarianism. For the African therefore, the empirical basis of knowledge has immediate practical results in agriculture, herbal medicine, crime prevention and remedy among others. Gyekye alludes to this when he says,

> our ancestors whose main occupation in farming knew of the systems of rotation of crops; they knew when to allow a piece of land to lie fallow for a while; they had some knowledge of the technology of food processing and preservation; and there is a good deal of evidence about their knowledge of medial potentialities of herbs and plants- the main source of their health care delivery system long before the introduction of western Medicine. (Even today, there are countless testimonies of people who have received cures from traditional healers where the application of western therapeutics could not cope). (1997: 26, 27)

Scientific studies have begun to recognize the positive role that endogenous knowledge plays in the formulation and implementation of sustainable development policies and projects in developing countries. World Commission on Environment and Development (1987: 12) has acknowledged this fact when it states that,

> Some traditional lifestyles are threatened with virtual extinction by insensitive development over which the indigenous people have no participation. Their traditional rights should be recognized and they should be given a more decisive voice in formulating policies about resource development in their areas (particularly in complex rain forest, mountain and dry land ecosystems).

While it is clear that the concept of traditional ecological knowledge does not exclusively belong to indigenous peoples of Africa, its agricultural script has been so altered by foreign policies on agro-ecological resource management thus making research on African indigenous ecological knowledge relevant. The following reasons have been advanced.

i. The long-term generation and transmission of knowledge of the local ecosystem offers a unique historical perspective into indigenous risk adjustment options.
ii. Modern scientists involved in the management and conservation of areas that may be ecologically fragile or marginal, or that contain genetically important plant or animal biodiversity, may benefit greatly from alternative knowledge.
iii. That African endogenous knowledge does not separate humans from natural environment; that the world is interconnected and interdependent.

This chapter sets out to distinguish between the abstract tradition of the west (scientific ecology) and historical tradition signposted by traditional systems of knowledge that is encoded in rituals and in cultural practices of everyday life. The object of this study is to show among other things that this traditional Ecological knowledge system possessed by people outside western scientific tradition has implication for African agrarianism. That in Africa, this knowledge system not only contributes to conservation of biodiversity, it promotes agro-ecological resource development, use and management. The chapter acknowledges the relevance of traditional knowledge and management systems and the use of local ecological knowledge in the cultivation of the soil and goes further to interrogate African concept of development from the perspective of agrarianism. It argues further that Africa is yet to appreciate the repertoire of its knowledge systems and heritage resource and the imperative of foregrounding its agrarian development on these knowledge systems. We argue for a recalibration of the agrarian system on the continent to allow for a critical appraisal of African Knowledge Systems and how such systems can be leveraged for development.

16.2 Conceptual Labyrinth

i. **African Agrarian Philosophy (AAP)**

Agrarianism is a heritage resource that goes back to the ancestors of the living great apes around 7 Million years ago when all humans on earth fed themselves exclusively by hunting wild animals and gathering wild plants. This ancient practice changed when people turned to what is termed food production, domesticating wild animals and plants and eating livestock and crops (Diamond 2017). In Africa today, this ancestral tradition has transformed humanity and social conditions making plant and animal domestication much more food and hence much denser human population consequently, there is food surplus, development of settled, politically centralized, socially stratified, economically complex, technologically innovative societies (Diamond 2017: 88).

Today therefore, Africans consume food that they produced themselves or that which is produced by other nations of the world. It is this historical dynamics that Africans have developed the agrarian culture, hence evolved an agro-ecological practices and food systems. As the philosophical ideas, African agrarian philosophy belongs to a political and social philosophy that relates to the ownership and use of land for farming, or to the part of a society or economy that is tied to agriculture (Merriam 2020; Collins 2020). It is a system of ideas relating to the ownership and use of land, especially farmland, or relating to the part of a society or economy that is concerned with agriculture. Thus, agrarian philosophy is a discipline devoted to the systematic critique of the philosophical frameworks (or ethical world views) that are the foundation for decisions regarding agriculture (Taliaferro and Carpenter 2010). This system of ideas regulates agricultural practices and food systems as well as advocates for rights and activities of farmers in urban and rural communities. It should be noted that there are many schools of thought within agrarianism though a recurring feature of agrarians has been a commitment to supporting and protecting the rights of small farmers and poor peasants against the wealthy in society. In the African parlance, local agrarian practice values rural society as superior to urban society and the independent farmer as superior to the paid worker and stresses the superiority of a simpler rural life as opposed to the complexity of city life.

ii. **Endogenous Knowledge Systems (AEKS)**

African Knowledge Systems are a highly contested area of knowledge in terms of terminology, location, identification and significance. Thus, African Endogenous Knowledge Systems (AEKS) is variously referred to as Local Knowledge Systems (LKS), Traditional Knowledge Systems (TKS) or Traditional Ecological Knowledge (TEKS). Specific definitions also differ depending on the different authorities. This essay shall use endogenous as umbrella references to ways of knowing that are traditional and encapsulated with local beliefs, practices, customs and world views that are helpful in managing environmental, spiritual and socio economic challenges. Endogenous Knowledge thus encapsulates traditional folk knowledge of plants and animals, belief systems, cultural and historical artifacts that supports human existential conditions.

The International Council of Science (2002: 3) defines Traditional Knowledge as 'cumulative knowledge, know how, practices and representations maintained and developed by people with extended histories of interaction with natural environments … encompassing languages, naming and classification systems, resource use practices, ritual, spirituality and world views' what this means is that, endogenous Knowledge factors in cultural experiences, epistemologies and empiricisms implicated in ecology, subsistence systems, medical practices, music, oral traditions and technology. This explains the conceptualization of the term by Eyong (2007: 122) to include a set of interactions between economic, ecological, political, and social, environments within a group or groups with a strong identify, drawing existence from local resources through patterned behaviours that are transmitted from generation to generations to cope with change. Houtondji (1977) has however expanded the concept to entail a corpus of elaborate knowledge in oral cultures. He rejects attempts to qualify AEKS with 'traditional' and 'indigenous' in favour of endogenous which he argues is more dynamic and less derogatory. In some, endogenous knowledge encodes the philosophical foundations that undergird physical and spiritual development in the different cultures on the continent (see Gundu 2010). Such knowledge systems are created, decoded and protected through components like language, heritage resource, folklore, cultural practices and cultural resource use and are distinguishable from western knowledge systems in five ways namely,

(a) It is transmitted through language and is also collectively owned. This distinguishes it from Western ways of knowing where collation and transmission is through writing. Western Knowledge Systems are also 'personal' commodities that are protected for personal and corporate gain.
(b) It has a characteristic attachment to particular places as opposed to the 'universality' of science. As multiple knowledge systems, they can best be understood within their cultural context.
(c) Compared to Western ways of knowing, African Knowledge systems are fore grounded on a holistic worldview where knowledge intertwines different facets of the cosmos to interconnect living things, the dead and the landscape through obligations and responsibilities.

(d) It also uses oral traditions and demonstration for preservation instead of documentation (see Gorjestani 2004).
(e) It also has an inward focus on local conditions around health, security, environment, economy, survival and religion.

It is thus safe to state here that, endogenous knowledge is a philosophical exploration aimed at a salutary initiative in the quest for achieving the much needed epistemic autonomy in all spheres of traditional African community. In essence, African endogenous knowledge systems are apparently the building blocks of human development. This type of traditional knowledge is acquired by the indigenous people over hundreds of years through direct experience and contact with the environment by means of indigenous stewardship method (ISM) through the physical, spiritual, mental, emotional, and intuitive relationship of indigenous peoples with all aspects and elements of their environment (Whyte 2013).

iii. **Sustainable Development**

As a concept, development is often times misunderstood by even the most sophisticated minds. It is most commonly conceived as a process of growing larger, fuller, more mature or becoming organized; thereby getting better results through better methods. It conceptual understanding includes economic growth, industrialization, progress of technology and Gross National Product (GNP) or Gross Domestic Product (GDP). Evandro Agazzi, thus defines development as "a set of interconnected changes, which are goal-oriented and produce a certain global result that is positive". He states,

> ...when a process is originated or promoted by man, a consideration of the intended goal becomes entirely obvious, and in this case it seems that with a positive connotation only if a process promoted is expected to lead to "good" results (Agazzi 1993: 31).

It follows, from the above, that, "development" as a concept contains an implicit teleological flavour. As a set of interconnected changes which are goal-oriented, and which produce a certain result, which is good, "development" is said to have a value-side added to its general idea. It refers to a harmonious growth of a multi-dimensional complex structure, which realize an intrinsic dynamism, a plan, a kind of ought to be. Understandably, every development has some structure of growth but not every structure of growth is development. This mean that we must be conscious of the different significations that the growth of each one of the numbers of values under considerations reveals.

It is very clear that the concept of development includes purpose and meaning in life, which means the possibility of displaying man's potentialities at the different levels of self-realization. It means fellow-feeding and respect for other human beings within the totality of their beings. This entails a progressive economic and social engineering of human society through maintaining the security of livelihood for all peoples and by enabling them to meet their most urgent needs, together with a quality of life in accordance with their dignity and well-being, without compromising the ability of future generations to do likewise (Sands 1993: 102). In relation to agro-ecology development relates to the wise management of land resources with the intent at enriching humanity and enhancing sustainable human well-being for the greater good of humanity. It suffices to say here that "development" is all about the human person in a given social milieu; of his rights, his essential desires; how much of them have been satisfied, of his entire quality of life, his entire security of livelihood etc. It is all about humanism, a concept which in itself needs clarification.

16.3 African Ecological Knowledge System

The ecosystem view of many indigenous African societies is reflected in the following traditional management practices which encompass individual and community wisdom and skills (Atteh 1989). In our indigenous ecological knowledge system, this knowledge system finds expression in what has come to be called Indigenous Ecological Knowledge (IEK). It refers to the knowledge base acquired by indigenous and local peoples over many hundreds of years through direct contact with the environment. It includes an intimate and detailed knowledge of plants, animals, and natural phenomena, the development and use of appropriate technologies for hunting, fishing, trapping, agriculture, and forestry, and a holistic knowledge, or "world view" which parallels the scientific discipline of ecology. It further includes knowledge of indigenous soil taxonomies, indigenous knowledge for potential use of local plants and forest products, and animal behavior and acquired hunting skills, local knowledge of important tree species for agro-forestry, firewood, integrated pest management, the control of soil erosion and soil fertility, and fodder management, indigenous agronomic practices such as terracing, contour bonding, fallowing, organic fertilizer application, crop-rotation and multi-cropping as well as indigenous soil and water conservation and anti-desertification practices.

In Nigeria for instance, Tiv farmers played an important role as agents of social and economic change through the dissemination of improved technology and techniques of food production. Varvar (2011) compliment this point when he says that Tiv migrants were agents of change and innovation in the traditional system of agriculture. Such techniques and innovations were evolved to cope with the agricultural realities of the areas intent at good land management. While not engaging in the unending debate about what constitute traditional as opposed to the western, indigenous ecological knowledge can very well be understood to be a cumulative body of knowledge and beliefs, handed down through generations by cultural transmission, about the relationship of living beings (including humans) with one another and with their environment.

For the Tiv therefore, God's benevolence at creation is meant that the creator did not want us to starve or die of hunger. What this means is that humanity is under obligation to work as a team for greater production and acquisition of wealth. It can thus be taken be an attribute of societies with historical continuity in resource use practices. It is here that humanity cannot be distinguished from endogenous knowledge system. Thus, agro-ecological epistemology acknowledges that land, inclusive of the soil, water, air and other resources is one of the most fundamental resources that support the essence and existence of the human person.

The development of endogenous knowledge, or the change in the application of acquired ecological knowledge, is predicated upon conscious efforts by both individuals and the local community. Although most innovations in the application of indigenous knowledge are typically regarded with caution by traditional societies in Africa, under special circumstances they may be readily accepted by the entire community. Such circumstances may occur when local ecological and climatic conditions have dramatically changed, or have become stressed to the point of threatening collective and individual survival, Famine caused by drought, deforestation, desertification or topsoil erosion, and declining productivity are some circumstances which may encourage or necessitate the acceptance of innovation.

In some of Africa's most ecologically fragile and marginalized regions, knowledge of the local ecosystem simply means survival. As so much is at stake in changing traditional natural resource management practices, any proposed change is usually based on an informal evaluation and consultation process among key community members (usually a peer group involving elders). By sharing and comparing knowledge of key indicators that describe ecological responses to change or the prediction of environmental trends, the community can weigh the long and short-term costs and benefits of change related to any new innovation or application of local ecological Management systems. Adande has confirmed this much when he says,

> the means and methods used by Blacksmiths in Burkina Faso to detect or identify underground veins and determine the quality of the *Ore*. According to him the methods includes, using long iron bar, examining certain visual, tactile or taste clues and using certain plants. This is basically founded on the indigenous belief that, *Ore* were like tremendous snakes moving about underground and that a person needed special skills to locate them. (Adande 1997: 79)

In his Traditional Ecological Knowledge in Perspective, Berkes (1993) records the practical significance of indigenous ecological knowledge. According to him, the preservation of Indigenous Ecological Knowledge (IEK) is important for social and cultural reasons. IEK is a tangible aspect of a way of life that may be considered valuable. For the rest of the world, there are also tangible and practical reasons why IEK is so important, quite apart from the ethical imperative of preserving cultural diversity. This has been the set mission of the International Union for Conservation of Nature. (IUCN); "to influence, encourage and assist societies throughout the world to conserve the integrity and diversity of nature and to ensure that any use of natural resources is equitable and ecologically sustainable" (https://www.iucn.org, IUCN 1986) The union's programme on Traditional Knowledge for Conservation chronicles a wide range of imperatives that is African for our consideration as follows:

Traditional knowledge for new biological and ecological insights: New scientific knowledge can be derived from perceptive investigations of African environmental knowledge systems, as in the case of life cycles of tropical reef fish (Johannes 1981).

Traditional knowledge for resource management: Much traditional knowledge is relevant for contemporary natural resource management, in such areas as wetlands.

"Rules of thumb" developed by traditional African resource managers and enforced by social and cultural means, are in many ways as good as Western scientific prescriptions.

Traditional knowledge for protected areas and for conservation education: Protected areas may be set up so as to allow resident communities to continue their traditional lifestyles, with the benefits of conservation accruing to them especially where the local community jointly manages such a protected area, the use of traditional knowledge for conservation education is likely to be very effective.

Traditional knowledge for development planning: The use of traditional knowledge may benefit development agencies in providing more realistic evaluations of environment and natural production systems. Involvement of the local people in the planning process improves the chance of success of development (Warren et al. 1995).

Traditional knowledge for environmental assessment: People who are dependent on local resources for their livelihood are often able to assess the true costs and benefits of development better than any evaluator coming from the outside. Their time-tested, in-depth knowledge of the local area is, in any case, an essential part of any impact assessment.

In these and through local knowledge, Africans are capacitated and enhanced to appreciate their cultures and life ways in which such knowledge system can advance their beingness and so, used as a tool for qualitative social changes in local communities. Nakashima represents a good example here when he says that the ecological knowledge of the Northern Canadian indigenous peoples provides insight into the life of the people of the community (Berkes 1993). Thus, ethno-science as made available through indigenous knowledge has paved the way for the acceptability of the validity of traditional knowledge in a variety of fields. Traditional African ways of knowing started to receive currency in several disciplines, including ecology. Various works showed that many indigenous groups in diverse geographical areas from the Arctic to the Amazon had their own systems of managing resources (Posey 1985). In Africa, the feasibility of applying Indigenous Ecological Knowledge to contemporary resource management problems was gradually recognized and used in agricultural practices in traditional rural areas in Africa. This has been the story of African Agrarian philosophy that is understood as agriculture and food systems that is built on Moral Ethics, Creativity, and Innovation, and which, unequivocally, identifies with the needs and aspirations of Africans. As stated in Our Common Future:

> Tribal and indigenous peoples "... lifestyles can offer modern societies many lessons in the management of resources in complex forest, mountain and dry land ecosystem (WCED 1987: 12). These communities are the repositories of vast accumulations of traditional knowledge and experience that link humanity with its ancient origins. Their disappearance is a loss for the larger society, which could learn a great deal from their traditional skills in sustainably managing very complex ecological systems. (WCED 1987: 114–115)

Indigenous Knowledge among the Coast Sami on fishing, hunting and gathering is here acknowledged. The economic adaptation of the Coast Sami is based on a wide range of ecological knowledge. Firstly, definitions of which components of nature are "resources" are based on such knowledge. Secondly, there is knowledge about how these resources can be utilized, and thirdly, there is knowledge about ecosystem functions, relations between species and sustainability of different resources. The moral element can be found in norms and unwritten rules about resource use which are more readily availed through indigenous wisdom passed from generation to generation. Berkes (2018) explores the importance of this local and indigenous knowledge as a complement to scientific ecology, and its cultural and political significance for indigenous groups around the world, and asks how we can learn from this knowledge and ways of knowing.

16.4 African Endogenous Knowledge and Modern Scientific Knowledge

African endogenous knowledge and modern scientific knowledge are similar and dissimilar. Both are based on an accumulation of observations though, endogenous knowledge differs from western science in some fundamental ways. Articulating the distinction between these two traditions of thought, Claude Levi-Straus posits that, "these two ways of knowing are two parallel modes of acquiring knowledge about the universe; the two sciences were fundamentally distinct in that the physical world is approached from opposite ends in the two cases: one is supremely concrete, the other supremely abstract" (Claude Levi-Straus 1962: 269). In a similar vein, the philosopher Paul Feyerabend (1987) distinguished between the two traditions of thought by stating that "abstract traditions (to which scientific ecology belongs) and the historical traditions, which include western science, knowledge that often becomes encoded in rituals and in the cultural practices of everyday life". Other scholars like Fikret Berkes, Johan Colding, and Carl Folke (2000: 1251), and Agrawal (1995a, b), have however cautioned against overdramatizing the difference between Western science and traditional knowledge and questioned if the dichotomy is real.

However, there are no doubt fundamental differences between the two knowledge paradigms. In Africa, the method of acquisition of traditional knowledge is ethnocentrically based. Such is why African Knowledge paradigms are characterized by an old African proverb which states "when a knowledgeable old person dies, a whole library disappears." As practitioners, guardians and educators of indigenous knowledge, the death of key elders (along with the current disinterest of youth to learn traditional ways and languages) can severely limit and threaten existing sustainable livelihoods. Unlike the documented scientific system, much of the remaining traditional ecological knowledge in Africa exists only in oral form, passed on from knowledgeable individuals through shared practice and story-telling.

Endogenous knowledge systems were altered and disrupted in Africa during the colonial period. This disruption is currently perpetuated by the inequitable north–south political and economic system, where endogenous knowledge systems are often ignored, under-valued or replaced by colonial, state practices. An important question facing development organizations that are interested in learning from indigenous knowledge is how to evaluate this alternative body of knowledge. Despite the inherent differences between traditional and scientific knowledge systems, innovative mechanisms are being sought by scientists such as anthropologists and development planners, to integrate both systems effectively in order to facilitate sustainable natural resource management planning. To achieve this, it is necessary to document, and consequently gain credibility and respect for the existing body of indigenous knowledge for the four agro-ecological regions in Africa (i.e., humid equatorial/ coastal lowlands, sub-humid).

An understanding of African endogenous knowledge systems within the cultural framework can help a development planner to understand more fully the dynamics of the local ecosystem. This approach can help establish a more flexible atmosphere through joint cooperation between the development planner and the affected indigenous community. For example, indigenous ecological knowledge may be utilized to suggest project site alternatives or mitigating measures which could help avoid or reduce inadvertent long and short-term damage to the ecosystem and traditional culture. As well, indigenous management practices and appropriate technology innovations that are implemented in partnership with development organizations and indigenous societies can also be adapted to help solve ecological problems faced by other societies in similar agro-ecosystems. It is pertinent to argue from hindsight that the utilization of local indigenous knowledge helps in protecting and sustainably utilizing the natural resource management strategies in the maintenance of biological diversity, biological and crop pest control strategies, recycling and fixation of soil nutrients, strategies to conserve soil and water.

16.5 A Maze of African Agrarianism

It was Gyegye who said that, "growth of knowledge, increases human power, but human power can either be positive or negative and that, scientific research can thus either increase a man's power to cause pain or misery to others or promote his ability to improve human welfare. In Africa, land provides food, shelter, herbs for medicines and water for existential purposes. This means therefore that humanity must consciously act in the fulfillment of its sustainable existential needs. In what follows, we discuss sustainable ways of African ethno-local agrarian practice in some select African ethnic groups.

(a) **Indigenous Fertilizer**

Since the 1960s scientists have recognized the validity of the indigenous bush-fallow system associated with shifting cultivation or slash-and-bum agriculture. Agricultural experts and extension workers have since developed a low-cost and labour intensive farming system called alley cropping, an adapted technique which capitalizes on the beneficial attributes of bush fallow, yet overcomes some of its limitations (Lai 1990).

With alley cropping, food crops are grown in wide rows that alternate with hedgerows of nutrient-producing trees and shrubs (for example, *Leucaena* and *Acaciaalbida*). The hedgerows are pruned periodically, and the nitrogen rich material is returned to soil as mulch, which inhibits weed growth and retains soil moisture. The hedges are usually planted along the contours of sloping land in order to act like terraces by decreasing water runoff velocity and subsequent soil erosion.

Between 1984–1988, the International Livestock Center for Africa (ILCA) and the International Institute of Tropical Agriculture (IITA) were involved in various on-farm research projects that introduce *alley cropping* to indigenous Nigerian farmers, particularly women. This was encouraged through the use of local theater and songs for promoting ecological knowledge and resource management skills. This innovative technique fund useful practice in African agrarian projects by overcoming not only the suspicion and hesitance of local farmers but also offering them total independence and self-sufficiency in food production.

The Tiv of Nigeria is not unaware of the fact that some unwholesome human activities negatively affect plants and animal life. Gaseous emissions and excessive production of greenhouse gases, carbon monoxide or methane and small droplets of acid in the atmosphere (acid rain) constitute a nuisance to the biotic components of the ecosystem. Knowledge of traditional science and technology for example has informed the adoption of self-purification method of the natural environment by the local Tiv population. Agricultural wastes like yam and cassava peels, and banana leaves are dumped into the village forest and fallowed farmland to manure the land. This waste is then converted into decomposing waste through a process of biodegradation to manure crops and plants for high productivity. Paul Bohannan argues similarly that, the fallow practice among the Tiv is intended to revitalize the land for higher productivity. In his words,

> the fallow land that is cleared for planting of fallow crops, grass and small trees are allowed to dry and burned. The ashes, then, are mixed with the soil when mounds or ridges are made … some Tiv say this increases the yield. Others say that burning the grass and bush is the easiest way to get rid of it, adding that if you wanted it for fertilizer you would bury it rather than burn it. (1969: 19)

Tiv agrarian practice is intensive though, soil management to maintain fertility is a priority hence aa deliberate indigenous manuring approach is adopted to keep the farm field fertile for successive phases of crop production with little or no intervening fallow period. This explains why the African gives the earth (land) the symbol of motherhood that evokes warmth, security, fruitfulness and protectiveness that are a value in life in general. When the African speaks of mother therefore, it is not in the Nietzschean tradition that connotes her inferior status as the inferior quarter and not the better half of man (Oscar Levy 1906: 55), but as a being that inspires almost all moral as well as deep passions in man. In managing the land, the African seeks the recreate humanity, to protect it to serve humanity sustainably.

(b) **Indigenous Pharmacopoeia**

The human environment, it can be said is encyclopedic in the sense that it harbours every secret of human knowledge intent at sustainability. Thus, human environment from the earliest antiquity opened up itself for human appropriation, to treat the sick as well as to communicate supernatural forces. Thus, the pharmacopoeia located in our human environment for our purpose can be said to include, fruits, seeds, leaves, rhizomes, barbs, roots and minerals extracted from the earth, and shells and the remains of animals, vegetables as well as vertebrates and invertebrates (Souza 1997: 191). In Africa, an interface of magico-medical and ritual incantations has the maximum potency of balancing the environment as they help the individuals psychologically ready and receptive, whether they are patients expecting to be cured, or simply individuals seeking to be told their fortunes. In so doing, use the various organic and inorganic ingredients culled from the environment to heal rather than kill the biotic life. In Africa therefore, knowledge of medicinal elements in plants is a universal distribution in Africa. Every family in Africa can boast of at least one family member who can identify medicinal plant seed or fruit.

(c) **Pastoralism**

This United Nations Research Institute for Social Development (UNRISD) case-study investigates some of the impacts on the traditional land management practice of the Barabaig, a semi-nomadic pastoral group in Tanzania, imposed by a large-scale agricultural development scheme.

Over many generations, the Barabaig have learned to sustainably exploit various foraging regimes based on sophisticated seasonal grazing rotations. The forage regime of most importance to the Barabaig is *muhajega*, a highly nutritious mix of grasses and herbs which grow on fertile soils that collect in depressions on the Basotu plains in Hanang district. The muhajega is high valued by the Barabaig for its capacity to produce high milk yields and stimulate cattle growth, as well as to improve the recuperative powers of livestock suffering ill health from stresses involved with dry season and droughts (Lane 1990).

In this sub-Saharan region of Africa, the availability of water is the most limiting factor in the sustainable use of the common property. The Barabaig recognize that their use of land is limited to the right of usufruct, which permits the use of common land only when it is not denuded beyond recovery or when other users are not disadvantaged (Lane 1990). Partly in response to an expected increase in the demand for wheat in Tanzania and the inherent fertility of the muhajega,' the Tanzanian government appropriated large tracts of land, including much of the fertile Basotu plains to implement a large-scale and controversial foreign-aid wheat scheme called the Tanzania Canada Wheat Program (TCWP) (Lane 1990). Much of the rationale involved in the appropriation of traditional Barabaig muhajega was the appearance to development planners and scientists that Barabaig land was often left vacant or "lying idle" (Young 1983). Lane (1990) assumes that these descriptions probably meant that the land was perceived as underutilized and that it could be used for more productive purposes.

In reality, this is representative of Barabaig traditional knowledge of the dynamics of seasonal grazing regimes and the need to let the ecosystem recover through fallow periods. The Barabaig have learned that, to make efficient use of natural resources, access to grazing needs to be controlled to prevent exploitation past the ecosystem's carrying capacity. All Barabaig are subject to strict and complex restrictions developed and enforced by a hierarchy of Barabaig jural institutions that control the use of land, interpret customary rule, and adjudicate in rare conflicts over rights and duties (Lane 1990).

It has been reported that a powerful women's council threatened those involved in the illegal farming with a curse. Threatening a curse is effective because all Barabaig believe that a curse will bring ruin to people's lives (Lane 1990). This helps the Barabaig to protect and sustain all common resources for equitable benefit. For example, although surface water is universally accessible to all Barabaig, routes to and from water sources are not to be restricted by homestead construction, and shared water sources must not be overly used, diverted or contaminated (Lane 1990).

Along with some of the adverse erosion such as gullies and sheet erosion resulting from ecologically inappropriate mechanized mono-cropping of wheat on the Basotu plains, and increased Barabaig reliance on the remaining forage regimes (i.e., unsustainable grazing rotations due to excessive grazing demand and hoof traffic), the overall carrying capacity of the forage ecosystems has been significantly lowered. This has serious implications for both the Barabaig people and the long-term fertility of the Hanang plains. Existing traditional common land tenure systems should, in future, be recognized for their efficient or sustainable land-use regimes based on accumulated indigenous knowledge and local culture.

(d) **Indigenous Bio-pesticides**

Although traditional pest control systems were once widely used in tropical countries, their use has been severely disrupted by the introduction of modern agro-chemicals. This dependence on expensive modern pesticides, apart from posing a potential threat to the health of the poor traditional farmer, is often poisonous to the local ecosystem. Many reported cases of human poisoning by Gamalin 20 (Lindanne) at University College Hospital Ibadan attest to the harmful practice of Agro-chemicals. Similarly, 20 public health field workers were poisoned by Malathion in Ondo State of Nigeria. It was also reported that a family of 5 people died after eating meals contaminated with pesticides (Heeds 1991).

The earliest known mention of poisonous plants having bio-pesticide properties is found in the Indian Rig Veda (2000 B.C.). Today, there are some 1600 plant species which have been reported to possess such properties. The neem tree *Azadirachtaindica* is one of the most promising (Hoddy 1991). The neem tree, or Indian lilac, is a hardy and fast-growing deciduous tree which is drought and salt tolerant. It can be grown on marginal soils with low fertility due to its powerful root system which can extract nutrients from deep layers of badly leached and sandy soils (Heeds 1991).

In Africa, indigenous farmers have known about the insecticidal properties of the neem tree for centuries, In Nigeria and other parts of Africa, farmers have long observed the immunity of its leaves to desert locust attack (Emsley 1991). It is also an effective antibiotic for fever and other related ailments. Although not as powerful as synthetic pesticides, the neem extract ingredients, which makes it difficult for any insect pest to develop a resistance to them all (Hoddy 1991).

Heeds (1991) reports that, some indigenous farmers in Africa are using scientific assistance to develop a neem spray made from the seeds of the fruit. It works as a repellent and anti-feedant to many chewing and sucking insect pests in the larva or adult stages, including desert and migratory locusts, rice and maize borers, pulse beetles, and rice weevils. It also upsets the insect's hormone balance so that it becomes permanently incapacitated.

In a seminar organized jointly by the Natural Resources Institute (NRI) of British Overseas Development Administration, African farmers were availed of the opportunity of exchanging views on ways of reducing crop losses due to pests. The NRI, working on the Mali Millet Project, described how indigenous farmers in north-western Mali placed leaves of the neem tree under the millet heads when they lay them on the ground to dry. This practice discourages insect infestation (Pickstock 1992). A project funded by USAID recently brought together a team of entomologists and social scientists from Niger and the University of Minnesota to promote the exchange of indigenous knowledge on the uses of neem products in improving the sustainability of traditional agriculture in Niger (Warren and Pinkston 1998). Similarly, the Tiv of Nigeria has developed a formula for preserving grains through the use of pepper and neem leaves (Ihuah 2013).

Emsley (1991) is also on record as saying that, chemists in 1988 determined the chemical structure of the neem tree extract, *azadirachtin*. Currently, over a dozen companies inindustrialized countries are working on commercial neem products. In 1983, the American Environmental Protection Agency registered a commercial neem pesticide for marketing under the name "*Margosan-O*" (Hoddy 1991). There are many other properties in Africa that qualify for similar recognition. It 1 s thus concluded that the challenge of Africa and Africans is to discover and patent a chemically modified version of *azadirachtin* that is stable and as effective as naturally occurring neem. This is the challenge of sustainability and development that makes a contribution to world science and food security.

16.6 Tiv (Nigerian) Agrarian Practices

Africans are indigenous farmers with Yams (*discorea* spp.), millet (*penisetum* spp.), and guinea corn (*sorghum* spp.) as their main crops. The Tiv of Nigeria for example have one word for Farms and produce (*Yiagh)* to represent their ideas of wellbeing, economy and work. In this word is found the meaning of subsistence in its broadest sense; food in the field, in the granary, in the pot, and in the belly; the farms that produce the food; the work that is required to plant, tend, and harvest; and the whole range of activities that centre around physical nourishment and comfort (Paul and Laura Bohannan 1968: 39).

For many Africans, cultivation of the soil not only an equivalent of sustainable living, it creates harmony between humanity and nature. This occupation offers for them total independence and self-sufficiency as opposed to the busy technological urban life and capitalism that destroys inter-dependence and dignity of the human person. Furthermore, the technological urban life fosters vice and weakens the social system. This explains why agrarian practice rests on solid radar in African communities. Besides, it brings order and a sense of identity, historical and religious tradition, a feeling of belonging to a concrete family, place, and clan, which are psychologically and culturally beneficial. This instantiates harmony within his life and in turn, checks the encroachments of a fragmented alienated modern society. This is the practice that has been lived by the African for centuries.

In Nigeria, farming is more of a way of life. Cultivation of the soil and other aspects of agrarian practices are common in Nigeria though, the Middlebelt territory as a whole is the most arable agricultural block. Its ambient climate is suitable to virtually every subsistence or commercial/export crop that can be cultivated in Nigeria. Bala Takaya authoritatively reports that,

> the region is a natural home to, among others, all grains (rice, maize, sorghum, millet and even wheat and barley); all legumes (like all varieties of beans, groundnuts, peas, bambara, etc); all fruit and tree crops (e.g everything citrus, mango, apple, pineapple, coffee, cocoa, tea, etc); all root crops (e.g yams, cassava, cocoyam, and their temperate varieties like beetroots, raddish, parsnips, etc); as well as everything vegetable (e.g tomatoes, water melons, cabbages and even exotic temperate veggies like broccoli, cauliflower, fennel, sage, basil, mint, etc.) and all livestock grazing and other forms of animal husbandry; whether commercial or pastime animal rearing (cattle, goats, sheep, piggery, pack animals, all forms of poultry, or even rabbitery, snails and grass-cutters). Above all, the territory is amenable to year-round crop cultivation by both rain-fed and *fadama* irrigated farming. (Takaya 2017: 4, 5)

It is thus safe to argue here that, African agrarian practice connects the practitioner with the community and nature, and stimulates a positive spiritual good that offers the virtues of honour, self-reliance, courage, moral integrity, and hospitality in the practitioners. It is sufficient to state here that the professional agro-ecological practice enhances harmony and sustainability in the eco-system. The Tiv Farming practice of making mounds (*avom*) is an example in question. The mounds are made by the men with heavy hoe (*Ikyar/gbar*) and covered with dried or fresh grass/leaves left on the field after clearing. The purpose of this according to Bohannan (1969: 17), "is both to prevent erosion and crumbling of the mounds and to afford from the sun for the sprouting yam during the first few days it is above the surface". The fallow system practiced by the Tiv is another agro-ecological practice that not only strengthens the knowledge and scholarly networks in African agriculture, it is an enabler for multiple pathways that creates new knowledge, intelligence, science, and foresight capabilities in relationship to agricultural crops and environment. This point has been well captured by Uchendu for whom agriculture has been described as *staff of life* for the Igbo. Like other African tribes, land for the Igbo "is an important factor in the construction of their social identity, the organization of religious life, and the production and reproduction of culture" (Uchendu 1965: 30).

The point of attraction here is that, traditional knowledge is in use in this agro-ecological friendly practice. As it were, this local knowledge prioritises yam farming after the fallow is broken, to starts the rotation cycle followed with millet and guinea corn and other farm products in a two to three year cropping season. The local knowledge system also capacitates the African farmers to know when a piece of fallow land is ready for cultivation by the type of grass which it produces. One tradition holds that after a piece of land has been cropped, the grass which it grows is predominantly sword grass (*ihila-imperatacylindrica*). According to an authority on Tiv knowledge system, *agom* grass begins to replace it for the next few years, again clumps get larger and heavier until they are about a foot in diameter at the base and the sword grass is replaced with *apel* grass—this is an indication that the fallow is ready to be farmed again. In the exact words of Bohannan (1969: 19), "trees and earth-worms are used as indicators for rich soil large undergrowth trees and large size grass chumps, with plenty earth-worm moulds, mark fallow which is ready for cultivation". We may say then that, multiple management of traditional farming practice has thrown up a number of new agro-ecological adaptations of new knowledge that have been generated, accumulated and transmitted to the community. Nyityo has availed us one of such farming practice in Tivland that,

> Agriculture in *Ikyumbur* is closely linked up with the Tiv concept of a clearly defined land area that surrounds or adjoins the residence of a famer. Provision is made here largely for grain crops such as maize and millet. In some cases yams Tobacco and ginger are also planted on this plot which is commonly refers to as *akongu* or homestead field. (Nyityo 2011: 99)

Here understood, ownership and use of land for farming not only defines the African rural farmer, it shores his commitment to his community and traditional values. Besides all these, farming for the African superintends the urban society and that their independent farming, simple rural life is not only opposed to the complexity of city life, it is superior to the paid worker. For the African therefore, farming as a way of life is a profession that shapes the ideal social values of the distressed world today. This is captured in *Ala*, the most important deity in Igbo cultural life, concept of Earth Goddess, the spirit that increases the fertility of man and the productivity of the land, and fecundity; the primordial mother.

16.7 The Evolution of African Agrarian Philosophy

It is the claimed conviction of Aristotle that virtue requires a certain level of wealth that reflects a version of agrarian values that stresses the importance of leisure time that can be devoted to philosophy. What this assertion suggests is that any householder that owns a farm is not only a model for the state, he/she is identified as the most valuable citizens. This classical thinking also holds true of the Africans who reason that cultivation of the soil is the basis of community life. Thus, there is a relationship between the locations of the farms and stability of the polis. This is because the farmers are inherently tied to a particular geographic location they develop a loyalty to the polis as a place. In his words,

> their investment in tree and vine crops requires multi-generational stability, they exhibit both a loyalty and a proficiency in the arts of defense and citizenship that mercenaries and those whose professions can be relocated will never have. Hanson argues that of all the city states, Athens' development of sea power created a class of traders whose economic interests were

> not so closely tied to the land, leading to the political upheaval we associate with time of Socrates and the eventual clash with the more firmly agrarian Sparta. (Hanson 1995)

Studies in agricultural systems of the Ancient world avails us three typologies namely, that which was owned by the sovereign and centrally managed. Herein, practical knowledge was limited to members of the priesthood and a few overseers. Secondly, there is in contrast, the agriculture of the Peloponnesian Peninsula which relied upon relatively small production units controlled and managed by the head of the household. The third system was true *subsistence farming* found on the margins of the ancient world. This system was not productive enough to sustain the division of labor needed to support the military and economic institutions of civilization (Mazoyer and Roudart 2006; Barnhill et al. 2018).

In Africa and for Africans, the third typology applies for a number of reasons. Apart from the acknowledged approval of farming by the forefather that virtue requires everyone born of a woman a certain level of wealth to feed the household. In Africa therefore, household agrarianism is not simply a blanket praise of farmer's virtue, but is more properly understood as a philosophy that takes agriculture to be especially significant for larger questions in social ethics and politics. This explains why access to customary land and secure rights in Africa is a burning political question. With the rapidly growing population, the African sees access to cultivable land as a security in an insecure world; a sure way to human sustainability and continuity. For as the Tiv say *shagba dugh shin nya* (wealth is generated from the soil/land/earth). This does not mean only that land (*Inya*) has an immense power over individual and social life, it controls the entire being of the human person and his/her existence and takes necessary action to ensure the community of communal life and in collaboration with the ancestors, and she establishes prohibitive conduct (Ihuah 2013). This is an argument in the direction of advocating a more ecologically friendly approach in agro-ecological stewardship in place of drawing upon uncompensated harms or violations of human (or non-human) rights (Jackson et al. 1984).

African ontological categories situate land as second after God. This finds expression among the Tiv that *Aondo hemba, Inya dondo* (Ihuah 2013). The Igbo of Nigeria express similar sentiments when they say, "land (*Ala*) is the messenger of *Chukwu* (the great creator of the great God) and the most important spirit after Him" (Anyanwu 1983: 109) This is the underlying idea in African agrarianism, as a mental attitude of the African people, as a mode of perception and interpretation of experienced reality and as a mode of value. Land provides the standard as well as the mechanism for affecting personal and social adjustment. What this entails is that the power of social integration which land encapsulate is based on the feeling and emotion of the ideal not on analytic logic or on empirical fact.

This is the land tenure system that informs the Tiv cultural frame work encoded in cultivation of the soil through a shifting farming process. This farming regime involves the pulling down grass (*ihyande*) for yam farm followed with *avom* mounds culminates into planting (which becomes *sule*, i.e., yam field (*itiev*)). When the yam farm is harvested, the farm becomes an old field or (*akuur*). Guinea corn, Barbara nuts (*igbough-ahi*) ground-nuts (*bum-ahi*) are selectively planted on the *akuur* in the third round of planting after which the *tsa* is allowed to fallow for three years. This cultural framework commonly practiced in Africa is a cultural framework for resource management. This cultural script is an internalized plan consisting of routine steps with alternative subroutines decision modes, and room for experimentation derived from experiences and local knowledge of Tiv farmers over generations.

What is clear here is that, African agrarianism is a deeply religious one, regulated by codes of behaviour and customs approved by ancestors, and enforced by the gods in respect of land use. The socio-moral value of land here therefore lies in the emphasis on community, its wellbeing, and its ordered existence as well as on the strengthening of all vital relations (Anyanwu 1983: 111). For the African therefore, life consists in the mutual interdependence between natural and supernatural forces in the journey of peaceful living.

The use of symbolic breaking of the kola-nut by Africans (the Igbo in particular) to call all beings and forces to communion by saying: "He who lives above, the giver of life, we thank you, *Ani* (the earth Goddess) come and eat kolanut, *Amadioha* (God of thunder), come and eat kolanut, may the river not dry up and may the fish not die; we shall live" (Momoh 2000: 372) is all intent at rejuvenating the land to prosper for humanity inherit the earth. For the African therefore, cultivating the soil has within it a positive spiritual benefits namely, acquisition of a virtuous life of honor, self-reliance, courage, moral integrity, generousity and hospitality (Inge 1965). By having direct contact with nature and, through nature, the farmer not only creates a closer relationship to God, he recreates order out of chaos in the image and likeness of God.

This is a philosophy of sustainable human environment that is informed by a nonviolent and gentle attitude towards nature. It is a consciousness of the limits in which we must live in order not to degrade natural and human environment. This philosophy guarantees a humane and more human friendly development which improves the quality of human life on earth. Devendra Kumar advices us here to learn from the bees the manner we serve nature in order to get its sustenance simultaneously. According to him, "The more the honey it collects from flowers, the more it serves in the propagation of the plants by helping in their fertilization. We could emulate the bees by fulfilling our needs through a similar symbiotic relationship with nature" (Kumar 2001: 2). As explicitly captured elsewhere, sustainable human development is a journey inward the essential human nature and the integral well-being of man in his material and spiritual life. It involves shifting the balance of human development towards improving the quality of human life on earth (Ihuah 2010a, b).

This in no way suggest the superiority of humans over other lesser beings, it only encourages them to co-operate rather than compete with the lower creatures and to simply enlarge *the boundaries of the community to include soils, waters, plants, and animals, or collectively: the land*" (Leopold 1949: 239). This is to say that the spirit world of Africa is intimately connected with the daily life of the people. K.C. Anyanwu alludes to this when he says,

> the relationship man and th spiritual world is maintained through many channels. Obedient to the codes of behavior and the customs approved by the ancestors and enforced by the earth goddess through priests and titled elders, and the heads of various extended families, is the most important channel. (1983: 109)

Thus, African Agrarian Philosophy here interpreted as rural agro-ecological practices avails the theoretical knowledge of man in setting him properly on the road to his dignity and destiny; of sustainable living in the biotic community. It rests upon the idea that every person and everything is related to every person and everything, and that human nature is good and instinctively seeks the divine and that humans only become dysfunctional when they grow up in a sick culture which produces violent and damaged humans. Humanity is organically embodied in a series of associations and relationships with nature which appears to have full value only in those close ties. African Agrarian culture is today at its lowest ebb needs to seek nourishment from traditional African system to remain relevant and functional. The need of moment therefore is to recalibrate African Agrarian Philosophy through African endogenous knowledge systems and critical heritage resource. To achieve this, we need to decolonize indigenous languages. This is the point eruditely made by Wiredu that we need to embark on a critical and comparative exercise of both (foreign and indigenous) conceptual frameworks in our own indigenous languages in order to further evaluate the intelligibility of these theories and concepts. He says then that,

> ....Try to think them through in your own African language and, on the basis of the results, review the intelligibility of the associated problems or the fallibility of the apparent solutions that have tempted you when you have pondered them in some metropolitan language. (Wiredu 1995: 24)

The giant strides made by the Asian Tigers in recent years are directly attributable to their use of local languages in their educational system up to the university level. The use of local languages at all levels of education in these societies has also 'freed' these patches of indigenous knowledge which was hitherto locked up due to lack of use (Roy-Campbell 2006). Considering the oral nature of African cultures, language can play an important role in cultural transmission and knowledge dissemination. This role is also tied to African identity. Appropriate promotion of African languages devoid of foreign idiosyncrasies can actually lead to a better understanding of rural agriculture and local knowledge known and owned (intellectual property) by Africans. In many ways than one, the power of African endogenous knowledge can be enhanced through language which will in turn promote African agrarianism for sustainable living. In many parts of Africa, ignorance of endogenous knowledge in the areas of biodiversity, herbal medicine and intangible heritage has exposed it to piracy which continues to date. The inability of African Governments to recognize, protect and preserve this knowledge system as intellectual property has slowed the development of its agro-cultural economy. We must therefore strive for development of endogenous science and knowledge says Claude Ake. He adds further that,

> we cannot fully emancipate ourselves…even though the principles of science are universal, its growth points, applications and the particular problems which it solves are contingent on the historical circumstances of the society in which the science is produced. (Ake 1986: ii–iv)

Such is the idea of sustainable human development that guides African agro-ecological community. It is informed by a nonviolent and gentle attitude towards nature. It is a consciousness of the limits in which we must live in order not to degrade our environment and ourselves. This represents the spirit of the *African humanistic heritage*, an African moral philosophy that ensures care, concern, co-existence and communal responsibility that ensures a harmonious relationship with other members of the biotic community. It is the African spirit of self-reliance i.e., of making sense of the human soul through the human body, the appropriation of endogenous knowledge for African Agrarian Philosophy.

Thus, African agrarianism entails a conscious effort at inventing, improving and changing agricultural practices through the simplest tools on earth produced by men relying on selves to make man be himself to himself through a process of self-reltrieval of the self. This is an African spirit that makes the call for the restitution of Africa more actual, acute and urgent (Ogundowole 2007: 18). This is a call for self-reliance in agro-ecological practice; a principle of progressive evolution by man/woman relying on the self in him/her to make a change in farming practices that improves the African environment in food production and processing through the self (i.e. self-retrieval of the self). Self retrieval of the self for the African is the needed impetus to resolve the challenge and trigger human and land resource development beyond imagination. Relying on the self in him/her man/woman can instantiate a system of activities or satisfactory ways of resource management. Ogundowole's explicit point clarifies the matter better. He avers,

> the self, thus, is more than a mere summation of the social-historical and anatomic-biological in the human, not just the socio-biological, it is the totality of all these plus the correlation between them which thus constitutes the intrinsic life-force propelling the human towards creative, innovative purposive activity distinguishing the human from all other living beings, nay other animals. (Ogundowole 2007: 23, 24)

Thus, self-reliacism, i.e., realization of the principle of self-reliance here argues for self-retrieval of the self. This means that, the self is a propelling life-force that engineers the socio-cultural material and immaterial, spiritual possibilities to so act for the positive change directed towards self-realization in all human endeavours. This philosophy seeks to reconcile agriculture and local communities with natural processes for the common benefit of nature and livelihoods. It employs a variety of methods to understand the varied components of the environment.

One point that must be bone in mind is that, in the cultivation of the soil, AAP must keep science relevant and future focused, it must strengthen knowledge and scholarly networks in traditional ecological knowledge system to create new knowledge, intelligence and foresight capabilities to guide the direction of agro-ecological resource management to reduce hunger and poverty. AAP connotes agriculture and food systems built upon moral ethics, creativity and innovation and which unequivocally identifies with the needs and aspirations of African in thought and action.

16.8 Conclusion

Our discussion thus far reveals that African Agrarian philosophy is a lens with which Africa can implement Sustainable Development Goals (SDGs). This can be done by strengthening of endogenous knowledge agricultural practices that enables multiple pathways to create new knowledge, intelligence, science, and foresight capabilities. Our local challenges and the shortcomings of our industrial and technological economy notwithstanding, Africans are enjoined to not only focus on farming, they should endeavour to develop practices and policies that promote the health of land, community and culture hence the African survival is inextricably linked. Africans must seek to sustainably nourish the earth and protect the right to food for the greatest number of the Africans even as there is the increasing growth of human population and the decline of soil and water resources. While the arguments for and against the full use of technology in food production rages and the race for genetically modified crops are being advanced, the urgent need for establishing the morality of farming in Africa becomes sacrosanct. Indeed, it is not an exaggeration to say that a creative and explicit African Agrarian philosophy is important in the varied actions of individuals and groups in African agriculture. As rightly asserted by Barnhill et al. (2018) "if there truly is anything special about farming, the time to consider that question is now". The urgent need of the moment in African agro-ecology is an ethical dialogue; an open-ended, evolutionary and arduous process of intercultural debate and consensus-building revolving around the basic issues of relationship with nature, human fulfillment, and respect for the cultural rights of the individual and community, and justice for all.

References

Adande, A.B.A. 1997. Traditional Iron Metallurgy in West Africa. In *Endogenous Knowledge: Research Trails*, ed. P. Houtonji. Dakar, Senegal: CODESRIA.

Africa Part VIII, ed. Daryll Forde. London: Stone and Cox ltd.

Agazzi, E. 1993. Philosophical Reflections on the Concept of Development. In *Ideas Underlying World Problems*, vol. 1, ed. I. Kucuridi. Ankara Moloksan Co. Ltd.

Agrawal, A. 1995a. *Indigenous and Scientific Knowledge: Some Critical Comments*. Indigenos Kowledge and Develoment Monitor.

Agrawal, A. 1995b. Dismantling the divide between indigenous and scientific knowledge. *Development and Change, 26*(3):413–439. https://doi.org/10.1111/j.1467-660.1995.tb00560.x

Agrawal, A. 2009. Why "Indigenous" Knowledge? *Journal of the Royal Society of New Zealand* 39(4). https://doi.org/10.1080/03014220909510569

Ake, Claude. 1986. Editorial: Raison D'etre. *African Journal of Political Economy/Revue Africained' Economie Politique*, 1 (1 Southern Africa in Crisis).

Anyanwu, K.C. 1983. *The African Experience in the American Marketplace: Smithtown*. New York: Exposition Press.

Atteh, O.D. 1989. Indigenous Local Knowledge as Key to Local-Level Development: Possibilities, Constraints and Planning Issues in the Context of Africa. In *Seminar on Reviving Localself-Reliance: Challenges for Rural/Regional Development in Eastern and Southem Africa (Unpublished)*.

Barnhill, Anne, Mark Budolfson, and Tyler Doggett (eds.). 2018. *The Oxford Handbook of Food Ethics*. https://doi.org/10.1093/oxfordhb/9780199372263.001.0001/oxfordhb-9780199372263-miscMatter-3.

Berkes, Fikret. 2018. *Sacred Ecology: Traditional Ecological Knowledge and Resource Management*. London: Routledge.

Berkes, Fikret, Johnan Colding, and Carl Flke. 2000. Rediscovery of Traditional Ecological Knowledge as Adaptive Management. *Ecological Applications* 10 (5): 1251–1262 (The Ecological Society of America).

Bohannan, P. 1969. *Tiv Farm and Settlement*. Ibadan: The Ibadan University Press.

Bohannan, Laura, and Paul Bohannan. 1962. *The Tiv of Ceentral Nigeria, Ethnographic Survey of Africa, West Africa Part VIII* (ed. Daryll Forde). London: Stone and Cox Ltd.

collinsdictionary.com paraphrased Archived Wayback Machine, 2020.

Claude Levi-Straus, C. 1962. *The Savage Mind*. Chicago, IIllinois: University of Chicago Press.

Diamond, J. 2017. *Guns, Germs and Steel: The Fates of Human Societies*. New York: W.W. Norman & Company lIndependent Publishers.

Emsley, J. 1991. Piecing Together a Safer Insecticide. *New Scientist* 132 (1798): 24.

Eyong, C.T. 2007. Indigenous Knowledge and Sustainable Development in Africa: Case Study on Central Africa. *Tribes and Tribals* 1: 121–139.

Feyerabend, P. 1987. *Farewell to Reason*. Verso, London, UK.

Gorjestani, N. 2004. Indigenous Knowledge: The Way Forward. *Local Pathways to Global Development* 45–59. www.http://www.worldbank.org/afr/ik/ikcomplete.

Gundu, Z.A. 2010. Endogenous Knowledge Systems: The Challenge of Salvaging a Strategic Heritage Resource in Nigeria. *Or-Che Uma: African Journal of Existential Philosophy* 1 (2): 115–126 (Benue State University).

Gyekye, K. 1997. Philosophy, Culture and Technology in the Post-colonial Africa. In *Post Colonial African Philosophy*, ed. F.C. Eze.

Hanson, Victor Davis. 1995. *The Other Greeks: The Family Farm and the Agrarian Roots of Western Civilization*. Berkeley: University of California Press.

Heeds, A. 1991. Botanical Pesticides. *Alternatives* 18 (2): 6–8.

Hoddy, E. 1991. Nature's Bitter Boon: The Neem Tree—A Substitute for Pesticides. *Development and Communication* 5: 29–30.

Holt, Rinehart, and Winston World Commission on Environment and Development (WCED). 1987. *Our Common Future*. New York: Oxford University Press.

Hountondji, P.J. 1977. Introduction. In *Endogenous Knowledge: Research Trails*. Dakar: CODESRIA Book Series.

Ihuah, S. Alloy. 2010a. *The Ethics of Science and Technology: An African Perspective*. Saarbrucken: LAP Lambert Academic Publishing AG & Co. KG.

Ihuah, S. Alloy. 2010b. *Philosophy and Human Existence: Critical Essays in Philosophical Discourse*. Saarbrucken: LAP Lambert Academic Publishing AG & Co. KG.

Ihuah, S. Alloy. 2013. African Indigenous Ecological Knowledge and Sustainable Development. In *Studies in Culture, Gender and Education in Africa*, ed. Maurice Nyamanga Amutabi, 9–30. Nairobi: The Catholic University of Eastern Africa (CUEA).

Inge, M.T. (Ed.). 1965. *Agrarianism in American Literature*. New York: Odyssey Press.

IUCN. 1986. List of Threatened Animals Prepared by The IUCN Conservation Monitoring Centre Cambridge U.K. International Union for Conservation of Nature and Natural Resources 1986.

Jackson, Wes, Wendell Berry, and Bruce Coleman. 1984. *Meeting the Expectations of the Land*. San Francisco: North Point Press.

Johannes, R.E. 1981. Working with Fishermen to Improve Coastal Tropical Fisheries and Resource Management. *Bulletin of Marine Science* 31: 673–680.

Kumar, D. 2001. Excerpt from his Award Lecture, the Indian National Science Academy (INSA) Annual B.D. Tilak Award for Rural Development. Available at http://www.insa.org.

Kwasi Wiredu. 1995. *Conceptual Decolonization in African Philosophy: Four Essays* (ed. Olusegun Oladipo Ibadan). Hope Publications.

Kyle Powys Whyte. 2013. On the Role of Traditional Ecological Knowledge as a Collaborative Concept: A Philosophical Study. *Whyte Ecological Processes* (*A Springer Open Journal*) 2: 7. http://www.ecologicalprocesses.com/content/2/1/7. Accessed June 15, 2022.

Leopold, Aldo. 1949. *Sand County Almanac*. Random House Digital Inc.

Mazoyer, Marcel, and Lawrance Roudart. 2006. *A History of World Agriculture: From the Neolithic Age to the Current Crisis* (Tr. James H. Membrez). New York: Monthly Review Press.

Merriam-Webster.Com. 2020.

Momoh, C.S. 2000. Substance of African Philosophy. Africa Philosophy Projects Publications.

Nyito, S.G. 2011. A Preliminary Ethno-Archaeological Study of Agricultural Practices in Ikyumbur Area of Tivland. In *Archeology, History and Environment in the Middle-Benue Valley Nigeria: Essays in Honour of Bassey Wai Andah*, 91–107, ed. Alloy S. Ihuah, and Zacharys Anger Gundu. Saarbrücken: VDM Verlag Dr. Müller GmbH & Co. KG.

Ogundowole Ezekiel Kolawole. 2007. *Inexhaustability of Self-Reliance*. Lagos: University of Lagos Press.

Oscar Levy. 1906. *The Revival of Aristocracy*. Translated by Leonard A. Magnus. London: Probsthain and Co.

Pickstock, M. 1992. Small Farmer Knowledge and Pest Control. *International Agricultural Development* 12 (1): 5–7.

Roy-Campbell, Z.M. 2006. The State of African Languages and the Global Language Politics: Empowering African Languages in the Era of Glbalization. In *Selected Proceedings of the 36th Annual Conference on African Languages*, ed. Olaoba F. Arasanyin, and Michael A. Pembertu, 1–13. Summerville: Cascalla Proceedings Project.

Sands, P., ed. 1993. *Greening International Law*. London: Earthscan Publications Ltd.

Senghor, L.S. 1964. *On African Humanism*. New York: Praeger.

Taliaferro, C., and S. Carpenter. 2010. Farms. In *Life Science Ethics*, ed. Gary L. Comstock.

Takaya J. Bala. 2017. *Pastoralists and Farmers Conflicts in Central Nigeria: Towards a Holistic Analysis of Causal Factors. In Herders and Farmers Conflict in Central Nigeria: Learning from the Past*. Makurdi, Centre for Research Management, Benue State University, Makurdi.

Uchendu, Victor. 1965. *The Igbo of Southern Nigeria*. New York.

Varvar, Toryina Ayati. 2011. Discrimination and Innovations in Indigenous Yam Production Technology and Techniques in Tivlannd: The Role of Migrant Farmers. In *Archeology, History and Environment in the Middle-Benue Valley Nigeria: Essays in Honour of Bassey Wai Andah*, 129–143, ed. Alloy S. Ihuah, and Zacharys Anger Gundu. Saarbrücken: VDM Verlag Dr. Müller GmbH & Co. KG.

Warren, D.M. and J. Pinkston. 1998. Indigenous African Resource Management of Tropical Rain Forest Ecosystem: A Case Study of the Yoruba of Ara, Nigeria, 158–189. In *Linking Social and Ecological Systems Management Practices and Social Mechanisms for Building Resilience*. ed. F. Berkes and C. Folke. Cambridge, UK: Cambridge University Press.

Warren, D.M., Slikkerveer, L.J., and Brokensha, D. 1995. The Cultural Dimension of Development: Indigenous Knowledge Systems. Intermediate Technology Publications, London, Retrieved 3rd June, 2022.

Alloy S. Ihuah is Professor of Philosophy, Benue State University, Makurdi-Nigeria. He is a Member of many Learned Societies including World Council on Values, African Studies Association (ASA), USA, Council for Research in Values and Philosophy (CRVP), International Federation of Philosophical Societies, Association of Philosophy Professionals of Nigeria (APPON)which he is its current National President. Professor Ihuah is also a member of the Nigerian Academy of Letters (MNAL) and a Member of the Knight of Columbus (KofC). Professor Alloy Ihuah is winner of the 2014 Asante award for outstanding research of the University of Georgia, USA. He researches and publishes in African/Inter-Cultural Philosophy, Philosophy of Science and Existentialism. He is married to Maureen Mbafan Ihuah and are blessed with two girls and two boys.

Chapter 17
The Shona People's 'Zunde raMambo' (King's Granary) as a Model for Social Responsibility: A Task for Higher Education Systems

Erasmus Masitera

Abstract For the Shona people of Zimbabwe, social caring and responsibility is a virtue, a moral calling which is inherently instilled in one's social schooling. Social schooling involves both practical and theoretical teaching and learning, the emphasis though being on enthroning humanising interactions. Some humanising perceptions were to be reflected in agricultural activities. The activities were culminated in a community welfare in which the less privileged received assistance. Through the 'Zunde raMambo' (King's granary), Shona communities ensured that food security, poverty and hunger were reduced, good health and social well-being strengthened. At the same time ethical values such as communal responsibility, sharing and caring, communing and common good ensued thereof. The informal safety nets that 'Zunde raMambo' provides reveal the African conception of community development which contemporary society ought to learn from especially the formation of a mental attitude of collaborative work for the good of the community. I argue that this is achievable through an educational system that reflects African ethos and training that encourage African values that humanise interactions that promote communal care and livelihood.

Keywords Shona people · Zunde raMambo · Social responsibility

17.1 Introduction

The concept Zunde raMambo is part of the Shona people's history and future as well. Ringson (2017), Muyambo and Marashe (2020) note that the idea is being revived in contemporary Zimbabwean society. The 'Zunde raMambo' concept is value-laden, an axiological expression of how Shona people preserve, promote and maintain every

E. Masitera (✉)
Great Zimbabwe University, Masvingo, Zimbabwe
e-mail: lmasitera@gmail.com

M. J. Tosam and E. Masitera (eds.), *African Agrarian Philosophy*, The International Library of Environmental, Agricultural and Food Ethics 35,
https://doi.org/10.1007/978-3-031-43040-4_17

member of the community's life. Members contribute to the nutritional well-being of the community through offering various forms of labour that contribute to the 'King's granary'—the king's reserve granary for the needy. The granary is filled through communal planning, performance and procedure of working together as a way of supporting each other (Mahohoma and Muzambi 2021: 1); that is aiding to the collective storage that assist the needy. In fact Zunde raMambo is the practice of communal working and producing together, for everyone's good. It is an indirect and direct way of responding to the Sustainable Development Goal #1 expectation that says no poverty, though in this sense the reference is to eradicate material (nutritional) poverty. There is a sense of ensuring food security for every member of the community, and bettering the lives of community members through redistribution of resources.

While a considerable number of authors have said something on the 'Zunde raMambo' practice, their writings lack the application of axiological perspective or thinking to the agrarian issue. In that respect, my chapter intends to fill that lacuna. I will philosophise on values such as communal living and common living, among others, and their relation to agrarian philosophy. Beyond that I argue that educational institutions are one of the ideal platforms through which such traditional practices are to be imparted and reimported to society; this is another aspect that is not reflected upon by philosophers who delve in agrarian philosophy such as Zvavanyange (2016a, b), Thompson (2008), Chibvongodze (2017). I should also mention here that there are a few scholars who have delved into African agrarian philosophy yet most would have discussed related issues to agrarian philosophy such as environmental philosophy (c.f. Museka and Madondo 2012; Chemhuru 2019, 2020). Others such as Mahohoma and Muzambi (2021), Mandikwaza (2018), Zvavanyange (2016a, b) and Ringson (2017) limit their discussion on Zunde raMambo to issues such as the concept's usefulness in sustainable development, food security, and social security. Important as these may be, there is a lack of focus on Zunde raMambo as an axiological vehicle for strengthening and reawakening a public good and a distributive pattern that is to be disseminated via educational institutions.

In order to argue my case I have divided this chapter into six sections. The first section introduces the Shona people and the concept of 'Zunde raMambo', the second section focuses on justifying the revival of the practice and its connection to educational discourses. Third section, reveals the ethical implications of 'Zunde raMambo' and the fourth section discloses some educational lessons that may be drawn from the practice. In the fifth section attention is on how higher education may achieve or successfully fulfil its duties in disseminating the ethical perceptions that have to be drawn from the practice. And lastly, the sixth section makes a conclusion in which the author brings together the ideas that African agrarian practice promote ethical practices.

17.2 Who Are the Shona People, and What is 'Zunde raMambo'?

The Shona people are inhabitants of the greater part of present day Zimbabwe. The term 'Shona' refers to a variety of indigenous languages such as the Karanga, Zezuru, Manyika, and Korekore. More to that the Shona people share ethical and religious beliefs; they share the Ubuntu (a humanising philosophy) ethical practice and thinking. Among the thinking of Ubuntu is the idea of communal living through strengthened bonds of relational living.

Community living is reflected through caring and sharing basic needs and guaranteeing social security. Social food security, one of the social securities, is one of the many public goods that the Shona people seek as necessary for social development (Muyambo and Marashe 2020: 232). 'Zunde raMambo' is an indigenous system and method of ensuring that communities survive through a shared practice of guaranteeing food security. According to Ringson (2017), the concept of 'Zunde raMambo' is translatable to the English word 'king's granary'. Literally the king's granary is for everyone in the community, it is a social food bank to which those in need and deemed by the community in need appeal to for assistance. Through careful planning, 'Zunde raMambo' is a process of producing and storing food (Mahohoma and Muzambi 2021: 1). This will involve some form of work parties that contribute to the production of food through availing labour to produce grain, and offer assistance in the identification and distribution of the reserves to the needy in the community (Stathers et al. 2000: 5; Muyambo and Marashe 2020: 235). The produce is safely stored under the custody of the village head (Stathers et al. 2000: 5). It is important to mention here that each village has a communal field to which members work in and there is a village granary reserved for storing the produce (c.f. Stathers et al. 2000).

'Zunde raMambo' is both a method of producing agricultural produce and a system of ensuring communal food security through communal storing. Further to that the practice of 'Zunde raMambo' has the ethical implications of promoting public/common good, through working together, caring and sharing labour and produce, and being responsible for each other's well-being. Importantly, the practice seamlessly fit into the developmental goals of eliminating poverty (SDG #1) through safeguarding communities against hunger and shameful conditions associated with poverty or hunger. It is without doubt that among the Shona people, 'Zunde raMambo' is a way through which the nutritional well-being of the community is assured. In addition to that 'Zunde raMambo' is a way through which members of a community show concern, and care for each other through being sensitive to each other's needs and working towards bettering each other's situations. There is thus a sense in which other-regarding values are treasured among the Shona people.

17.3 Justifying Reviving Zunde raMambo in Contemporary Educational Discourse[1]

This section will reveal the link that exist between agrarian practices and the axiological issues such as those drawn from ubuntu. The link is that the agrarian practices directly impact the day to day lives of the people and they shape and reshape of society. While the concept of 'Zunde raMambo' was a traditional method of assisting the less fortunate in terms of food security, it also served to show the Shona way of distributing societal benefits and burdens. 'Zunde raMambo' directly and indirectly prompted the ideas of responsibility, caring and sharing as foundations for human living among the Shona people. More to that, it begs questions that have to do with distributive justice—that is how benefits and burdens in society are to be shared by communities. This means that the practice of 'Zunde raMambo' has something to do with social justice and how to address social inequalities. The benefit and the burdens in this regard refer to the use of social goods such as land, and natural resources–how are these are used for the benefit of the community.

I also view the revival 'Zunde raMambo' as a decolonial project. By decolonial project, I am referring to a system that intends to reveal, first, the ability of the indigenous people to recognise that they are have an organised way of living. Such a realisation is a rejection of colonialism and its imposed narratives (c.f. Ranger 1967: 2; Kohn 2012: 4). Colonialism negated everything that had to do with indigenous organisation—these ranged from social, economic and political ways of surviving. To his amusement, Idowu (2006) views colonialism as a self-centred thinking that was termed as 'civilised system and mission'; Idowu (2006) concludes that the civilising mission was a way of demeaning native societies and establishing social control. This social control implies knowledge or ability to arrange the means to survive, meaning that the indigenous people were capable of arranging their life. As such the concept of 'Zunde raMambo' epitomises the ability of the local people at organising their lives to ensure that nutritional values are attained by members of society and that they are responsible for each other's well-being.

Second, by decolonial, I am referring to the endeavour of liberating the African mind from the colonial bondage. In this sense, the decolonial project is the self-affirmation, and recognition of a people's liberty as opposed to the Euro-centric thinking that was imposed by the colonial system(s) (Mendiata 2020; Ndlovu-Gatsheni and Ruhanya 2020: 2). In fact this is epistemological liberation. Mendiata (2020) actually views the decolonial project as a means by which societies liberate themselves from colonial methods and systems of oppression. There is a sense in which revitalisation and reaffirming social practices is one of the aims of decolonisation. In this regard, I view 'Zunde raMambo' as fulfilling that aim.

In connection to the above, and especially the first perception, 'Zunde raMambo' stands out as a means of showing the Indigenous Knowledge System's contribution

[1] While the practice of 'Zunde raMambo' is widely recorded as a traditional exercise, there have been in recent years the efforts to revive the scheme by the Zimbabwean government. Little success has been noted though.

to epistemic entrepreneurship. Indigenous knowledge had been neglected, negated and excluded from the world epistemic categories of knowledge (c.f. Matsika 2012: 35–37). Indigenous knowledge was considered dangerous and not fitting into the European system of education. What is important is that there was no effort to understand the indigenous forms of education and perhaps there was a premeditated move to obliterate it and replace it with that of the colonisers. By replacing the knowledge system, the colonisers were in a way working at ensuring control of the locals (Atkinson 1972: 11). The imposition of western forms of knowledge generally implied the negation of local knowledge systems and resulted in the appreciation of the colonisers' educational systems and methods. Notably Matsika (2012: 19–39) opines that the indigenous systems that were both practical and theoretical (embedded with ethical, spiritual ideas) were replaced by rudimentary education. This is a kind of education that limited locals to servitude, an education that was not holistic. The education was limited to specialising in specific areas. Whereas the indigenous knowledge system was holistic, it instructed people on all aspects of life. As for 'Zunde raMambo', there is a sense in which people are taught how to live with each other both practically and theoretically. The revival of such a practice acknowledges and points to the fact that knowledge about society is passed on through doing and communing. Through these, individuals become aware of what life is like and what is expected of them in community living.

Closely connected to the last point in the preceding paragraph, the African practice of work parties is located within the agrarian philosophical practice of agronomics. Agronomics involves the practice of agriculture, and speaks of the normative institutions that are associated with the practice of agriculture (Thompson 2008: 529). The interrelation that exists on the practice and social institutions is crucial and is of importance to this chapter. Accordingly, Thompson (2008: 527–528) argues that:

> Agrarian philosophy stresses the role of nature, soil and climate in the formation of moral character as well as social and political institutions.. *in addition*[2]
>
> ... the term 'agrarianism denotes a class of philosophical views on human culture and practices as they relate to the broader environment. Agrarian views are distinguished by their emphasis on the role of material practices in the formation of norms, values and social institutions.

The formulation of norms and social institution is important in as much as it shows the influence that the environment has on human lives. To this end, Zvavanyange (2016a, b) posits that agrarian philosophy is the use of moral ethics in building agriculture and food system. In a sense, the needs and aspirations of specific people are adhered and reflected in agricultural practices (c.f. Zvavanyange 2016b: 8). In the African sense this includes providing for the protection, prolonging life and provision of needed nutrition; additionally the use of localised norms in agricultural enterprises fits within the logic of contemporary development ethos (c.f. Thompson 2008: 531; c.f. Zvavanyange 2016a).

In short agrarian philosophy is important as it ensures that social groups and communities survive or are sustained through meeting material (food) needs. Such

---

[2] Own addition so that the statement makes sense.

expectations are realised through use of norms that enforce and maintain social cohesion and cooperation of members of a community. Beyond all this, it is important that nature (biosphere), is cared for and protected and promoted by humans (human practices ought to promote environmental well-being) aided by social institution (norms, and values (morality)). Clearly, agrarian philosophy, and indeed African agrarian philosophy, is a situated philosophy, a philosophy that responds to agrarian existential realities. That is, situations that are African centred in celebrating African agrarian successes and addressing African agrarian challenges. In this regard, I now turn to identify some of the values that are pertinent in African agrarian practice of 'Zunde raMambo' and their link to higher education.

African morality is centred upon the good of every one that is the well-being and livelihood of everyone is considered a public good that ought to be promoted. For this reason I note that communal living and common good are fundamental in developing an agrarian philosophy that is authentically African centred. Communal living or community life is connected with the adherence to or living within the community's moral systems. Noteworthy is that "within the communal moral system, individuals are expected to conform to the demands of the community, in as much as individuals have played a part in the formation of that community" (Masitera 2019: 104). Conforming to communal norms is always aimed at producing the common good (Gyekye 1997). Common good in this sense is the establishment of a good that benefits everyone in a community through harmonising, maintaining and promoting human interests and needs (Wiredu 1992; Gyekye 1997, 2010; Ndwandwe 2018). Composta (2008: 162) sums this up by saying that the common good is the *end* aim of any society that, is seeking to promote order and progress of society (material and immaterial desires of people). A question that may be posed is how this is to be achieved in face of the contemporary social practice which is individualistic and mostly against communal living. It is at this juncture that I think that educational institutions, and in particular higher education institutes have a role to play. Since ancient times, Plato claimed that educational institutions have a duty to impart knowledge and morality among the citizens of the country. The same applies today. However the moral practices in this case are of African origin. Higher education ought to have an obligation to teach and impart knowledge about African living systems and in particular African ethics that relate to humane living.

17.4 Ethical Implications of 'Zunde raMambo'

Humanness is the aim of living within an African community (Metz 2019). This is to be exhibited through prizing communal relations. Communal relations are extremely important as they assist in sustaining and maintaining communities in meeting their material and immaterial needs (c.f. Thompson 2008: 528). The needs and aspirations of the community are easily identifiable and expressed through community acting in unison so as to diffuse and eliminate difficulties, and to celebrate achievements (c.f.

Mandikwaza 2018; Mahohoma and Muzambi 2021: 1). In this sense the distribution of communal difficulties and joys of members is a communal concern.

In matters that have to do with material needs of people in communities, there is a sense in which issues related to responsibility, caring and sharing are invoked. These ethical considerations reveal the obligations and duties that communities have towards each other, that is, the aspect of other-regarding responsibilities which are at the heart of communal activities. Material needs in this case refer to both physical and biological requirements that ensure the survival or well-being of people in society. Importantly, seeking community well-being or security is the main aim (Mandikwaza 2018: 2). The foregoing invites discussion pertaining to what is prioritised in African ethics—is it the individual or community, individual rights or community obligations? My view to this is that the individual is prior to community; I base my argument on Gyekye, Oyowe, and Metz's thinking which argue that a community is an association of autonomous beings who are self-determining, can be accorded responsibility and have the capacity to commune with others (Gyekye 2010; Metz 2012; Oyowe 2014). Such a view opposes the view that is broadly shared by Menkiti (1984) and Molefe (2018) who argue that community and obligations are a priority in African societies. I contend that the latter position denies individuals the capacity to move towards each other, it imposes duties and thus denies individuals the right to liberty. Further, the position allots conformity to community rather than assign the individuals the ability to make choices and change community thinking and behaviour especially in cases of abuse. Having said this, I advance the thinking that caring, sharing and responsibility are showcased within communal activities of working for the good of everyone, it further illustrates the individual commitment to do good and seek good from others when need arises. In the same sense, by participating within communal activities such as 'Zunde raMambo' the Shona people demonstrate individual commitment to collective living.

In connection with the idea of collective living is the African conception of communing or communal living exhibited through the agrarian practice of 'Zunde raMambo'. Importantly the role of the individual and that of the community obviously becomes a central issue, I shall not belabour this issue as it has already been discussed above. Yet close to the same is the idea of socialising individuals into communal expectations such as communal responsibilities that are meant to eliminate selfishness and self-centeredness and ensuring the dominance of communal welfare (c.f. Mandikwaza 2018: 2). There is a sense in which these two (selfish and self-centeredness) promote and significantly lead to disharmony in society as bonds of connection are dishonoured through self-interestedness and in the process neglecting factors that bring people together such as understanding and caring for others. As such communal living involves the idea of other-regarding responsibilities. Other-regarding responsibilities is the capacity to empathise and sympathise with others, or as Hoffman and Metz (2017) put it, the ability to be in a relationship or being in a relation with others. Being in a relation with others is the movement towards each other, that is, the removal of barriers that prevent inclusivity and togetherness in society. Communal living promotes togetherness, communal allegiance and surely communal responsibility. In the case of 'Zunde raMambo' the idea of lessening or removal of (communal) poverty, social inequalities and social exclusion is a communal activity that is derived from being responsible, being interconnected through love and friendship, and through being in a relationship with others. Characteristically community or communal living is promoted through working together or participating in the process of filling the King's granary and in identifying needy members.

In economic affairs the practice of communal living as exemplified by 'Zunde raMambo' brings with it ethical considerations such as transforming moral economy of the community by doing away with selfishness and promoting food security. The idea of cooperating with others and being socially responsible is key to eliminating incidences that mark individualism or selfishness. Individualism in this sense refers to egoism that is aimed at self-aggrandisement and being inconsiderate to others. Such kind of living is incompatible with the African way of living and is destructive to ethos of community building. The idea here is that African ethical systems encourage working together and uplifting everyone in society. In addition to that, caring and being responsible for each other are greatly advocated for. The sum total of caring and being responsible for each other is shown through the establishment of a communal system that support food security for members of society. That is members who are found wanting and in need are taken care of by the community through community food stored in a communal barn which the local leader presides over. The idea is that members of the community and even those beyond, need generosity every now and then for them to survive; and in that case the community food storage becomes handy.

Feeding members of the community from a common food trough ensures that the nutritional needs of a community are achieved. Beyond that, the exercise falls within the ethical expectation of African ethics that is stimulating common good. Common good in African thinking is considered as having shared values (Gyekye 2010; Masitera 2018) that a community develop and respect. Some of the values that are communally established are aimed at establishing communal harmony, stability and solidarity (Bongmba 2018) ultimately intended at forming mutuality, reciprocity, and a sense of self-sacrifice for the individual's good and that of the community. Furthermore, there is the founding of respect, dignity, security and satisfaction (Gyekye 2010; Bongmba 2018). The main idea in common good is that the focus is building a better society, a society where everyone fits in and plays a role by contributing to communal development in which the individual finds fulfilment.

Implied in the foregoing paragraph is that common good flows from having respectable communal relations. As such cementing social relations is a normative expectation in African ethical practise (Menkiti 1984; Wiredu 1992; Gyekye 2010; Ndwandwe 2018: 105). Social relations are necessary in as much as they guarantee supportive systems and structures which humans badly require. The sociality includes the different webs of social, economic and political interactions which empower or and disempowers them. In the African sense, social interactions are meant to promote, protect, and enhance the way(s) of living of individuals. The social interactions are based on mutuality and reciprocity which are the foundations for cooperation and the bedrock for stability, harmony and solidarity. 'Zunde raMambo' epitomises a practice were these expectations are expressed.

17.5 Educational Lessons

The discussion in the preceding section is only informational but is well articulated and disseminated through educational systems. Education in this case is a process of enlightening and a platform for transmitting and renewing cultural perceptions (Adeyemi and Adenyika 2003). The idea is that education opens up avenues of interchange and integration of perspectives so as to enrich human interactions. In a sense education is a catalyst for social development through efforts of liberating the mind and through advancing new knowledge that is inclusive and that capacitates the human mind.

Flowing from the above then, the quest for authentic liberation through education is a mission that Africans in general seek. Since the colonial times, the demand for freedom has been deafening. Moving away from colonial hegemony which came through the imposition of foreign ways of living while neglecting African systems (Matsika 2012) is a clarion call. Liberation in this sense refers to formulating and living by one's own aspirations (Nasongo and Musungu 2009: 111), and reflecting independently on what people want. This will include epistemic freedom that is freedom to share knowledge, understand its implications and implement it for the benefit of the world and more importantly local individuals. Liberation is therefore an effort at attaining various forms of freedoms which include physical, political, social and economic freedoms. According to Nyerere (1967), liberation points in the direction of self-sufficiency, self-reliance and self-definition as authenticating and achieving a decolonial project. This thinking is well presented and argued for by Molefe (2019: 106) highlighting the importance of other-regarding responsibilities. Nurturing communal living is the basis of a social insurance that ensures that the needs of the community are realised; I am here thinking of members of a community who are vulnerable, for example, the aged, people living with various forms of disability, the young and marginalised members who are unable to feed and protect themselves. Such members are assisted through communal assistance systems which are embedded within the 'Zunde raMambo' practice whereby members of a community work, produce, and reserve the produce for the unfortunate members of community. In this sense, 'Zunde raMambo' becomes a kind of social welfare through which social burdens and benefits are shared by the community and are addressed by the same community. Ibhawoh and Dibua (2003: 70) posit that such communal activities reveal that social cooperation and collective production are self-help mechanisms that build common well-being of the community. Further to that, in a way, social inequalities are also reduced through the distribution of (nutritional) material resources (Ibhawoh and Dibua 2003: 71; Otunnu 2015: 19). A form of egalitarian distribution and obligation to work is also nurtured by communal engagement in producing and helping unfortunate members of society (Otunnu 2015: 19). These perceptions are all important in that they indicate the moral benefits of communal practices and they reveal the Nyerere (1967) expectation of self-reliance, and they are essential in identifying and addressing challenges that communities encounter (Thompson 2008: 531). They are instrumental in giving identity as well. My contention is that all these expectations are best dispersed through an educational system that reflects African ethos.

Education in the African traditional structure aimed at initiating people into the communal way of living (Adeyinga and Ndwapi 2002; Matsika 2012). The educational system aimed at inculcating societal expectations, duties and responsibilities. Education in that sense had the moral role of orienting people into that which a particular community expects (Plato; Adeyinka and Ndwapi 2002: 19). Adeyinka and Ndwapi (2002: 19) succinctly posit that the role of education is to:

> train young people for their … roles; to inculcate into them a feeling of belongingness and interdependence between all members of the community; to mould their morals and character in a socially-acceptable manner, and, finally, to equip them with the skills needed to engage themselves in various occupations satisfactorily.

This means that education had mostly an instrumental function of introducing people to social mores, and to prepare them for participation in a society though not specifically stated in the above quotation, practicing communal living through engaging in community oriented and focused activities. Communal spirit is a holistic and lifelong type of education (Adeyemi and Adeyinka 2003: 425) that aims at cultural transmission and even renewal of some practices. In this endeavour, education and specifically higher education becomes a transmitter of social morals. In saying this, higher education is considered by Adeyemi and Adeyinka (2003: 426, 428) as a vehicle through which social expectations are handed to others. Through transmitting ideas and in this case African moral ideas that are connected to 'Zunde raMambo', people learn communal expectations.

Apart from being the transmitter of values, higher education also is a process of training towards an end that society seeks (Metz 2009: 181). By the term 'end', I am referring to the ultimate aims of engaging in an activity. Some of the aims of higher education include: facilitating and ensuring the development of human society such as promoting and prolonging human life. Steering human development has become an urgent requirement of higher education, development in this case is used in an expansive way to include issues such as ensuring well-being (from Capability Approach) and economic well-being (propounded by materialists). In saying all this, I imply that higher education has the role of fostering human emancipation and empowerment by training or guiding people into a mind-set that is development and peace oriented. According to Mandikwaza (2018) the African collaborative work systems such as 'Zunde raMambo' were engaged in as a way of promoting sustainable peace and building community cohesion; these aspect become crucial in human living and especially in African communities which are diverse and mostly divided according to race and cultural differences. In this sense, educational institutions become instrumental in building those senses of belonging and of uniting different people.

This rhymes with Walker's thinking. Walker (2005: 105) opines that higher education aims at establishing social justice through the creation of enabling environments, that is, environments that promote opportunities and liberty for all. This comes about through personal and institutional conditions and contexts in which freedom may be achieved, in other words, universities are liberated zones through which individuals exercise freedom to be analytic and critical about life in general. The way in which higher education institutions are made is such that individuals have the opportunity to broaden their intellectual horizons by inter- and intra-personal interactions (Walker 2005: 105). Beyond that, individuals also encounter and experience unlimited regulation in regards of activities they engage in. They learn valuable life lessons on what to and what not to do. This is essential experience in the formation of individual values which one uses in suggesting or contributing to communal ideas on how to live together. Further to that Walker (2020) argues that higher education is the vehicle through which epistemic injustice is addressed; this assertion is relevant in this case, though I am arguing that traditional practices and theories deserve to be revived and revitalised. For Walker (2020) epistemic justice includes acceptance and widening of both practices and theories of traditional knowledge (she however was referring to Black Consciousness). Through disseminating ideas that are linked or that come out from Zunde raMambo practice, there is a chance of including African moral values and also promoting them through higher educational organizations.

17.6 How Higher Education May Achieve the Expectations

The deliberation above pointed to the importance of disseminating values that are considered important for African societies. My suggestion is that there is need for curricula development to reflect the African ethos. Curricula development implies changing and or refocusing (Posillico et al. 2021) the university curriculum so that it reflects an African flavour. As noted Matsika (2012), most African (university) education is mere replication of Western education. This Western education is the one which at colonisation replaced the African educational system. However through the initiatives that decolonial education demands, reintroduction of African pedagogy is expected. According to Shawer (2017), this refers to designing, developing and implementing a curriculum that is contextualised. Regarding the need to have university training incorporating and reflecting African values. It becomes necessary to revise modules/courses so that they include and reflect African values. Through revising curricula the chances are that African standards will be introduced or reintroduced to students in African universities. In addition to that, chances are that such values will be inculcated into the lives of students. In this way the African values are revived in practice and theory. Notably there will be transformation in the way people think and act.

Apart from reviewing curricula, there is also need to focus on the actual teaching of values. Curricula only contain the expectation, yet actual teaching is the practical orientation into thc expectations. Passing on of the actual information and knowledge takes place through teaching. Teaching involves enlightening students while at the same time allowing them the opportunity to make assessments concerning materials brought before them. Teaching is not merely a kind of indoctrinating, but ought to promote critical individuals. When critical assessment is permitted, students are granted the freedom to choose to follow, alter and most importantly make informed decisions concerning the kind of life they may want to follow and even improve decision making up to choosing a value relevant for a particular event. Teaching is therefore a way of introducing and broadening students' horizons concerning different ways of life, and it is also a way of offering students a chance to actively be involved in shaping their lives and that of the community. It is my opinion that through the teaching of African values and especially those that have been discussed in this chapter, students will be accorded a chance to appreciate, critique, and improve the African norms and practices. This is possibly a contribution that African morals have to offer to the world.

17.7 Conclusion

In this chapter I have shown that the Shona concept of Zunde raMambo translated as the King's granary is a process or procedure that is carefully planned by a community as a social security means for communal well-being. I have also argued that while 'Zunde raMambo' is an agricultural and food security system it is at the same time built upon moral ethics of the Shona people. In that regard, I have reflected on ethical standards such as responsibility, common good, communal living and sharing. These I have noted as the basis of forming the Shona (African) collective livelihood and importantly the basis of the African distributive system. At the end of the chapter, I have made reflections of the importance of education in the dissemination of African morals. I opined that higher education in particular is a vehicle for strengthening and reawakening the African values associated with 'Zunde raMambo' and the mentality that promotes common good and collective living in general.

References

Adeyinga, A.A., and G. Ndwapi. 2002. "Education and Moralitry in Africa", *Pastoral Care in Education* 20 (2): 17–23.

Adeyemi, M.B., and A.A. Adeyinka. 2003. The Principles and Content of African Traditional Education. *Educational Philosophy and Theory* 35 (4): 425–440.

Atkinson, N.D. 1972. *Teaching Rhodesians.* Harare: Longman.

Bongmba, E. 2018. Communitatianism in African Thought-Gyekye and Moderate Communitarianism? science.jrank.org/pages/8772/communitarianism-in-African-Thought-Gyekye-on-Moderate-Communitarianism.html. Accessed October 26, 2021.

Chemhuru, M. 2019. The Moral Status of Nature: An African Understanding. In *African Environmental Ethics*, ed. M. Chemhuru, 26–46. Cham: Springer.

Chemhuru, M. 2020. The Place of Ontology-Based Environmental Thinking in Africa's Higher Education. In *African Higher Education in the 21st Century: Some Philosophical Dimensions*, ed. A. Ndofirepi, and E.T. Gwaravanda. Brill/Sense. https://doi.org/10.1163/9789004442108_003.

Chibvongodze, D.T. 2017. Ubuntu is Not Only About the Human! Analysis of the Role of African Philosophy and Ethics in Environmental Management. *Journal of Human Ecology* 54 (2): 157–166.

Composta, D. 2008. *Moral Philosophy and Social Ethics*, Rome, Urbanian University.

Gyekye, K. 1997. *Tradition and Modernity: Philosophical Reflections on the African Experience.* New York: Oxford University Press.

Gyekye, K. 2010. African Ethics. In *Stanford Encyclopedia of Philosophy.* https://plato.stanfordedu/entries/african-ethics/. Accessed October 06, 2021.

Hoffman, N., and T. Metz. 2017. What can the Capability Approach Learn from an Ubuntu Ethic? A Relational Approach to Development Theory. *World Development* 97: 153–164.

Ibhawoh, B., and J.I. Dibua. 2003. Deconstructing Ujamaa: The Legacy of Julius Nyerere in the Quest for Social and Economic Development in Africa. *African Association of Political Science* 3 (1): 59–83.

Idowu, W. 2006. Against the Skeptical Argument and the Absence Thesis: African Jurisprudence and the Challenge of Positivist Histography. *The Journal of Philosophy, Science and Law* 6: 34–49.

Kohn, M. 2012. *Colonialism.* http://plato.stanford.edu/entries/colonialism.

Mandikwaza, E. 2018. Utilizing a Traditional Practice in Community Peacebuilding. *Conflict Trends* 2018 (2): 1–8.

Mahohoma, T., and P. Muzambi. 2021. Nhimbe as a Model for Reinvigorating Sustainable Socio-economic Development in Zimbabwe and Africa. *Theologia Viatorum* 45 (1): 1–8.

Masitera, E. 2018. Economic Rights in African Communitarian Discourse. *Theoria: A Journal of Social and Political Theory* 65 (157): 15–36.

Masitera, E. 2019. Traditional Communal Understanding of Crime and the Role of Social Therapy: Ideas from African Philosophy. *Filosofia Theoretica: Journal of African Philosophy, Culture and Religions.* 8 (3): 101–114. https://doi.org/10.4314/ft.v8i3.7.

Matsika, C. 2012. *Traditional African Education: Its Significance to Current Educational Practices with Special Reference to Zimbabwe.* Gweru: Mambo Press.

Menkiti, I. 1984. Person and Community in African Traditional Thought. In *African Philosophy: An Introduction*, ed. R. Wright. New York: United Press of America.

Metz, T. 2012. An African Theory of Moral Status: A Relational Alternative to Individualism and Holism. *Ethical Theory and Moral Practice.* 15 (3): 387–402.

Metz, T. 2019. *The African Ethic of Ubuntu.* 1000wordphilosophy.com/2019/09/08/the-african-ethic-of-ubuntu/. Accessed October 15, 2021.

Metz, T. 2009. "The Final ends of Higher Education in light of an African Moral Theory", *Journal of Philosophy of Education*, 43 (2): 179–201.

Mendiata, E. 2020. Philosophy of Liberation. In *Stanford Encyclopaedia of Philosophy.* plato.stanford.edu/entries/liberation/.

Molefe, M. 2018. Personhood and Rights in an African Tradition. *Politicon: South Africa Journal of Political Studies* 45 (2): 217–231.

Molefe, M. 2019. Ubuntu and Development: An African Conception of Development. *Africa Today.* 66 (1): 97–115. https://doi.org/10.2979/africatoday.66.1.05.

Museka, G., and M.M. Madondo. 2012. The Quest for a Relevant Environmental Pedagogy in the African Context: Insights from Unhu/Ubuntu Philosophy. *Journal of Ecology and the Natural Environment* 4 (10): 258–265.

Muyambo, T., and J. Marashe. 2020. Indigenous Knowledge Systems and Sustainable Development: The Case of Zunde Ramambo (Isiphala Senkosi) as Food Security in Chipinge, Zimbabwe. *INDILINGA—African Journal of Indigenous Knowledge Systems* 19 (2): 232–244.

Nasongo, J.W., and L.L. Musungu. 2009. The Implications of Nyerere's Theory of Education to Contemporary Education in Kenya. *Educational Research and Review* 4 (4): 111–116.

Ndlovu-Gatsheni, S., and P. Ruhanya. 2020. Introduction: Transition in Zimbabwe: From Robert Gabriel Mugabe to Emmerson Dambudzo Mnangagwa: A Repetition Without Change. In *The History and Political Transition of Zimbabwe: From Mugabe to Munangagwa*. Cham: Palgrave-Macmillan.

Ndwandwe, S. 2018. The Common Good and a Teleological Conception of Rights. *Theoria: A Journal of Social and Political Theory* 157: 100–122.

Nyerere, K.J. 1967. Education for Self-Reliance. *Cross Currents* 18 (4): 415–434.

Otunnu, O. 2015. Mwalimu Julius Kambarage Nyerere's Philosophy Contribution, and Legacies. *African Identities* 13 (1): 18–33.

Oyowe, O.A. 2014. An African Conception of Human Rights? Comments on the Challenges of Relativism. *Human Rights Review* 15: 329–347.

Posillico, J.J., D.J. Edwards, C. Roberts, and M. Shelbourn. 2021. Curriculum Development in the Higher Education Literature: A Synthesis Focusing on Construction Management Programmes. *Industry and Higher Education*. https://doi.org/10.1177/09504222211044894. Accessed November 24, 2021.

Ranger, T.O. 1967. *Revolt in Southern Rhodesia 1896–7*. London: Heinemann.

Ringson, J. 2017. Zunde raMambo as a Traditional Coping Mechanism for the Care of Ophans and Vulnerable Children: Evidence from Gutu District Zimbabwe. *African Journal of Social Work* 7 (2): 52–59.

Shawer, S.F. 2017. Teacher-Driven Curriculum Development at the Classroom Level: Implication for Curriculum, Pedagogy and Teacher Training. *Teaching Ad Teacher Education* 63: 296–313.

Stathers, T., T. Sibanda, and J. Chigariro. 2000. *The Zunde Scheme, Chikomba District, Zimbabwe*. https://assets.publishing.service.gov.uk/media/57a08d78e5274a31e0001898/R7034d.pdf. Accessed September 24, 2021.

Thompson, P.B. 2008. Agrarian Philosophy and Ecological Ethics. *Science and Engineering Ethics* 14: 527–544.

Walker, M. 2005. Amartya Sen'ss Capability Approach and Education. *Educational Action Research* 13 (1): 103–110.

Walker, M. 2020. Failures and Possibilities of Epistemic Justice, with some Implication for Higher Education. *Critical Studies in Education* 61 (3): 263–278.

Wiredu, K. 1992. Moral Foundations of an African Culture. In *Person and Community: Ghanaian Philosophical Studies*, ed. K. Wiredu, and K. Gyekye, 193–206. Washington DC: Council for Research in Values and Philosophy.

Zvavanyange, R.E. 2016a. *An African Agrarian Philosophy and Sustainable Development Goals: Nurturing Creativity in Science and Society*. Gfar.net/sites/default/files/THEME%203%20Zvavanyange%20Wed.pdf. Accessed September 06, 2021.

Zvavanyange, R.E. 2016b. *Philosophy and the Logic of Collective Action in African Agriculture: Perspectives I Africa's Transformation*. LAP.

Erasmus Masitera (June 3 1979–March 1 2022) was a senior lecturer in Philosophy in the Department of Philosophy and Religious Studies at Great Zimbabwe University, Masvingo, Zimbabwe. At the time of his demise, he was a postdoctoral fellow at the University of the Free State, South Africa. His research areas revolved around the connections of Ethics, Ubuntu, land reform and social justice.

Chapter 18
The Practice of African Indigenous Medicine and Agrarianism in Madamombe Area (Chivi District-Zimbabwe)

Tasara Muguti

Abstract African traditional medicine has been used by the African people since time immemorial. It has been used to deal with different livelihood challenges such as human and animal ailments, and bio-diversity conservation, among others. Among the Karanga people of Chivi, a Shona subgroup, this practice has persisted in modern times despite its denigration by both colonial authorities and missionaries in colonial Zimbabwe. The chapter examines the extent to which traditional medicine is relied upon by the Karanga people of Chivi. It mainly focuses on how traditional medicine has been used and continues to be used in crop enhancement, protection and animal husbandry. While other traditional institutions can be used to deal with agrarian issues in Madamombe area, it is believed that this precious Indigenous Knowledge System is being lost and should be preserved as contemporary challenges faced by the African people warrant a multi-faceted approach. As such, the chapter endeavours to problematise the efficacy of traditional medicines in promoting sustainable agriculture in Madamombe area. The chapter draws from face to face interviews and existing related literature.

Keywords Agrarianism · Indigenous medicine · Indigenous religion · Indigenous knowledge systems · Traditional healers

18.1 Introduction

Since prehistoric times, the African people have relied on their Indigenous Knowledge Systems (IKS) to deal with numerous livelihood challenges. For instance, African indigenous medicine (AIM) has been used to deal with an array of health challenges such as human and animal ailments, and biodiversity conservation, among

T. Muguti (✉)
Great Zimbabwe University, Masvingo, Zimbabwe
e-mail: tmuguti@gzu.ac.zw

M. J. Tosam and E. Masitera (eds.), *African Agrarian Philosophy*, The International Library of Environmental, Agricultural and Food Ethics 35,
https://doi.org/10.1007/978-3-031-43040-4_18

others. Among the Karanga people of Chivi, a Shona subgroup, this practice has persisted in modern times despite its denigration by both the colonial authorities and missionaries in colonial Zimbabwe. The chapter seeks to examine the extent to which traditional medicine is continuously and constantly relied upon by the people of Madamombe area in dealing with agrarian challenges. It examines the different ways by which indigenous medicine is used in diverse aspects of agrarian practices ranging from crop enhancement, processing and preservation of seed and crop yield, storage of harvest to animal husbandry.

While it is acknowledged that there are other traditional institutions that can be used to deal with agrarian issues, it is believed that this precious Indigenous Knowledge (IK) is fading and should be preserved. The study also attempts to problematise the efficacy of traditional medicines in promoting a sound agrarian system in the Madamombe area. Furthermore, today's agrarian challenges can not to be resolved by scientific interventions alone. As traditional healers and community elders pass on, this knowledge system is threatened with extinction; hence there is need to document it for posterity. The study will be qualitative in nature. It will draw from face to face interviews and existing related literature.

Madamombe area is located in the Chivi district of Masvingo province and is inhabited by the Karanga, a sub-group of the Shona people of Zimbabwe. It is situated on the south-western part of the province. Its southern boundary is the Nyarutedzi River and it elongates to the Shashe-Tukwi rivers confluence in the north-east. Its western boundary is the Zvinyamani hills which separates Masvingo from the Midlands province. The eastern boundary extends from Tokwe Bridge, Kilmanock and Mhandamabwe Townships up to Maramba turn off which are all situated along the Bulawayo-Mutare highway. Maramba village is the only part of Madamombe area which is sited on the eastern side of the highway.

Madamombe is home to several villages under headman Madamombe of the Mhari clan (Chiondegwa, 25 September 2021). It is located in an area with low agroecological potential due to frequent droughts resulting in crop failure in many seasons. Despite receiving erratic rains and being drought prone, the major economic activity in the area is subsistence farming. Like most parts of the province, Madamombe falls under agroecological region 4 and 5 with semi-arid conditions. As a result, off farming activities which have the potential to sustain livelihoods are limited in the area.

18.2 Indigenous Knowledge Systems, Traditional Institutions and Agrarianism

Indigenous Knowledge (IK) has been variously interpreted by different scholars (Sillitoe 1998; Berkes 2018: 8). It has been termed as indigenous knowledge, traditional knowledge, local knowledge, popular knowledge, folk knowledge, among others. However, as Sillitoe (1998: 223) argues, the lack of harmony in terminology "intimates the flux that characterises this fast-moving and exciting field in development practice". Johnson (1992) perceives IKS as knowledge that has been acquired by a specific group of people over succeeding generations as they interact with nature. Berkes (2018: 8) defines indigenous knowledge as a "a cumulative body of knowledge, practices, and belief, evolving by adaptive processes and handed down through generations by cultural transmission, about the relationship of living beings (including humans) with one another and with their environment."

While indigenous knowledge has been variously described by scholars, what is common among all these definitions or interpretations of IK is that it refers to traditional or localised knowledge systems that have sustained local communities since the earliest times. Today, there is mounting convergence between scholars and development partners that IKS, which have been marginalised since colonial times, have a lot to offer for sustainable development to take place in Africa (Ajayi and Mafongoya 2017; Kanu 2021: xi; Senanayake 2006: 91). According to Kambu (2010: 257) Indigenous Knowledge:

> Embodies knowledge, techniques and wisdom that cut across every thematic area and all aspects of social, ecological, economic and political aspects of indigenous and local communities. The local know-how to manage the environment, use plants and animals for therapeutic and medicinal purposes, preserve seeds for food and agriculture, and conserve and use water resources sustainably are a few examples of the loaded nature of IK…TK can be regarded as the code of life for many indigenous and local communities…TK is often very pragmatic, as it is developed and applied to solve specific and real problems that indigenous local communities grapple with in their daily lives.

Indigenous Knowledge has been used by many people worldwide across time. A disturbing trend has been the fact that power relations over knowledge has been a cause for concern since colonial times. Nonetheless, there has been a growing realisation since the last quarter of the twentieth century that scientific knowledge alone cannot effectively solve local problems and improve the livelihoods of local communities. For Ajayi and Mafongoya (2017), IKS can be treated as social capital for the poor as they may be used as an asset to ensure survival, food production and secure livelihoods. In the same vein, Senanayake (2006: 90) states that:

> This knowledge offers new models for development that are both ecologically and socially sound…development activities that work with and through indigenous knowledge have several important advantages over projects outside them…the critical strength of indigenous knowledge is its ability to see the interrelation of disciples, and then integrate them meaningfully. The holistic perspective and resulting synergism shows high levels of developmental impact, adaptability and sustainability than Western knowledge.

In this regard, it should be energetically advanced that Western scientific methodology should not be viewed as the only way of knowledge acquisition and the sole provider of solutions to diverse challenges to humanity today (Obiora and Emeka 2015). In challenging the notion of a single objective reality, George Kelly developed a philosophy which he termed 'collective Alternativism' (Mokgobi 2014). He argued that even though reality exists, the way it is created, interpreted and understood is multi-faceted. Thus, it then becomes problematic when one reality construction is given an elevated status and universalised since the world is extremely large and complex. Indeed, virtually everything including science is just but a matter of opinion, since it is not easy to authenticate anything indisputably (Ibid).

Furthermore, in recent times, climate change has invariably affected some vulnerable communities across the world particularly in rain-fed agric-based economies in Africa. As a result, these economies naturally become highly susceptible to climate change and its associated challenges. Thus, there is no doubt that decision makers should not only rely on scientific interventions but on the best available local knowledge in their quest to improve the livelihood of rural communities. Given this exigency, the use of traditional medicine in agriculture can play a crucial role in the search for a synergistic approach to agrarian challenges facing rural communities today. It is within these parameters that the contribution of traditional medicine to agrarianism in Madamombe area will be examined. Agrarianism refers to agricultural practices. In this chapter, I argue that there has been widespread use of traditional medicine in food production, storage and preservation by the Karanga people in Madamombe.

Among the Karanga people of Madamombe, traditional institutions interlock in sustaining peoples' livelihoods in a harmonious manner. The traditional leaders, spirit mediums (*masvikiros*) and traditional healers complement each other in many ways. As such, these institutions, which define the Karanga people should be preserved and protected for posterity. Perhaps it is prudent to make a cursory survey of how traditional leaders and spirit mediums contribute to agrarianism among the Karanga. Literature search demonstrates that traditional leaders and spirit mediums have always complemented each other in their work for the sustenance of their communities (Chemhuru 2010). In pre-colonial Africa, traditional leadership insisted on good agrarian management by making sure that community members compiled with specific customary laws, sanctions and taboos. During colonial times, there was serious disregard for traditional institutions as resource management was removed from communities. This had far reaching effects on rural economies which culminated in the distortion of the socio-economic and cultural development of many communities (Magoro 2008). Nonetheless, in post colonial Zimbabwe, the powers of the traditional leaders have been restored in their communities (Mazarire 2008). Traditional leaders are viewed as custodians of the land under their jurisdiction. They distribute land to their subjects, resolve land tenure disputes and preside over other issues that affect their immediate communities. The chiefs, headmen and village heads work closely with spirit mediums (*masvikiros*) in dealing with issues that affect their communities. Besides, traditional medical practitioners (TMPs) also play an integral role in their communities as levers to a wide range of indigenous agricultural farming practices.

In the Karanga worldview, religion and spirituality are treated as very important aspects. This worldview incorporates the visible and the invisible worlds. These two worlds are inseparable. The Karanga believe in the existence of a Creator God (*Musikavanhu*) who is approached through ancestral spirits (*vadzimu*). The *masvikiros*, who work closely with chiefs, are also an important part of Karanga traditional leadership structures. The *masvikiros* are the link between the living and the dead and as such are highly respected and revered (Risiro et al. 2013).

This Karanga religious hierarchy, which was exceedingly venerated before the advent of colonialism in Zimbabwe, is still venerated today. It is within the context of medico-spirituality that the contribution of AIM to agrarianism will be problematised.

18.3 African Indigenous Medicine and Agrarianism

The use of AIM among local and indigenous communities around the world is as old as humankind. Traditional medicine can be defined as:

> The totality of all knowledge and practices, whether explicable or not, used in diagnosis, prevention and elimination of physical, mental or social disequilibrium and relying exclusively on practical experience and observation handed down from generation to generation, verbally or in writing. (WHO 2000)

While this definition may appear to be exclusively focused on the social, physical and mental health of individuals, it should be acknowledged that AIM can be used beyond the wellness of human beings. Although African philosophy and culture have been disparaged and perceived in condensing terms, the failures of western style culture, has made the Africans to realise that they must realign their own culture by incorporating valuable western practices with functional indigenous components (Makinde 1988). In the African philosophy, traditional medicine is regarded as a holistic medical system that encompasses indigenous herbalism and African spirituality (Izichukwu 2019). Indeed, the philosophy of the practice of AIM is deeply rooted in the African world view (Onwuanibe 1979).

As noted elsewhere in this chapter, AIM can be used to deal with many livelihoods challenges including agriculture. The people in Madamombe area use a wide range of AIM in sustaining their crops and in animal husbandry. Despite the demonisation and spirited attempts to annihilate the practice of AIM by white settler authorities in colonial Zimbabwe, the practice has survived. In colonial Zimbabwe, IK was relegated to the periphery as western education in sciences and technologies took centre stage. This was in a bid of creating an epistemological supremacy that favoured the global north (Nyoni 2017).

In Madamombe area, the use of AIM in agriculture remains deep-seated. This IK form cannot be easily erased or relegated to the dustbin of history as it is not accidental to the Karanga people. As Nyoni (2017: 238) argues, "it constitutes the Africaness of the African, thereby his/her true identity which cannot be concealed during times of need and crises". It is in this context that the study endeavours to delve into how AIM can significantly contribute to sustainable agriculture among the people in Madamombe area. African Indigenous Medicine has been used in an array of activities that can enhance agricultural productivity. Indeed, across sub-Saharan Africa, indigenous plants have been successfully used in preventing and curing diseases in plants and animals (Risiro et al. 2013). Besides using AIM to promote human wellness, the Karanga people also use it to enhance, preserve and protect their agricultural produce. With limited livelihood opportunities outside agriculture among the Karanga, the locals who heavily rely on their crops and animal husbandry for survival would do everything possible to protect their sources of livelihoods. This is within a context in which the government is struggling to provide the necessary support systems for the rural folk.

Besides, there are known limitations of scientific interventions in dealing with some agrarian challenges facing subsistence farmers. The failures and inadequacies of scientific interventions in dealing with some agrarian challenges force the Karanga to continue to rely on AIM. Indeed, there are certain crop and animal diseases that the Karanga believe can only best respond to AIM. This has forced many people to resort to the use of AIM in agriculture. Indigenous medicine has always been part and parcel of the African past and the African people cannot be easily separated from it. It has been used by the African people for thousands of years and it can still be effectively used to curb food insecurity. As Mpepereki (2015: 16) argues:

> Africans must be careful, your ancestors discovered valuable sciences and technologies despised by the western as 'indigenous knowledge'. The stigmatization of African scientific knowledge and practices resulted in the erosion of the productive base of African communities as the Africans were forced to abandon the wisdom of their forefathers (sic) and found themselves with the few working technologies accepted by the western colonizers. Thus, western sciences and practices have caused environmental damage and proved to be unsustainable. While African science has stressed harmony with nature, western science has caused disharmony and environmental pollution. Instead, of working with, it has worked against nature.

The Karanga people depend on AIM to deal with different agrarian challenges which threaten food security. In New Millennia Zimbabwe, the use of AIM has accelerated due to the socio-economic challenges facing the country. While the post-colonial government worked strenuously to ensure that medical veterinary facilities and other services were accessible during the 1980s and early 1990s, the situation has significantly deteriorated over the years. This has been a result of mismanagement of the economy by the ruling elite, corruption and contested election outcomes particularly since 2000.

Indeed, unimaginable poverty, rising inflation as the local currency continues to shed value and the rising cost of living has exacerbated the lives of rural communities to the extent that they are forced to rely more and more on their surroundings for their survival. In this context, the use of AIM in agriculture can then become a practical alternative to Western agrarian approaches. Perhaps, it is critical at this juncture to examine specific aspects in which AIM has been used in agriculture in Madamombe area.

Indigenous African Medicine has been effectively used in animal husbandry. Among the Karanga people of Madamombe area, livestock play numerous roles in fighting poverty and sustaining people's livelihoods. For Mtshali et al. (2014: 195) "livestock does not only serve as a family and cultural asset, but is also a vital source of food, rich in protein for many poor inhabitants". Livestock are also a source of inorganic manure which can be used as a substitute for fertilisers by the poor communal farmers in the face of a comatose economy.

In Madamombe, the people have always used traditional medicine to deal with an array of animal diseases. The high temperatures in the area make livestock susceptible to an array of diseases. This is within the context of a regime which is struggling to maintain a viable and sustainable health delivery system for both humans and nonhumans. The depressed microeconomic environment in the country has seriously curtailed the government's capacity to provide essential veterinary services to the detriment of many rural farmers who traditionally rear diverse livestock. In Madamombe area, the cattle are not regularly dipped due to decaying infrastructure and shortage of appropriate veterinary medicines. The government is struggling to supply communal farmers with constant supplies of amitraz tick buster dip powder. Cattle in the area frequently die from tick-borne diseases which can be easily controlled by dipping them regularly (Simon Shumba, 11 September 2021). As such, domestic animals are exposed to some contagious diseases and the people in the area are forced to rely on their local knowledge of animal husbandry in looking after their livestock. An elderly man from Maramba village had this to say:

> In colonial Zimbabwe, cattle were dipped once a week. The Ian Smith regime would arrest anyone who failed to dip their cattle as it was treated as a serious offence. In post colonial Zimbabwe, cattle were dipped once a month during the First Republic under Robert Mugabe. When the Second Republic took over the reins of power under the leadership of Emmerson Mnangagwa, cattle are dipped once in three months or are never dipped at all due to shortages of dipping chemicals. At times, the villagers are forced to dig deep in their pockets to pull together financial resources to purchase the dipping chemicals. (Simon Shumba, 11 September 2021)

As a result, the people in the area, in order to avoid the loss of their livestock, especially cattle which many perceive as a form of wealth, draught power, source of pride and status symbol, are now more and more relying on traditional medicine to sustain their livestock. Nonetheless, many people in the area constantly loss their cattle to diseases such as anthrax, black leg and foot-and-mouth. A villager from Nyamadzawo narrated how some residents in the area and surrounding villages lost large numbers of cattle in 2021 due to theileriosis or January disease (James Mukono, 20 October 2021). Some people lost their entire herd of cattle. Mukono himself lost 15 herd of cattle and had to sell the remaining five at ridiculously give away prices. Interestingly, the Karanga people used to keep a lot of health cattle, sheep and goats before the advent of colonialism in Zimbabwe. This was made possible through their reliance on AIM. As the economic situation rapidly declines, the Karanga people are now even more and more relying on AIM to treat their livestock and maintain healthy domestic animals. The medicines, which have been used for a very long time, are embedded in customs and tradition and have been transmitted from generation to generation through oral mediums like stories and folklore.

It was established during the study that the elderly people, especially those over 60 years generally have comprehensive knowledge about the medicines that can be used to treat animal diseases. The youngsters are also socialised early in the use of these medicines as they at times accompany elders when they go to collect these medicines. The people in Madamombe use amarula tree (*mupfura*) bark, aloe (*gavakava*), pepper (*mhiripiri*), canthium huillense (*muvengahonye*), among others, to treat different animal ailments. The traditional herbs act as effective antibiotics in treating livestock. As one informant noted, nearly all the drugs that are prescribed by traditional medical practitioners (TMPs) to treat human beings can also be used in treating livestock diseases (Dzoro Living, 15 November 2021). The plants parts that are commonly used are dependent on the type of disease to be treated. These include stems, leaves, fruits, bark, seeds and flowers (Mudzengi et al. 2014).

Some of these traditional herbal remedies are not confined to the treatment of particular type of livestock but can be used in treating diverse livestock. The above tallies with Chavunduka's (1976) assertion that there are a wide range of indigenous veterinary medicines which can be used to treat cattle, pigs, goats, sheep, dogs, and chickens. These traditional indigenous veterinary drugs comprise those for the treatment of worms, constipation, burns, painful udder, diarrhoea, eternal parasites, wounds, snake bites and painful eyes (Dzoro, 15 November 2021). As one informant demonstrated, if a communal farmer is well acquainted with traditional herbs that prevent and cure livestock diseases, he/she may incur minimal or no expenses at all without the fear of losing their livestock.

A variety of traditional remedies are used in treating chicken. Locals use *gavakava, mupfura* bark, *mhiripiri*, among others. Almost every homestead in Madamombe area keeps a reasonable number of domestic chickens (roadrunners) which are treated as special meat. These are normally slaughtered during special family occasions. For instance, a communal farmer may keep large numbers of roadrunners (free range birds) that can freely move around with little care as they scavenge for food. These birds which do not need a lot of food, take up to 13 to 16 weeks to mature and produce eggs resistant to diseases. As one prominent farmer at Mhandamabwe Business centre puts it:

> The roadrunners which can be kept in large numbers do not need too much food, care and attention. These can be used for domestic consumption and can also act as a stable source of income for a family as they can be sold on the domestic market. If you keep plants like gavakava and mhiripiri around the homestead, the road runners easily treat themselves as they feed on these plants as they scavenge for food. (Mandebvu Robert, 02/07/2019)

Traditional medicine is also used by the locals to protect their livestock from thieves. Traditional medicines are planted in cattle pans to protect them from thieves. It is also used to recover stolen livestock. This was confirmed by headman Madamombe and village heads in the study area during a focus group discussion. The traditional leaders attested that traditional healers are regularly consulted by the people to recover their stolen domestic animals.

While traditional healers were denigrated as witches and charlatans by the colonial authorities and missionaries in colonial Zimbabwe, the practice has survived in post colonial Zimbabwe. Indeed, almost every village has a traditional healer who performs multiple roles. In their day to day work, traditional healers are guided by certain myths, taboos, cultural norms and values that regulate their operations for the sustainability of their communities. Besides, traditional healers have been known to be the custodians of cultural values and counsellors in their communities. Traditional healers act as caregivers, community therapists, family counsellors, health educators and also perform wider community functions as diviners, ritual specialists, teachers, community leaders, priests, moral and ethical guides (Payyappalli 2010).

Dissimilar to biomedical practitioners, traditional healers are also well known for providing preventive and promotive health care aspects in their communities. It is significant to note that traditional healers have often been perceived in both negative and positive terms in the development discourse. They are at times perceived as quacks and witches but are also, on a positive note, regarded as gatekeepers and custodians of indigenous knowledge and cultural identities (Ibid). As such, they do not only practice good environmental practices but make these practices cascade to the wide community through interactions with community members. They also work hand in hand with traditional leaders in ensuring that the members of the public practice sustainable environmental practices. It should also be noted that the mere practice of traditional medicine necessitates biodiversity conservation as the practitioners largely depend on plants and forests for their medicines.

Furthermore, the THPs are consulted by the people to provide expert services and to bring about social justice to victims of criminal misdemeanours perpetrated against other unsuspecting community members (Humbe 2018). Some THPs are renowned for being specialists in dealing with different animal ailments. While veterinary scientists cannot treat barren livestock to conceive, some healers in Madamombe area are specialised in this. Whereas minor livestock ailments and other agrarian challenges are managed at household level, there are other challenges that need the intervention of THPs. A traditional healer confirmed that people in the area paid her frequent visits seeking redress to problems associated with different agrarian challenges such as poor crops, crop and animal theft, livestock diseases, among others (Mbuya Mutawi, 20 October 2021). In the hyper inflationary environment in which accessing basic veterinary services is becoming a mirage, people in rural areas tend to rely on traditional medicine and the services of traditional healers who are easily accessible in their communities.

18.4 Crop Management: Planting, Enhancing, Protecting, and Postharvest Storage

The people in Madamombe area also extensively rely on traditional remedies based on their inherited local knowledge in crop management. Crop management involves planting, enhancing harvest, protecting crops in the fields, preserving and storing seeds and harvest. At the commencement of the planting season, some communal farmers in the area made sure that certain rituals are performed. The performance of certain agricultural rituals has always been an integral part of the Karanga people's way of life. Rituals such as the rainmaking ceremony (*mukwerere*) were meant to facilitate a successful farming season. In a related study carried out in rural Sri Lanka, it was also noted that eco-friendly crop protection practices such as rituals have survived since they are effective and "if these had no real effect they would have disappeared long ago" (Senanayake 2006: 91).

Most often, farmers make appeal to indigenous medico-ritual practices to enhance their crop yields. As confirmed by some informants, African traditional medico-spiritual potency is used by some farmers in the area to reverse misfortune on crop yields. For instance, *Divisi* is widely used for crop enhancement by the different sub Shona groups across Zimbabwe. In Madamombe area, the Karanga people are also known to have traditionally used *divisi* as a strategy to enhance their yields. According to several informants who were familiar with the practice, *divisi* comes in diverse forms and dimensions. *Divisi* is defined by some as a type of witchcraft that can be used to enhance agricultural productivity among the rural poor. This practice is widespread among many Shona people and is interpreted differently by people in different localities. Humbe (2018: 269) contends that "agricultural heightening is powered by practising incestuous relationships, particularly between parents and their natural children who are popularly known as *vana vembeu* (children of the seed)". This practice is associated with the use of charms "to protect his or her family from danger, to protect his property and livestock from intruders. This is known as *kuromba.* In genuine African spirituality, *kuromba* is very positive and socially acceptable" (Ibid).

Another type of *divisi* which is meant for crop enhancement is when an individual transplants some seedlings from someone's fields and replant them in his own field. It is believed that when this happens, an individual will get an enhanced crop yield since all the potential yield from his victim will be transferred to his field. There are also reported instances when some individuals have been caught red-handed rolling stark naked in other people's fields with potentially good harvests (Mutenhe Kennias, 15 November 2021). The belief is that when one does this, the victim's impending good harvest will be transferred to the person with this kind of *divisi* (Ibid). Indeed, these practices have resulted in constant friction between the communal farmers in the area. Some of the conflicts emanating from these practices have ended up at the village heads or headman's traditional courts.

Another informant narrated how his uncle, would in a bid to boost agricultural productivity, collect and mix different seedlings (*divisi*) such as cereals, groundnuts, maize and others of a similar nature. The seeds were mixed with some indigenous medicines and stored for planting. At the commencement of the rain season, he would take a black pot regularly used by his wife and broadcast the seedlings in the field in anticipation of an enhanced harvest (Mupani James, 20 October 2021). A village head also confirmed that AIM is frequently sprayed in cattle pans and fields in Madamombe area to boost livestock and crop productivity (Maramba village head, 25 October 2021).

Furthermore, some *hurudza* who kept large herds of cattle were known to use *divisi.* Several informants intimated that in actual fact, among these perceived cattle, most of them would actually be baboons. The 'cows' would give birth annually without fail and the milking was done by designated members of the family only. The milk was consumed within the family and not given to outsiders in order to conceal the *divisi* (Mutenhe Kennias, 15 November 2021).

In protecting their crops from thieves, the Karanga use *rukwa* (fencing). According to an informant, if an individual attempts to steal from one's field, he will be found roaming about in the field as they would get confused and fail to exit the field (James Makura, 20 October 2021). As testified by several informants in Madamombe area, there are individuals whose fields are systematically avoided by thieves as these people are well known to use AIM in protecting their crops. There are numerous stories narrated by elders in the area in which some people were found wandering about in the fields after attempting to steal from some people's fields. The law enforcement agents such as the police force cannot be everywhere to protect people's property which forces the locals to rely on indigenous remedies. Besides, as advanced by one informant:

> At times we are frustrated by the actions of the corrupt police officers that procrastinate in dealing with theft cases or release thieves without charge. As a result, we take it upon ourselves to protect our crops and livestock. (Makoro Jameson, 22 October 2021)

The People of the Madamombe area have always practiced certain indigenous ways of preserving foodstuffs and seed for prospective seasons. In this regard, they use a variety of traditional medical remedies to conserve their harvests. These methods have proven to be very effective in dealing with pests and diseases. In spite of their importance, some of the indigenous agrarian practices are slowly disappearing or experiencing rapid transformation. This is because people are gradually gravitating towards the cultivation of modern crops such as maize which has become the staple food for many. Indeed, up to the end of the twentieth century, the people in the area were known for growing crops such as millet, sorghum, rapoko, round nuts, groundnuts and maize. They treated the harvest with indigenous remedies such as manure and wood ash mixture as preservative. The ash concoction is sprinkled on the grain before being thoroughly mixed. As one informant reiterated, ash from trees is generally bitter. This bitterness protected the grain from rats and pests for a prolonged period which is much more effective than modern chemicals (Mototi James, 20 October 2021). At times, the local people mix modern chemicals with ashes in preserving maize. They believed it was not only cheaper and secure to do this but that it was a very effective remedy in maize preservation.

Women in particular, due to their perceived patience, were known to possess superior skills of identifying, processing good seed varieties of grain cobs. These cobs were tied and hanged inside kitchen. Smoke wafted from the hearth to the grain to protect it from borers and rats because of a bitter taste from the soot (soot preservation). Sorghum and millet seeds tassles are also slashed, dried and coated with smoke in kitchens awaiting the farming season. These were hanged in kitchens were they were smoke-coated. This has always been a very effective and cheap way of deterring pests since smoke produces a pungent taste. In this regard, there is no doubt that the African people have, over the years, developed detailed and elaborate methods of preserving seeds that modern scientists have a lot to learn from.

While it is a fact that the cultivation of traditional small grain crops such as sorghum and millet has lost traction in the last two decades or so, there is no doubt that people in the area are slowly gravitating towards their revival because of climate change and recurrent droughts. The recurrent droughts induced by cyclones such as EL NINO in the 2019 farming season urgently demand that subsistent farmers in drought prone areas embrace the growing of drought resistant small grains such as rapoko, sorghum (*mapfunde*), finger millet and the rearing of adaptive animal genus such as the hard Mashona type in order to sustain their livelihoods. As one subsistent farmer observed, the growing of small grains have numerous advantages including the fact that they act as an adequate buffer against hunger and starvation. Finger millet does not necessarily need any pesticides to preserve it (John Mandaza, 20 October 2021). Besides, finger millet is known to possess medical properties that can be used in treating ulcers. Though sorghum is prone to pesticides, the use of spirostactiys (*mutovhoti*) branches which produce a very strong scent can be effective in sending away beetles (*zvipfukuto*) and grain borers (Ibid).

Due to the prevailing socio-economic challenges in the country today, many people are finding it extremely difficult if not impossible to purchase treated seeds from reputable companies such as SEEDCO due to their prohibitive costs. At times, farmers miss the planting season as they fail to purchase seeds in time due to financial constraints. This has forced the poor people to rely on Presidential inputs which are usually distributed on partisan lines. A Movement for Democratic Change-Alliance (MDC-A) supporter lamented:

> We have always been struggling to access agricultural inputs such as seeds and fertilisers. These are distributed by our local Member of Parliament and councilors who are members of the ruling ZANU-PF party. These people always make it a point that members of the opposition are deprived of these inputs despite the fact that the President is on record saying that the inputs should not be distributed on partisan lines. (Machuma Jane, 20 October 2021)

Nonetheless, some informants who benefited from the maize seed complained that those treated seeds had a very low level of germination. As one informant retorted:

> We are right now in the middle of replanting using seeds we had stored from the past farming season. These seeds from government are substandard. On the other hand, we cannot afford seeds from reputable seed companies which are exorbitant. (Peter Chitayi, 20 December 2021)

The foregoing challenges are forcing people in Madamombe area to embrace indigenous agrarian approaches which have been inherited from generation to generation.

Traditional medicine is also used in preserving foodstuffs such as meat when it was obtained in abundance for future consumption. Dried meat (*chimumukuyu*) is preserved by cutting meat into long stripes which are then sprinkled with salt or ash of certain plants used in place of salt. The meat is then hung over a fire to absorb smoke and left to dry. This practice is particularly common among the Karanga during hunting safaris where they sometimes spend days away from home. Meat from domestic animals is also processed in the same manner though it may usually be boiled first before being dried.

The same process is used in preserving fish, mice and mushroom. Other plants were also used as catalyst to ripen or soften food. In this regard, there is no doubt that IKS can still play a fundamental role in food and seed preservation at minimal cost as it is easily accessible, affordable and effective. Besides providing food security for the local people, a lot of income can also be generated by the locals as they can sell the various dried food products to sustain their families. This is particularly true of communities in Madamombe area where livelihood opportunities are limited.

In Madamombe area, post harvest storage, preservation of seeds and food has been mostly the preserve of elderly women whose understanding of indigenous knowledge systems has always been comprehensive through years of trial and error. The same sentiments are echoed in other studies with regards to other Zimbabwean rural communities were the selection of appropriate seeds was mostly done by elderly women (Makamure et al. 2001; Matsa and Mukoni 2013; Parawira and Muchuweti 2008).

After the grading and selection processes, the grain for future use was protected from weevils by using indigenous medicine. As noted earlier, smoke coating and ash mixtures/soot have traditionally been the predominant chemicals the peasant farmers use in the storage of most seeds and grain crops for use in the long term as they are stored in granaries. Among the Karanga, the women who had ordinarily perfected the skills of classifying the best seed varieties collect maize cobs and grain, tie and hang them inside their kitchens directly under the fire place. The grain would then be coated by the smoke wafting from the hearth. As Mapara (2009: 150) observes:

> In this manner, the grains were protected from grain borers and rats because of the bitter taste that would result from the soot. By employing this method, the people ensured that they could use the seed even after two or three seasons. When this method is compared to current ones that are used by seed houses, it can be observed that it is very difficult to keep today's seed for more than one season. The short life span of today's seed is an indicator that Western science does not always benefit those who are interested in long term planning, especially those who want to stock seeds for more than two seasons.

While keeping grain in granaries is no longer as pervasive as in the past, some farmers, especially those involved in the cultivation of sorghum, rapoko and millet still keep grain in granaries which are carefully constructed. Over succeeding generations, the Karanga people in the area have developed elaborate scientific knowledge of identifying suitable sites for granaries, for example, on top of rocks to be free from moisture and termites. The inside of the granaries were wrinkled with cow-dung. This further protected the grain from borers.

An informant, Kennias Mutenhe highlighted that people in the area at times place gum tree leaves on the floor of the granary, on the middle and on top of the harvest to protect the grain from pests (Mutenhe Kennias, 15 November 2021). The granaries were then tightly sealed or closed as people wait for the next rain season. Tightly sealing the granaries ensured the suffocation of the pests already in the granaries due to lack of oxygen. This ensured food security among the locals. The practice resonates with other studies that have been carried out on other Zimbabwean indigenous people (Nyota and Mapara 2008; Matsa and Mukoni 2013).

Nonetheless, the use of AIM in the public health care system and in agriculture has resulted in the endangering and extinction of some plant species in Madamombe area. As the species and genetic diversity is disturbed, the ecological stability of the habitats in which these are found is affected as well. Magoro (2008: 5) states that:

> The loss of biodiversity from habitat destruction and unsustainable harvesting practices often means that a range of medicines will no longer be available to both rural or urban THPs, and increasing pressure on diminishing wild stocks of plants in conservation taboos being ignored and ultimately lost.

What is also significant is that habitat loss is more often than not associated with the loss of indigenous knowledge. While THPs harvest plant medicines in a sustainable manner, traditional medicine is used by almost everyone in dealing with human and non human ailments. The non THPs sometimes use unsustainable ways of harvesting thereby posing a serious threat to habitats. In this regard, it then becomes critical for both traditional healers and the ordinary people to "understand traditional medicine from a traditional ecological perspective and the knowledge to protect threatened species" (Ibid: 6). In protecting the medicinal plants that benefit communities, the enforcement of restrictions that guard against the overexploitation of these plants species becomes of essence. In Madamombe area, traditional leaders and community policemen continue to play a critical role in this regard. Strengthening and enforcement of indigenous resource utilisation and management practices can result in minimal ecosystem destruction and better disease control.

18.5 The Future of the Use of AIM in Agriculture

Contemporary agrarian challenges induced by climate change and variability, among other factors, calls for the embracing of indigenous knowledge practices of processing, preservation and storage of crops and other food types by communal farmers. In this regard, documenting these indigenous agrarian practices then becomes imperative. If this is not done immeasurable knowledge of indigenous agrarian practices that may be beneficial in improving food security and income generation may be lost in the fast modernising world. As argued by many decolonial scholars, it is about time that policy makers, development practitioners and researchers realise that there is no knowledge system which is superior to the other. All knowledge systems developed in particular environments and protracted their communities since time immemorial. Nkondo (2012) avers that all knowledge systems have initially been localised but were universalised through the process of subjugation and colonialism. Therefore, as Vilakazi (1999: 203), cited in Maila and Loubser (2003) puts it:

> Africans need to acknowledge that their indigenous processes of knowing and knowledge production may not only enhance and sustain them as a people, but could also contribute to the global pool of knowledge in search of sustainable solutions to global challenges such as climate change…This implies that Africa cannot be excluded from global influences; neither should Africa be guided only by her past because no civilisation...can manage to develop in isolation from 'others'.

Indigenous knowledge base should be reinforced and improved through further research by academics for the benefit of many Third World countries which "have survived on it for centuries and will continue to do so for the foreseeable future" (Dhewa 2008: 6). As Dhewa (2008: 11) further states "traditional medicine is…a part of Africa's development resource not well-studied, not adequately appreciated and developed". Certainly, thoughtless would be a person who despises his own knowledge forms. When people pay little attention to their knowledge system, it may be easily threatened with erasure or extinction or may be effortlessly misappropriated (Mkhize 2021). Deconstructing the fact that western knowledge is superior to indigenous knowledge can be facilitated through intra-generational transfer of IKS farming practices.

Furthermore, the teaching of IKS should be incorporated into the school curricula at primary, secondary and tertiary levels. Universities and agricultural colleges in particular should prioritise the teaching of IKS. Higher education in Africa should be used to debunk the notion that western approaches are always the best. The carrying out of research on medicinal properties of plants in institutions of higher learning will indisputably increase the knowledge of their applicability in agricultural practices. As Mothibe and Sibanda (2019) assert, while there has been a lot of research going at different levels, the efforts are largely uncoordinated which significantly hinders the acceptance of traditional agrarian practices at the official level. Growing the knowledge on the use of traditional medicine in agriculture will facilitate the exchange of local and western knowledge as stigma against the former fades away.

IKS has often been perceived as pagan, inferior, primitive, backward and unchristian and has been looked down upon (Chanza and Mafongoya 2017). To add to that, Christianity has poisoned people's minds to the extent where they regard traditional knowledge as a myth rather than a source of production. To a greater extent, higher education in Africa has been used to strengthen rather than to break the bond with western scientific ways of doing things. Integrating or synthesising IKS with modern agrarian approaches will go a long way in improving food security and people's livelihoods especially among the most vulnerable communities. If the teaching of IKS is prioritised in schools, this will help in dispelling the myths and prejudices that are associated with the preservation of this alternative knowledge system which has sustained African communities since immemorial times. Matsa and Mukoni (2013: 244) observe that:

> Overlooking indigenous knowledge of farming practices, processing, preservation and storage of food crops may create negative attitudes not only towards agricultural extension officers who implemented modern methods. They may be viewed by indigenous societies as officials who are bent on destroying their IKS as viewed by post-colonial theorists. If agricultural extension officers get to understand IKS and use correct vernacular terms used in local environments, this would boost understanding of farming practices by both subsistence communities and officials themselves.

Thus, the preservation of Karanga agrarian practices anchored on the medico-spiritual realms then become imperative as these practices have sustained them for thousands of years.

18.6 Conclusion

In conclusion, it can be observed that while some of the Karanga indigenous agrarian practices are shrinking as some elderly indigenous knowledge holders perish, there is still widespread use of AIM for sustainable agriculture in Madamombe area. As traditional healers pass on, this knowledge system is threatened with extinction; hence there is need for its documentation for posterity. There is a lot to learn from the IKS of the local people. In agriculture, the use and application of suitable IK can significantly alleviate rural poverty through increased agricultural output. Indeed, IK can act as a community's armour in mitigating and adapting to environmental shocks. It provides local communities the opportunity to tackle agricultural challenges by embracing endogenous solutions at their disposal.

Indeed Africans should appeal to their indigenous medico-spiritual assets in sustainable agriculture to diminish the overdependence on western-centric protocols in addressing African needs. It could be a bail out of the wider dependency syndrome that is characteristic of Black Africa. The use of traditional medicine in agriculture can proffer more affordable and sustainable solutions to agrarian challenges facing communal farmers in Zimbabwe today. In that regard, institutions of higher learning, policy makers and planners should prioritise the teaching of IKS in their curricula as the acknowledgement that all knowledge systems are culturally embedded continues to gain currency.

References

Ajayi, O.C., and P.L. Mafongoya, eds. 2017. *Indigenous Knowledge Systems and Climate Change Management in Africa*. CTA: Wageningen.

Berkes, F. 2018. *Sacred Ecology*. New York: Routledge.

Chemhuru, M. 2010. Democracy and the Paradox of Zimbabwe: Lessons from Traditional Systems of Governance. *The Journal of Pan African Studies* 3 (10): 180–191.

Dhewa, C. 2008. Is Traditional Medical Practice in Africa still Community Property? Lessons from Zimbabwe. In *Governing Shared Resources: Connecting Local Experiences in Global Challenges. 12th Biennial Conference of the International Association for the Study of Commons, Cheltenham, England.*

Humbe, B.P. 2018. *Divisi* Witchcraft in Contemporary Zimbabwe: Contest Between Two Legal Systems as Incubator of Social Tensions Among the Shona People. In *Religion, Law and Security in Africa*, ed. C. Green, T.J. Gunn, and M. Hill, 269–283. ACLARS, African Sun Media.

Johnson. M. 1992. *Lore: Capturing Traditional Environmental Knowledge*. Ottawa: IDRC.

Kambu, A. 2010. Bridging Formal and Informal Governance Regimes for Effective Water Management: The Role of Traditional Knowledge. In *Traditional Knowledge in Policy and Practice: Approaches to Development and Human Well-Being*, ed. S.M. Subramanian, and B. Pisupati, 252–266. New York: United Nations University Press.

Kanu, I.A. (ed.). 2021. *African Indigenous Ecological Knowledge Systems: Religion, Philosophy and Environment*. Maryland: APAS.

Magoro, M.D. 2008. Traditional Health Practitioners' Practices and the Sustainability of Extinction-prone Traditional Medicinal Plants. Masters dissertation, University of South Africa.

Maila, M.W., and C.P. Loubser. 2003. Emancipatory Indigenous Knowledge Systems: Implications for Environmental Education in South Africa. *South African Journal of Education* 23 (4): 276–280.

Makamure, J., J. Jowa, and H. Muzuva. 2001. In *Liberalisation of Agricultural Markets*. SAPRI Zimbabwe, Harare.

Makinde, M.A. 1988. *African Philosophy, Culture and Traditional Medicine*. Ohio Centre for International Studies.

Mapara, J. 2009. Indigenous Knowledge Systems in Zimbabwe: Juxtaposing Postcolonial Theory. *Journal of Pan African Studies* 3 (1): 139–155.

Matsa, W., and M. Mukoni. 2013. Traditional Science of Seed and Crop Yield Preservation: Exploring the Contributions of Women to Indigenous Knowledge Systems in Zimbabwe. *International Journal of Humanities and Social Science* 3 (4) [Special Issue–February 2013]: 234–245.

Mazarire, G.C. 2008. 'The Chishanga Waters have their Owners': Water Politics and Development in Southern Zimbabwe. *Journal of Southern African Studies* 34 (4): 757–784.

Mkhize, N. 2021. African/Afrikan-centered Psychology. *South African Journal of Psychology* 51 (3): 422–429.

Mokgobi, M.G. 2014. Understanding Traditional African Healing. *African Journal for Physical Health Education, Recreation and Dance* 20 (Sup-2): 24–34.

Mothibe, M.E., and M. Sibanda. 2019. African Traditional Medicine: South African Perspective. *Traditional and Complementary Medicine* 1–27.

Mpepereki, S. 2015. Food Insecurity: Result of Abandoning Sound African Science*:* In *The Patriot, Celebrating Being Zimbabwean*. Harare (October 30–November 5, 2015).

Mtshali, M.N.G., T. Raniga, and S. Khan. 2014. Indigenous Knowledge Systems, Poverty Alleviation and Sustainability of Community Based Projects in the Inanda Region in Durban, South Africa. *Studies of Tribes Tribals* 12 (2): 187–199.

Mudzengi, C.P., E. Dahwa, J.L.N. Skosana, and C. Murungweni. 2014. Promoting the Use of Ethnoveterinary Practices in Livestock Health Management in Masvingo Province, Zimbabwe. *A Journal of Plants, People and Applied Research* 12: 397–405.

Nkondo, M. 2012. Indigenous African Knowledge Systems in a Polyepistemic World: The Capabilities Approach and the Translatability of Knowledge Systems. In *The Southern African Regional Colloquium on Indigenous African Knowledge Systems: Methodologies and Epistemologies for Research, Teaching, Learning and Community Engagement in Higher Education*. Big Spring: Howard College.

Nyoni, B. 2017. Abundant Life and Basic Needs: African Religions as a Resource for Sustainable Development, with Special Reference to Shona Religion. Ph.D. thesis, Theologische Fakultät der Universität Rostock.

Nyota, S., and J. Mapara. 2008. Shona Traditional Children's Games and Play: Songs as Indigenous Ways of Knowing. *Journal of Pan African Studies* 2 (4).

Obiora, A.C., and E.E. Emeka. 2015. African Indigenous Knowledge System and environmental Sustainability. *International Journal of Environmental Protection and Policy* 3 (4): 88–96.

Parawira, W., and M. Muchuweti. 2008. An Overview of the Trend and Status of food Science and Technology Research in Zimbabwe over a Period of 30 Years in Scientific Research and Essays. *Academic Journals* 3 (12): 599–612.

Payyappalli, U. 2010. Supplementary Feature Knowledge and Practitioners: Is There a Promotional bias? *Traditional Knowledge in Policy and Practice* 194–208.

Risiro, J., D.T. Tshuma, and A. Basikiti. 2013. Indigenous Knowledge Systems and Environmental Management: A Case Study of Zaka District, Masvingo Province. *International Journal of Academic Research in Progressive Education and Development* 2 (1): 19–38.

Senanayake, S.G.J.N. 2006. Indigenous Knowledge as a Key to Sustainable Development. *The Journal of Agricultural Sciences* 2 (1): 87–94.

Sillitoe, P. 1998. The Development of Indigenous Knowledge: A New Applied Anthropology. *Current Anthropology* 39 (2): 223–252.

WHO. 2000. *Traditional Medicine: Definitions*. Available at http://www.who.int/medicines/areas/traditional/definitions/en/. Retrieved August 10, 2021.

Informants

Chiondegwa Upenyu, Chiondegwa Village, 25 September 2021.
Chitayi Peter, Makambe Village, 20 December 2021.
Dzoro Living, Chingovo Village, 15 November 2021.
Machuma Jane, Makambe Village, 20 October 2021.
Makoro Jameson, Nyamadzawo Village, 22 October 2021.
Makura James, Interview, Makambe village, 20 October 2021.
Mandebvu Robert, Mhandamabwe Business Centre, Chivi District, Interview, 02/07/2019.
Mandiva John, Muzvidziwa Village, 20 0ctober 2021.
Maramba Village Head, Maramba Village, 25 October 2021.
Mbuya Mutawi, Maramba Village, 23 October 2021.
Mototi James, Mutsauri Village, 20 October 2021.
Muguti Moses, Makambe Village, 23 October 2021.
Mukono James, Interview, Nyamadzawo Village, 20 October 2021.
Mupani James, Interview, Gwapedza Village 20 October 2021.
Mutenhe Kennias, Gwapedza Village, 15 November 2021.
Shumba Simon, Maramba Village, 11 September 2021.

Tasara Muguti is a Senior lecturer in the History, Archaeology and Development Studies department at the Great Zimbabwe University. He holds a Bachelor of Arts (Hons) in Economic History, a Master of Arts in African Economic History and a Graduate Certificate in Education, all obtained from the University of Zimbabwe. He is currently a registered Ph.D. Student with UNISA. He has several published book chapters and journal articles to his credit. His research interests are in Indigenous Knowledge Systems, with special emphasis on African traditional medicine, land reform, human rights, democracy and many other topical issues on contemporary Southern African studies.

Part V
Contemporary Agrarian Issues in Africa

Chapter 19
Henry Odera Oruka's Parental Earth Ethics as Ethics of Duty: Towards Ecological Fairness and Global Justice

Pius Mosima

Abstract The current global ecological crises have, *inter alia*, led to an upsurge in massive migrations, food crises, diseases, pandemics, increased conflict and war especially in African societies. Most of those affected are the small farmers in rural communities who depend on agriculture. This crisis has not only raised concerns about the extent of the damage humans and human activities are causing to the natural environment but has also ignited discussions about the urgent necessity for a change in human behavior and our obligations towards a new agrarian philosophy. In this chapter, I argue that Henry Odera Oruka's Parental Earth Ethics could be adopted as a new principle of global justice and for grounding global environmental ethics. This principle invites us towards rethinking ecological fairness and global justice. Specifically, I opine that Oruka's Parental Earth Ethics could be a veritable resource in our search for: (i) promoting global justice in the use and equal distribution of wealth and world resources for the common good and security of all, and (ii) global environmental ethics. It is basically driven by the ethical principle of the human minimum, which ought to inspire us humans to be more ethically and politically responsible in their use and distribution of world resources for the good and security of all.

Keywords African Agrarian philosophy · Eco-philosophy · Parental earth ethics · Global environmental ethics · Global justice · Wellbeing

19.1 Introduction: Philosophy Must Be Made Sagacious

The name Henry Odera Oruka (1944–1995) commands a lot of respect in the history of contemporary African philosophy. His influence and respect do not only come from his attempt to order and structure the discourses on the nature and possibility of

P. Mosima (✉)
Department of Philosophy, University of Bamenda, Bamenda, Cameroon
e-mail: piusmosima@yahoo.com

M. J. Tosam and E. Masitera (eds.), *African Agrarian Philosophy*, The International Library of Environmental, Agricultural and Food Ethics 35,
https://doi.org/10.1007/978-3-031-43040-4_19

an African philosophy (hence with a platform from which to understand and interpret the discourses), but he also made an original contribution to these discourses. His insights and brilliance have been universally sanctioned, inter alia, as being synonymous with the concept and Trend of Sage philosophy/Philosophic sagacity.[1] Yet his contributions to practical philosophy, with issues related to discourses on justice, poverty and freedom, underdevelopment, environmental ethics which are crucial in his works (Oruka 1997), are lesser known and have not been explicitly addressed.[2] Anke Graness (2012, 2015), for example, points at the exclusion from the academic discourse of a scholar like Oruka, and thinks that the debate on global justice, a debate which is at the core of global ethics, is largely being conducted by European and American scholars from different disciplines without taking into account views and concepts from other regions of the world, particularly, from the Global South. The lack of a truly intercultural, interreligious, and international exchange of ideas provokes doubts whether the concepts of global justice introduced so far are able to transcend regional and cultural horizons.[3] How to tackle problems of global justice and ecological fairness have puzzled and provoked thinkers for ages. There is the need for careful, systematic and critical thinking from a broader perspective that characterizes philosophy. Philosophy, in its many forms and guises, has a broader relevance than many people seem to realize. Oruka was convinced that philosophy has a special mission of enhancing the socioeconomic wellbeing of people without causing havoc to the environment. Oruka's famous maxim was that "philosophy must be made sagacious" (Graness and Kresse 1997: 253–254); meaning, philosophy ought to be practically relevant in creating a society where everyone feels ethically responsible toward the wellbeing of the 'other', a society that privileges no one and excludes no one. That would translate into improvement in a global society that is both egalitarian and communitarian oriented.[4] One area that Oruka's practical philosophy could be relevant and conveniently applied is in agrarian philosophy. Agrarianism concentrates on the fundamental goods of the earth, on communities of more limited economic and political scale than in modern society, and on simple

[1] For more on Oruka's philosophic sagacity, see, for example; Azenabor (2009), Graness and Kresse (1997), Janz (2009), Kalumba (2002, 2004), Kresse (1993, 2007), Masolo (1994, 1997, 2005), Mosima (2016, 2018), Ndaba (1996), Ochieng'-Odhiambo (1994, 1996, 1997, 2002a, b, 2006, 2007), Oduor et al. (2018), Oseghare (1985), Presbey (1996, 1997, 1999, 2000, 2002, 2007, 2012),Tangwa (1997), Van Hook (1995).

[2] It would be instructive to note that only two of Oruka's (1990, 1991) books explicitly address issues concerning the nature and possibility of an African philosophy. The other books focus on what he calls practical philosophy. Following Immanuel Kant's distinction between theoretical reason and practical reason, Oruka opines that the former treats issues about the fundamental principles of knowledge and the metaphysics of reality, while the latter addresses principles of ethics and the rules of their application in the social, political, religious and legal life of humankind (Oruka 1997: xi).

[3] African/Africanist scholars have been discussing this especially towards environmental problems Chimakonam (2017), Okeja (2018) and Boogaard (2019).

[4] Oruka states his passion for practical philosophy in these words "But I was more interested (coming from science) in philosophy that would be useful for understanding the problems of Africa" (Oruka 1997: 212).

living, even when the shift involves questioning the "progressive" character of some recent social and economic developments. As a social and political philosophy it promotes subsistence agriculture, smallholdings, egalitarianism by supporting the rights and sustainability of small farmers and poor peasants against the wealthy in society. Agrarian philosophy stresses the role of nature, soil and climate in the formation of moral character as well as social and political institutions. One thing Oruka acknowledges is the scale of our environmental crises which is undeniably global. Environmental problems do not seem to respect conventional boundaries drawn by nation-states or communities and so need a global ethics. Following Oruka, I refer to African agrarian philosophy to an area of critical enquiry into the nature and justification of values and norms that are global in kind and into the various issues that arise such as world poverty and international aid, environmental problems, and human rights. Oruka thinks that earth is a common good for all and that there are or should be rights and obligations that obtain among the inhabitants of the earth in relations to the use and distribution of the resources of the earth. The rights and duties should be such that they ensure the preservation of human life as a fundamental right as well as furthering the enrichment of human life and all of nature. To illustrate that the earth is a common wealth, Oruka uses the analogy' of "Parental Earth Ethics" (Oruka 1993). In this analogy, the earth or the world is analogously equivalent to a Family Unit, in which members have kith and kin relationship with one another. The earth is a common wealth and hence every human being has a right to share in the resources of the earth. It is an ethic of duty towards man and the environment, because, according to him, there is a link between the manner of thinking our link to the universe, the cosmos, and the manner of thinking a fair society.

In this chapter, I argue that Oruka's Parental Earth Ethics is a starting point for developing a more truly ecological orientation to Agrarian philosophy as it could be relevant in fostering global justice and ecological fairness. This chapter seeks to provide answers to these two different but interrelated questions: What are the main arguments in Oruka's Parental Earth Ethics and how relevant are they to current discourses on global justice and ecological fairness? Could the principles of Parental Earth Ethics provide a practical basis on which to formulate a new conception of global justice and ecological fairness?

19.2 Grounding African Agrarian Philosophy on Oruka's Parental Earth Ethics

Oruka grounds his ethical and political theories on human realities and existential facts about human beings and the human condition. These realities, according to Oruka, include poverty, underdevelopment, freedom, socio-economic injustice, inhumanness, and environmental degradation stem from our global irresponsibility (Oruka 1997: xi). Even though he argues from an African base with a strong appeal for bodily needs and political freedom, he thinks these issues are of great global concern and could be of intercultural relevance in discourses on global justice (Graness 2015). For Oruka, it is a moral imperative for philosophers to adopt practical ways of improving the wellbeing of the community. This would require developing new methods of solving classical metaphysical paradoxes (Oruka 1997: 99). Oruka referred to this requirement as "the ethical minimum" which is an absolute moral principle which is based on a universal right with a corresponding universal duty. Oruka asserts that the right to a human minimum is the right of every moral agent, which the world owes him or her, to a life with the dignity of a human being. It is the very minimum a human being demands from the world so that he or she may be in a position to understand and recognize the rights of others. Otherwise he or she would become morally blameless if he refuses to acknowledge the right of anybody or nation to anything however legal (see Oruka 1993: 23). This right does by definition impose a duty on every moral agent (no matter from which part of the globe) to help ensure that there are no human beings who lack the means to fulfill this right. Oruka saw it as the pivotal guiding principle that can establish an egalitarian global society, a more humanized life on earth (Oruka 1997: 85) a society that will ensure a genuine practice of justice (i.e., egalitarian and ecological fairness) at global level beyond national borders and not charity or humanitarian aid. In "Parental Earth Ethics", a paper first published in 1993 in the journal *Quest* (Vol. VIII, No. l, June), and later published as a chapter in the revised edition of his book *The Philosophy of Liberty* (1996) and *Practical Philosophy: In Search of an Ethical Minimum* (1997) respectively,[5] Oruka responds to an article by Garrett Hardin "Lifeboat Fthics: The Case Against Helping the Poor" which was first published in 1974 (*Bioscience*. Vol. 24, No. 10, October 1974).[6] Let us examine Hardin's argument below.

19.3 Hardin's Argument

In the article "Lifeboat Ethics: The Case Against Helping the Poor", Hardin (1974) questions policies such as foreign aid, immigration and food banks. He introduces the Life Boat Ethics to help him defend a position that finds it senseless and suicidal for the rich nations to offer charitable or humanitarian aid to the poor nations. His argument is most likely a response to two basic directions in the modern practice of

[5] In this chapter I am using the 1993 version.

[6] Garrett James Hardin (1915–2003) was an American ecologist who warned of the dangers of overpopulation. He focused on human population growth, the use of the Earth's natural resources, and the welfare state. He argued that if individuals relied on themselves alone, and not on the relationship of society and man, then the number of children had by each family would not be of public concern. A utilitarian, Garrett Hardin in his Lifeboat Ethics, a metaphor he used for resource distribution, argues that an international state should refrain from sharing resources with and providing help for other states to maximize its people's welfare. The global resources are finite and states ideally should share it equally for maximum collective interest. The version I refer to is printed in The Dialectics of Third World Development, I. Volger and A. De Souza, Rowman and Allenheld Publishers, USA, pp. 171–185.

foreign aid. First there is the environmentalist ethics with an emphasis on the notion of the Spaceship Earth or the Finite Global Truth; and second, the international philanthropists who are out to eradicate poverty.[7] Hardin compared the lifeboat metaphor to the Spaceship Metaphor of resource distribution, which he criticizes by asserting that a spaceship would be directed by a single leader which the Earth lacks. Hardin asserts that the spaceship model leads to the Tragedy of Commons.[8] In contrast, the lifeboat metaphor presents individual lifeboats as rich nations and the swimmers as poor nations.

Hardin argues that the notion of a Spaceship Earth would make sense only if the world had a governing sovereign. Such a sovereign would need to make sure that within the Space ship those on board have their rights only in so far as they have corresponding responsibilities. He notes that:

> What about Spaceship Earth? It certainly has no captain and no executive committee; The United Nations is a toothless tiger because the signatories of its charter wanted it that way. The spaceship metaphor is used only to justify spaceship demands on common resources without acknowledging corresponding spaceship responsibilities.

In the second place there are the usual International philanthropists who are inclined to the eradication of hunger and general poverty as an understandable responsibility for any reasonable moral agent. At the time he was writing, Hardin asserted that approximately two-thirds of the world's population was "desperately poor" and the remaining one-third was "comparatively rich" before launching his metaphor of each rich country being in a full lifeboat, while poor countries were in "much more crowded lifeboats"; he described emigration as a continuous process where "the poor fall out of their lifeboats and swim for a while in the water outside, hoping to be admitted to a rich lifeboat, or in some other way to benefit from the 'goodies' on board" before asking what the rich lifeboat should do. He argues that wealthy people should not be responsible for the poor and that the consequences of feeding the poor are detrimental to the environment and to the society as a whole. The rich or affluent have the obligation to help only themselves and their posterity; to ensure their survival and wellbeing. According to him, the poor are too many to be helped by the rich without a threat to the very survival of the rich. Consequently, rich nations should stop giving foreign aid to the poor nations that are in need. The well-developed nations, including the United States of America and other European countries are known for the aid they offer whenever a country is in need. However, Hardin claims that giving a foreign aid to other countries in need will be detrimental to the rich nations' economies. Hardin is very skeptical when it comes to foreign aid and he encourages the rich nations to stop helping those in need so as to them a lesson on

---

[7] It is also referred to as the Spaceship Earth or Spacecraft Earth or Spaceship Planet Earth. It is a worldview encouraging everyone on earth to o act as a harmonious crew working toward the greater good.

[8] The tragedy of the commons refers to a situation in which individuals with access to a shared resource (also called a common) act in their own interest and, in doing so, ultimately deplete the resource. This leads to over-consumption and ultimately depletion of the common resource, to everybody's detriment.

how to control their population and better their living conditions. To say that "it is time to refuse to give aid in the form of food to needy countries that do not accept responsibility for limiting their population growth. He explains in many occasions throughout the text that programs such as "food for peace program", will only do harm than benefit the countries eventually. Because of the failure of the food for peace program, Hardin assumes that the newly developed program called the World Food Bank, will also fail because of human greed and it will do no good but harm those who are in need.

In Hardin's view, many of the richer countries are seen as which is only capable of carrying so many people. People in poorer countries are "in the water" and want to get into the lifeboat which represents the rich countries. By letting more people on the lifeboat than the boat can handle will drown everyone.

Hardin illustrates this argument by using an analogy of a lifeboat with the full capacity of 50 and a safety factor of 10 people having many desperate people, more than its full capacity, swimming towards it. That means that there are already 40 people in the lifeboat. The captain has three options of action. The first option is, following the Christians and Marxist maxims which would teach us to help all those who approach us for help, to admit as many people as the lifeboat can take until there is no more space left that is until it sinks. This is what he refers to as complete justice, complete catastrophe. The second option is to admit 10 more people, on first come first admission, to take up the safety factor and lose the safety factor. Hardin asserts the reduction in safety factor to zero will be paid for dearly and asks "which ten do we let in? Would we not feel guilty for admitting some and leaving out others? Those admitted will not themselves feel guilty for their luck, for if they are, they would not in the first place accept the offer. The third option, which Hardin sees as the way out, is to admit nobody and ensure the safety of all those already aboard. "Don't admit anymore to the boat and preserve the safety factor. Survival of the people in the life boat is then possible" (Hardin 1974: 173). So, according to Hardin, that is the realistic truth that should prevail in the modern world even if many see it as unjust. He notes further that the aid from the richer to the poorer nations is made absurd and unsolvable because the former control their population growth, while the latter are breeding exponentially and often even in direct proportion to the aid which they receive from outside. Without foreign aid, Hardin observes, they would sooner be checked in their human growth by pestilence, wars, earthquakes etc. But with foreign aid they will continue to multiply and the doomsday is postponed. Yet, when it comes, it will be doomsday for the aid receivers, but also for the aid donors.

19.4 Oruka and Parental Earth Ethics: The Earth as a Common Wealth

Oruka acknowledges that when he wrote his 1986 paper "The Philosophy of Foreign Aid: A Question of the Right to a Human Minimum" he was apparently not aware of Garrett Hardin's paper. But after reading Hardin's paper, he had to update his argument. Oruka earlier considered international charity, international trade and historical rectification as the main reasons for the practice of foreign aid. Yet he later dismissed all three reasons as insufficient and substituted the right to a human minimum as the basis for the practice of foreign aid. So "Parental Earth Ethics" was a development of the earlier argument for the right to a human minimum and a response to Hardin's argument.

Oruka articulates and proposes an ethics of distributing world resources that would guarantee, at least, the fundamental universal human rights. This is the ethics of "the rights to a human minimum". This is the right that every moral agent demands from the world in order to live with dignity as a human being, and to recognize the rights of other human beings. This right, according to Oruka would not make sense if one does not assume some common wealth for all human beings, which, of course, exists. The earth and the resources therein are common wealth for all human beings. Even though Hardin argues against the rich helping the poor, it seems there are no debts or common wealth between the boaters and the swimming millions. He does not interrogate the hegemonic and exploitative relationship between the rich countries and the poor countries. In this relationship, which Oruka considers unjust, part of the riches of the rich were gained from this unjust relationship which also contributed to the poor getting more impoverished. For Oruka, at the beginning all boats were poor. Then a number of the sailors of the now rich boats sailed to the now poor boats and. by all means possible, plundered the wealth of many of those boats and used the gain to cause economic and safely disparity between the boats (Oruka 1993: 24). This relationship partly has contributed to the riches of the rich and the poverty of the poor and the stark disparity in wealth between nations. Moreover, whereas Hardin thinks there is just one boat, Oruka thinks in the real world there are many life boats—a few are affluent while the large majority of boats are poor. If there are many rich countries, then that makes it easier for the rich to help the poor without endangering the very survival of the rich. Oruka judges that the rich boats owe their part of their current self-preservation to the gains brought to them by the inter-boat pipes. If indeed all poor boats would sink, eventually all the rich boats would also sink. This explains Oruka's concern for a basic ethics for global redistribution, i.e. aid. But what ethics is relevant for our global environment and redistribution?

19.5 The Earth as a Family Unit

In the "Parental Faith Ethics" Oruka uses an analogy of a family with six children two of whom are relatively rich while four are generally poor. Among the rich, one is extremely rich while of the poor four, three are very poor. The reasons for the differences in wealth have to do partly with family history, partly with personal luck, and partly with individual talents. The children find out that from time immemorial this family finds itself guided by two main unwritten ethical principles; (1) the parental debt (bound) principle (PP), and (2) the individual luck principle (IP).

(1) **The Parental Debt Principle**

The parental debt (bound) principle is comprised of four related rules; (i) The family Security Rule, (ii) The Kinshipshame Rule, (iii) The Parental Debt Rule, and iv) The Individual and Family Survival Rule.

(1a) The Family Security Rule: The family security rule slates that the fate and security (physical or welfare) of each of the members ultimately bound up with the existential reality of the family as a whole. Any one of the six members, may, for example, be arrogant and have enough to claim self-sufficiency and independence from the rest. But sooner or later he or his own children or his grandchildren will experience a turn of events which must surely make them desperately in need of protection from the Family Tree! Therefore, family gives members security.

(1b) The Kinshipshame Rule: This rule states that the life conditions of any one member affect all of them materially and emotionally, so none of the members can reasonably be proud of his situation however 'happy', if any member of the Family Tree lives in decadence. In other words, given the shared humanity, no normal human being can feel happy in his riches when some fellow human being lives in extreme deprivation.

(1c) The Parental Debt Rule: This rule states that whoever in the family is affluent or destitute partly owes his fortune or misfortune to the parental and historical factors inherent in the development of the Family. Hence, within the Family no one is alone fully responsible for his affluence, nor for his misfortune.

(1d) Individual and Family Survival, which states the ethical rule that anyone member of the family has, given (1a), b, c, no moral obligation to refrain from interfering with the possessions of any other affluent or destitute brother or sister who ignores the obligation to abide by the rules of the Family ethics. This rule allows the disadvantaged to demand assistance from the affluent, but it allows creative and hardworking members of the family to repossess undeveloped possessions of the idle relatives and develop them for use to posterity (Oruka 1993: 24–27).

The PDP above stresses the importance of an ethics of distribution which takes into consideration an ethical minimum for our collective survival. The poor or disadvantaged can only demand assistance from the affluent when the assistance would be necessary for individual or group survival. This principle does not justify laziness by some idle people on the back of creative and hardworking people. This explain why Oruka talks about the right of the poor to demand assistance from rich, or the creative and hardworking to repossess the undeveloped resources from the lazy relatives under "the individual and family survival rule". The repossessing of the undeveloped resources is only morally justified if it is intended for the common good; "for posterity". It cannot be done simply for the individual to become richer and enjoy more comfort.

(2) **The Individual Luck Principle**

This principle is made up of three constituent rules namely; (i) the personal achievement rule, (ii) the personal supererogation rule, and (iii) the public law rule.

(2a) The Personal Achievement Rule: This rule states that what a member possesses is due mainly to his or her special talents and work.

(2b) The Personal Supererogation Rule is a corollary of the Personal.

Achievement rule or it assumes the truth of the personal achievement rule. It states that every member of the family has the right to do whatever he/she wishes with his/her possessions.

(2c) The Public Law Rule states that any member of the family who contravenes the right of another member as given by the second principle will be subject to the Family Public Law, and punished or reprimanded to restore justice.

19.6 Priority Order

In order to avoid a conflict of rights, Oruka makes it clear that The Parental Debt Principle is prior to the Individual Luck Principle. Hence, if any of the rule sin 1 comes into conflict with any of the rules in 2, then then 1 takes priority. This is because, according to Oruka,

> … The ethics of common sense shows that when in any given community matters of common wealth and security conflict with matters of personal possession, luck or achievement, the former must prevail over the latter. There is no country in which, for example, one would accept a wish or a will from one of its citizens which stipulates that upon death all his achievements, however dear to the country, should be exterminated or kept out of use by anybody. The reason for such will would be that those achievements are personal and hence the personal supererogation rule is to prevail. The objection to this will can be supported by invoking issues of common origin, common security and common wealth of the community of which the person was a member. (Oruka 1993: 26)

The point that Oruka is reinforcing is that the right to a human minimum cannot be overridden by the property right such as the right of the first or prior occupation, or claim to territorial sovereignty. The property right is not or should not be treated as absolute which gives one an exclusive right over the resources which find themselves in one's possession. The individual luck principle deals with individual right to property; one's possessions. This therefore makes it a secondary right and not a basic or fundamental right and should not be treated as so because it presupposes the right to self-preservation and it is based on the right of the first occupation. This means that there are some other values or considerations of greater moral significance that may override it. For instance, the right to life is prior to it and is of greater moral significance. Therefore, when one's right to life is in conflict not with another person's right to life, but with a right to property, then the right to life should take precedence. The point Oruka is making with the PDP is that the individual has a place but ethical obligations give the reason to maintain family security.

19.7 Parental Earth Ethics as an Eco-philosophy

From his Parental Earth Ethics, Oruka makes a claim that the earth or the world is a Family Unit, in which members have kith and kin relationship with one another. It is an ethic of duty towards man and the environment. The earth is a commonwealth to all humanity, in which we are all interconnected; hence we need mutual care, humility. He is advocating for a kind of global ethics in which the whole Earth is conceived as a global society. It is a sort of "organic unity" governed by the principles of interdependence and ethical responsibility (Graness and Kresse 1997: 257). Hence, from the point of an environmental ethics, Oruka makes it clear that the earth is a "complex web of being" with symbiotic relation, where everything seems to depend on each other for survival; and nothing or no one is supposed to be more important than the other. It is an eco-philosophical approach which recognizes the totality of (spatial, temporal, spiritual and other) inter-linkages in nature because "there is a need for a shift towards a new epistemological outlook in which humankind is viewed as part of a complex and systematic totality of nature (Oruka and Juma 1994: 115).

Eco-philosophy can therefore be regarded as the totality of the philosophy of nature, and, hence, it is conceived to be broader than subjects such as environmental studies or environmental ethics (Oruka and Juma 1994: 119). Eco-philosophy is 'ecological' in the sense that it sees humanity as one with nature, as an integral part of the process of evolution which carries the universe onward from matter to life, to consciousness, and ultimately to the divine.

There are two conceptions of nature, which imply two conceptions of ethics, one—the Western one—is anthropocentric, and the other—the African one—is ecocentric. The dominant discourses in global environmental ethics have been from Western philosophical perspectives with a solid grounding on anthropocentrism, meaning a "human-centered" worldview. It assigns "special value or worth to human beings and locates the human person at the center of the environment. With his Parental Earth Ethics, Oruka challenges anthropocentricism and proposes ecocentrism, which prescribes to recognizing the importance of relation. This invites us to think about a relational ontology and a relational cosmology.

According to Oruka and Juma, we have to "adopt a holistic outlook in which everything is related to everything else. This inter-relatedness requires a corresponding philosophical approach that looks at nature in its totality and derives from it ethics that reflect this outlook (Oruka and Juma 1994: 117). This is a way to construct an eco-philosophy. Contrary to environmental studies which "have so far, restricted themselves to the study of the earth and atmosphere", eco-philosophy "must include the totality of both human-made as well as non-human-made philosophy about nature and the totality of the universe" (Oruka and Juma 1994: 119).

The eco-philosophical approach invites us to found a new ethics, which would take into account the complexity, and totality of nature. This would be a parental earth ethics. This implies to take care of the human beings as well as the non-human beings. Parental earth ethics is, for Oruka "a basis ethics that would offer a motivation for both a global environmental concern and a global redistribution of the wealth of nations (Oruka and Juma 1994: 128).

19.8 The Principle of Ethical Human Minimum as the Basis of Global Justice

Oruka argues for global justice in two key articles, *John Rawls' Ideology: Justice as Egalitarian Fairness* (1981) and *The Philosophy of Foreign Aid: A Question of the Right to a Human Minimum* (1989). In these articles he exhorts thinkers to take the responsibility and apply the principles of justice to a global level on a global scale. Oruka engages with the philosophy of one of the leading philosophical defenders of the modern democratic capitalist welfare state, John Rawls. According to Rawls, the principles of justice emerge from 'the original position' and are chosen by people who are under 'a veil of ignorance' and are denied knowledge of their own particular history and identity. Rawls presents his two principles of justice as universal principles that people would agree to under certain impartial circumstances and which would therefore withstand rational and critical scrutiny. The first principle states that "Each person is to have an equal right to the most extensive total system of equal basic liberties compatible with a similar system of liberty for all" The second principle states that "Social and economic inequalities are to be arranged so that they are both: (a) to the greatest benefit of the least advantaged, consistent with the just savings principle, and (b) attached to offices and positions open to all under conditions of fair equality of opportunity. According to Rawls, the first principle is prior to the second. This means that the demands of the first principle cannot be overridden by the imperatives of the second. The first principle, namely, the liberty principle, pertains to everyone having a maximum and equal degree of liberty, including all the liberties traditionally associated with democracy such as the right to vote and to hold public office, freedom of speech and assembly, freedom of thought and conscience, and the right to hold personal property. The second principle, which Rawls refers to as "the difference principle" and Oruka calls "the socioeconomic principle" (Oruka 1997: 116), concerns the distribution of wealth and income. Rawls argues that the inequalities in wealth and income should be such that they are to the highest benefit of the least advantaged members of the society in such a way that they benefit more than they would by their effort.

Nevertheless, Oruka finds the Rawlsian approach global justice as inadequate for two main reasons: First, the order of Rawls' principles ought to be reversed so that the second is prior to the first. Most of the rights that comprise Rawls' first principle are political entitlements, while his second principle concerns economic rights. Oruka argues contrary to Rawls, that economic rights are more fundamental than political rights; thus endorsing his case for right to a human minimum. Rawls' theory, according to Oruka, may be relevant to people in richer than poorer countries. Second, such a rendering of theory of justice as universal, absolute justice can neither be egalitarian nor just because of the disparity between rich and poor countries (Ourka 1997: 117–119).

Consequently, Oruka contends that it is difficult to formulate a universal theory of social justice would be naïve and dangerous. For such an abstraction to be relevant, as in this case or an African agrarian philosophy, it needs to take into account the level of economic advancement, historical traditions, ideological realities and experiences of the societies for which it is meant. Yet, could helping the poorer countries or global redistribution not solve the problem?

In the paper "The Philosophy of foreign Aid: A Question of the Right to a Human Minimum" Oruka examines the three possible rationales for the current practice of foreign aid or assistance to the poor by the rich, and finds all of them morally deficient as it cannot bestow or safeguard the life and dignity of the recipient of aid as the rich design this aid in their favour. He makes a distinction between "global justice" and "international justice." According to him, egalitarianism would be more in line with global justice than with international justice. Global justice requires equal distribution of the world's wealth among its population regardless of the national, racial, technological, or geographical differences, which is to say that it requires the total eradication of inequality in the world. International justice, on the other hand, requires an internationally recognized law that would ensure that everyone has a right to a minimum standard of living: it is, at best, for the elimination of abject poverty. Thus, international justice is open to large inequality as long as everyone has a right to a minimum standard of living (Oruka 1997: 118). Consequently, Oruka thinks that when Rawls exalts the Liberty Principle above the Difference Principle, it would allow for a Society of Unbalanced or Wild Justice (SUWJ), in which, for example, it would be acceptable for a few wealthy members to live three times longer than their poor compatriots, as long as the poor compatriots had access to the basic requirements of life (Oruka 1997: 118 ff.).

For Oruka, assuming the ethics of egalitarian fairness, a SUWJ can be shown to be unjust for two main reasons: First, great inequality in wealth and income, even in services and benefits derived from these items, is in conflict with the nature of equality required by egalitarian social existence. Part of the aim of egalitarian fairness is to suppress and eradicate, as a matter of cardinal ethical principle, any development toward inequality in wealth and liberty. Equality in egalitarian terms is an end in itself and inequality an evil to be eliminated, even at a high price. Second, the possibility for some people (a minority) in one society to acquire the means for such a good life while others (the majority) cannot afford such means, would ensure serious disharmony, envy and distrust in the society. Yet, a just society, in communitarian terms, must be free of such problems; social harmony and mutual trust and understanding between the fortunate and the unfortunate must be a condition of justice treated as fairness (Oruka 1997, 120).

For these reasons, Oruka re-structured the two principles of Rawls in these words:

(1) Social and economic differences are to be arranged so that they are both (a) to the greatest benefit of the least advantaged, consistent with the just savings principle, and (b) attached to offices and positions open to those whose ethical inclination is to advance the requirement in (a).
(2) Each person is to have an equal right to the most extensive total system of equal basic liberties compatible with a similar system of liberty for all (Oruka 1997: 123–124). Hence, when Oruka reformulates the Rawlsian principles of global justice and enunciates on the principle of the right to a human minimum, he is not only proposing an ethics of distributing resources among the citizens of the globe, but also prescribing a principle of global justice which would guarantee, at least, the fundamental universal human right for all the citizens of the world. This right, as earlier explained above, is to exist, to live, and to function as a human being, a being with dignity, a person with moral worth. The implications is that the world should be humanized (Ouka humanism) and focus more on global justice which are insulated on the principles of Parental Earth Ethics than what prevails under the practice of the international justice which cannot guarantee and ensure the survival and freedoms for all or most people In the world. It does not stipulate what justice is as pertains to the relationship among the citizens of the world; international justice docs not have principles of rights and duties that hold among the citizens of the globe.

19.9 Conclusion

Throughout this chapter, I have argued for the relevance of Oruka's Parental Earth Ethics, in our quest for global environmental ethics and the promotion of global justice. It calls for the use and equal distribution of wealth and world resources for the common good and security of all. It is basically driven by the ethical principle of the human minimum, which ought to inspire us humans to be more ethically and politically responsible in the way we relate to other humans and the rest of nature. While challenging the positions of Western scholars in the liberal tradition like Garret Hardin and John Rawls, Oruka opines that there is a link between the manner of thinking our link to the universe, the cosmos, and the manner of thinking a fair society. He is not only skeptical about the global acceptability and applicability of their claims, but stresses the importance of the ethical human minimum which points to some universal values and norms with claims and responsibilities or obligations that are global in scope. In his Parental Earth Ethics he articulates his claims for a non-anthropocentric principle for nature and together with Juma (Oruka and Juma 1994), they argued against the Judeo-Christian view of nature, which they saw as promoting a form of possessive individualism that disrupts the complex web of being of which humans are a part. They contended that it is this Judeo-Christian anthropocentric advocacy for the supremacy of human beings that has led to the wanton global destruction of the environment to gratify selfish human interests. It would be instructive to note that Oruka wrote in the 1990s with the dominance of an egalitarian form of democratic liberalism in the US, and most of Europe was under the influence Rawls etc. But we need to go back to Oruka seriously because liberalism is crashing and the poor, who are never part of it, can hardly survive in the liberal system. This is where Oruka sees things differently and tries to formulate a new position which I find very interesting. Yet, why is it important for Africans to reflect on agriculture from a philosophical point of view? There seems to be uneasiness within the frame of development in agriculture from an African perspective, as agriculture is seen as primitive, should be brought to global standards with fertilizers, trade, to alleviate poverty etc. but is it the right way to do justice? Agriculture is also a cultural treasure in Africa, with its extensive rapport with nature; is it not time for us explicitly interrogate this relationship?

References

Azenabor, G. 2009. Odera Oruka's Philosophic Sagacity. *Thought and Practice (New Series)* 1 (June): 69–86.

Boogaard, B.K. 2019. The Relevance of Connecting Sustainable Agricultural Development with African Philosophy. *South African Journal of Philosophy* 38 (3): 273–286.

Chimakonam, J. 2017. African Philosophy and Global Epistemic Injustice. *Journal of Global Ethics* 13 (3): 1–18 (September 2017). https://doi.org/10.1080/17449626.2017.1364660.

Graness, A. 2012. What is Global Justice? Henry Odera Oruka's Contribution to the Current Debate. *Journal on African Philosophy* 6: 31–46.

Graness, A. 2015. Is the Debate on 'Global Justice' a Global One? Some Considerations in View of Modern Philosophy in Africa. *Journal of Global Ethics* 11 (1): 126–140. https://doi.org/10.1080/17449626.2015.1010014.

Graness, A., and K. Kresse, eds. 1997. *Sagacious Reasoning: H. Odera Oruka in Memoriam.* Frankfurt: Peter Lang.

Hardin, G. 1974. Lifeboat Fthics: The Case Against Helping the Poor. *Bioscience* 24 (10) (October 1974).

Janz, B.B. 2009. *Philosophy in an African Place*. Lanham: Lexington Books.

Kalumba, K. 2002. A Critique of Odera Oruka's Philosophic Sagacity. *Philosophia Africana* 5 (1).

Kalumba, K. 2004. Sage Philosophy: Its Methodology, Results, Significance, and Future. In *A Companion to African Philosophy*, ed. Kwasi Wiredu, 274–282. Malden: Blackwell.

Kresse, K. 1993. Interview with Professor Henry Odera Oruka. *Quest—Philosophical Discussions: An International African Journal of Philosophy/Revue Africaine Internationale de Philosophie* IX (2)/X (1): 22–31.

Kresse, K. 2007. *Philosophising in Mombasa: Knowledge, Islam and Intellectual Practice on the Swahili Coast.* Edinburgh: Edinburgh University Press.

Masolo, D.A. 1994. *African Philosophy in Search of Identity*. Bloomington/Indianapolis: Indiana University Press/Edinburgh: Edinburgh University Press.

Masolo, D.A. 1997. Decentering the Academy: In Memory of a Friend. In *Sagacious Reasoning: Henry Odera Oruka in Memoriam*, ed. Anke Graness, and Kai Kresse. New York: Peter Lang.

Masolo, D.A. 2005. Lessons from African Sage Philosophy. *Africa e Mediterraneo* 53 (December): 46–53.

Mosima, P.M. 2016. *Philosophic Sagacity and Intercultural Philosophy: Beyond Henry Odera Oruka, African Studies Collection 62.* Leiden/Tilburg: Tilburg University.
Mosima, P.M. 2018. Henry Odera Oruka and the Female Sage: Re-evaluating the Nature of Sagacity. In *African Philosophy and the Epistemic Marginalisation of Women*, ed. J.O. Chimakonam and L. du Toit, 22–41. London: Routledge.
Ndaba, W.J. 1996. Odera Oruka's Sage Philosophy: Individualistic vs. Communal Philosophy. In *Beyond the Question of African Philosophy: A Selection of Papers Presented at the International Colloquia, UNISA, 1994–1996*, ed. A.P.J. Roux, and P.H. Coetze. Pretoria: University of South Africa (UNISA) Press.
Ochieng'-Odhiambo, F. 1994. The Significance of Philosophic Sagacity in African Philosophy. Ph.D. thesis, University of Nairobi.
Ochieng'-Odhiambo, F. 1996. An African Savant: Henry Odera Oruka. *Quest: An African Journal of Philosophy/Revue Africaine de Philosophie* IX (2)/X (1) (December 1995/June 1996): 12–15.
Ochieng'-Odhiambo, F. 1997. Philosophic Sagacity Revisited. In *Sagacious Reasoning: Henry Odera Oruka in Memoriam*, ed. Anke Graness, and Kai Kresse. New York: Peter Lang.
Ochieng'-Odhiambo, F. 2002a. The Evolution of Sagacity: The Three Stages of Odera Oruka's Philosophy. *Philosophia Africana* 5 (1): 19–32.
Ochieng'-Odhiambo, F. 2002b. Some Basic Issues About Philosophic Sagacity: Twenty Years Later. In *Perspectives in African Philosophy: An Anthology on 'Problematics of an African Philosophy: Twenty Years Later*), ed. Claude Sumner, and Samuel Wolde Yohannes. Ethiopia: Addis Ababa University Press.
Ochieng'-Odhiambo, F. 2006. The Tripartite in Philosophic Sagacity. *Philosophia Africana* 9 (1): 17–34.
Ochieng'-Odhiambo, F. 2007. Philosophic Sagacity: A Classical Comprehension and Relevance to Post-colonial Social Spaces in Africa. *Quest: An African Journal of Philosophy/Revue Africaine de Philosophie* XXI (1–2).
Oduor, R.M.J., Oriare, N., Owakah, F.E.A. 2018. *Odera Oruka in the Twenty-First Century.* Washington, DC: The Council for Research in Values and Philosophy.
Okeja, U. 2018. *African Philosophy and Global Justice: Critical Essays*, 1st ed. London: Routledge.
Oruka, H.O. 1981. Rawls' Ideological Affinity and Justice as Egalitarian Fairness. In *Justice: Social and Global*, ed. Ericsson O. Lars, H. Ofstad, and G. Pontara, 77–88. Stockholm: Gotab.
Oruka, H.O. 1989. Philosophy of Foreign Aid: A Question of the Right to a Human Minimum. *Praxis International* 8 (4): 465–475.
Oruka, H.O. 1990. *Trends in Contemporary African Philosophy*. Nairobi: Shirikon.
Oruka, H.O. 1991. *Sage Philosophy*. Nairobi: Acts Press.
Oruka, H.O. 1993. Parental Earth Ethics. *Quest* VII (1) (June 1993): 20–27. "Rejoinders". *Quest* VII (I) (June 1993): 106–109.
Oruka, H.O. 1996. *The Philosophy of Liberty* (Revised and Reprinted). Nairobi: Standard Textbooks Graphics and Publishing.
Oruka, H.O. 1997. *Practical Philosophy. In Search of an Ethical Minimum*. Nairobi: East African Educational Publishers.
Oruka, H.O., and C. Juma. 1994. Eco-philosophy and Parental Earth Ethics (On the Complex of Being). In *Philosophy, Humanity and Ecology: Philosophy of Nature and Environmental Ethics*, ed. H. Oruka, 115–129. Nairobi: ACTS Press.
Oseghare, A.S. 1985. Relevance of Sagacious Reasoning in African Philosophy. Doctoral dissertation, University of Nairobi.
Presbey, G.M. 1996. African Sage-Philosophers in Action: H. Odera Oruka's Challenges to the Narrowly Academic Role of the Philosopher. *Essence: An International Journal of Philosophy (Nigeria)* 1 (1): 29–41.
Presbey, G.M. 1997. Who Counts as a Sage? Problems in the Future Implementation of Sage Philosophy. *Quest—Philosophical Discussions: An International African Journal of Philosophy/Revue Africaine Internationale de Philosophie* XI (1–2): 52–66.
Presbey, G.M. 1999. The Wisdom of African Sages. *New Political Science* 21 (1): 89–102.

Presbey, G.M. 2000. On a Mission to Morally Improve Society: Odera Oruka's African Sages and the Socratic Paradigm. *International Journal of Applied Philosophy* 14 (2): 225–240.

Presbey, G.M. 2002. African Sage Philosophy and Socrates: Midwifery and Method. *International Philosophical Quarterly* 42 (2): 166.

Presbey, G.M. 2007. Sage Philosophy: Criteria That Distinguish It from Ethnophilosophy and Make It a Unique Approach within African Philosophy. *Philosophia Africana* 10 (2): 127–160.

Presbey, G.M. 2012. Kenyan Sages on Equality of the Sexes. *Thought and Practice: A Journal of the Philosophical Association of Kenya (PAK) (New Series)* 4 (2): 111–145.

Tangwa, G.B. 1997. Sagacious Reasoning: Henry Odera Oruka in Memoriam (ed. Anke Graness, and Kai Kresse. Frankfurt am Main: Peter Lang Review). *Quest—Philosophical Discussions: An International African Journal of Philosophy/Revue Africaine Internationale de Philosophie* XI (1–2): 175–182.

Van Hook, J. 1995. Kenyan Sage Philosophy: A Review and a Critique. *Philosophical Forum* XXVII (1): 54–65.

Pius Mosima teaches Philosophy at the University of Bamenda, Cameroon and is Research Fellow at the Vrije Universiteit Amsterdam, the Netherlands. His research interests include African/Intercultural Philosophy; Globalization and Culture, Moral and Political Philosophy. He has several peer-reviewed articles and book chapters, supervised research, has presented papers in many conferences, facilitated many workshops and have given guest lectures in Summer Universities on these topics in many African, European and Asian universities. He is the author of *Philosophic Sagacity and Intercultural Philosophy: Beyond Henry Odera Oruka (2016) and editor of A transcontinental career: Essays in honour of Wim van Binsbergen (2018).*

Chapter 20
Food Security as a Fundamental Human Right: A Philosophical Consideration from Africa

Maduka Enyimba and Victor C. A. Nweke

Abstract In this paper, we argue that among the three basic human needs, namely, food (air and water), shelter and clothing; food is the most significant. This is the case because food is a necessary requirement for the preservation of human life. We argue that if food is a prerequisite for human life, the right to life suggests the right to food security. And since the right to life is a fundamental human right, food security should therefore be a fundamental human right. By food security we specifically refer to equal access to the necessary qualitative food for proper human flourishing. We adopt the critical and argumentative methods of philosophy to buttress our claim. Accordingly, we present a critical analysis of the relationship between food security and the right to life. We also expose the relationship between food security and crime in some African states. Using Nigeria as a point of reference, we maintain that there is food insecurity in many contemporary African countries. And that it is orchestrated by poverty—the lack of economic and financial resources needed to access quality food in its right proportion. We demonstrate how food insecurity leads to crime in Nigeria and in sub-Saharan African states. We conclude by suggesting that food insecurity can be confronted if the government can invest more in agriculture and also create more employments with reasonable income level.

Keywords African philosophy · Crime · Food security/insecurity · Human rights · Right to life · Agrarianism

M. Enyimba (✉)
University of Calabar, Calabar, Nigeria
e-mail: enyimbamaduka@unical.edu.ng

V. C. A. Nweke
University of Koblenz-Landau, Mainz, Germany
e-mail: vcanweke@uni-koblenz.de

M. J. Tosam and E. Masitera (eds.), *African Agrarian Philosophy*, The International Library of Environmental, Agricultural and Food Ethics 35,
https://doi.org/10.1007/978-3-031-43040-4_20

20.1 Introduction

The philosophy of agriculture is generally perceived as the study of the relationship between human society, agriculture, environment and food. Such study can be done from an ethical, anthropological, sociological or even environmental perspective as it concerns food related issues (Suchara et al. n.d.). It also goes by the name agrarianism. In this sense it is seen as an ideological movement that places premium on farming and farmers over the rest of a nation's economy (Thompson 2010, 2008). Agrarian philosophy extends its investigation to food production, its safety and stability, as well as factors affecting food production such as climate change, nature, political and economic systems. Zravanyange argues that African agrarian philosophy is concerned with agriculture and food systems that are built on moral values, creativity and innovation in a way that identifies with the needs and aspirations of Africans and those of its development partners (2016). The point being made here is that agrarian philosophy or philosophy of agriculture broadly conceived raises moral, political, economic, environmental, religious and social concerns as it relates to food production and factors influencing it.

In this paper, using Nigeria as a reference point, we investigate the nature of agrarian philosophy in Africa from the perspective of food security/insecurity and human rights, and their interface with factors such as poverty and crime. Most Sub-Saharan African countries lack the means to adequate qualitative food. Poverty induced by gross unemployment and underemployment makes it impossible for most people in these African countries to be able to have access to adequate food through legitimate means. This situation often compels people to embrace crime to escape from the looming starvation, hence, the high crime rate and continual exodus of people to other parts of the world through whatever means possible. The fact is that every human being should be entitled to a basic income that will enable her/him to have access to adequate qualitative food.

There is food insecurity in many contemporary African countries of which Nigeria is among and this is mainly due to lack of awareness of the relationship between right to life and right to food security. We intend to draw attention to such relationship in the essay. Food security is the presence or availability of balanced food, as well as easy access to it by the citizens for at least three times daily. Anything to the contrary would amount to food insecurity, which is the absence or lack of easy access to balanced food by the citizens of a given society for at least three times in a day. This is the idea of food security expressed by Afolabi and others (2018). According to them, food security can be defined as a condition when people can no longer adequately feed, either due to the unavailability of food or increased prize of the little that is available, which in turn diminishes the quality of their life and leads to malnutrition and eventual death (Afolabi et al. 2018). In other words, the absence of food insecurity is the presence of food security. Food security generally entails a condition where people have adequate physical and economic access to sufficient food to meet their dietary needs for a productive and healthy life (Olayinka 2017: 1–2). In the same vein, the Food and Agricultural Organization refers to Food security as equal access to necessary qualitative food for proper human flourishing (FAO 2017: 29). It has also been variously described as the right to adequate food and/or the right to be free from hunger and starvation (FAO 2005: 5, 37). This suggests that food security should be seen as a human right.

The universal declaration of human rights have been described as the highest level of expression of the natural law theory that sets a universal standard for rights to be respected without exceptions (Corradetti 2013: 8). This international human right document makes provision for the protection of the right of persons to have access to sufficient food (UN 1948). So also does the constitutions and legal documents of different nations of the world provide for the need for adequate food for the human person. However, these aspects of the provisions of these legal documents seem to have been so frequently overlooked that it has become necessary to draw attention to them. It is against this backdrop that we aim in this work to demonstrate that food security is an essential aspect of fundamental rights of the human person in Africa and indeed across the globe. It follows thus that food insecurity is an evidence of the denial of right to food, which is a denial of human right, and which gives way to crime in its various forms.

In order to achieve the aim of this work, we employ the method of critical analysis and argumentation to first, examine the relationship between food security/insecurity and human rights, then we attempt to show the relationship between crime and food insecurity in Nigeria and other African countries. Some of the causes of food insecurity and their attendant crimes are also highlighted in this section. Finally, we submit that governments and stakeholders in African countries in general and Nigeria in particular, should take food insecurity seriously by investing more in agriculture and creating gainful employment for their citizens if there is going to be relative peace, harmony and progress in the continent.

20.2 Food Security/Insecurity and Fundamental Human Rights

Today human rights records have become the criteria for the assessment of the quality of the governance of any nation and its government (Corradetti 2013: 8). Iwe buttresses this point by stating that by virtue of their natural characteristics, human persons are the foundation and source of fundamental human rights (2000: 5). If Iwe's opinion is to be carefully scrutinized, it would imply that an attempt at denying any of the fundamental rights of the human person would amount to a direct attempt at dehumanizing, debasing and destroying the dignity of the human person. In other words, the human person has emerged as a bearer of intrinsic and inviolable rights. This emergence of the human personality with fundamental natural equality with fellow humans and with the recognition of the inviolability of natural powers and rights is the most decisive of all historical land-marks (Iwe 1986: 64; Laski 1966: 93, 141). At this point, one is reminded of Jeffery Flynn's attempt at balancing the past and contemporary definitions of the notion of human rights by referring to the historical studies on human rights. In doing this, he observes a disruption of the pattern between the meanings of the past and those of the present. He suggests that a more inclusive definition of human rights needs to be provided, and that a work of historical and conceptual clarification is needed. This submission stems from his conviction that the inherent legal nature of human rights, and the revolutionary founding of nation states defended by scholars such as Habermas, cannot explain the contemporary conception of human rights as a language of moral protest (Flynn 2013).What Flynn does here is to draw attention to the legal and moral connotation of the notion of human rights.

In their attempt to clarify the notion of human rights, Cruft et al. (2015: 1–3) question the nature and justification of human rights. They reveal the diverse positions and approaches to the issue of human rights with no prevailing philosophical view on them. According to them, this is a sign of intellectual, cultural and political fertility of the notion of human rights (2015: 1–3). Following this, they define human rights as those rights which human beings possess by virtue of their humanity and which can be identified by the use of ordinary moral reasoning (Cruft et al. 2015: 4). In other words, human rights, though associated with natural reasons are opposed to the sort of conventional reasons created within particular social or institutional contexts. The central focus of this conception of human rights is that it is possessed by all human beings everywhere in the world by virtue of the fact that they are human beings. Similar conception of the nature of human rights has been expressed by scholars such as Griffin (2008: 54) and Simnons (2001: 185) among others.

Cruft etal (2015: 4) drew some significant conclusions from the above renditions of the nature of human rights which we tend to leverage on in this work. They suggest that human rights can be linked to natural rights. In so doing, it will be seen to connote equal moral rights of all human beings. Moreover, because fundamental human nature has not changed, all peoples across the globe have legitimate claim to rights such as food, education and so on (2015: 4).

It is worthy of note here that scholars such as Rawls (1999), Beitz (2009) and Raz (2010) among others have given a political conception to the notion of human rights. According to these scholars, human rights are to be understood in relation to their role or function in contemporary international political practice. What this means is that human rights are not based on any quality or characteristics shared by all human beings anywhere in the world. While this political conception of human rights might seem to be opposed to our moral definition of human rights, it is nevertheless an important aspect of human rights that is worth consideration. However, it will be beyond the scope of this work to dive into the discussion of the relationship between the two conceptions of human rights.

One of the theories of justification of human rights which makes clearer the sense of human rights that we employ in this work is the instrumental justificatory theory of human rights. According to this theory, human rights are justified because they are necessary and useful means of realizing the values and essential features of human lives. Such features include the notion of human agency, the idea of good life, and the notion of basic needs (Cruft et al. 2015). This justificatory theory of human rights is proposed by those who are of the view that human rights are rights possessed by all humans specifically because they are human beings. Griffin (2008) and Cruft et al. (2015) argue that unlike the non-human animals, human animals as agents have the ability to form concepts of a good life. They also have the ability to pursue such conception of a good life they might have chosen for themselves. It is this capacity that is believed to confer human dignity on them. In other words, to have human dignity is to pursue a given course of life without external interference. This means that human rights are justified on the grounds that they protect human dignity. The implication here is that when human rights are violated or denied, human dignity is betrayed. This is the case also when basic human need like food is denied.

Many scholars have identified human rights to include; right to life, right to freedom of speech, right to freedom of association, right to freedom of religion, right to freedom of expression and right to self-determination (Uduigwomen 2006: 136; Omoregbe 1997: 95–9). This identification coheres with Osita Eze's definition of human right as the demands or claims which individuals or groups make on society, some of which are protected by law and become part of the justifiable legal norms, while others remain aspirations to be attained in the future (1984: 5). Human rights are those inviolable and inalienable moral powers of every human being to have, to do, to require from others, to possess or give something (Ibanga 2000: 185; Iwe 1986: 156). For Ukavwe, human rights are those rights to which an individual is entitled by virtue of his status as a human being (2012: 183). The implication of the foregoing definitions is that human rights are naturally inherent in the human person and therefore a denial of it is a debasement of the human person. This substantiates, Pegels' position that in contemporary terms, human rights are based on the idea that the individual has rights, claims upon the society, or against society, that these rights which must be solely recognized, on which it is obliged to act, are intrinsic to human beings (1979: 9). This explains why they are also called fundamental rights. Human rights consist in a moral power of a person to claim something as one's own. They point to the general goodwill and conscience. Even when a person may seem to have no physical power to enforce his claims it does not invalidate his rights. (Chukwudozie 2000: 5).

Food insecurity is closely related to food security and the fundamental rights of human persons. Food insecurity according to Olayinka (2017: 3), occurs when a country's system does not match its demand with sufficient quantity and quality. Food insecurity exists when there is uncertainty about future availability and access to food. It is the insufficiency in the amount and kind of food required for a healthy lifestyle (Afolabi et al. 2018). What is meant here is that where there is uncertainty, insufficiency, unavailability, inaccessibility and non-utilization of food in the right proportion in a given household, community or society, there is a case of food insecurity. An assessment of two major forms of food insecurity reveals the possible causes of food insecurity in any given society. The first is *chronic food insecurity*. As a long term or persistent insecurity of food, it occurs when people cannot meet their minimum food requirement over a sustained period of time. This type of food insecurity is believed to ensue from an extended period of poverty, inadequate or outright lack of access and financial resources to cater for oneself or one's household

(Adeoti 1989). The important point to note here in line with the trajectory of this work is that poverty is a fundamental indicator/cause of food insecurity. We will return to this in subsequent section of this work.

The second type of food insecurity that we would wish to underscore is known as *seasonal food insecurity*. This type is believed to be predictable and follows a sequence of known events. It lasts for a limited time and it is also very recurrent. Seasonal food insecurity occurs when there is a cyclical or recurrent pattern of inadequate availability and access to food occasioned by climate change, irregular or lack of work opportunities and/or cropping pattern (FAO 2002). A closer look at the nature of this type of food insecurity would reveal the place of unemployment and/or underemployment in the pervasiveness of food insecurity in the society. The point that needs to be noted here is that food security is intertwined with the phenomenon of food insecurity. An adequate guard against the occurrence of food insecurity would ensure the enduring presence of food security, and a careful maintenance of food security would in turn eradicate the menace of food insecurity and its attendant challenges.

Food security refers to a state where food is available at all times, to which all persons have means of access, and when there is nutritionally adequate supply of it in terms of quantity, quality and variety, and is acceptable within a given culture (Clover 2003: 7). The Human Development Report (HDR) of the United Nations 1994 whose focus was on the issue of eradicating poverty throughout the world, specifically listed food availability as one of the important components of security in the nations of the world. What this implies is that food insecurity constitutes a threat to social security across the globe. A society can be said to be secured when there is freedom from fear and want of basic human needs of which food is one (Afolabi et al. 2018: 1).

As Olayinka (2017: 1–2) rightly observes, the provision and availability of food to citizens are as important as the right to life among other human right provisions. This is so because the right to food is a fundamental provision under article 22 and 25 of the Universal Declaration of Human Rights (1948). It contains provisions for food and human dignity. This is also provided in article 11 of the Covenant on Economic, Social and Cultural Rights (1990), and articles 24 and 27 of the United Nations Convention on the Rights of the Child (2009). It is stated that the state shall direct its policies towards ensuring that suitable and adequate food is provided for all citizens in the 1999 constitution of Nigeria section 16 (2) (d). The main focus of Nigeria and other African countries should be to embark on agricultural revolution through the creation of more jobs, and by making agriculture the hub of economic growth in the country and in the continent. This will enhance the achievement of a hunger free country, which is a country that is food secured (see Nwozor et al. 2019: 9).

Our argument is that among the fundamental human rights is the right to food security or the right to freedom from food insecurity. This can also be described as the right to freedom from hunger. Hunger, it must be noted is a by-product of poverty. And poverty is one of the factors that dehumanize human persons and in most cases force them into one crime or the other. Referring to poverty as a common enemy of humanity, one must note that poverty is responsible for most of the topical challenges in the world yesterday and today. Therefore, whoever cares about the fate of humanity should seek for ways to reduce the rate of poverty in the world. The point being made here is that if poverty is one of the causes of food insecurity or hunger and starvation (Ejiogu 2014: 98), then poverty is one of the many factors that deprive the human persons of their fundamental human rights. This fact goes further to reveal the relationship between human right and food security. In subsequent section of this essay we shall discuss the role of poverty in the relationship between crime and food security/insecurity in Nigeria and few other African countries.

What many authors and human right activists seem to gloss over is the implication of the right to life for food security in any country like Nigeria and as imbedded in the definition of human rights. The fundamental human right to life presupposes the right to food security or the right to freedom from hunger or the right to adequate food. Thus, the defenders of human rights to life ought to recognize themselves as defenders of human rights to adequate food or freedom from hunger. The right to life refers simply to the freedom a person has to choose to live or exist and to live a good life (Okafor 1992: 77, 106–107; Njoku 2009: 20, 41–42; Okon 2005: 114). If humans have the right to choose to live and they ought to eat in order to live, then it follows that the right to adequate food (food security) is a fundamental and inviolable human right intrinsically embedded in the fundamental human right to life.

Therefore, to deny a person access to safe and adequate food is to subject the person to hunger and starvation which may lead to the person's death. This will logically suggest that any denial of adequate food to a person, is a denial of the person's right to freedom from hunger (right to food security) and since hunger and starvation can lead to loss of life, then, it is equally a denial of the person's right to life. It follows that food security, right to adequate food or right to freedom from hunger is a fundamental human right. Accordingly, as Jacques Diouf rightly pointed out "the right of everyone to have access to safe and nutritious food, consistent with the right to adequate food and the fundamental right of everyone to be free from hunger was reaffirmed in 1996, at the world food summit" (2005: 111).

20.3 Poverty, Food Insecurity and Crime in Nigeria

After the Second World War, a new agricultural system was developed in most of the industrialized nations. This led other countries of the world to begin a new approach to agriculture in order to attain self-sufficient levels of production, and prevent food shortages for their growing population. This led to the introduction and use of chemicals in the production of crops, and this drew attention to the need to ensure the safety of the food (Slater et al. 2014; Rizzuti 2020: 2). This incident led to the first ever world food conference where the notion of food security was first conceptualized as the availability and price stability of basic food stuffs at the international and national levels (FAO 2006).

In his analysis of food theory and technology, Kent (2013: 4) disclosed how access to food might influence social skills and behaviors. According to him, there is a significant relationship between food insecure children and juvenile crimes. As a result of this finding, Kent suggests that juvenile crime can be prevented if food security is guaranteed for the populace and for the children in particular (Kent 2013: 15–16). The point is that food security, which is access to food, has direct and positive influence on social skills and behaviors. If this is the case, then it follows that the lack of access to food (food insecurity) would also directly and negatively affect or influence social skills and behaviors. In other words, social ills such as stealing, robbery, corruption etcetera, do also result from the prevalence of food insecurity.

Food insecurity as stated earlier, refers to the lack of access to good and adequate food that is necessary for proper human flourishing. It also consists of the inability to access a safe and nutritious food which is consistent to the right to adequate food. Food insecurity is opposed to food security already enunciated earlier in this essay. Food insecurity is an existential reality in Nigeria and most African countries, and this is not without a cause. Although there may be varied causes of food insecurity in Nigeria, Africa and the world over, in this essay we focus on one major cause, namely, poverty.

The World Bank defines poverty as the level of income below which it is possible to say that a person is poor (1986). This definition does not reveal much as it excludes quite a lot of the dimensions of poverty. Amartya Sen defines poverty as the absence of a sufficient level of fundamental entitlements such as the right of access to essential goods, such as the right to what is being produced (Sen 1981). This, of course, would include food. Poverty is linked to low or unequal capacity of choice possessed by an individual in his/her attempt at accessing the necessary goods and services for human flourishing. Human Development Report as produced by United Nations Development Programme since 1990 shows that poverty is one of the major causes of food security for poor households. For instance, in a low income family, once expenditure on basic necessities such as energy, clothes and shelter has been deducted, the resources left are not sufficient to meet the family's other needs such as food (UNDP 1994). Poverty can then be conceived as both a cause and a consequence of under nourishment resulting from severe food insecurity (Broca 2002).

Poverty is a major cause of food insecurity in most African countries like Nigeria. The rate of poverty unveils the rate of crime in the society, and most of the crimes prevalent in contemporary African nations are traceable to food insecurity. Thus, there is a link between poverty, crime and food insecurity in Nigeria and in the continent of Africa at large. Poverty is a state of impoverishment that renders one unable to have access to basic human needs. It is the quality or state of being poor or indigent. A person is thus, said to be poor, when he/she is in a state of want or has scarcity of means of subsistence. Any deficiency or lack of basic elements or resources that are needed to preserve life or make life meaningful and worth living constitutes poverty. Poverty is one major factor that constitutes grave challenges to food security in Nigeria. Nigeria like most African countries is blessed with abundant human and natural resources, yet its people are poor. According to Kempe, poverty is a particularly disturbing phenomenon in Africa. Despite being one of the most richly endowed regions in the world, the continent remains one of the poorest (2002: 389). The World Bank recently declared that about 112 million Nigerians are deemed extremely poor. This figure is measured by World Bank parameter of those living on less than US$ 1.25 per day. The nation is ranked among the world's poorest countries. According to United Nations Development Programme (UNDP 2009), in Nigeria, hunger exhibits its ugly face in most homes where the average citizens contends with a life of abject poverty. The consequence of this is that the poor masses become easily brainwashed, their rights and choices are terribly manipulated, and this sometimes leads them to violence and crime.

The notion of crime has been defined as "a conduct that is forbidden by law and for which punishment is prescribed" (Mberu 1999: 186; Ndubisi 2009: 64). Crime refers to any activity publicly proscribed by the written laws of a society (McGuire 2004: 3). It includes such acts as theft, fraud, murder, corruption etcetera. Crime is a social phenomenon and as such can be induced by socio-economic condition of a society. Some of the major causes of crime have been identified as unequal distribution of wealth and power, social and environmental conditions, and unemployment, among others (Ozdemir and Oner-ozkan 2017: 353). The fact is that environmental factors such as physical, social, economic, cultural and political environments tend to shape or influence people's behavior. As rightly observed by Ozdemir and Oner-ozkan, poor physical, social and family environment often strengthen the intention toward criminal acts (2017: 353). This suggests that poverty and negative family experiences such as lack of access to good and enough food may encourage criminal intents and actions.

Ndubisi enunciated two kinds of crime; social and religious crimes. According to him, social crimes are those directed against individuals and which ultimately upset the harmony in the society. Examples are stealing, adultery, fighting, lying and egocentricism (2009: 64). Religious crimes are those committed primarily against the gods but having effect on human life. Such crimes are capable of incurring disaster on both the individual and the community at large (Ndubisi 2009: 64). These forms of crime include murder, incest, suicide and termination of pregnancy. In addition to these types of crime, Mberu identified two other kinds of crime, namely core crimes and white collar crimes (1999: 187–189). For this scholar, core crimes are the centre of criminal law, example robbery, whereas white collar crimes refer to illegalities carried out by persons of respectable and high social standing in the course of the discharge of their duties. Examples of such crimes include bribery and corruption, insurance fraud etc. Crime is thus, part of the broader phenomenon of social deviance (Enyimba 2003: 58). According to Sarki, the most common white collar crime is when an employee steals from the employer or cheats customers. It may also occur in a situation where a government official uses his/her office to obtain undue personal benefits (Sarki 2019: 8).

Other forms of crime include corporate crime, organized crime, victimless crime and conventional crimes. Corporate crimes are committed by a business entity or individuals representing or acting on the behalf of that business entity or corporation. Such crimes occur when there is a violent act by an organization against its workers or consumers; or when there is price-fixing, false advertising and tax evasion, etcetera (Payne 2016; Gobert and Punch 2003). Organized crime is a structured conspiracy in carrying out a criminal act with the use of fear, corruption and violence (Wright 2013;

Galeoti 2014; Sarki 2019). Organized crime involves murder and buying of illegal goods and services, among others. Victimless crimes are committed by adults who willingly exchange strongly demanded, but legally prohibited goods and services. It involves the acts proscribed by law in which no third party is hurt or involved (Yi 2008; Kadish 1983). Conventional crimes are crimes that are common and more visible in the society. They are common because they are offences which draw the attention of law enforcement agencies like the police more frequently. These types of crimes are mostly committed by the less privileged in the society, and constitute a large percentage in crime statistics. They include theft, rape, murder, burglary, arson, assault, battery, robbery, and domestic violence (Sarki 2019: 12).

The point we wish to make with the above breakdown of different forms of crime is that these crimes are directly or indirectly influenced by the intrinsic or extrinsic desire by the perpetrators to have a good life and improved condition of living. This includes access to adequate resources that will enable them gain access to basic human needs such as food. In other words, no matter the type of crime involved, the ultimate end or driving force for any crime is to have access to good life, better living standard of which food and nutrition is among. It is to be able to provide for one's own and for one's posterity such that they may not have to suffer the lack of access to basic human needs such as good food.

Thus, poverty is one of the root causes of food insecurity and food insecurity contributes to most crimes in Nigeria and other countries of Africa. The struggle for survival in contemporary Africa is the struggle to live, which is in turn the struggle to have access to safe, adequate and nutritious food. This is why most citizens of Nigeria and Africa at large, who may feel that their rights are denied as a result of poverty may take up arms against the government or take to crime in other to regain access to their fundamental human rights (Ofoegbu 2014: 53) of which food security is one. This is evident for instance, in the Niger Delta region of Nigeria, where the youths take up arms against the state for what has been described as impoverishment and deprivation of their legitimate means of livelihood (Tamuno 2011: 145–173).

The report of the International Society and Development Center to the Food and Agricultural Organization on December 22 2016 identified a close relationship between food security and violent crime and conflict across the globe. Using Somalia and Ethiopia as reference point, the report shows that policy makers may benefit from the discovery that food security tends to engender crime-free society, thereby ensuring peace and stability in the society. This shows that there is a significant relationship between food insecurity and violent crimes and conflicts in the society (Bruck et al. 2016: 2). Coughron (2016: ii) corroborates this finding when he demonstrates in a study that food insecurity has been shown to have a negative impact on the health and psychological wellbeing of those who suffer from it. According to him, there is a significant relationship between food insecurity and violent crimes and conflicts at both country and global levels; these are induced by low income levels in these countries of the world (Coughron 2016: 4).

For example, in Somalia, Ethiopia as well as Nigeria it has been shown that there is an intrinsic relationship between violent crime or conflict and decreased agricultural production. Food security indicates that crimes and conflicts have a significant relationship with food wastages and shortages (Bruck et al. 2016: 3). As Bruck et al. (2016: 14) further observes, and in line with the argument of this work food insecurity presents both material and non-material reasons or factors that induce individuals to engage in some form of behavior that threatens the peace and stability of a given society. This means that some acts or social behaviours which can be referred to as crimes, tend to ensue from individuals who suffer or experience food insecurity. This is one of the reasons why some corrupt politicians in some African countries like Nigeria can induce citizens with food items and money to either vote against their consciences or to commit such crimes as arson and murder all in the bid to gain some political advantage.

Besides, factors that bring about poverty are mostly artificially induced by the economic conditions of the states, and the individuals. Some of these poverty inducing factors in Nigeria and other African countries include bad governance, low income, unemployment, underemployment etcetera. These factors of poverty prevent the people from accessing adequate, safe and nutritious food (food security), hence they take to violence and crime of all sorts in order to regain, maintain and preserve their fundamental right to adequate food which of course, amounts to preservation of their inviolable right to life. In order to confront these challenges, governments of Nigeria and other African states need to invest massively in agricultural production and also ensure the creation of employment with reasonable income level. This will grant the populace adequate and equal access to good food and other basic needs. Moreover, employing food assistance as one of the strategies of sustainability in world food programme, food insecurity will become a thing of the past. Food assistance is a process of ensuring food security by understanding of people's long-term nutritional needs and of the diverse approaches required to meet them.

20.4 Conclusion

The argument in this chapter has been that food security otherwise known as the right to safe, adequate and nutritious food is constitutive of the right to freedom from hunger, which in turn is incorporated into the larger labyrinth of fundamental human rights. If the right to life presupposes the right to choose to live, then the right to food security is the right to life, since humans need food to live. Denying a person access to food, amounts to denying the person's fundamental human right to life, and by extension, a denial of the person's dignity. A relationship was further established between, poverty, crime and food insecurity. It was argued that poverty is one of the major sources of food insecurity in most African countries and the absence of food security is the cause of most crimes in many African countries today. Thus, poverty which is induced by the general economic conditions of the states also affects the individual's economic status. The governments of these states should then be able to invest more in agriculture and promote employment for its citizenry with such minimum income that can enable them gain access to adequate, safe and nutritious food. This is believed to be a viable way of ensuring peace and stability in the society.

References

Adeoti, J.V. 1989. Economic Crisis in Developing Countries: The Food Dimensions. *Ilorin Journal of Business and Social Sciences.*

Afolabi, M.B., A.A. Ola, B. Oyinloye, and P.N. Egbogu. 2018. Food Security: A Threat to Human Security in Nigeria. *Journal of Health and Social Issues* 7 (2): 1–30.

Beitz, C. 2009. *The Idea of Human Rights.* New York: Oxford University Press.

Broca, S.S. 2002. *Food Security, Poverty and Agriculture: A Concept Paper.* ESA Working Paper No. 02–15. www.fao.org/es/esa. Retreived June 22, 2022.

Bruck, T., N. Habibi, C. Martin-Sheilds, A. Sneyers, W. Stojetz, and S. Weezel. 2016. *The Relationship Between Food Security and Violent Conflicts: Report to FAO.* International SXecurity and Development Center. Berlin: gGmbH. www.isd-center.org.

Chukwudozie, C.M.N. 2000. *Introduction to Human Rights and Social Justice.* Enugu: Ochumba Printing and Publishing Company.

Clover, J. 2003. Food Security in Sub-Saharan Africa. *African Security Review* 12 (1).

Constitution of the Federal Republic of Nigeria, 1999.

Corradetti, C. 2013. Philosophical Dimensions of Human Rights: Some Contemporary Views—Introduction. *SSRN Electronic Journal.* https://doi.org/10.2139/ssrn.2211203.

Coughron, J.R. 2016. An Examination of Food Security and its Impact on Violent Crimes in American Countries. *All Theses* 2565. https://tigerprints.clenson.edu/all_theses/2565.

Covenant on Economic, Social and Cultural Rights, Comment No. 3, Paragraphs 2, 4. 1990.

Cruft, R., S.M. Liao, and M. Renzo. 2015. The Philosophical Foundations of Human Rights: An Overview. In *Philosophical Foundation of Human Rights*, ed. R. Cruft, S.M. Liao, and M. Renzo, 1–41. Oxford: Oxford University Press.

Ejiogu, E.A. 2014. *The Problematic of Human Dignity in Africa.* Owerri: Living Flames Resources.

Enyimba, Maduka. 2003. *Democracy, Politics and Society: A Philosophical Appraisal.* Calabar: Iyke Press.

Flynn, J. 2013. Human Rights in History and Contemporary Practice: Source Materials for Philosophy. In *Philosophical Dimensions of Human Rights: Some Contemporary Views*, ed. C. Corradetti.

Food and Agriculture Organization of the United Nations. 2002. *The State of Food Security in the World*, 4th ed. Retrieved from http://www.fao.org.

Food and Agriculture Organization of the United Nations. 2005. *Voluntary Guidelines to Support the Progressive Realization of the Right to Adequate Food on the Contest of National Food Security, Rome.*

Food and Agricultural Organization of United Nations. 2006. *Food Sercurity Policy Brief 2.* Available online: http://www.fao.org/fileadmin/templates/faoitaly/documents/pdf/pdf_food_security-concept-note.pdf. Accessed June 29, 2022.

Food and Agriculture Organization of the United Nations. 2017. *The State of Food Security and Nutrition in the World, Rome.*

Galeoti, M. 2014. *Global Crime Today: The Changing Face of Organised Crime.* Routledge.

Gobert, J., and M. Punch. 2003. *Rethinking Corporayte Crime.* Cambridge: Cambridge University Press.

Griffin, J. 2008. *On Human Rights.* Oxford: University Press.

Human Development Report of United Nations. 1994.

Ibanga, Michael. 2000. Democracy as a Foundation for Human Rights Protection. *Ndunode: Calabar Journal of the Humanities* 3 (1).

Iwe, N.S.S. 1986. *The History and Contents of Human Rights.* New York: Peter Long Publishers.

Iwe, N.S.S. 2000. *The Dignity of Man as the Foundation of Human Rights: A Message for Nigerians.* Calabar: Seasprint.

Jacques, Diouf. 2005. *"Foreword" Voluntary Guidelines to Support the Progressive Realization of the Right to Adequate Food in the Contest of Natural Food Security.* Rome: F.A.O.

Kadish, S.H. 1983. *Encyclopedia of Crime and Justice*, vol. 4. Free Press.

Kempe, R.H. 2002. From Crisis to Renewal: Towards a Successful Implementation of the New Partnership for Africa's Development. *African Affairs* 101: 389.

Kent, B.D. 2013. *Food Insecurity as a Factor in Felonious or Misdemeanor Juvenile Crimes*. Doctoral dissertation, Walden University.

Laski, Harold J. 1966. *A Grammar of Politics*. London: George Allen and Unwin.

Mberu, B.U. 1999. *Social Structures and Institutions: Nuclear Themes in Sociology*. Abakiliki: Willy Rose and Applesead Publishing Company.

McGuire, J. 2004. *Understanding Psychology and Crime: Perspectives on Theory and Action*. Berkshire: Open University Press.

Ndubisi, F.N. 2009. Crime and Punishment of Ancient African Morality. In *From Footmarks to Land Marks on African Philosophy*, ed. A.F. Uduigwomen. Calabar: Jochrisam Publishers.

Njoku, Francis O. C. 2009. *Igbo Jurisprudence: An African Exercise in Legal Coherentism*. New Jessey: Goldine and Jacobs Publishers.

Nwozor, A., J.S. Olanrewaju, and M.B. Ake. 2019. National Insecurity and the Challenges of Food Security in Nigeria. *Academic Journal of Interdisciplinary Studies* 8 (4): 9–20.

Ofoegbu, Francis C. 2014. *Live and Let Live Nigeria: A Philosophical Cum Historical Reflection on Nigeria @ 100*. Owerri: Applause Multi-sects.

Okafor, F.U. 1992. *Igbo Philosophy of Law*. Enugu: Fourth Dimension Publishers.

Okon, Etim E. 2005. The Universal Declaration of Human Rights: Legal and Philosophical Appraisal. *Sophia: An African Journal of Philosophy* 7 (2).

Olayinka, O.F. 2017. The Right to Food in Nigeria: What is the Impact of University Education and Food Security? In *Issues in Curriculum and Language Education*, ed. F.V. Falaye, and J.A. Adegbile, 411–429. Ibadan: Ibadan University Press.

Omoregbe, Joseph. 1997. *An Introduction to Philosophical Jurisprudence: Philosophy of Law*. Lagos: Joja Press.

Osita, Eze C. 1984. *Human Rights in Africa*. Lagos: Institute of International Affairs.

Ozdemir, F., and B. Oner-ozkan. 2017. The Nature of Crime: Different Approaches Towards the Causes of the Criminal Act. *Nesne Psikoloji Dergisi (NPA)*. 5 (11): 345–361.

Payne, B.K. 2016. *White Collar Crime: The Essentials*. Sage Publications.

Pegels, E. 1979. The Roots and Origins of Human Rights. In *Human Dignity: The Internationalization of Human Rights*, ed. A.H. Henkin. New York: Aspen Institute for Humanistic Studies.

Rawls, J. 1999. *The Law of the Peoples with the Idea of Public Reasons Revisted*. Cambridge Mass: Harvard University Press.

Raz, J. 2010. Human Rights Without Foundations. In *The Philosophy of International Law*, ed. Samantha Besson, and John Tasioulas. Oxford: Oxford University Press.

Rizzuti, A. 2020. Food Crime: A Review of the UK Institutional Perception of Illicit Practices in the Food Sector. *Social Sciences* 9 (112): 2–11. https://doi.org/10.3390/socsci9070112.

Sarki, Z.M. 2019. Types of Crime: The Three Clusters. In *Reading in Criminology*, ed. S.A. Abdullahi, 1–15.

Sen, A. 1981. *Poverty and Famines: An Essay on Entitlement Deprivation*. Oxford: Clarendom Press.

Simnons, J.A. 2001. Human Rights and World Citizenship: The Universality of Human Rights in Kant and Locke. In *Justification and Legitmacy: Essays on Rights and Obligations*, 179–196. Cambridge: Cambridge University Press.

Slater, R., S. Kay, and W. Steve (eds.). 2014. Food Security. In *The Companion to Development Studies*, 3rd ed, 151–156. New York: Rutledge.

Suchara, T., M. Akitsu, and K. Ohishi. (n.d.). *Philosophy of Agricultural Science: Social and Ethical Studies of Food, Life and Environment*.

Tamuno, Tekena N. 2011. *Oil Wars in the Niger Delta: 1849–2009*. Ibadan: Sterling-Holden Publishers Ltd.

Thompson, P.B. 2008. Agrarian Philosophy and Ecological Ethics. *Science and Engineering Ethics* 14 (4): 527–544.

Thompson, P.B. 2010. *The Agrarian Vision: Sustainability and Environmental Ethics*. Lexington: University Press of Kentucky.

Uduigwomen, A.F. 2006. *Introducing Ethics: Trends, Problems are Perspectives*. Calabar: Jochrisam Publishers.

United Nations. 1948. *Universal Declaration of Human Rights*. New York: United Nations.

United Nations Convention on the Rights of the Child. 2009. *General Comments, No. 12*.

United Nations Development Programme (UNDP). 1994. *Human Development Report: New Dimensions of Human Security*. New York.

United Nations Development Programme (UNDP). 2009.

Ukavwe, H.O. 2012. A Philosophical Discourse on Human Rights. *Abuja Journal of Philosophy and Theology* 2.

World Bank. 1986. *Poverty and Hunger: Issues and Options for Food Security in Developing Countries*. Washington D.C.

Wright, A. 2013. *Organised Crime*. New York: Routledge.

Yi, X.I.A.O. 2008. Defining the Concept of the Victimless Crime in View of Criminal Policy. *Journal of Capital Normal University (Social Sciences Edition)* 6.

Zvvavanyange, R.E. 2016. *An African Agrarian Philosophy and Sustainable Development Goals: Nuturing Creativity in Science and Society*. A Paper Presentation at the Third Global Conference on Agricultural Research for Development, 5th–8th April 2016, Johannesburg, South Africa.

Maduka Enyimba (Ph.D.) is a senior lecturer at the University of Calabar. His teaching and research interests include; Epistemology, African Philosophy, Environmental Ethics, Philosophy of Education, Development Studies, Philosophy of Literature, and Philosophical Anthropology. Some of his publications have appeared in journals such as, *Filosofia Theoretica, African Symposium, South African Journal of Philosophy, Dialogue and Universalism* etc. Maduka is a member of Conversational School of Philosophy (CSP), and employs the methods and principles of conversational thinking to interrogate and engage with other scholars on the idea of sustainability, inclusiveness and development across borders. Presently, he is working on two projects; (1) the conversational idea of —Creative Struggle|| and Conflict Management in Sub Saharan Africa (2) a theory of Philosophical anthropology that grounds humanness in African place through the methodology of conversationalism.

Victor C. A. Nweke is an academic philosopher and political theorist. His major research interests include post-colonial African philosophy, global political theory, intercultural philosophy, applied logic and normative ethics. He has presented invited lectures on topical issues in these areas and related subjects in academic conferences and workshops. He has also published his ideas in reputable academic platforms. His peer-reviewed publications in journals and edited books attempt to interrogate and reconceptualize articulate perspectives on right conduct and human flourishing in a complex world using ideas in African philosophy as a focal point of departure. Victor is a foundational member and affiliate researcher of the Conversational School of Philosophy (CSP). He is also on the editorial board of *Filosofia Theoritica: Journal of African Philosophy, Culture and Religions*. He is currently a researcher/Ph.D. candidate at the institute of Cultural Studies University of Koblenz-Landau, Germany.

Chapter 21
Rethinking Shangwe Traditional Philosophy in Resolving Agrarian Conflicts in Contemporary Gokwe Communities

Elvis Tsvangirayi Siziva

Abstract Agrarian disputes and contestations have culminated into a plethora of far-reaching developmental challenges in Shangwe communities. It is unfortunate that these agrarian conflicts have remained a thorn in the Gokwe region because traditional law codes have not been sufficiently used in conflict resolution. Agrarian wrangles are inevitable in any society especially in Africa where agriculture is the backbone of national economies. Agrarian conflicts in the Gokwe region usually emanate from land pressure, farmer-grazier contestations, and crop and stock theft. The Zimbabwean government has relied on contemporary law in dealing with land contestations and other agrarian issues. It is important to note that indigenous Shangwe people developed sophisticated codes for conflict resolution. These codes and laws are embedded in several linguistic of genres including proverbs, folklores, riddles, idioms, myths, legends and taboos. These were used by sage during indigenous court proceedings as a form of reprimanding wrongdoers. The essence of the chapter therefore is to problematize the efficacy of Shangwe jurisprudence in resolving agrarian conflicts in contemporary Gokwe communities. However, a nexus between contemporary law and indigenous customs should be established in order to efficiently and rigorously solve agrarian wrangles. Contemporary courts and indigenous Shangwe courts should operate in-loco-parentis. They should complement each other in terms of handling agrarian matters.

Keywords Shangwe · Traditional philosophy · Agrarian conflicts · Jurisprudence

---

E. T. Siziva (✉)
Great Zimbabwe University, Masvingo, Zimbabwe
e-mail: elvissiziva38@gmail.com

M. J. Tosam and E. Masitera (eds.), *African Agrarian Philosophy*, The International Library of Environmental, Agricultural and Food Ethics 35,
https://doi.org/10.1007/978-3-031-43040-4_21

21.1 Introduction

Agrarian disputes and contestations have culminated into a plethora of far-reaching developmental challenges in Shangwe communities. Against this background, the Zimbabwean government has over time pursued contemporary law in dealing with issues of land tenure, crop theft, livestock and other related agrarian disputes as a panacea to the seemingly insurmountable problems. Indigenous law has not been adequately applied in addressing such issues and that has posed some serious issues on rural communities in Gokwe, and Zimbabwe as a whole. Advocates of Western styled agrarian laws argue that customary laws are ambiguous, uncertain and loose in dealing with land tenure and agrarian issues. The colonialists introduced new dimensions to land ownership, title, and management, as well as wider rights and responsibilities related to land and natural resources (Kasimbazi 2017: 7). During the colonial period a new set of laws regarding land ownership were introduced and left a legacy that influenced land policies in many African countries.

As a result of colonialism, a system of land tenure based on freehold and leasehold was introduced. In most cases, existing forms of customary land tenure were either ignored or overridden (Ibid). Traditionally land was considered as common property, controlled by lineage heads and chiefs From the same perspective, Makanyisa et al. (2012: 176) also argue that in pre-colonial Zimbabwe, land was communally owned and placed under the custody of traditional chiefs whose mandate was to delineate pastures from arable lands. From the same perspective, Peters (2004: 274) argues that land in Africa is usually vested in collectivities such as chiefdoms or clans. The principle of communal land ownership became the basis for solving land disputes in rural Zimbabwe.

Agriculture and livestock production is practised in the entire Gokwe region which result in scarcity of land resources managed under unsecured customary land ownership and communal grazing. As such agrarian wrangles continue to rock the Shangwe communities. Some of the conflicts have degenerated in physical conflicts, witchcraft and death. Due to pressure on land, grazing land becomes a major predicament in the region and livestock end up devouring crops resulting in farmer-grazier conflicts. In some cases, some people due to food insecurity resort to stealing crops and livestock as a solution to food insecurity. Ingawa et al. (1999) opines that the main causes of farmer/grazier conflicts include access rights, whereby traditional access rights to communal grazing and water resources are being obstructed by the individual tenure system of arable farmers. The essence of this chapter is to problematize the efficacy of Shangwe indigenous customs in managing agrarian wrangles.

There is no denial that traditional laws, which regard land as a communal resource, are relevant in dealing with rural issues like land tenure contestations, crop and livestock theft and conflicts associated with environmental degradation. Traditional sages who are the custodians of indigenous law have successfully employed these laws in dealing with agrarian conflicts for several centuries. Under traditional land tenure systems land given to a clan member by the traditional leader belongs to the clan. The Shangwe traditional law emphasises justice in terms of access to land.

Moreover, the Shangwe people believe that nature or the Earth is sacred and must be treat with reverence. It is for this reason that the Shangwe people observe sacred days (*zvisi*) where the land is not cultivated. In the case of the Nemangwe community, two sacred days are observed, every Thursday and 28th day of every month people are not allowed to cultivate the fields. However, other Gokwe communities only observe every Thursday. Gombe (1998: 195) asserts that it was the role of the chief to ensure that people were adhering to the *chisi* principle. The king or chief, was a representative of the spiritual world on earth, he was the guardian of the fundamental values of life and was largely responsible for the prosperity of his people (Bourdillon 1976: 111; Tosam, in this volume), of which land was considered as the basis of life. Therefore, it was also believed that disobeying the chief was tantamount to disobeying God (Ibid). This translates to the fact that the chief had a firm hand on agrarian matters since his decisions are considered to be God's decisions. Shangwe traditional law also emphasises respecting land boundaries created by traditional leaders, village heads, headmen and chiefs.

21.2 Historical Background to the Shangwe Ethnic Group: Revisiting the Conundrum Associated with the Term Shangwe

This section unravels the identity of the Shangwe ethnic group. The term Shangwe has always been a subject of critical debate between the *Madheruka* migrants, most of whom came from Chilimanzi area and the Shangwes. The debates associated with the term in question sometimes degenerate into conflicts between the two groups. This chapter seeks shed light on the myth associated with the term Shangwe. The term Shangwe is derived from Shangwa meaning poverty. When the original Shangwe people came into the Mafungabutsi which later became Gokwe they lamented that they had come to a poverty stricken country (Nyika ye Shangwa). As such they started to call themselves the Shangwe people meaning people living in a poverty stricken domain. When the Madheruka came into the scene during the 1950s they misconstrued the term Shangwe to mean backwardness.

There are two categories of the Shangwe people, namely, the Karanga Shangwe and the Korekore Shangwe. Siziva (2019: 11) argues that the Karanga Shangwe were the first to settle on the Mafungabutsi plateau, which is present day Gokwe. According to oral sources, Nyamusasa was the paramount ruler while Mubvuma, Mufunga, Mupare and Manyowi were his sub rulers (Ncube 2004). These were the first to adopt the term Shangwe. The Karanga Shangwe under Nyamusasa remained in control of the Mufungabutsi until the eighteenth Century when they were defeated by Chireya, the Korekore leader (Beach 1980: 282). The Korekore under Chimera, later Chireya, came from Guruuswa area (Ibid). This translates to the fact that the Korekore Shangwe was part and parcel of the Rozvi Kingdom and became victims of Rozvi succession squabbles. When Chireya destroyed the Karanga dominance the

Korekore Shangwe became the new feuds of the Gokwe region. When Chireya and his people came into the Mufungabutsi area they also adopted the term Shangwe (Siziva 2019: 11).

Against the above history of the Shangwe people, the Madheruka gave the term Shangwe a completely different and denigrating meaning. The 1950s–1960s saw an influx of the *Madheruka* migrants into Gokwe region as a result of the Native Land Husbandry Act of 1951. Nyambara (2002: 288) observes that colonial officials regarded the migrants as the embodiment of 'modernization' and as the model farmers to be emulated. They therefore concentrated their extension work among the migrants, giving them the necessary technical advice and credit facilities. As such, the migrants began to denigrate indigenous Shangwe as backward and primitive people (Ibid). On the contrary, the Shangwe claim that the term Shangwe refers to a place rather than to their ethnic identity (Ibid). Even today the Madheruka people still disparage, though to a limited extent, the indigenous Shangwe as lagging behind the dynamics of technological development. However, the denigrative interpretations of the term Shangwe were a product of colonial mentality and not a true reflection of the Shangwe people. The Shangwe, like any other Shona societies, back then, had sophisticated life coping strategies which enabled them to solve their agrarian and other wrangles prior to colonisation and Madheruka influx.

21.3 The Causes, Effects and Dynamics of Contemporary Gokwe Agrarian Conflicts

This section explores the factors which culminated into agrarian conflicts and the nature of the wrangles associated with agrarianism in the Gokwe region. It articulates economic, political and social causes of agrarian wrangles in the entire Gokwe region and how the wrangles have impacted on the lives of the local people.

Agrarian wrangles are inevitable in any society especially in Africa where agriculture is the backbone of national economies. According to Diao et al. (2010: 1), since almost all rural households depend directly or indirectly on agriculture, and given the sector's large contribution to the overall economy, it might seem obvious that agriculture should be a key sector in development. Competition for agrarian resources inevitably leads to turbulence in Gokwe communities. However, as suggested by Chivasa and Mutsvangwa (2014: 167) the inevitability of conflicts suggests that they should not be viewed as a negative force but as a resource for social change and development. However, at the end of the day, conflicts need to be resolved. The Shangwe society has experienced some serious agrarian conflicts since the nineteenth century which demand appropriate conflict management mechanisms. These conflicts emanate from economic, political and social factors, as examined below.

21.4 Insecure Land Tenure Systems in Rural Gokwe Communities

In Africa, wrangles over land emanate from boundaries to ownership of the whole plot of land; they also range from intra family or clan to inter family and clan conflicts. Land squabbles are an increasingly common feature of life in Africa, exacerbated by land tenure insecurity, land pressure and land degradation.

Among the Shangwe people land is distributed customarily; using customary principles. Under customary law land is not privately owned but communally owned and individually worked in order to plant food crops for the family (Jacobs 2000). However, communal land in Zimbabwe is owned by the state. No individual has title deeds under communal tenure system. Moyo (2007) suggests that the radical title to communal land is vested in the state through the President who holds it in trust for the communities. Therefore, no-one is allowed to dispose the land the way he/she sees fit except the state. Such land tenure is legally inferior to the extent that the state can displace people from such land. In Gokwe some residence have become victims of displacement by government tiers creating bitter squabbles between those concerned and the government. Land is allocated by chiefs to household heads who are principally males. There was no individual titling but there was consensus that if one was allocated land it belonged to him (Ibid). This to some extent explains why African land was taken away by the white settlers who erroneously interpreted this to mean that land in Zimbabwe (as in most African colonies) did not belong to anyone since the Africans had no title deeds (Ibid). In post-colonial Zimbabwe, land insecurity remains a major cause of tension in rural communities such as Gokwe. The Chemagora, Sikombela, Bhadha and Sibusiso incidents are typical examples of conflicts which occurred as a result of lack of security of tenure in Gokwe. A couple of families were evicted from Chemagora by the government of Zimbabwe. According to News Day (2019) Three thousand Gokwe families in Chemagora, Sikombela, Bhadha and Sibusiso under chief Njelele were left homeless after being evicted by armed soldiers and police.

The villagers were reportedly settled by Chief Njelele and Zimbabwe African National Union Patriotic Front (ZANU PF) politicians during 2008, 2013 and 2018 election periods (Ibid). The Zimbabwe National Army (ZNA) and the Zimbabwe Republic Police (ZRP) demolished property belonging to the settlers. The evictions caused protracted wrangles between the victims and the Chief as well as the politicians concerned and the government. The evictions led to the arrest of Chief Njelele who was accused of unlawfully parcelling out land to get cattle and money (Ibid). The irony is that politicians were not affected by the matter, a factor which seemingly entails that the arrest of Chief Njelele was just a smokescreen meant to blindfold the affected people, civil society organisations, opposition political parties, international community as well as the general public. There is no doubt that the people in these areas became victims because their lands were not secured by title deeds. As such the government and politicians could violate these people's rights anyhow in terms of land ownership.

21.5 Rampant Land Pressure: A Key Factor of Agrarian Conflicts in Gokwe

Nyambara (2001) postulates that the post-independence cotton boom which attracted more migrants; the effects of the Economic Structural Adjustment Programme (ESAP) initiated in the early 1990s resulted in massive retrenchments of workers. The majority of the retrenchees found their way into Gokwe villages. As a result, signs of land pressure began to manifest in the form of ubiquitous land disputes among various claimants. The situation was aggravated by the large influx of *Madheruka* into Gokwe district. Additionally, multitudes of unemployed youths in Gokwe today are clamouring for pieces of land resulting in acute land conflicts in the study area.

Due to land pressure in Gokwe, the residents adopted the practice of share cropping. Sharecropping in Gokwe villages usually takes place between on one hand households that have accumulated land but are unable to sufficiently utilize it due to resource constraints and on the other hand the landless in-migrants (Ibid). It is however, not uncommon for some land rich farmers to enter into share cropping arrangements with other land rich farmers for the purpose of maximising their production capacity, especially during good cotton seasons (Ibid).

According to a village head in Nemangwe community many people in Gokwe region are practising share cropping mainly because of land pressure. This has culminated into some serious wrangles over the years. Conflicts occur when one of the parties fails to comply, for example when the tenant fails to fulfil his or her obligations at the end of the season. It may also arise if the land lord refuses to grant the tenant permission to cultivate the land in the next season. Conflicts of this nature are common in contemporary Gokwe communities and have always been prevalent since the coming in of the *Madheruka* during the 1950s–1960s.

Land pressure resulted in farmer-grazier conflicts as village heads tended to allocate land to young adults on land that was formerly designated as pasture lands. According to Mufutau et al. the conflict between farmers and herders is one of the social problems that provoke serious security challenges and obstruct with severe threat to the entrepreneurship and unity of the people. Farmer-grazier conflicts are a perilous feature of the Shangwe communities emanating from land pressure. Muhammed also reiterates that disagreements over the use and allocation of food and resources such as grazing areas between herders and local farmers are intense because of the interest of both parties. Pastoralists and crop farmers are intertwined, sharing land, water, fodder and other resources. As a result, there are several problems bordering on the relationships between pastoralists and crop farmers, foremost of which is the perennial conflict over resources (Shettiman and Tar 2008: 64). Farmers require large pieces of land for crop cultivation while herders also need vast land to graze their animals resulting in an environmental dilemma between the two fraternities. Such a situation is rampant in the Gokwe region.

Mutasa (2011: 9) suggests that it is almost impossible to talk about Zimbabwe's agricultural development without touching on its thorny land issue that has punctuated the country's life since the turn of the millennium.

Land pressure in Gokwe has also led to cultivation of fragile areas like stream banks. According to Mlowoka (2012: 1) cultivation of riparian zones of rivers lead to massive land degradation. Land degradation has created tension between traditional authorities and community members. It has caused bitter conflicts among community members in Gokwe. Stream bank cultivation along major rivers in Gokwe such as Sasame, Sengwa, Ume, Gwave, Svisvi, Bopoma and Tare has destroyed a lot of arable land creating tension between the perpetrators of land degradation and the owners of the affected arable lands.

21.6 Politics of Patronage and Agrarian Wrangles in Gokwe Communities

Interview with a former Movement for Democratic Change (MDC) Councillor of Nemangwe Ward 11 established that Zimbabwe's political landscape has also contributed to massive agrarian wrangles in Gokwe region. Following the emergence of the Movement for Democratic Change (MDC) in 1999 Zimbabwe was polarised into rural ZANU PF and urban MDC (Harrold-Barry 2004). Such a scenario appears to have culminated into a plethora of agrarian wrangles in Zimbabwe. The Centre for Conflict Management (2019) maintains that the politicisation of development initiatives also causes conflict, for instance, the selective distribution of food or seed relief such as the presidential seeds and the grain loan scheme. There are also intra and inter party conflicts, hate speech, malice, exploitation of youth, violence and politicisation of service delivery. ZANU PF became the dominant party in Gokwe and gave members of the party in the region the powers (illegitimate) to manipulate agrarian benefits in their favour. For example, one village head echoed that government input scheme mostly favoured ZANU PF party supporters. Thus, partisan service delivery by government culminated into dire conflicts in Gokwe communities.

Again when ZANU PF faced a serious defeat at the referendum of February 2000 Mugabe and his party officials resorted to Fast Track Land Reform Programme (FTLRP) as a backdrop to rejuvenate their party (Mugabe 2001: 120). Moreover, some people from all parts of Gokwe became excited and joined the *jambanja* land reform hoping to benefit chunks of land. They even forfeited their communal land ownership rights in Gokwe. When they failed to get the land elsewhere they returned home only to find their pieces of land ceded to other people by the local authorities. Such a scenario brewed tense relations between the local leaders and the returnees and between returnees and the new beneficiaries of the land.

21.7 Social Implications of Agrarian Conflicts in the Gokwe Region

Most agrarian conflicts in Gokwe are socially grounded. During bad seasons crop and stock theft becomes a cause for concern. Gokwe is a drought prone region. Droughts expose people to some untold predicaments. One village head in Gokwe Nemangwe also opined that due to a series of droughts, many people in Gokwe experienced precarious lives which at times caused some to engage in atrocious activities such as stealing and witchcraft. When the area is rocked by a serious drought, many people find it difficult to secure food and as such some resort to stealing grain and livestock from those who managed to produce something. This is yet another area that creates agrarian wrangles.

There is no doubt that the economic slump which is seemingly a permanent feature of Zimbabwe today has created acute poverty in many contemporary Gokwe communities. These social ills have created tensions among members of the region in question. Tinga (1998) asserts that witchcraft is a plot to inflict harm or undermine the progress of other people. A witchcraft is the use of supernatural power for nefarious purposes (Kugara 2017). Most Africans believe that witchcraft cause unusual phenomenon like accidents, conflicts, death, domestic and public aggression, poverty, sickness and failure (GechikoNyabwari and NkongeKagema 2014). Research shows that a serious conflict that occurred between chief Nemangwe and one of his subjects in his neighbourhood. It is alleged a certain farmer at Halfway area near chief Nemangwe's residents used magical powers to make people work his fields during the night, the *divisi* juju.

Unfortunately the Chief was one of the victims of the practice. It is also alleged that the recalcitrant farmer made a special tool for the chief, a huge mattock which he called the '*bhidha ramambo*', the mattock for the chief. When an unforeseen circumstance occurred which the people thought was a result of witchcraft the chief summoned a **traditional healer** for assistance. The traditional healer exposed the farmer and the farmer admitted his evil practice. One informant echoed that the chief and the cunning farmer were very close friends but the incident dented their relations permanently.

Still on the same vein, a certain young man from the Zarova area in Gokwe Nemangwe articulated how witchcraft tarnished his relations with his biological parents. According to the man his little daughter died mysteriously. This prompted him to consult a traditional healer (n'anga) after seeking permission from the village head. The traditional healer alleged that the aggrieved man's parents killed the sibling and transformed her into a goblin in order to boost crop yields. The scenario became a bitter pill for the young man and his wife and created a gulf between the couple and the man's biological parents. As such, the man and his wife parted ways with the bride's family. It is important to note that some agrarian conflicts in Gokwe emanated from social disorders sometimes degenerating into permanent family fragmentation.

21.8 Gender Dimension and Agrarian Wrangles in Gokwe Communities

After attaining independence in 1980, Zimbabwe has remained largely a patriarchal society. The patriarchal nature of the Zimbabwean society has shaped and perpetuated gender inequality and female subordination (Kambarami 2006: 9). According indigenous custom, men are the heads of the families and according to customary law land is allocated to the head of the family. Because land is owned by men, women only gain access to land through marriage. This means that women's access to resources like land is limited.

Moreover, this implies that female-headed households in Gokwe find it difficult to own land. In case of death of the husband the wife of the deceased loses land to a male relative of the late husband, a situation which has caused continuous wrangling between men and women in the Gokwe region. Research has shown that in Gokwe North and South cases of this nature have been dealt with by traditional leaders such as village heads, headmen and chiefs. However, families without elder male children usually lose the case. This idea is accentuated by Goebel (2005) who posits that most Southern African societies have a patrilineal system in which land tenure is most frequently in the hands of males and generally the eldest son or uncle inherits title to land. It is only in exceptional cases, if there is no husband's brother or son to inherit the land, that we find widows inheriting title to land, provided they remain in the family and if they had children with the deceased husband (Gaidzanwa 1995).

Customary law treats women as minors and gives rights to land only to the male members of the family. Land is allocated by traditional leaders who are mostly male and who strive to protect men's rights and access to land. Thus, usufruct rights to land are the most prevalent phenomenon in Zimbabwe. Despite feminist agitations, women remain marginalised in land ownership; they still have no secure access to land. Their rights and access to land is still mediated by their male relatives. Moyo (1995: 71) asserts that this is because of the government's rhetorical emphasis on redressing historical imbalances. According to Jacobs (2000: 2), women's demands for land have been neglected by the state which does not want to lose support among its male-dominated political party structures. This was clearly brought out by the then Vice President, Joseph Msika, who when asked at a press conference why women did not have land rights said, 'because I would have my head cut off by men if I gave women land…men would turn against the government…giving wives land, or even granting joint titles, would destroy the family.' Such sentiments have aggravated gender discrimination in Zimbabwe. Thus, in Gokwe women do not own land and that has caused women to quarrel even with their own sons over land left behind by their husbands.

Furthermore, men and women in families have been involved in wrangling over choice of crops. The patriarchal nature of Shangwe societies has always placed women on the peripheries when it comes to decision making. Hence, during the planting season men usually dictate the crops and varieties to be planted. Since cotton is considered as the golden leaf, men prioritise cotton at the expense of other crops such as ground nuts, cowpeas and round nuts. According to Jacobs (2000: 2), customarily, a wife is entitled to own a ground nut plot known as *tseu*. As such women would prefer their ground nuts and related crop varieties to be grown earlier than any other crops but men take advantage of their customary status as heads of families and ignore the wishes of the women. This has also caused acute tensions between men and women.

21.9 Effects of Agrarian Squabbles in Gokwe Region

This section delves into the effects agrarian wranglings in Gokwe communities. Agrarian squabbles have impacted heavily on the development of Gokwe and Zimbabwe as a whole. There is no denial that agrarian wrangles have affected the contemporary Gokwe communities. The effects of the wrangles were not always detrimental to the region in question. Wrangles are normal events and there is no reason to "pathologise" them as it is too often the case. Development institutions tend to see conflicts as negative phenomena to be solved. Sometimes agrarian wrangles help to develop the agriculture sector by bringing new and sophisticated agrarian order in the region in question.

However, Marongwe (2002: 112) has identified some detrimental effects of conflicts. He suggests that local struggles and local conflicts by their very nature have a tendency of retarding developmental efforts. Their effects have devastated many communities; deep divisions exist among members of the same community, making them unable to work together for their own development. Over and above all, divisions lead to negligence and poor management of land resources in an area. In the Gokwe region, for example, some people, because of land disputes, are now working the land recklessly leading to massive land disintegration, a situation which is bound to affect future generations. Erosion is now a menace in most areas of agrarian conflicts. The Centre for Conflict Management and Transformation (CCMT) (2013) also observed that instability in an area leads to reduced economic opportunities. Depletion of local resources such as land creates a donor syndrome. Residents in Gokwe seem to have developed a huge dependence on non-governmental organisations. Failure to manage resources due to conflicts has impoverished the region to the extent that the region became a home for non-governmental organisations.

Agrarian wrangles have also culminated into very tenuous and fragile relationships among community members in Gokwe. Broadly, small-scale land squabbles have sometimes degenerated into widespread civil wars, posing threats on national security. Such situations caused victimisation of some members in the region leading to a tense atmosphere. Agrarian conflicts in Gokwe communities have affected the region politically, socially and economically. Most people in Gokwe are living precariously, languishing in dire poverty because of conflicts associated with agriculture.

Conflict constitutes the major explanatory factor for famine, hunger and malnutrition given the complex nature of the humanitarian crisis that results from conflict. Conflict is disruptive and has strong implications for rural development (Lund 2003). Conflicts in Gokwe have undermined the competitiveness of the region in terms of agriculture. Gokwe was one of the agricultural giants in the country especially in the 1990s. The region was well known for its cotton production but currently the level of production has drastically deteriorated, mainly because of agrarian conflicts.

There were many harvest suicides in Gokwe in 1997 where 153 women committed suicide because their husbands had squandered the money from the proceeds of the land (Chingarande 2004). This demonstrates that women are left out in decision making concerning with the money they have earned from their labour on the land (Ibid). Women are also very vulnerable to domestic violence during harvest times in Gokwe. One Gokwe woman complained bitterly on how. During harvesting, men quickly forget that women contributed or did most of the work during the course of the agricultural season. They sometimes batter their wives for demanding a share of the money obtained from farming. For this reason some women get depressed to the extent of committing suicide.

Agrarian wrangles in some Gokwe communities have attracted witchcraft. The Police Outpost (2016) confirmed an incident where a Chief instructed two Village Heads to conduct a cleansing ceremony with their subjects. At one of these ritual ceremonies, a more mysterious event occurred. It is alleged that in full view of the villagers, a big eagle flew past and picked up a full grown dog from the homestead of Village Head Pauro where the cleansing ceremony was taking place. The ceremony was immediately abandoned.

> In another case, a woman from Sidojiwe Village who was pregnant is said to have gone into labour before giving birth to a frog-like creature in a case largely linked to witchcraft. Though these sound **like** scripts extracted from a horror movie, they are a clear demonstration of the prevalence of witchcraft in Chief Njele's area. These cases, according to Chief Njelele, emanate from marriage disputes, infidelity, and adultery. He also indicated some conflicts normally occurred during the ploughing season when people fight over farm boundaries as well as harvesting time (Ibid).

While it is true that agrarian wrangles have impacted negatively on Gokwe communities, it is important to note that at times these wrangles are necessary in as much as people would want to progress in terms of agriculture. Agrarian conflicts have helped the government and other stakeholders in the agriculture sector to come up with meaningful strategies that have helped to improve the sector. For example, the Pfumvudza scheme. Through the Pfumvudza most people who joined the scheme and abide by it were provided with seeds.

Again, to ease up boundary conflicts the Agritex department is pegging contours for each farm. The contours also serve as boundary markers, a strategy which was designed because of boundary conflicts. When conflicts arise they bring in new constructive ideas. As such conflicts are not always negative. They at times promote change of systems and the much needed development.

21.10 Harnessing Shangwe Knowledge Systems in Conflict Resolution: Reflections on How the Shangwes Handle Agrarian Conflicts in Gokwe Communities

Here I examine the effeicacy of Shangwe knowledge systems in resolving agrarian conflicts. The essence of the section is to discuss the applicability of Shangwe jurisprudence in handling conflicts associated with agriculture.

The Shangwe conflict management system depends on the ethnic group's didactical philosophy. It should be appreciated that the Shangwe people have a rich history which is embedded in their oral traditions. The Shangwe ethnic group has oral traditions Which is a repository of their history, world view and wisdom. Newberry (2004) observes that oral tradition is a source of wisdom. The community is a web of mythological symbols which are transmitted orally (Ngara et al. 2014). Apart from mythologies the Shangwe have a diverse cultural heritage of folktales, proverbs, riddles, praise poetry lyrics, similes, songs, myths and taboos. These are carefully and appropriately utilised to solve agrarian wrangles among members of the society. Agrarian wrangles in Gokwe have been dealt with successfully using traditional philosophy for a very long time.

Agrarian squabbles in Gokwe communities can be successfully managed without any intervention of contemporary jurisprudence. The indigenous conflict management systems apply not only to minor inheritance or boundary disputes, but also to extended land use conflicts concerning common property and natural resources. According to Moyo (2007), land conflicts can be resolved with minimal state intervention and even without any state intervention at all. The Shangwe are capable of handling their agrarian wrangles without necessarily involving contemporary state institutions like the Agritex department, Environmental Management Agency (EMA) and the contemporary court system.

21.11 Rethinking the Efficacy of Shangwe Indigenous Jurisprudence in Resolving Agrarian Conflicts

Agrairian disputes form an important part of Shangwe history and the people the Shangwe have effective conflict management systems that have been used to address such conflicts. This section attempts to explore indigenous laws employed by Shangwe people in dealing with agrarian conflicts in the Gokwe region.

Shangwe communities like any other communities in Zimbabwe have customary court systems which deal with all sorts of conflicts. The traditional Shangwe court system (*dare*), which is still significant in Shangwe communities today, shows a unique approach to jurisprudence. For the Shona, an individual crime is said to affect the whole community, hence the social and moral dimension of crime. Hence the dictum, "I am because we are, and since we are, therefore I am" (Mbiti 1989). This implies that the Shangwe work as a community in resolving agrarian conflicts, a factor which is bound to bring informed judgements on agrarian issues. At these traditional courts the Shangwe depend on their rich traditional customs which is enshrined in their oral traditions such as in their folktales, proverbs, riddles, or poetry lyrics, and songs. These are carefully and appropriately utilised to solve conflicts among members of the society.

For example, proverbs during the court session are used to prosecute the alleged offender (Mandova and Chingombe 2014). They question the logic of one's engagement in certain acts. This is because proverbs have the power to invoke one's imagination by clarifying the situation under discussion (Ibid). Nandwa and Bukenya note: A proverb ... fascinates us by calling our attention simultaneously to the general reality around us to the particular reality of the situation in which the proverb is used (Nandwa and Bukenya 1985). Shangwe people have a rich reservoir of proverbs that can be used for arbitration of court cases. The elderly Shangwe man and women are well versed in Shangwe philosophy.

One common proverb used at Shangwe court sessions is '*mhosva hairovi*' meaning crime does not lapse in the passage of time. This proverb is used to warn people against committing crimes unnecessarily thinking that they will get away with it. It has been noted that some people after their animals have devoured a neighbour's crops they appear before a Shangwe court and they are charged a fine they choose not to comply. This proverb assures such recalcitrant elements that the court will follow them up until they comply. Proverbs are a fundamental genre of literature in of Shangwe indigenous jurisprudence. They play a pivotal role in conflict resolution and are commonly used in traditional courts system, that is, from family level court to chief's court.

One other important aspect of Shangwe philosophy is their emphasis on the importance of using amicable methods of resolving disputes, including agrarian disputes. They believe that violent methods do not work. The following proverb underscore this idea:

> Ngoma hairidzwe nedemo"
>
> Do not beat a drum with an axe.

The Shangwe people use the above proverb at the start of the court session as a way of advising the disputing parties to avoid getting emotional during the court proceedings so that their peace building agenda becomes a success.

According to Mandova and Chingombe (2014) African philosophy emphasizes the spirit of communality through extension of hospitality to others. As such the Shangwe court system encourages communality in terms of pasture lands. In case of wrangling over grazing land, the Shangwe employ proverbs:

> Mombe inopfuura haipedzi uswa
>
> A passing ox does not finish the grazing

By the above proverb members of the conflicting parties are encouraged to share grazing areas with affected areas.

Of all the various forms of oral genres, proverbs are the most important instruments employed in conflict resolution among the Shangwe communities in Gokwe. Okpewho (1992: 235) argues that proverbs are the storehouse of the wisdom of the society. This is because most proverbs have a philosophical mileage which is the result of an appropriate and sensitive observation of human interaction and experience. In case of agrarian disputes in Shangwe communities, proverbs are treated with respect and authority. They are regarded as truth tested by time, and they are often used for resolving conflicts and other problems between members of the community.

The Shangwe people are well known for their magical conflict management mechanisms. They at times use magic to deter recalcitrant elements in the society especially regarding stock and crop theft. Magic is a powerfull device for averting agrarian conflicts among the Shangwe.

In case of agrarian wrangles associated with witchcraft, the Shangwe consult a *tsika mutanda*, a fortune teller. It is alleged that one old woman bewitched her two year old granddaughter to death in order to use her as a goblin for boosting her productivity in agriculture. Research has it that the mother of the child accused her mother in law of witchcraft immediately after the death of her child and that created a sharp gulf between the two women. The mother in law reported the matter to the chief. She accused her in law of defamation of character. The chief advised the two parties to consult a forth teller through the village head. That was duly done and the forth teller claimed that the mother in law had bewitched her daughter in law for agricultural purposes. The mother in law was found guilty before the chief and was fined two herds of cattle, one for the chief and the other one for her daughter in law.

Furthermore, some Shangwe songs reprimand wrongdoers. It is common in Shangwe communities that women commit suicide during harvest times because of rogue men who misappropriate family funds after marketing their produce especially cotton. Usually the Shangwe sing the song,

Jembere Guru	elderly woman
Raigara musango	who stayed in the forest
Raisa musoro musaidhoni	put her head in poison

According to one respondent, the song was composed when a certain Njelele woman committed suicide when her husband squandered money from agricultural produce. It is reported that the woman spent much of her time in the fields working extraordinarily hard. As such the song is meant to remind men to respect their hardworking wives so that they do not commit suicide.

If a community member is caught stealing either livestock or crops the Shangwe people force the juvenile thief to sing a mockery song which says:

Kurima musana wandirwadza	during the farming season my back aches
Pakudya ndomera manhenga	In terms of consuming I develop feathers

This song discourages laziness becausxe this may result in crop and stock theft. Obliging the culprit to sing the song is a way of embarrassing him or her in front of other villagers or community including children.

One other Shangwe philosophical genre which is essential in resolving agrarian wrangles is folklore. To warn people against crop theft the folklore entitled, Tsuro na Nzou (Hare and Elephant) is often used among Shangwe societies in Gokwe. The folktale reads:

> Hare had his field with a lot of big pumpkins. Elephants used to go and steal the pumpkins while the Hare was asleep. In the next morning Hare would only see footprints. One day he designed a wise plan. He cut one of his biggest pumpkins and got inside with a very sharp knife in his hand. As usual the Elephant came. He preferred those big pumpkins. The elephant swallowed the pumpkin which was housing the hare. The hare then cut the elephant using his knife and the Elephant died on the spot.

The moral implication of this folklore is that stealing other people's crops may lead someones in a serious tragedy, including death. As such the folklore is often used to deter people from stealing other people's produce. In most cases Shangwe people depend on magic as their security measure against crop and animal theft, the *rukwa juju. Rukwa juju* sometimes leads to death of the victim.

Like any other Shona community in Zimbabwe, the Shangwe also use taboos to manage agrarian disputes. Chemhuru and Masaka (2010: 123) define taboos as 'avoidance rules' that forbid members of the human community from performing certain actions, such as eating some kinds of food, walking on or visiting some sites that are regarded as sacred, cruelty to nonhuman animals, and using nature's resources in an unsustainable manner. Among Shangwe communities a certain taboo is often used to settle boundary disputes. Taboo forbid farmers to get into a neighbour's or any member's field without his or her permission.

21.12 Conclusion

Shangwe people have a rich reservoir of moral codes and customs that can be used for arbitration of court cases and also to avert agrarian conflicts. Community-based customs and mechanisms for conflict resolution in Gokwe communities are still relevant although they are largely undermined in contemporary jurisprudence. Moreover, colonial authorities and contemporary policies have stripped traditional authorities of some of the powers they formerly enjoyed such as expelling witches from the community. No stringent and deterrent measures can be taken against recalcitrant elements in the society, a factor which now makes conflicts a permanent feature of Shangwe communities.

There is need to take into consideration indigenous agrarian values and methods of conflict resolution in the resolution of agrarian conflict in contemporary Gokwe legal institutions and indigenous jurisprudence to avoid contradictions in terms of enforcing the law. Indigenous courts should work in liaison with contemporary courts. The traditional African codes should be employed alongside contemporary codes in resolving agrarian conflicts.

References

Beach, D.N. 1980. *The Shona and Zimbabwe 900–1850: An Outline of Shona History*. Gweru: Mambo Press.

Bourdillon, M. 1976. *The Shona Peoples: An Ethnography of Contemporary Shona, with Special Reference to Their Religion*. Gweru: Mambo Press.

Centre for Conflict Management Transformation. 2013. *Conflict Transformation: A Multi-dimensional Task*. Hugh Miall.

Chemhuru, M., and D. Masaka. 2010. Taboos as Sources of Shona People's Environmental Ethics. *Journal of Sustainable Women Access to Development in Africa* 12 (7): 121–133.

Chingarande, S.D. 2004. *Land in Context of Fast Track Land Reform Programme*. Brief Policy Prepared for the African Institute for Agrarian Studies.

Chivasa, N., and P. Mutsvangwa. 2014. An Examination of the Role of Shona Folktales in Promoting Peacebuilding Among Modern Communities in Zimbabwe. *International Journal of Humanities Social Sciences and Education (IJHSSE)* 1: 161–169. ISSN 2349-0373 (Print), ISSN 2349-0381 (Online). www.arcjournals.org.

Diao, X., P. Hazell, and J. Thurlow. 2010. *The Role of Agriculture in Development: Implications for Sub-Saharan Africa*, https://researchgate.net/publication/223557221

Gaidzanwa, R. 1995. Land and the Economic Empowerment of Women: A Gendered Analysis. *Southern Africa Feminist Review: The Gendered Politics of Land* 1 (1): 1–2.

GechikoNyabwari, B., and D. NkongeKagema. 2014. The Impact of Magic and Witchcraft in the Social, Economic, Political and Spiritual Life of African Communities. *International Journal of Humanities Social Sciences and Education (IJHSSE)* 1 (5): 9–18. ISSN 2349-0373 (Print), ISSN 2349-0381 (Online).

Goebel, A. 2005. Zimbabwe's 'Fast Track' Land Reform: What About Women? *Gender, Place and Culture* 12 (2): 145–172.

Gombe, J.M. 1998. Tsika DzaVaShona 2nd ed. Harare College Press.

Harrold-Barry, D. 2004. *Zimbabwe: The Past is the Future: Rethinking Land, State, and Nation in the Context of Crisis*. Avondale: Harare Weaver Press.

Ingawa, S.A., A. Ega, and P. Erhabor. 1999. *Farmer-Pastoralist Conflict in Core States of the National Fadama Project*. Abuja: FACU.

Jacobs, S. 2000. Zimbabwe: Why land reform is a gender issue, *Sociological Research online* 5 (2). https://www.socreonline.org.uk/5/2/jacobs/html.

Kambarami, 2006. *Culture, Femininity and Sexuality and Culture: Patriachy and Female Subordination in Zimbabwe*, University of Fort Hare.

Kasimbazi, E. 2017. *Land Tenure and Rights for Improved Land Management and Sustainable Development*. United Nations Convention to Combat Desertification.

Kugara, S.L. 2017. *Witchcraft Belief and Criminal Responsibility: A Case Study of Selected Areas in South Africa and Zimbabwe*. South Africa: Venda University.

Lund, M. 2003. *Prevention Policy and Practice in Pursuit Theory. The Institution for Poverty, Land and Agrarian Studies*. http:// anothercountrysidewordPress.

Makanyisa, et al. 2012. The land tenure system and the environmental implications on Zimbabwean society: Examining the pre-colonial to post-independent Zimbabwean thinking and policies through out history and philosophy. *Journal of Sustainable development in Africa* 14 (6).

Mandova, E., and A. Chingombe. 2014. The SHONA Proverb as an Expression of UNHU/ UBUNTU. *International Journal of Humanities Social Sciences and Education (IJHSSE)* 1: 161–169. ISSN 2349-0373 (Print), ISSN 2349-0381 (Online). www.arcjournals.org.

Marongwe, N. 2002. *Conflicts Over Land and Other Natural Resources in Zimbabwe*. Harare: ZERO Publications.

Mbiti, J. 1989. *Introduction of Africa Religion*. Oxford: Heinemann Educational Publishers.

Mlowoka, C. 2012. *Relationship Between Stream Bank Cultivation and Soil Erosion in Dedza*. Malawi.

Moyo, Sam. 1995. *The Land Question in Zimbabwe*. Harare: SAPES Trust.

Moyo, S. 2007. *Emerging Land Tenure Issues in Zimbabwe*. Monograph Series, Issue No. 2/07. Harare: African Institute for Agrarian Studies (AIAS).
Mugabe, R.G. 2001. Inside the third chimurenga, Harare Publishing House.
Mutasa, M. 2011. *Taming the Beast: Vulnerability to, Coping and Adaptation with Drought Impacts in Rural Zimbabwe*. Paper Prepared for the Initiative on Climate Adaptation Research and Understanding Through the Social Sciences (ICARUS-2) Meeting at the University of Michigan (5–8 May 2011) Themed, Vulnerability and Adaptation: Marginal Peoples and Environments.
Nandwa, J., and A.L. Bukenya. 1985. The Role of Proverbs in the Judicial System. *International Journal of Asian Social Science*.
Ncube, G.T. 2004. *A History of North-Western Zimbabwe, 1850–1960*. Gweru: Mambo Press.
Newberry, D. 2004. *Contradictions at the Hearts of the Canon: Jan Vansina and the Debate Over Oral Histriography 1960–1985*. Cambridge University Press.
News Day. 2019. *300 Families Evicted from Chemagora*.
Ngara, R., and J. Rutsate, and V. Mangizvo. 2014. Shangwe Indigenous Knowledge Systems: An Ethnometrological and Ethnomusicological Explication. *International Journal of Asian Social Science*.
Nyambara, P.S. 2001. The Politics of Land Acquisition and Struggles Over Land in the Communal Areas of Zimbabwe. The Gokwe Region in the 1980s and 1990s. *Africa: Journal of the International African Institute* 71(2): 253–285.
Nyambara, P.S. 2002. Madheruka and Shangwe: Ethnic Identities and the Culture of Modernity in Gokwe, Northwestern Zimbabwe, 1963–79. *The Journal of African History* 43 (2): 287–306.
Okpewho, I. 1992. *African Oral Literature, Backgrounds, Character and Continuity*. Bloomington: Indiana University Press.
Peters, P.E. 2004. Inequality and Social Conflict Over Land in Africa. *Journal of Agrarian Change* 4 (3): 269–314.
Shettima, A.G., and U.A. Tar. 2008. Farmer–Pastoralist Conflict in West Africa: Exploring the Causes and Consequences. *Information, Society and Justice* 2: 63–84. ISSN 1756-1078 (online). https://doi.org/10.3734/isj.2008.1205.
Siziva, E.T. 2019. *Shangwe Chieftainship: Succession Disputes in the Nemangwe Chiefdom and the Quest for Legitimacy cc19th–21st*. Great Zimbabwe University. http//ir.gzu.acc.zw.8080/xmlui/handle/123456789/337.
Tinga, K.K. 1998. *Cultural Practice of the Mijikenda at Cross roads: Divination, Healing Witchcraft and the Saturday Law*. AAP.

Elvis Tsvangirayi Siziva holds a Master's Degree in History and a Post Graduate Diploma in Higher and Tertiary Education from the Great Zimbabwe University. He is currently a Secondary School Teacher at the Nemangwe Secondary School. His research interests are ethnicity and Politics.

Chapter 22
Murimi Munhu: A Quest for Decoloniality in Black African "Small Scale" Subsistence Farmers in Rural "Reserve" Zimbabwe

Joseph Pardon Hungwe

Abstract I employ Oliver Mtukudzi—the late Zimbabwean musician's *Murimi munhu* lyrical composition—to highlight the coloniality embedded in "small scale" subsistence farmers in Zimbabwe. Arguing from a decolonial perspective; this chapter seeks to achieve two things. Firstly, in deploying the decolonial theory, attention here is focused on confronting the negative stereotypes and marginalising practices that are systematically expressed and practiced against "small scale" subsistence farmers in rural "reserve" Zimbabwe. By their very nature, dehumanising stereotypes and exclusionary practices are not only a negation of African communal values, but also perpetuate and entrench coloniality. Secondly, by addressing coloniality as the basis of dehumanising tendencies towards small scale subsistence farmers, the chapter endeavours to conceptually restore the humanity of subsistence farmers in rural Zimbabwe. Therefore, the central argument here is that decoloniality is imperative to countervail dehumanising practices and orientations as highlighted in the *Murimi munhu* song.

Keywords Smallscale farming · Decoloniality · Dehumanising · Subsistence farmer · Coloniality · African communal values

22.1 Introduction

Murimi munhu, which literally means "a farmer is a human being", is a famous Shona song by the late Afro-Jazz musician, Oliver Mtukudzi. The song aptly captures and problematises the dehumanising practices and attitudes that are expressed towards small scale subsistent farmers in Zimbabwe. On the basis that "music not only reflects society's hopes and aspirations, it also articulates issues facing the populace and is

J. P. Hungwe (✉)
College of Education, University of South Africa, Pretoria, South Africa
e-mail: josephhungwe@gmail.com

M. J. Tosam and E. Masitera (eds.), *African Agrarian Philosophy*, The International Library of Environmental, Agricultural and Food Ethics 35,
https://doi.org/10.1007/978-3-031-43040-4_22

a mirror of the political, (social and economic) landscape and temperature" (Maguraushe 2020: 2), this chapter employs the *Murimi munhu* song to articulate the coloniality perpetrated by both the state and general populace in Zimbabwe. In cognisance of the dehumanising practices, I deploy the decoloniality theory towards elimination and combating the colonial vestiges, which continue to shapesubsistence farming in Zimbabwe. The historical injustice to land access and distribution in Zimbabwe is a matter that has received scholarly attention (Hungwe 2021).

It is recorded that small scale farmers make a substantive contribution towards food security and sustainability in Zimbabwe. For instance, before the 2002 farm invasions and the consequent disruptions, black African small scale farmersproduced more than half of the national maize tonnage. Nevertheless, black African small scale farmers continue to be systematically dehumanised through derision, negative stereotyping, squalid living conditions and government's general systematic marginalisation. Succinctly, dehumanising practices and attitudes are manifested on two levels. Firstly, the unfair economic market forces deny small scale subsistence farmers the power to determine the selling prize for their produce. Moreover, small scale farmers are not unionised, and cannot therefore, bargain collectively forthe selling price of their agricultural commodities. Secondly, the "urbanised" populace derides small scale subsistence farming as an engagement for the "unemployable"and uneducated persons (Hungwe 2021). In fact, it is common in Zimbabwe that rurality/*ruzevha* as representative of a place occupied by subsistent farmers is negatively regarded asa primitive place occupied by primitive people (*ibid*). In this chapter I locate these dehumanising tendencies in the colonial arrangement, in which black Africans were confined to reserved areas that are colloquially known as *ruzevha* (rural areas).

The primary concern of this chapter is that the colonially entrenched dehumanising practices and attitudes in the reserve area, which currently is referred to as smallscale subsistence farmers, has continued in post-independence Zimbabwe. Deploying decoloniality as a theoretical framework, I seek to argue that smallscale farmers are dehumanised in several forms. This chapter is, therefore, important because it brings a paradigm shift in the pertinent surrounding land matters in Zimbabwe and Africa at large. Most literature on land matters in Zimbabwe focus on land appropriation under colonial establishment and redistribution in the post-independence Zimbabwe (Moyo 2011; Tom 2015). In this respect, the chapter juxtaposes subsistent farmers and decoloniality. To adequately appreciate the embedded coloniality in subsistence farmers in the rural reserve context, this chapter is divided into six sections. Firstly, I discuss the historiography of land issue in Zimbabwe. This discussion is important because land was/is not only an economic, but also a racialised issue in Zimbabwe. The second section is an exposition of theoretical understanding of land, while the third section explores the contextual lyrical relevance of the *Murimi munhu* song. The fourth section theorises rurality, and the fifth section expounds on the notion of decoloniality. Finally the sixth section appropriates theoretical tenets on decolonising the small scale farmer sector in Zimbabwe.

22.2 Historiography of Land Issue in Zimbabwe: Colonial Partitioning of the Country

The contentious issue of land became evident in the colonial era. In 1890, about 196 pioneer column members arrived in what was to become Rhodesia, a country that was later renamed Zimbabwe (Chitiyo 2000). The company had come on the assumption that there was abundant gold, but when it became apparent that there had been an exaggeration of the gold, diamond and other mineral wealth, attention was shifted to agriculture. Large tracts of fertile and arable land in areas with high annual rainfall were confiscated from black Africans (Shumba 2018). Consequently, black Africans were forcibly removed and restricted to drought prone areas with poor soil for agricultural purposes. The agriculturally poor land that was allocated to black indigenous Africans were known as "reserves". In semantics terms, areas that were allocated to both black Africans and wild animals were known as native reserves and national park reserves respectively. The land that was forcibly taken from black Africans was given to pioneer column members as compensation for their efforts in the colonisation project. While under the traditional setup land was communally owned under the tutelage of chiefs and other traditional leaders, the colonial establishment instituted private land ownership. According to Tshuma (1998: 78), "in the pre-colonial Zimbabwe, family and individual land use rights were mediated through membership of political communities". Within communities, families and individuals created communally recognised use rights over specific process of land by investing their labour. The traditional community structures regulated access and use of land and people were protected against infringements of land ownership.

The first reserves for Africans were designated in 1894 by British South Africa Company (BSAC) and these were arid Gwai and Shangani areas. These areas were badly watered, sandy and unfit for settlement. By 1905, the BSAC had created about 60 reserves that occupied only 22% of the new colony; the settler community appropriated the bulk of the land (Shumba 2018). By 1922, 64% of all Africans were required to live in reserves and this land expropriation was designed to force Africans to seek formal employment. For Nyandoro (2019), legislative foundation culminated in the compulsory eviction, migration and forced resettlement of the black African people in squalid conditions and on semi-arid land. The culmination of the establishment of reserve areas was crystallised by the Land Appointment Act of 1930. Accordingly, this Act "put definitive limit on land available for black through a legalized division of the country's land which prohibited members of either racial group from owning land in areas assigned to other. By 1930, 50.8% of the total land had been declared European while 30% had been reserved for the African population" (Shumba 2018: 61). Resettling black Africans in agriculturally unproductive and overcrowded areas reduced them into deep economic poverty. By this Act, the country's natural resources were racially divided with whites having more access to fertile land and natural minerals.

Furthermore, Moyo (2011) notes that the dispossession of indigenous Zimbabweans of land and natural resources from the 1890s led to the highly skewed agrarian structure and racially discriminatory land tenure, crowding the majority into marginal Tribal Trust Lands known as communal areas. Shumba (2018) further observes that the colonial regime began to pass regulations, assigning specific parcels of land to blacks while reserving the best land for Europeans. By 1910, 24.3% of land had been appropriated for whites and 26% had been declared Native Reserves, later known as Tribal Trust Land. In 1951, the promulgated Native Land Husbandry Act was premised on the imperative to "revolutionise" African agriculture. The development of settler capitalist agriculture and the underdevelopment of the African agricultural sector were intimately linked to settler land expropriation. To graphically illustrate the land imbalance, it is stated that "the Land Tenure Act of 1969 reserved 15.5 million hectares, largely in the most productive areas for 6000 farms owned by both individual white farmers and large estates, 16.4 million hectares for 700,000 black families and 1.4 million hectares of 8500 black small-scale farmers" (Southall 2011: 91).

It is instructive to state that under colonialism, land grabs were "legitimatised" through laws such as the Native Reserves Act, Land Apportionment Act of 1930, and the 1969 Rhodesia Constitution (Shumba 2018; Mlambo 2010). As a result, the cumulative consequence of land grab is that an imbalance of land ownership between black Africans and white farmers began to emerge. White farmers owned almost half the arable land, while black Africans were congested within agriculturally unproductive areas. As a consequence of the skewed land redistribution, black Africans who had economically relied on land for subsistence farming became impoverished while white settlers farmers accumulated wealth (Hungwe 2021).

One important aspect to note here is that land became a demarcation signifier of race. Racial discrimination between Europeans and black Africans was concretised through skewed land distribution. Agricultural and pastoral land allocated to European settlers had title deeds, whereas customary tenure governed land was occupied by black Africans. For Tshuma (1998), native reserves were created in the aftermath of the unsuccessful Ndebele and Shona liberation war of 1896. The Native Reserve Commission of 1914 recommended that land should be set aside for the use of and occupation by Africans. The demarcation between native reserve and European settler land was done in such a way that a reserve did not encroach over to the fertile and productive land set aside for Europeans. Africans in the reserves were not granted titles to the land. The "justification" for allocating arid and unproductive land to black Africans to be subsistent farmers in rural/*ruzevha* areas was drawn from the misconception that they were incapable of appropriate farming. As Africans were considered to be uncivilised, irrational and irresponsible, they were excluded from individual property rights in terms of land ownership. Specifically, it was considered an economic risk to allocate commercial fertile land to the black Africans because traditional methods of farming were considered as archaic and unproductive. By 1905, half of the population of black Africans was living in the designated reserve areas. The colonialists stereotyped black Africans as "wasteful and irresponsible" farmers, judging their farming methods as causing soil erosion (Chitiyo 2000). Dispossession of land entailed that black Africans who could not economically sustain their livelihood were forced to supply their labour to the white farmers in exchange of token monthly "salary". The Native Affairs Department, whose primary remit was to "oversee" the welfare of Africans, ensured that there was economic underdevelopment in the native reserves (Hungwe 2021). Apparently, both forced resettlement and the colonial state's monopolisation of land compromised black Africans' sense of citizenship. Basic individual and communal freedom granted to whites were constrained for black Africans. For black Africans, there was a simultaneous loss of cultural and social connection with land, and traditional structures were disrupted. In other words, reserves were and have remained areas of enforced habitation. It is for this reason that even in post-colonial Zimbabwe, subsistence farmers are often recipients of government-assisted food parcels. Consequently, an appropriate conceptualisation of subsistence farming would account for and eliminate cultural, political and economic disempowerment initiated by colonial administration and left unresolved by the post-colonial government.

In the post-colonial Zimbabwe, the notion of small scale farmers—which is often used interchangeably with smallholdering farmers—are famers in communal areas who practice subsistence farming, producing maize, cotton, small grains such as millet, sorghum, groundnuts and some cash crops. In most cases, subsistence farmers sell surplus to the Grain Marketing Board and other private grain and cotton buyers. It is noted that subsistence farmers produce food for the nation and generate foreign revenue for the nation (Kang'ethe and Serima 2018). Despite the point that small scale subsistence farmers in rural areas greatly contribute towards food security and the general economy in terms of "occupation", the rural reserve area has remained literally underdeveloped in all aspects. For instance, the absence of good roads in the reserve areas meant that Africans could not easily facilitate the marketing of their agricultural products. In addition, while white commercial farming enjoyed government subsidies, African farmers were denied access to both bank loans and government financial grants (Tom 2015). Astactics for suffocating and disrupting the black African indigenous farming activities, white farmers could raid the cattle and burn down crops on the African farms. The introduction of preposterous taxes, such as cattle and hut taxes, consequently forced Africans to offer their services as cheap labour in order to pay the prescribed taxes. However, there is a need to understand how land is conceptualised in the African perspective.

22.3 Theorising Land: The African Perspective

I am using the term, Africa, in its generic sense in reference to black people of sub-Saharan Africa. In the traditional African set up, land had three specific roles: the social, economic and spiritual aspects. In the social aspect, peoples' land was occupied along ethnic lineage. In Zimbabwe, you would find that the Ndebele would settle as a collective group in one part of the country, and this goes for Shona people and all other ethnic groupings. Consequently, still today, there are areas officially and socially referred to as Matabeleland, Mashona-land and Manica-land. Socially, land denotes collective identity and bestows a sense of belonging. In the traditional set up, land is a communally owned land under district administration and traditional leadership. So land in this context remains a community asset or resource upon which members are supposed to draw economic, social and political benefits. Economic aspect refers to sustenance and income generating activities, which are made possible through access to land. From the land, community members can till and grow crops for food, rear cattle, and build and extend homes. So in a nutshell, the land serves two primary functions, namely, economic and residential. Land was/is a common community resource. Land was central to the livelihood of black African people.

Finally, the spiritual aspect is a distinctive African perspective of land. Perhaps there are certain similarities in social and economic aspects between Africans and other racial and ethnic groups across the globe. However, the spiritual aspect is deeply embedded in the perspective on land among Africans. In this aspect, land belongs to the livingdead (ancestors), the living and the forthcoming generations. So one of the distressing experience during colonially-forced land dispossession and the consequent relocation to native reserves was that, for Africans, there was a sense of separation from the persons "lying in graves". Ultimately, colonisation was simply a racial encounter between white and black Africans. On the contrary, white settler colonialist perceived colonisation as occasioning the meeting point of white as symbolically representing civilisation, while black Africans represented cultural and social primitiveness.

22.4 The Notion of Rurality in Zimbabwe

This brief paragraph seeks to do two things. Firstly, it is important to conceptualise rurality on the basis that this concept is a construction of both colonial and post-colonial Zimbabwe. Secondly, this section unpacks and exposes the material and economic conditions of rurality in Zimbabwe. This second point is important because subsistence farmers are located and conditionally determined by the conditions that characterise rural areas.

So from a conceptual perspective, rurality is contested. There are some suggestions that the appropriate descriptive definition of rurality can be drawn from juxtaposing rural and urban. In most cases, the concept of rurality is juxtaposed to urbanity in as far as infrastructural development is concerned. In Zimbabwe, rurality is perceived as an area of anti-progression and cultural conservatism. Arguably, since rurality is a contextually and historically defined concept, there is a multiplicity of definitions. For instance, in the context of the United Kingdom, Cloke states that ruralityis a fluid space that is characterised by conservatism and, in some instances, oppressive conditions. While there is a dominant perception on rurality, there are also scholars who highlight numerous advantages of rural over urban areas. Comparatively, rural dwellers have bigger space/land, a more peaceful environment, there is a lesser need to spend money on daily basic commodities, and there is less working-hours pressure that the urban counterparts experience (Masinire and Ndofirepi 2020). In accounting for the divergent conceptual viewpoints on rurality, it suffices to note that rurality and urbanity are characteristically demarcated in countries.

On the second point concerning rurality material and economic conditions, it is the case that in Zimbabwe, colonialism and coloniality account for deliberate underdevelopment in rural areas. Materially, rurality is described as a place of deficit, scarcity and unavailability of means for decent basic living needs. Mlambo (2010) states that the overwhelming majority of Zimbabweans live in the rural areas, which is often referred to as the countryside. Colonialism ensured that rurality is characteristically deficient in many aspects. The poor, or in some cases, the absence of proper road networking implied that black African subsistence farmers could not easily access markets for their cash crops. In fact, the few transporters who would avail themselves to transport agricultural products would charge exorbitant transport fees to black African subsistence farmers. Consequently, the negative implication is that financial profit was highly compromised. Currently, statics indicate that about 64% of Zimbabweans live in the rural areas where there are no rural industries and most rely on rainwater instead of irrigation. While rural areas are prone to diseases such as malaria and bilharzias in addition to frequent life-threatening venomous snake bites, it is the case that health services are usually few and distant.

In Zimbabwe, the living conditions in rural reserves have remained difficult because of limited social services such as education, healthcare, electricity and public transport (Chigonda and Chazireni 2018). In most rural reserves, people use open water-holes to access water for both human and animal consumption. For Chitiyo (2000), small scale farmers are impoverished, alienated and landless. In fact, one of the prime reasons small scale substance farmers are derided is that they generally live in abject poverty. In the rural areas, land enables the provisioning of food. For O'Flaherty (1998), about 57% of the Zimbabwean population reside and work in communally-held reserve rural areas. Communal areas used to be called Tribal Trust Land, so the country was partitioned between European and African agricultural farming. As a colonial construct, rural areas were and continue to be deprived of facilities like mechanised farm equipment and funding through bank grants and loans. It was and continues to be regarded as a business risk to lend money and agricultural inputs to subsistence farmers. It is within this context that *the Murimi munhu* song advocates for decoloniality of the small scale subsistence farmer sector in rural reserves in Zimbabwe.

22.5 *Murimi Munhu*: Lyrical Contextualisation

Oliver Mtukudzi, who died in 2019, was a famous Zimbabwean Afro-jazz musician whose popularity extends beyond Zimbabwe to Africa and the world at large. Though he predominantly sung in chiShona a language spoken by the majority in Zimbabwe, there are several songs he sung in isiNdebele and English. Perhaps what made Mtukudzi's songs popular in Zimbabwe is that the songs had contextual messages that resonated with the economic, social and political context. Arguably, "African artists are not simply entertainers or cultural educators but also fully acknowledged law-makers. Through song and dance, musicians create and disseminate principles that form the foundation of a people's system of thought and action" (Maguraushe 2020). It is in this respect that the song, *Murimi munhu*, succinctly captures the historicity and contemporarily contentious issue of subsistence farmers in the rural reserve areas of Zimbabwe.

Before analysing and theorising the lyrical context of *Murimi munhu*, it is imperative to state two affirmations in this chapter. Firstly, owing to the colonial arrangement and consequent legacy, farming is a largely racially construed concept. As discussed in the foregoing sections of this chapter, the colonial administration established a dual farming system in which white farmers were allocated large hectares of productive land. Besides access to productive land, white farmers had access to bank loans and grants. So farming became a lucrative business that guaranteed economic profits. On the other hand, black Africans who were allocated poor land in arid areas drew only basic economic livelihoods from land in the reserve area. In line with *Murimi munhu*, farming became and has remained a matter of basic survival for black Africans. Secondly, though subsistence farmers are not a homogeneous group in rural Zimbabwe, it is the case that on the average, this model of farming does not enable farmers to accrue economic benefits. So subsistence farmers live in povertystricken and dehumanising circumstances.

In the view of the foregoing observations, the *Murimi munhu* song pointedly depicts subsistence farming as a demeaning and dehumanising undertaking. For instance, Mtukudzi states that "*murimi tora badza urime, nehunyanzvi hwako torarama*" (the farmer should continue to use his/her agricultural expertise to feed the nation). This is a plea that despite the colonial and coloniality stereotypes, a subsistence farmer must persevere because his or her agricultural activities ensure the provision of food. Accordingly, "*kuva mudealer kudya zvemurimi*" (entrepreneur feeds on food from the farmer) or "*kuva mudzidzisi kudya zvemurimi*" (teachers feed from the farmer's products) are lyrics that point to the indispensability of the subsistence farmer. The lyrics of this song are pleading the case towards the restoration of the humanity and dignity of farmers. The background of such as plea is founded on the acknowledgement that small scale subsistence farmers are systematically dehumanised in both the colonial and post-independence Zimbabwe.

Paradoxically, *Murimi munhu* solicits for respect of subsistent farmers on the basis that they "feed the nation", yet this model of farming is not designed to commercially supply the national granaries. In fact, by its very design, subsistence farmers are supposed to harvest crops and store for their own consumption, and not for sale. Nevertheless, it is an escapable fact that in the post-Fast Track Land Reform programme, commercial farming was disrupted and in most cases, destroyed. The destruction of commercial farming as manifested in the vandalisation of farm equipment and the flight of experienced white commercial farmers (Hungwe 2021) entailed that subsistence farmers have, to a large extent, filled the gap of commercially supplying and guaranteeing food security in the country. In reference to under mechanised agricultural equipment used by small scale farmers in reserved areas, Mtukudzi persuasively suggests that "*murimi iwe tora kapadza*" (subsistence farmers take up the hoe to plant and cultivate crops for food production). It is instructive to note that even after many years of political independence, the subsistence farmers in the colonially-established reserve areas employ traditional methods of agriculture. Such methods, which used to be referred to as "primitive", include tilling the land using donkey or cattle draught power. In most of the rural reserve areas, subsistence farmers rely solely on seasonal rainwater to irrigate their crops. So it means that fundamentally, agricultural farming activities are only carried out in the four-month rain season. Pointedly, subsistence farmers are located in the rural impoverished and remote areas that have no basic infrastructure and rely on waterbodies like dams to install and sustain a mechanised "year-round" irrigation scheme. One other constraint encountered by subsistence farmers is the fact that they are not unionised. This implies that they cannot negotiate and bargain with the companies that buy their commodities for a fair price. For instance, cotton farmers in Gokwe and other areas have, for the past few years, been paid in groceries instead of cash (Moyo 2021). Even now, people who live in the reserve areas do not have title deeds to the land. In fact in Zimbabwe, the state's efforts towards the promotion of agricultural activities in the reserve areas have been suspected to be more political than economic endeavours. There are programmes such as Command Agriculture in which the state provides agricultural seeds, farming equipment/implements and expert knowledge. To advance the central argument in this chapter, I deploy decoloniality as a theoretical framework.

22.6 Decoloniality: Some Conceptual Considerations

This chapter is theoretically couched in decoloniality. In order to adequately appreciate decoloniality, it is critical tooutline the conceptual difference between colonialism and coloniality. These terms are important in further appreciating the above-articulated distinction between decolonisation and decoloniality. Colonialism refers to the political, social, authority and economic relations that are established between the colonising andcolonised countries. As the term connotes, colonising of a country imposes and dominates the colonised country in terms of administrative discharge, political, economic and cultural relations. For coloniality, however, an acknowledgement is made that long after the official termination of colonialism, there are remnant colonial values that persist, albeit as underlying, to determine and shape social, political, economical and educational facets of a "post-colonial" country. The continuities of marginalisation and underdevelopment in rural areas are accordingly a manifestation of coloniality embedded in the agrarian scope of Zimbabwe. For this reason, Bardhan and Zhang (2017: 286) note that "coloniality is the continuation of the domination of power of the structuring systems such as race that were put in place during European colonialism".

Decoloniality has become a catchword that means different things to different people depending on historical, political and social context. In fact, this term has become popularised to the extent that it has now become a multidisciplinary concept.

Illustratively, decoloniality is used in politics, education, economics, agriculture, business and many other sectors. A distinction needs to be outlined between decolonisation and decoloniality. Accordingly, "decolonisation is the historical and ongoing project through which formerly colonised societies think through the legacies of colonial domination and work to extricate themselvesfrom it". Decolonisation is, therefore, the undoing or correcting the legacy of colonialism. On the political side, decolonisation served as a blue-print towards the ideal emancipation of Africans from the forms of discrimination, oppression and inhumane practices wrought by colonialism. In the case of education, Fredua-Kwarteng (2020) notes that, "decolonisation entail inclusion of African indigenous ways of knowing, knowledge and values in African university curricula and the exclusion of Western knowledge and values that are deemed irrelevant or destructive to African development or realities".

It would appear that the concept of decolonisation has been popular from a political perspective. In this respect, Hansungule (2000: 306)—arguing from a political perspective—opines that "de-colonisation in the sixties and early seventies was a total let down. With the 'wind of change' in the midst, most Africans looked to de-colonisation with great hope. However, de-colonisation failed to emancipate the African." Admittedly, political decolonisation, which bestowed political independence on the previouslycolonised nations, instilled great excitement and a sense of hope. But with contentious issues such as land still topical in Africa, it is evident that colonialism embedded values that persist to influence political, social and economic discourse. Decades after political independence, decolonisation has is a multidisciplinary concept that covers all facets of the society that was disrupted by colonialism. On the other hand, decoloniality is an endeavour towards curtailing the perpetuities of colonialism. To understand decoloniality, one has to appreciate the notion of coloniality. Basically, coloniality is the acknowledgement that while colonialism is a historical occurrence, coloniality refers to the continuous influence of values associated with colonialism. Fundamentally, decoloniality attends to issues of power and dominance. Perhaps the succinct conceptualisation of decoloniality is offered by Maldonaldo-Torres (2007: 243) who notes that, "coloniality refers to long standing patterns of power that emerged as a result of colonialism, but define culture, labor, inter-subjective relations and knowledge production well beyond the strict limits of colonial administration". It is my considered perspective that social negativity that characterises small scale subsistence farmers is a result of coloniality in Zimbabwe.

Conclusively, colonialism classified humanity in accordance to race, ethnicity and sex. Consequently, the colonial arrangement entailed that material and economic conditions were determined by the racial, ethnic and sex classification arrangement. To give an example, a sex hierarchy in which white males occupied pinnacle positions, followedby white females, black African males and African females, accordingly was characteristic in colonial Zimbabwe and perhaps the rest of Africa. The implication of this colonial arrangement is that white males would access economic benefits, such as preference in employment, and property ownership, such as land and getting financial grants from the banks. On the other hand, there were myriad impediments to economic resources for all other categories of classification. So subsistence farmers are financially excluded. From this classification of humanity, the Land Apportionment Act of 1930 drew rationalisation of the creation and establishment of reserves for black Africans. This is the reason why decoloniality is appropriate in this chapter, because it tends towards the restoration of humanity, in this context, of the small scale subsistence farmers. The next section dwells on this matter.

22.7 Appropriating Decoloniality to "Small Scale" Farmers in Rural Zimbabwe

A conceptually adequate decoloniality of the subsistence farming in the rural reserve context in Zimbabwe needs a multifaceted approach. In cognisance of the point that*the Murimi munhu* song entreats for a re-affirmation of the humanity of subsistence farmers, it can be argued that it is a decolonisation song. To this end, I expound decoloniality on two levels, namely, the ideological and materiality. I am aware that these levels are in some cases interwoven. However, the quest for decoloniality in "small scale" subsistence farming in Zimbabwe is premised on the observation that the colonial structures and values that occasion de-humanising marginalisation has continued in post-colonial Zimbabwe. Henceforth, Hughes (2010) suggests that "imperial colonizers do not seize land with guns and plows alone. In order to keep it, especially after imperial dissolution, settlers must establish a credible sense of entitlement. They must propagate the conviction that they belong on the land they settled."

At an ideological level, subsistence farmers are considered as primitive, uneducated and culturally underdeveloped because they reside in rural and remote areas. This dovetails with the observation that in Zimbabwe, and perhaps as is the case across Africa, rurality connotes economic, social and political deficiency (Masinire and Ndofirepi 2020). The origin of this ideology is traceable to colonial administration that ensured that the dual agricultural model favours the European settlers at the expense of black African subsistence farmers. As noted already in this chapter, European settler white farmers were subsidised by the state and could easily access markets for commodities. Additionally, rural reserves was an arrangement for constraint and confinement as black Africans were not permitted to live in the urban areas. During the colonial era, only men of working class age were permitted to reside in the urban township so that they could provide a labour force for the growing industries. The urban areas were developed while rural areas remained neglected.

On the basis of the pictorial impression of the above paragraph, the ideological decoloniality implies that rurality in which subsistent farmers are located has to be conceptually reconfigured. This can be done on two fronts. Firstly, education in general and higher education in particular need to establish ways of curtailing the coloniality stereotypes associated with rurality and subsistence farming. For instance, rural areas are perceived to be epicenters of witchcraft, the loitering of ghosts at night, *sangomas*, witchdoctors and other "evil" practices. These stereotypes feed into the mainstream agricultural activities of subsistence farming. Illustratively, among the Shona people, there is a perceived "witchcraft" practice called *divisi*. *Divisi* is said to be a magical wand that ensures that a subsistence farmer accumulates higher harvests than the rest of his/her villagers. The usage of the advantages that *divisi*is said to posses can also result in high cattle production in terms of cattle offspring and dairy products.

Furthermore, from an ideological perspective, subsistence farming in the rural reserve areas is stereotypically considered to be a pastime of the "academically-challenged" members of society. In other words, subsistence farming is said to be employment for people who would have failed school examinations and therefore have no opportunities of securing formal salaried jobs. It is instructive to state that Zimbabwe's education has been and continues to be orientated towards academics, with less emphasis laid on entrepreneurship and vocational skills. The graduate as an "end product" of such an educational system is expected to seek and gain formal employment in industrial and civil services, which are located in the urban areas. It is for that reason that civil servants such as teachers, policemen or prison guards who will be stationed in rural areas command more social respect and recognition than small scale subsistence farmers. The assumptions was/is that a certificated clerical job is more prestigious than the manually-orientated work demanded in small scale farming.

What then does decolonising the coloniality ideology of subsistence farming imply? If decoloniality is conceptually understood as confronting and eliminating the vestiges of colonialism in agriculture, then as I noted earlier in this section, higher education needs to play a critical role in the formative approach to subsistence farming. As important spaces for knowledge production in Africa, higher education can research and produce models of subsistence farming, which restores human dignity; for instance, the deficit narrative in which a rural area is perceived as a place of shortcomings. In pursuance of decoloniality, the gap between urban and rural needs to be narrowed. This will militate against the residual remains of colonialism that stifles economic development in the rural areas, thereby dehumanising subsistent farmers.

On the other hand, a materiality level of decoloniality requires that the material economic conditions of the rural areas, in which the majority of black African subsistence farmers reside, have to be wholly transformed. An emerging problem in the discourse on decoloniality is that it has become a hollow concept that is deficit of practicalities. So transforming the material conditions in rural reserve areas is a matter of decoloniality. This transformation might imply a branded model in which communal and individual ownership of land are concurrently instituted. Instructively, under colonial era, "tribal trust lands were established as a way of accommodation and keeping indigenous people away from the urban areas and would only be allowed into towns and cities to provide cheap labor in industry, mines and on farms most of which were in peri-urban areas" (Mapuva 2015: 144). Considerably, a decoloniality perspective is crucial in curtailing instances, practices and attitudes of coloniality against small scale subsistence farmers.

This chapter has already noted the point that a dually-established agrarian model during the colonial era led to concurrent infrastructural underdevelopment of rural areas, while noticeable development was concentrated in the urban areas. Since the colonial era, the rural areas have been neglected and marginalized—a situation that encumbers the subsistence farming model. Though colonial legislation was deliberately designed to curtail development, the post-colonial government has not done much to improve rural areas (Mapuva 2015). Most rural areas have no electricity, supply of clean running water, good roads, and reliable internet and telecommunication services. Apparently, the post-colonial Zimbabwe government has equally neglected the rural areas. Accordingly, a plethora of legislative initiatives such as the 1982 Communal Land Act, Growth with Equity Policy, First Five Year National Development Plan of 1985, Free Educational Policy and Land Acquisition Act of 1992, and the subsequent Fast Track Land Reform, have all been unsuccessful in redressing the skewed economic development imbalance between the rural and urban areas of Zimbabwe. In so many ways the skewed economic imbalance occasions dehumanising practices against subsistence farmers who reside in reserve areas. For instance, because of the absence of running water and flushing toilets, subsistence farmers resort to the bush as toilets to relieve themselves.

At a material level, decoloniality can also reinforce the notion of Growth Point in Zimbabwe. The primary objective of the Growth Point concept, which was introduced after the attainment of political independence in 1980, was to develop infrastructure in rural areas. A Growth Point would have industries that acted as absorption markets for commodities of subsistence farmers. For instance, Gokwe had a functioning ginnery industry that buys cotton products—a cash crop commonly grown in Gokwe. This is the case across all rural reserve areas, thereby disadvantaging small scale farmers in Zimbabwe.

Additionally, decoloniality should imply re-configuring the rural reserve areas in Zimbabwe. As noted already in this chapter, under colonial administration, black Africans were forcibly resettled on arid and droughtprone areas. Though post-colonial Zimbabwe subdivided the country into five regions in accordance with agricultural patterns, black Africans who were resettled by colonial administration in non-crop producing areas are still stuck in such areas. "Good productive farming" is still today associated with white males. The implication of remaining in these areas is that for such people, subsistence farming does not translate into acquisition of a subsistence crop harvest. Decoloniality would mean the resettlement of people into areas that suit their preferred forms of agricultural activities for sustenance.

22.8 Conclusion

In this chapter, I have outlined the imperative for decoloniality in view of the systematic marginalisation of small scale farmers in rural Zimbabwe. It has become apparent that the social negativity depicted on the small scale farmers in rural areas is a manifestation of coloniality. Rurality is a symbol of deprivation and therefore, small scale subsistencefarmers as rural inhabitants are equally perceived as people who live on the social periphery. The implication of this is that the agricultural contribution made by small scale farmers to national grain and produce is not recognised by the state and society in general. It can, therefore, be argued that one of the primary reasons as to why rural areas are underdeveloped emanates from the social negativity associated with rural reserve areas from the colonial to contemporary post-colonial Zimbabwe.

References

Bardhan, N., and B. Zhang. 2017. A post/decolonised view of race and identity through the narratives of US international students from the Global South. *Communication Quarterly* 65 (3): 285–306.

Chigonda, T., and E. Chazireni. 2018. Once Rural, Always Rural? Social Service Provision in Selected Rural Cases from Zimbabwe. *National Journal of Multidisciplinary Research and Development* 3 (1): 469–475.

Chitiyo, T.K. 2000. Land Violence and Compensation: Reconceptualising Zimbabwe's Land and War Veterans' Debate. *CCR* 9 (1): 1–26.

Fredua-Kwarteng, E. 2020. Lecturers are Key to Ending Colonial Epistemicide. *University World News: Africa Edition*. https://www.universityworldnews.com/post.php?story=202114502190.

Hansungule, M. 2000. Who Owns Land in Zimbabwe? In Africa? *International Journal on Minority and Group Rights* 7: 305–340.

Hughes, D.M. 2010. *Whiteness in Zimbabwe: Race, Landscape, and the Problem of Belonging*. New York: Palgrave Macmillan.

Hungwe, J.P. 2021. Ivhu Kuvanhu/Umhlabathis Ebantwini: The 'Violent' Ubuntu in the Fast Track Land Reform Programme in Zimbabwe. In *Philosophical Perspectives on Land Reform in Southern Africa*, ed. E. Masitera. Chan: Palgrave Macmillan.

Kang'ethe, S.M., and J. Serima. 2018. Exploring Challenges and Opportunities Embedded in Small Scale Farming in Zimbabwe. *Journal of Human Ecology* 46 (2): 177–185.

Maguraushe, W. 2020. Nuances of Political Satire, Advocacy for Peace and Development in Selected Oliver Mtukudzi Songs. *Mankind Quarterly* 61 (2): 1–19.

Maldonaldo-Torres, N. 2007. On Coloniality of Being: Contributions to the Development of a Concept. *Cultural Studies* 21 (2–3): 240–270.

Mapuva, J. 2015. Skewed Rural Development Policies and Economic Malaise in Zimbabwe. *African Journal of History and Culture* 7 (7): 142–151.

Masinire, A., & Ndofirepi, A.P. 2020. *Rurality, Social Justice and Education in Sub-Saharan Africa*, vol. 1. Chan: Palgrave Macmillan.

Mlambo, A.S. 2010. 'This Is Our Land': The Racialization of Landin the Context of the Current Zimbabwe Crisis. *Journal of Developing Societies* 26 (1): 39–69.

Moyo, S. 2011. Land Concentration and Accumulation After Redistributive Reform in Post-Settler Zimbabwe. *Review of African Political Economy* 128 (38): 257–276.

Moyo, J. 2021. *Zimbabwe's Cotton Farmers Sweating for No Gain*. Anadolu Agency. https://www.aa.com.tr/en/africa/zimbabwe-s-cotton-farmer-sweating-for-no-gain/2241607.

Nyandoro, M. 2019. Land and Agrarian Policy in Colonial Zimbabwe: Re-ordering of African Society and Development in Sanyati, 1950–1966. *Historica* 64 (1): 111–139.

O'Flaherty, M. 1998. Communal Tenure in Zimbabwe: Divergent Models of Collective Land Holding in the Communal Areas. *Africa* 68 (4): 527–559.

Shumba, J.M. 2018. *Zimbabwe's Predatory State: Party, Military and Business*. Pietermaritzburg: University of KwaZulu-Natal Press.

Southall, R. 2011. Too Soon to Tell? Land Reform in Zimbabwe. *Africa Spectrum* 3: 83–97.

Tom, T. 2015. Post Zimbabwe's Fast Track Land Reform Programme: Land Conflict at Two Farms in Goromonzi District. *Journal of Humanities and Social Sciences* 10 (1): 87–92.

Tshuma, L. 1998. Colonial and Post-Colonial Reconstruction of Customary Land Tenure in Zimbabwe. *Social and Legal Studies* 7 (1): 77–95.

Joseph Pardon Hungwe holds Ph.D. in Education from the University of Johannesburg (South Africa). He is currently affiliated to the University of South Africa as a Research Fellow. He has published several journal articles and book chapters. His research areas include decolonisation and internationalization of higher education in Africa. Additionally, he has researched and published on the politics of land in Africa.

Chapter 23
The Farm in Colonial and Postindependence Imagination: A Crisis of Continuity

Mbuh Tennu Mbuh

Abstract In the note of content for this book, emphasis was partly on the "limited resources at [the] disposal" of humans with which "to take care of the ever-growing population." The functional relationship between humans and their circumambient environment—where the former is an integral part of the latter—in its transitional state will be vital to my engagement here with the manner in which the farm (or farming) became, and has remained, a metaphor for survival, status, rituals, control, and identity; while conditioning thought patterns in family, communal, national, and even global contexts. In Pinyin, as in most, if not all villages of the North West Region of Cameroon, the government policy of "green revolution" has been undermined by a gendered determinant toward crop cultivation, harvesting, consumption, commercialisation, and ritual performance. While exploring the causality between colonial and postindependence agricultural cultures, it is difficult, ironically, to determine the boundary between alien and indigenous attitudes toward the task of farming especially since some of the popular crops were brought in from foreign lands. By analysing the various tasks related to farming, and based on practical and textual evidence, it will be obvious in the end that neocolonial and postindependence African economies have failed to decolonise the colonial prefiguring in this domain, leading to a continuity of redundancy in replicating mega policy and even of stagnation. The ongoing campaign for a cultural glocalism generally, and in farming specifically, will further signify the imperative to Africanise policy and praxis in this area.

Keywords Postindependence imagination · Green revolution · Agricultural continuity · Mega policy · Cultural glocalism

M. T. Mbuh (✉)
Department of English, The University of Bamenda, Bamenda, Cameroon
e-mail: mbuh.tennu@uniba.cm

M. J. Tosam and E. Masitera (eds.), *African Agrarian Philosophy*, The International Library of Environmental, Agricultural and Food Ethics 35,
https://doi.org/10.1007/978-3-031-43040-4_23

23.1 Introduction: A Worrying Determinism as a Global Scheme

In the note of content for this book, emphasis was partly on the "limited resources at [the] disposal" of humans with which "to take care of the ever-growing population". The functional relationship between humans and their composite environment—where the former is an integral part of the latter—in its transitional state will be vital to my engagement with the manner in which the farm (or farming or agriculture) became, and has remained, a metaphor for several rubrics, including status, rituals, control, and identity. We should bear in mind throughout this chapter that the establishment of colonial farms, as highlighted by Edward Said (1989: 156, 158; 2005: 90) in his appraisal of Austen's colonial vision, was a means of serving the colonist's family interests of subsistence and status in particular and those of the metropolis in general. From this terminal goal in which the colonial farm resolved subsistence matters in the metropolis, I propose to elaborate the continuities that influenced both independence and postindependence systems, together with the accompanying mismanagement, gendering, and othering in agricultural matters. The dramatic increase in food imports from colonies resulting from World War I (Janes et al. 2019), vindicated this colonial structuring of agriculture.

Beginning with the need to pacify the survival instinct, we should recall how in his proximate state, early man invented strategies that evolved from wild fruits to roots and then to crop farming. The existentialist status that was associated to this bears lasting indicators in every hierarchical social structure, ranging from caste subalternity to the realm of the lords and royalty. In the uneasy continuity between precolonial and colonial societies, social status and the nature of subsistence was reflected in one's seasonal produce: the case of Okonkwo is perhaps the most iconic, dramatized at the start of Achebe's (1995) *Things Fall Apart* as a factor of "his personal achievement". The barns of yam, his wives, and his aspiration for the highest title in the land are related to his ambition and status as a hero-farmer, who is diametrically opposed to his father, Unoka. Although the latter has been unfairly analysed in critical discussions,[1] it is the son's aggressive strides in social mobility that become interesting in our understanding of how precolonial farming did not really suffer from a productivity complex before the advent of colonialism. It is instead its commissioned continuities that have developed into a more polarised society, reflecting the way farming had evolved into a classist complement to capitalist and neoliberal determinism in the new world order of the twentieth century.

Another characteristic of the changing farming culture is the control of spheres of influence through it, specifically from the colony to the postcolony, and then in communal contexts. If in the past certain activities were attributed to men, it

[1] See my reaction (Mbuh 2015) against such reductionist generalization and its distortion of the already problematic transition from the precolonial into the colonial world, in "Art, the Artist, and Postcoloniality: Chinua Achebe's Unoka and Implications of Representation in *Things Fall Apart*" (*The Gong: An Interdisciplinary Journal of English Language and Literary Studies*, Vol. 1, No. 1, 2015 (104–121).

was mainly because of the physical strength and biological conditioning (say in the case of procreation by the female) that was required to accomplish them. These included hunting, trading, fetching of firewood, etc., through which identities were highlighted. Precolonial leadership in a family can best be understood generally by drawing on Abraham's misapplied personality. Traditionally described as a patriarch—and Okonkwo's communal ambitions lean in this direction—he was not (yet) the controlling male in feminist propaganda. Complementarity was key to male–female relationships, just as mutual respect was key to any form of engagement. Even so, farming has become a restricting occupation that conditions the female to a permanent category of dependency, with only symbolic gestures of good faith from the constituted male authority in both political and scholarly spheres.

The above variables fit into the now polarised logic of farming politics and then go on to condition thought patterns in family, communal, national, and even global settings. In Pinyin, as in most, if not all villages of the North West Region of Cameroon, the government policy of "green revolution" was undercut by the groomed biases of male prefiguring in the postindependence era, and has thus been further undermined by a gendered determinant toward crop cultivation, harvesting, consumption, commercialisation, and ritual performance. While exploring the causality between colonial and postindependence agricultural cultures, it is difficult, ironically, to determine the boundary between alien and indigenous attitudes toward the task of farming especially since some of the popular crops were brought in from foreign lands and their financial potentials were exploited by both colonial and postindependence authorities in favour of the family or community head or government. By analysing the various tasks related to farming, and based on practical and textual evidence, it will be obvious in the end that neocolonial and postindependence African economies have failed to decolonise the colonial prefiguring in this domain through an organic method—one that invests in the positives of the precolonial method, leading to a continuity of redundancy in replicating mega policy, and even of stagnation. The ongoing campaign for a cultural glocalism generally, and innovations in farming specifically, will further signify the imperative to Africanise policy and praxis in this area.

23.2 Understanding a Gift of Ants-Infested Faggots

I remember as a child when in the developing Aghem town of Wum in the North West Region of Cameroon, where my father served as a butcher, 'German Farm' was a household name in the heterogeneous community. A heritage of German expansionism till 1919 when the Treaty of Versailles dispossessed Germany of her colonies, the farm still served the postindependence elite right into the 1990s without any major improvement in its method of production. Wum was a space at the crossroads of rural–urban shifts in development, and, on the edge of both civilisations, we were caught in a whirlwind of emotions and for obvious reasons, our preference was for the ostentatious. The Germans had been forced out of Kamerun after 1919, but not after they had erected their forts which have stood the test of time—if only they were properly maintained. It has not been easy to erase those footprints by successive "masters" and leaders who introduce new ideologies. The natives had grown stout on the Kaiser's tutelage, and German monuments today attest to that faith, long after the English and the French came and left controversial imprints.

These marks are evident in postindependence policy in Cameroon today, and it will be understandable from this colonial baggage, why Anglophones are increasingly identifying with their colonial past, unearthing facts that destabilize the present arrangement; while the Francophones are more centripetal in their relationship with leadership. This also has a bearing on agriculture, with Anglophone regions constituting the bread basket of the country. Even the heritage of colonial farms can be enabling, whereas in the Francophone zone, individual initiative is subsumed into expectations from government hand-outs. In doing this, they also resurrect German landmarks that have been described by Kitts Mbeboh as having been abandoned to a decoration of cobwebs. Mbeboh is concerned with the symbol of German political manhood in Cameroon, including the farms that became parastatals and what is today described as the Prime Minister's Lodge in Buea, former capital of British Southern Cameroons. The strategic location of this edifice symbolizes both the colonial preference for a topographical vantage point and its complement of the overseer's gaze across the plantation slopes and plains. As the centre of colonial authority, the Lodge is a remnant of what initiated Cameroon's encounter with mega trends in agriculture, but instead of pushing forward with the initiative at independence, and customising it where necessary, the new elite was easily seduced by, and mimicked, colonial social models without the will to ensure the resources that will sustain it. Thus, if this historical symbol facilitated the Germanisation of Cameroon, and from which the farm is understood as a powerful metaphor of colonial authority, it is a typical postindependence characteristic of foreshortening our potentials of the private sector that has transformed the farms into state property. The mixed economy format that was permissible to this practice is no longer tenable in the late capitalist world and beyond, in which Bretton Woods institutions that were set up to ensure the perpetual hegemony of the global North continue to assist in the death throes of the global South. If we have to qualify this claim, it will only be because the IMF/World Bank SAP program is not a blanket deal and has been customised successfully elsewhere. The failure of countries like Cameroon is related to the indulgence of policy makers in maintaining systems that were long redundant elsewhere.

On a rebroadcast of BBC's "The Food Chain" of Sunday November 28, 2021, Gabriella D'Cruz, an Indian female was being showcased after winning the Corporation's Global Youth Championship award. In her experiment with seaweed farming in India "to change lives and revive the oceans using seaweed", we realise how new demographics and their reengagement with environmental awareness or concerns are revitalising the farming industry in ways that could not have been imagined, let alone encouraged, half a century ago. Significantly, while the winner is from the Indian middle class, her idea is an inspiration that cuts across social barriers and can serve for innovative experiments elsewhere. In this case, the middle class is no longer the crusoeistic stimulus of eighteenth-into-nineteenth century European colonisation as etched in Defoe's novel, *Robinson Crusoe*, but a reawakening that is concerned with uplifting economically marginalised peoples from the controlled stasis of mega economies: the middle class is no longer a status, but a motif for global balance, and while Crusoe's farm is a metaphor for emerging empire whose ultimate brunt India was to experience more than any other colonised space, D'Cruz's seaweed farm represents another shade of the empire farming back to the centre—to paraphrase Salman Rushdie's (1982) famous line in postcolonial hermeneutics—with or for global assurance. This positive mood brackets green discourse with a revivalism of the Spivakian subaltern, not as a condition, but a momentum. By going back to the roots and seeking what Prahalad (2005) describes as the fortune at the bottom of the pyramid, the Indian laureate was also cautioning against perverse globalisation and its bullying diplomacy. While Prahalad asserts that the green revolution "has made great strides in agricultural productivity in India" (321), he also describes the collective nature of the sacrifeces that ensured such success. This is what is lacking in Africa where, it can be argued, green politics has bypassed us, who should in fact be leading the debate, making us conspicuously absent during climate change protests and similar events all over the world.

The literature on the relationship between colonial agricultural methods and replication in the (post)colony follows a pattern long overwritten in Europe. The intention here is to engage some indicative measures toward an orientation that will be inclusive enough to liberate the postcolony from the stranglehold of neoliberalism. The eclecticism will hopefully facilitate the understanding of even symbolic boundaries between disciplines, while at the same time strengthening the advocacy for a less catch-up appeal.

23.3 From a Colonial Backcloth: The Africanisation of the Colonial Farm

Evidence from the English novel suggests that farmland was becoming insufficient for the agrarian interests of the population. Migration to the colonies for various reasons coincided with the establishment of colonial farms which necessitated the transfer of existing knowledge, only that this took the form of either condescension or exploitation. The sourcing of labour into the colonies is evident in D. H. Lawrence's early novels in which migration to the New World is largely based on the prospects offered by the vast agricultural land. But in analysising the aversion toward colonial space later in *Lady Chatterley's Lover* (1993), we meet Mellors who implicitly denounces the establishment of farms abroad especially when these structures become symbols of conquest. Such farms encouraged the emergence of aggressive capitalism through which imperialism achieved or even attempted to proffer justifications of its stranglehold on a global imagination.

Jean-Paul Sartre's (2001: 12–13) description of exclusive Frenchification of agriculture in colonial Algeria, in which it took "just a century to dispossess them [Algerians] of two-thirds of their land" typifies the overall situation in Africa: "Thus colonization has turned the Algerian population into an immense agricultural proletariat. It has been said of the Algerians that they are the same men as in 1830 and work the same land; only instead of owning it, they are the slaves of those who own it". First published in 1964, Sartre's work is contemporaneous to the eddying memories of World War II, when war-time food rationing in Germany and Britain justified Sartre's claims especially regarding the need for the colonial powers to intensify the control of food production in the colonies. No wonder too that even today, IDPs of the Anglophone Crisis in Cameroon are rationed food that comes mostly from a global food chain.

While the symbol of the colonial farm can be seen in literatures all over the continent, the motivation developed from its representation in English literature. The English initiative was probably a faithful account of the colonial establishment as worthy of the propaganda of its civilization. Defoe speculates on this apparently inoffensive portrayal and even attributes a benevolent gloss to his imperial imagination in *Robinson Crusoe* (1986). Defoe suggests that divisions of racist, class, and ideological assumptions never overruled an expansionist policy that targeted a global space. As we have seen, in Austen, the trade in territoriality translated into an individual and domestic spur to national consciousness. *Mansfield Park* suggests the need for a relationship in which the colonial farm, as Said reminds us, helps to sustain the home family: "What [Austen] sees more clearly than most of her readers is that to hold and rule Mansfield Park is to hold and rule an imperial estate in association with it. What assures the one, in its domestic tranquillity and attractive harmony is the prosperity and discipline of the other". According to Said, then, "no matter how isolated and insulated the English place is (e.g. Mansfield Park), it requires overseas sustenance" (Said 1989: 15, 158).

In Schreiner's (1977) *The Story of an African Farm*, however, the farm is more than a collateral for the metropolis against the rainy day: it has its own logic that does not abide by colonial policy—South Africa itself being a colony within a colony, thus challenging the notions of an "imagined community" (Anderson 1992) and of "internal colonialism" (Hechter 1975)—and in this way the characters figure out ways of self-assertion that demystify every form of constructed arrogance. Schreiner is indispensable to any discussion of how agriculture became a political tool for the imaginative representations of colonial heritage in Africa, and how her foresight inspires meaningful revaluations today of what has been essentialised on a masculinist turf. Schreiner's novel can be analysed as foundational in the task of writing the farm in African literature in the *fin de siècle* moment of the nineteenth century. The narrative is both prophetic of the conspiracies that are still being staged against the continent today, and affirmative in its advocacy for a more tolerant and therefore inclusive identity politics. Basically, the Africanist bias in Schreiner's vision reflects an anticolonial sentiment that characterises her foundational narration of the awkward binary which served colonial convenience. Her dedication of a prior text, *Dreams* (2003) "To a small girl-child, who may live to grasp somewhat of that which for us is yet sight, not touch" testifies to the unique foreview that her works offer, defying the cultural conceit which appropriated her to their condescending worldview. The difference between "sight" and "touch" is fundamental in our understanding of Schreiner's rejection of the Eurocentric conceit that tended to other those whose reality was taken for granted. This difference and Schreiner's attitude toward it provides ample evidence of her sympathetic representation of the indigenous worldview. In agricultural matters, and bearing in mind that South Africa was a complex Settler colony, she was already writing the country's composite version of the plantation and house slave experiences, overlapping as the farm changes ownership, and anticipating the tensions that will arise if the rhetoric of racial complexes was not at least mitigated. In her preface to *The Story*, this distinction is weighted against the glossy condescension of the English literati who, disadvantaged by distance and obviously buoyed by bias anthropological travel accounts, still have the illusion of authorising a colour essentialism in their encounter with Schreiner's world. Describing two types of life, one staged and the other lived, Schreiner was making a strong case for the contextual appreciation of cultural difference, even in agricultural responsibilities. The staged life—at the end of which "when the curtain falls, all will stand before it bowing" evokes the formal closure of a Shakespearean motif that imagines life as a performance; while the lived version is more functional:

> Here nothing can be prophesied. There is a strange coming and going of feet. Men appear, act and re-act upon each other, and pass away. When the curtain falls no one is ready. When the footlights are brightest they are blown out; and what the name on the play is no one knows … Life may be painted according to either method; but the methods are different (*Story* 2-3).

This is the recalcitrant intellectual energy which Schreiner exerts in identifying with the native canvass of her aesthetics, campaigning for the recognition of local rights and privileges in a context where "[t]he canons of criticism that bear upon the one cut cruelly upon the other". Ironically, those who negotiate cultural distortion have only a poorly mediated clue of the reality on the ground, and go on to certify what appeals to their whims as a universal given. The recognition of the distance between "sight" and "touch", the one grafted and the other factual (3) is thus crucial to our understanding of the farm metaphor here as also the appeal and possible misconception of distance to the centre. The repeated binary of black and white, night and day, far and near, and appearance and reality becomes the leitmotif in Schreiner's writing of farm politics as a colonial trope of Self and Other that grafts itself on the psyche of the natives and stymies every endeavour to achieve relative autonomy in future. The impasse can be resolved through a choice of either returning to the basics, if inoffensive, or strategizing to survive the stranglehold of emergent transnational benefactors in the sector through an inclusive diplomacy.

Having parcelled out arable lands in the UK to the point of relative scarcity, the colony became the new frontier in agrarian imperialism where innovative farming methods such as we observe in the British novel, were implemented, as depicted in Schreiner. In Hardy's *The Mayor of Casterbridge* and *Tess of the d'Urbervilles* for instance, a traditional approach to farming is seen as antiquated, and already being subsumed into the imperial alternative of subtle conquest. In other words, Hardy's narratives demonstrate how agriculture in England was already being influenced by technological drives from Scotland, embodied by Farfrae, who is on his way to the New World. A paternalistic comportment characterised this universalisation of farming initiatives, and persists till date. The mechanised transition in Europe thus inspired the experimentation in the colony with both forms simultaneously, implanting the systemic confusion over private and parastatal initiatives that have dominated the farming culture in the postcolony. While these farms were naturally transformed through new technologies, they were also reminiscent of the tension between innovative and conservative forces in Hardy's Farfrae and Henchard respectively. As Farfrae, the foreigner from Scotland, emerges victorious, already eyeing the American "new" world with a post-Columbus conceit, it is his former boss, Henchard, who degenerates both physically and psychologically, to the point of contemplating suicide, indicative of the fate of conservative mores when juxtaposed with liberal ideas. In a sense, the Henchard-Farfrae tension anticipated the dramatization of agro-diplomacy in Africa, where these binaries could hardly be sustained one way or the other.

In "The Birth of Colonialism in Malaya", Janes et al. (2017) point out that an intimate kinship existed between plantation enterprise and the establishment of colonies. This is a significant pointer to the capitalist inspiration behind the intimacy that also explains the overthrow of indigenous farming systems. This was a transformation that inspired the indigenes to begin aping the alien method, to the point of gendering farming through exclusive tags of cash and subsistent crops. The problem with postindependence policy makers was/is their endorsement of prescribed programs that eclipse(d) any possibility of innovation.

In Cameroon, farming in the rural areas has hardly evolved beyond the colonial sentence that encouraged primary production, caught as it were in the foiling of methods at independence. In fact, growing up in rural Cameroon in the early 1970s, farming was a habitual activity mainly because there were no other options. The family, most often polygamous, engaged in coffee farming as the main source of income that was managed by the husband. At school, we were taught the virtues of shifting cultivation, where fallowed lands were revisited after several years in the hope of a better yield. At the same time, we were drilled on how to obtain compost manure for use especially in gardening. It was only years later, in secondary school, that we were again taught advanced mechanised agricultural methods, including innovations in green house farming, to ensure variety and maximum productivity. Ironically, what we learned remained theoretical and was geared toward validating examination requirements beyond which there was no functional benefit. Meanwhile, government policy continued to prize agriculture as the backbone of the economy. Otherwise, the distinction between cash and subsistence crops reinforced colonial binaries within the domestic setting, with the husband controlling the coffee farms while his wives and children became *his* workers. In that local context, even the plantain was identified with the male, while the banana belonged to the female. At the national level, however, where farming was done on a semi-industrial scale, the gendering was neutralised and the colonial father-figure, whether as state or foreign ventures, was imposed and validated in the duplicities of international trade.

In the former West Cameroon, this transformation was represented by a colonial establishment like the German Farm which was later customised as the Wum Area Development Association (WADA). Even more sophisticated was the creation of the Produce Marketing Board (PMO) which was charged with the management of farmers' produce and ensuring favourable market deals on the world market, and in return paid relatively healthy and regular bonuses to the farmers. The rural economy experienced bumper decades, and politicians were quick to take the glory and reminded the citizens that before oil there was agriculture and that after the oil must have been exhausted, the country shall still be dependent on agriculture.[2] As far as propaganda held sway in the political economy of agriculture, it was a strategic bargain that guaranteed not just food self-sufficiency for the country but also served the Central African sub-region. While the PMO remain a farmers' association in the main, it was the Cameroon Development Corporation (CDC), that attained the status of a plantation even as it was to become vulnerable to the neoliberal arrangements of consensual indebtedness. As Konings and Nyamnjoh (2003: 14, 15) note, "as plantation agriculture expanded, it soon became evident that the local tenure system could not cope with the increasing flow of migrants that was leading to the development of a land market, land shortages and numerous land disputes". The situation please the British colonial masters who "acknowledged the existing land problems but did not introduce any structural changes. They did not want to disrupt the economic role of immigrants whose entrepreneurial spirit and hard work were said to compare favourably with the consumerist attitudes and laziness of the local population" Lamentably, more than sixty years after independence, we still propagate discussions of second generation agriculture, without any upgrade.

23.4 Writing the Farm: Olive Schreiner and Bessie Head

Perhaps owing to its middle class origins, it is understandable why the novel genre easily became a medium for narrating the farmer's story in the African version. From its rudimentary representation in *Things Fall Apart*, with no specific details about the technicalities of the process, we come to see in Schreiner's *Story* the introduction and interrogation of foreign models. As it will turn out, however, the change became part of the osmotic consciousness that was to determine colonial and postindependence development in Africa. It can even be argued that at this rate, African agriculture is suffering from the fallacy of replication, a strategic form of redundancy which stimulates the very dormancy that is nurtures, based not only on a desire to copy alien models without the necessary infrastructure, but also due to the paternalistic character of the global food program. While such a change may be contested, it is

[2] See Ndzana (1987).

important to note that the difference between lived and mediated experience determines our orientation; and that Schreiner's novel—like Head's (1977)—uses the farm as the background for the exploration of multiple themes, one of them being technological experimentation. When Bonaparte arrives at the farm, and is intrigued by Waldo's invention that will help in shearing the sheep, the former's compliments suggest how much such initiatives were in vogue, especially if performant and then validated by the colonial metropolis: "we must get you a patent. Your fortune is made. In three years' time there'll not be a farm in this colony where it isn't working" (*Story* 2). This possibility is complemented in *Rain Clouds*, where initiatives are expected to be implemented and assessed always as a mirror of superior European brands: "What was he looking for? What was he doing? Agriculture? The need for a poor country to catch up with the Joneses in the rich countries? Should superhighways and skyscrapers replace the dusty footpaths and thorn scrub?" (*Rain Clouds* 184). This query already invokes the complex landscape of developmentalism into which independence leadership blundered without heeding expert advice. Inevitably, we are confronted with jumpstarted development project in all fields because of resistance toward the conceptualisation of the decolonial project. The easy alternative of parachauting into the colonial realm of evolving globalism has resulted in a reactionary momentum that stagnates at best, and otherwise renders us into perpetual beggars of even what we own. In other words, the miasma of Pan-Africanism has never been disentangled into enabling vistas mainly because the new elite of the 1960s were more attentive to their godfathers than to their indigenous experts, a neglect which both Soyinka (1986) and Armah (1988) capture beautifully in *The Interpreters* and *The Beautful Ones Are Not Yet Born*, respectively.

And this is where the phenomenon of cooperative societies becomes another sign of poor strategic management. I remember how as kids in the village, my uncle's house was used as the packing store of the local cooperative society; but the farmers were continuously exploited, beginning from the local staff to the highest rung, until PMO brought relative stability which ironically attracted government parastatist interest, signalling the death toll that came barely a few years after its transformation into the National Produce Marketing Board (NPMB). The culture of cooperative societies is thus a symptom of the top-down adaption of in the postcolony to protect elitist status. The seduction that camouflages good intentions as a subtle way to ensure terminal dependency is reflected in the vision of Lessing's protagonist, Gilbert Balfour: "He looked at Makhaya with his eyes full of dreams, because Makhaya had proved himself the magician who could make tobacco cooperatives appear overnight. He wanted Golema Mmidi to be cooperative in everything as that was the only way of defeating the land tenure system in the tribal reserves and the only way of defeating subsistence agriculture which was geared to keeping the poor man poor until eternity" (Head 155–56). Here, change is comes in the form of conquest, already implying a hegemonic attitude that can only be resolved in one of two ways: resistance or submission. Overall, it is the latter that has become the normative praxis, with a tint of gloss that seeks to vindicate the controlling rhetoric.

Gilbert's approach contrasts that of Waldo's father, Oto, the kind, almost philanthropist German farm-keeper, who is reminiscent of near-patriarchal farm-owners in the English novel. The major difference is that whereas the English farms are hereditary, even if they were already being compromised by exclusive capitalist and mechanised priorities, those in the colony kept on changing ownership, indicative of the mutative control over resources that still burdens the continent. Head was aware of this, and her novel insists that "things were changing rapidly. The colonialists were withdrawing, and the change was not so much a part of the fashionable political ideologies of the New Africa as the outcome of the natural growth of a people" (*Rain Clouds* 34). Yet, Balfour's desire to "assist in agricultural development and improved techniques of food production" is the logical start of the transition which was to characterise agricultural transformation:

> The country presented overwhelming challenges, he said, not only because the rainfall was poor but because the majority of the people engaged in subsistence farming were using primitive techniques that ruined the land. All this had excited his interest. He had returned to England, taken a diploma in agriculture, and now had returned to Botswana to place his knowledge at the service of the country (*Rain Clouds* 23).

Once he acknowledges the "primitive" status of agriculture in Botswana, he simultaneously runs the risk of being assimilated into "civilised" methods uncritically, partly because of his own ego. Such concessions have been responsible for the dependency syndrome that characterises the state of farming in Africa, generally. Where the industrial-scale farms were succeeding in colonial outposts like South Africa, Zimbabwe, Cameroon, and Kenya—and indicates the extent to which a metropolitan heritage can serve postindependence imperatives when conceits are not at play—it was the fallacy of independence autonomy, that is, the poorly rehearsed rhetoric of territorial integrity, for instance, which upset the apple cart of hope and substituted it with a form of compromised continuity. The case of Zimbabwe is representative and instructive at the same time, and already being mirrored by Cameroon's where the metaphoric bread baskets of the respective regions quickly transformed into a spiralling inflationary nightmare ironically due to the fact that even the colonial-into-neoliberal paternalism had to be troughed through a gullible, self-serving leadership which privileges a deterministic ideological pull to the functional expectations of their constituents. Ultimately, even the cooperatives are displaced by a cartel of "cattle-speculating monopolies [which] were in the hands of a few white traders", leading to the understanding that "co-operatives was the dirtiest word you could use to the monopolists" (*Rain Clouds* 35). This was the start of the suspicion between the new leaders and native experts that created a chasm in developmental consciousness which still plagues Africa today.

Like Schreiner, Head's concern with the inequalities in colonial spaces is significant especially because of her double gaze in racial identities, and how it offers a more objective mapping of difference. This endeavour is a good example of using the novel as a platform for advocacy:

> He [Gilbert] felt that he had stumbled on to one of the major blockages to agricultural progress in the country. The women were the traditional tillers of the earth, not the men. The women were the backbone of agriculture while the men on the whole were cattle drovers. But when it came to programmes for improved techniques in agriculture, soil conservation, the use of pesticides and fertilizers, and the production of cash crops, the lecture rooms were

> open to men only. Why give training to a section of the population who may never use it but continue to leave it to their wives to erode the soil by unsound agricultural practices? Why start talking about development and food production without taking into account who is really producing the food?" (*Rain Clouds* 34).

The politicization of farming here bears masculinist marks of conceit and neglect and follows a pattern of postindependence machoism: the new elite of the 1960s, so adequately fictionalised from Fanonian white-face mimics to self-effacing caricatures, not only excluded women from participatory politics, but also feminised the masses as the new bulls of burden for the new elite. McClintock (1995) has used the symbol of the wagon as one of colonial fetishes to demonstrate the way in which male hegemony gained ascendancy in colonial imagination both against the native and European female; and was consequently inherited by the independence elite in the reinvention of stagnation. Whereas the colonial wagon rumbled ahead with the arrogance of authority, the mimic authorities in Africa have collapsed all blocs of social consciousnesss into just two, theirs and the rest. This latter is the amorphous constituency in which the farmer is deneutered and then controlled often by proxy. Farming has thus been synonymous to a feminised psyche that is controlled and adaptable to the whims and caprices of both donor agencies and their enablers in the postcolony. As such, the task which Gilbert faces is similar to that of Soyinka's interpretive iconoclasts; that of circumventing the programmed dependency on a form of technology whose code are never available to the end user.

23.5 Conclusion: Fate of the Green Revolution

As a Ph.D. researcher in the University of Nottingham and doing part-time with the Royal Mail, I was fond of taking coffee during break, but soon realised that my 33p only fetched me a cup that was actually one-third coffee, the rest filled with water. So I deliberately withdrew the cup when water started coming out, until I had my cupful of coffee. One day one of the tellers asked to know why I indulged in my coffee ritual. I explained to her that as the son of a coffee farmer back in Cameroon, I love to drink the pure beverage, not a two-third, water-diluted version. This experience triggered my interest in fair trade complicities in propagating win–win deals that are mere formality. The decision by Cote d'Ivoire and Ghana to process their cocoa before exporting will definitely ignite positive change into an emerging agricultural vision for Africa.

It becomes imperative, therefore, to Africanise the Green Revolution as a means to control our agriculture sector, otherwise the sonority of green discourse will only end in the formality of speeches and the accompanying connivance from global food promoters. In the introductory chapter of their edited volume, Larson and Otsuka (2016: 1) outline the focus of the book as "clarifying the importance of Green Revolution, identifying emerging challenges, and suggesting an effective strategy towards an African Green Revolution". Their conclusion hails the possibility of a Green Revolution in Africa with regards to rice and maize farming, coupled with "improved management practices". As valid as these are hopeful, especially given the global chaos that results from disruptions in the food chain, they still insinuate a follower, adaptive mentality that will hardly lead to the emergence of African agriculture on its own terms.

It is alsopossible to conceive of a shift from Green revolution to Green devolution—the idea here is to reveal how global machinations condemn Africa to remain at the primary stage, rather than reflect on ways that can actually destabilise the conspiracy. If we move from shifting cultivation to a phasing alternative, that is, correcting postindependence mishaps in the sector by prioritising rather than engaging a "wholesale" farming culture even at the grassroots, even alien technology will help us stem the dance of apes the music of second generation agriculture only takes us back to the period of the Industrial Revolution in nineteenth century Europe. Accordingly, the abstracts of casing and gendering crop cultivation will dissolve into a degendered bias for quality and increased productivity. Finally, large-scale farming adversely affected indigenes, and there will need for farming and settlement to be reviewed. The process, inherited by postindependence leadership, promotes the marginality on which a classist heritage thrived and became the new normal in the postcolony, with the farmer situated at the lowest rung. This resulted from a situation in which the traditional owners of land were pushed to its stony fringes, and their labour valuated as a pathetic wage-marker. Inevitably, dependency is constructed as part of government policy, and mirrors the legitimization of difference.

Even after independence, prescriptive farming methods, senseless wars, and hunger had been constructed into a neo-colonial variable of dependency. During the last years of the Cold War, the case of Ethiopia can be compared with the misplaced priorities that befell Zimbabwe after Mugabe became President in 1980, to understand how compromised leadership played farm politics without mastering the rules of conquest embedded in it. Ironically, today, an organisation like the World Food Program (WFP) has become a neoliberal leeway to ensuring continuous dependency and stigmatization of Third Worldhood on or against Western good faith. Large-scale agro-industrial establishments in the West have exploited the same colonialist tactics of finance and marketing and even though they have been criticised for indulging genetically altered technology, still exploit cheap labour in the underdeveloped world. Fair-trade talk is another side show that distracts attention from the horrendous situation. There can be no fairness when there is no mutuality in determining the market price worsened from free market to late capitalism of the Bretton Woods. Recently, the Kenyan government approved research in genetically modified goods and one is bound to see this more as surrender to mega interests in the agricultural business than even an inkling of interest in innovative research. If it were to be the latter, indigenous knowledge systems would have provided an alternative orientation.

We can be inspired by this reconnectivity to our agrarian roots not as a means to fulfil some Bretton Woods conditionality, but in the words of Lawrence's heroine in *Women in Love* (1987), "reculer pour mieux sauter"; or as more eloquently phrased by Achebe, to begin to speculate where the rains started beating us. The exploitation of agriculture as the politics of dependency must be considered as anathema in the twenty-first century, and African governments must also enable reforms that will guarantee a more functional autonomy today, especially after the global effects following Mr Vladimir Putin's crass invasion of Ukraine on February 24, 2022. We have realised that the global economy of over-dependency can have long-term effects if the supply chain is disrupted. And even as I write, renewed famine in Somalia, the worst in four decades, is a gruesome signal to how Africa's dependency on charity and prescribed farming practices has been a disservice to the agricultural policy on the continent as a whole. Ironically, strides made by Ethiopia after its own crisis in the mid-1980s are threatened by a more sinister crisis in Ethiopia, as in most African countries, based on an inability to respect internal structures of statehood since the 1960s. At this rate, Africa is doomed to remain the charity bowl of the world while her enormous potentials are wasted on the puerile diplomacy of the Internationals and their warlords.

References

Achebe, C. 1995. *Things Fall Apart*. New York: Knopf Doubleday Publishing Group.

Anderson, B. 1992. *Imagined Communities: Reflections on the Origin and Spread of Nationalism*. London: Verso.

Armah, A.K. 1988. *The Beautyful Ones Are Not Yet Born*. Oxford: Heinemann.

Defoe, D. 1986. *Robinson Crusoe*. New York: Penguin.

Head, B. 1977. *When Rain Clouds Gather*. Chicago: Academy Press.

Hechter, M. 1975. *Internal Colonialism: The Celtic Fringe in British National Development, 1536–1966*. London: Routledge and Kegan Paul.

Janes, L. et al. 2017. The Birth of Plantation Colonialism in Malaya. In *Planting Empire, Cultivating Subjects: British Malay 1786–1914*, ed. Lynn Hollen Lees, 21–61.

Janes, Lauren, et al. 2019. World War I and the Origins of the Modern Food System, Global Food History. *Global Food History* 5 (3): 224–238.

Konings, Piet, and Francis B. Nyamnjoh. 2003. *Negotiating an Anglophone Identity: A Study of the Politics of Recognition and Representation in Cameroon*. Leiden: Brill.

Larson, Donald F., and Keijiro Otsuka. 2016. *In Pursuit of an African Green Revolution: Views from Rice and Maize Farmers' Fields*, vol. 1.

Lawrence, D.H. 1993. *Lady Chatterley's Lover* (ed. Michael Squires). Cambridge: Cambridge University Press.

Mbuh, T.M. 2015. Art, the Artist, and Postcoloniality: Chinua Achebe's Unoka and Implications of Representation. *Things Fall Apart* (*The Gong: An Interdisciplinary Journal of English Language and Literary Studies*) 1 (1): 104–121.

McClintock, A. 1995. *Imperial Leather: Race, Gender and Sexuality in the Colonial Contest*. New York: Routledge.

Ndzana, V.O. 1987. *Agriculture, Pétrole et Politique au Cameroun: Sortir de la crise?* Paris: L'Harmattan.

Prahalad, C.C.K. 2005. *The Fortune at the Bottom of the Pyramid*. Saddle River, NJ: Wharton School Publishing Upper.

Rushdie, S. 1982. The Empire Writes Back with a Vengeance. *The Times*, July 3.

Said, E. 1989. Jane Austen and Empire. In *Raymond Williams: Critical Perspectives*, ed. Terry Eagleton. Cambridge: Polity Press.

Said, E. 2005. *Culture and Imperialism*. London: Vintage.

Sartre, J.-P. 2001. *Colonialism and Neocolonialism*. London: Routledge.

Schreiner, O. 2003. *Dreams: Three Works*. Birmingham: The University of Birmingham University Press.

Schreiner, O. 1977. *The Story of an African Farm*. Edgbaston, Birmingham: The University of Birmingham Press.

Soyinka, Wole. 1986. *The Interpreters*. London: Fontana.

Women in Love. 1987. (ed. David Farmer, Lindeth Vasey, and John Worthen). Cambridge: Cambridge University Press.

A Fulbright and Commonwealth Scholar, Mbuh Tennu Mbuh holds a Ph.D. in English literature from The University of Nottingham, UK. He is the Head of the English Department, Faculty of Arts in The University of Bamenda—Cameroon. He is also a writer and the President of the Anglophone Cameroon Writers' Association, ACWA. His research interests revolve around postindependence nuances in literary and geopolitical contexts and how these generate new interpretations of empire today; interrogating feminist advocacy from indigenous and Africanist perspectives; and reconsidering religious (especially Christianity's) claims for global ascendancy.

The manufacturer's authorised representative in the EU is Springer Nature Customer Service Centre GmbH, Europaplatz 3, 69115 Heidelberg, Germany. If you have any concerns regarding our products, please contact ProductSafety@springernature.com

Printed and bound by CPI Group (UK) Ltd, Croydon, CR0 4YY
13/08/2026
02196020-0005